COLLOIDAL BIOMOLECULES, BIOMATERIALS, AND BIOMEDICAL APPLICATIONS

SURFACTANT SCIENCE SERIES

1. Nonionic Surfactants, *edited by Martin J. Schick* (see also Volumes 19, 23, and 60)
2. Solvent Properties of Surfactant Solutions, *edited by Kozo Shinoda* (see Volume 55)
3. Surfactant Biodegradation, *R. D. Swisher* (see Volume 18)
4. Cationic Surfactants, *edited by Eric Jungermann* (see also Volumes 34, 37, and 53)
5. Detergency: Theory and Test Methods (in three parts), *edited by W. G. Cutler and R. C. Davis* (see also Volume 20)
6. Emulsions and Emulsion Technology (in three parts), *edited by Kenneth J. Lissant*
7. Anionic Surfactants (in two parts), *edited by Warner M. Linfield* (see Volume 56)
8. Anionic Surfactants: Chemical Analysis, *edited by John Cross*
9. Stabilization of Colloidal Dispersions by Polymer Adsorption, *Tatsuo Sato and Richard Ruch*
10. Anionic Surfactants: Biochemistry, Toxicology, Dermatology, *edited by Christian Gloxhuber* (see Volume 43)
11. Anionic Surfactants: Physical Chemistry of Surfactant Action, *edited by E. H. Lucassen-Reynders*
12. Amphoteric Surfactants, *edited by B. R. Bluestein and Clifford L. Hilton* (see Volume 59)
13. Demulsification: Industrial Applications, *Kenneth J. Lissant*
14. Surfactants in Textile Processing, *Arved Datyner*
15. Electrical Phenomena at Interfaces: Fundamentals, Measurements, and Applications, *edited by Ayao Kitahara and Akira Watanabe*
16. Surfactants in Cosmetics, *edited by Martin M. Rieger* (see Volume 68)
17. Interfacial Phenomena: Equilibrium and Dynamic Effects, *Clarence A. Miller and P. Neogi*
18. Surfactant Biodegradation: Second Edition, Revised and Expanded, *R. D. Swisher*
19. Nonionic Surfactants: Chemical Analysis, *edited by John Cross*
20. Detergency: Theory and Technology, *edited by W. Gale Cutler and Erik Kissa*
21. Interfacial Phenomena in Apolar Media, *edited by Hans-Friedrich Eicke and Geoffrey D. Parfitt*
22. Surfactant Solutions: New Methods of Investigation, *edited by Raoul Zana*
23. Nonionic Surfactants: Physical Chemistry, *edited by Martin J. Schick*
24. Microemulsion Systems, *edited by Henri L. Rosano and Marc Clausse*
25. Biosurfactants and Biotechnology, *edited by Naim Kosaric, W. L. Cairns, and Neil C. C. Gray*
26. Surfactants in Emerging Technologies, *edited by Milton J. Rosen*
27. Reagents in Mineral Technology, *edited by P. Somasundaran and Brij M. Moudgil*
28. Surfactants in Chemical/Process Engineering, *edited by Darsh T. Wasan, Martin E. Ginn, and Dinesh O. Shah*
29. Thin Liquid Films, *edited by I. B. Ivanov*
30. Microemulsions and Related Systems: Formulation, Solvency, and Physical Properties, *edited by Maurice Bourrel and Robert S. Schechter*
31. Crystallization and Polymorphism of Fats and Fatty Acids, *edited by Nissim Garti and Kiyotaka Sato*

S

32. Interfacial Phenomena in Coal Technology, *edited by Gregory D. Botsaris and Yuli M. Glazman*
33. Surfactant-Based Separation Processes, *edited by John F. Scamehorn and Jeffrey H. Harwell*
34. Cationic Surfactants: Organic Chemistry, *edited by James M. Richmond*
35. Alkylene Oxides and Their Polymers, *F. E. Bailey, Jr., and Joseph V. Koleske*
36. Interfacial Phenomena in Petroleum Recovery, *edited by Norman R. Morrow*
37. Cationic Surfactants: Physical Chemistry, *edited by Donn N. Rubingh and Paul M. Holland*
38. Kinetics and Catalysis in Microheterogeneous Systems, *edited by M. Grätzel and K. Kalyanasundaram*
39. Interfacial Phenomena in Biological Systems, *edited by Max Bender*
40. Analysis of Surfactants, *Thomas M. Schmitt* (see Volume 96)
41. Light Scattering by Liquid Surfaces and Complementary Techniques, *edited by Dominique Langevin*
42. Polymeric Surfactants, *Irja Piirma*
43. Anionic Surfactants: Biochemistry, Toxicology, Dermatology. Second Edition, Revised and Expanded, *edited by Christian Gloxhuber and Klaus Künstler*
44. Organized Solutions: Surfactants in Science and Technology, *edited by Stig E. Friberg and Björn Lindman*
45. Defoaming: Theory and Industrial Applications, *edited by P. R. Garrett*
46. Mixed Surfactant Systems, *edited by Keizo Ogino and Masahiko Abe*
47. Coagulation and Flocculation: Theory and Applications, *edited by Bohuslav Dobiáš*
48. Biosurfactants: Production • Properties • Applications, *edited by Naim Kosaric*
49. Wettability, *edited by John C. Berg*
50. Fluorinated Surfactants: Synthesis • Properties • Applications, *Erik Kissa*
51. Surface and Colloid Chemistry in Advanced Ceramics Processing, *edited by Robert J. Pugh and Lennart Bergström*
52. Technological Applications of Dispersions, *edited by Robert B. McKay*
53. Cationic Surfactants: Analytical and Biological Evaluation, *edited by John Cross and Edward J. Singer*
54. Surfactants in Agrochemicals, *Tharwat F. Tadros*
55. Solubilization in Surfactant Aggregates, *edited by Sherril D. Christian and John F. Scamehorn*
56. Anionic Surfactants: Organic Chemistry, *edited by Helmut W. Stache*
57. Foams: Theory, Measurements, and Applications, *edited by Robert K. Prud'homme and Saad A. Khan*
58. The Preparation of Dispersions in Liquids, *H. N. Stein*
59. Amphoteric Surfactants: Second Edition, *edited by Eric G. Lomax*
60. Nonionic Surfactants: Polyoxyalkylene Block Copolymers, *edited by Vaughn M. Nace*
61. Emulsions and Emulsion Stability, *edited by Johan Sjöblom*
62. Vesicles, *edited by Morton Rosoff*
63. Applied Surface Thermodynamics, *edited by A. W. Neumann and Jan K. Spelt*
64. Surfactants in Solution, *edited by Arun K. Chattopadhyay and K. L. Mittal*
65. Detergents in the Environment, *edited by Milan Johann Schwuger*

66. Industrial Applications of Microemulsions, *edited by Conxita Solans and Hironobu Kunieda*
67. Liquid Detergents, *edited by Kuo-Yann Lai*
68. Surfactants in Cosmetics: Second Edition, Revised and Expanded, *edited by Martin M. Rieger and Linda D. Rhein*
69. Enzymes in Detergency, *edited by Jan H. van Ee, Onno Misset, and Erik J. Baas*
70. Structure–Performance Relationships in Surfactants, *edited by Kunio Esumi and Minoru Ueno*
71. Powdered Detergents, *edited by Michael S. Showell*
72. Nonionic Surfactants: Organic Chemistry, *edited by Nico M. van Os*
73. Anionic Surfactants: Analytical Chemistry, Second Edition, Revised and Expanded, *edited by John Cross*
74. Novel Surfactants: Preparation, Applications, and Biodegradability, *edited by Krister Holmberg*
75. Biopolymers at Interfaces, *edited by Martin Malmsten*
76. Electrical Phenomena at Interfaces: Fundamentals, Measurements, and Applications, Second Edition, Revised and Expanded, *edited by Hiroyuki Ohshima and Kunio Furusawa*
77. Polymer-Surfactant Systems, *edited by Jan C. T. Kwak*
78. Surfaces of Nanoparticles and Porous Materials, *edited by James A. Schwarz and Cristian I. Contescu*
79. Surface Chemistry and Electrochemistry of Membranes, *edited by Torben Smith Sørensen*
80. Interfacial Phenomena in Chromatography, *edited by Emile Pefferkorn*
81. Solid–Liquid Dispersions, *Bohuslav Dobiáš, Xueping Qiu, and Wolfgang von Rybinski*
82. Handbook of Detergents, *editor in chief: Uri Zoller*
    Part A: Properties, *edited by Guy Broze*
83. Modern Characterization Methods of Surfactant Systems, *edited by Bernard P. Binks*
84. Dispersions: Characterization, Testing, and Measurement, *Erik Kissa*
85. Interfacial Forces and Fields: Theory and Applications, *edited by Jyh-Ping Hsu*
86. Silicone Surfactants, *edited by Randal M. Hill*
87. Surface Characterization Methods: Principles, Techniques, and Applications, *edited by Andrew J. Milling*
88. Interfacial Dynamics, *edited by Nikola Kallay*
89. Computational Methods in Surface and Colloid Science, *edited by Małgorzata Borówko*
90. Adsorption on Silica Surfaces, *edited by Eugène Papirer*
91. Nonionic Surfactants: Alkyl Polyglucosides, *edited by Dieter Balzer and Harald Lüders*
92. Fine Particles: Synthesis, Characterization, and Mechanisms of Growth, *edited by Tadao Sugimoto*
93. Thermal Behavior of Dispersed Systems, *edited by Nissim Garti*
94. Surface Characteristics of Fibers and Textiles, *edited by Christopher M. Pastore and Paul Kiekens*
95. Liquid Interfaces in Chemical, Biological, and Pharmaceutical Applications, *edited by Alexander G. Volkov*

96. Analysis of Surfactants: Second Edition, Revised and Expanded, *Thomas M. Schmitt*
97. Fluorinated Surfactants and Repellents: Second Edition, Revised and Expanded, *Erik Kissa*
98. Detergency of Specialty Surfactants, *edited by Floyd E. Friedli*
99. Physical Chemistry of Polyelectrolytes, *edited by Tsetska Radeva*
100. Reactions and Synthesis in Surfactant Systems, *edited by John Texter*
101. Protein-Based Surfactants: Synthesis, Physicochemical Properties, and Applications, *edited by Ifendu A. Nnanna and Jiding Xia*
102. Chemical Properties of Material Surfaces, *Marek Kosmulski*
103. Oxide Surfaces, *edited by James A. Wingrave*
104. Polymers in Particulate Systems: Properties and Applications, *edited by Vincent A. Hackley, P. Somasundaran, and Jennifer A. Lewis*
105. Colloid and Surface Properties of Clays and Related Minerals, *Rossman F. Giese and Carel J. van Oss*
106. Interfacial Electrokinetics and Electrophoresis, *edited by Ángel V. Delgado*
107. Adsorption: Theory, Modeling, and Analysis, *edited by József Tóth*
108. Interfacial Applications in Environmental Engineering, *edited by Mark A. Keane*
109. Adsorption and Aggregation of Surfactants in Solution, *edited by K. L. Mittal and Dinesh O. Shah*
110. Biopolymers at Interfaces: Second Edition, Revised and Expanded, *edited by Martin Malmsten*
111. Biomolecular Films: Design, Function, and Applications, *edited by James F. Rusling*
112. Structure–Performance Relationships in Surfactants: Second Edition, Revised and Expanded, *edited by Kunio Esumi and Minoru Ueno*
113. Liquid Interfacial Systems: Oscillations and Instability, *Rudolph V. Birikh, Vladimir A. Briskman, Manuel G. Velarde, and Jean-Claude Legros*
114. Novel Surfactants: Preparation, Applications, and Biodegradability: Second Edition, Revised and Expanded, *edited by Krister Holmberg*
115. Colloidal Polymers: Synthesis and Characterization, *edited by Abdelhamid Elaissari*
116. Colloidal Biomolecules, Biomaterials, and Biomedical Applications, *edited by Abdelhamid Elaissari*
117. Gemini Surfactants: Synthesis, Interfacial and Solution-Phase Behavior, and Applications, *edited by Raoul Zana and Jiding Xia*

ADDITIONAL VOLUMES IN PREPARATION

Colloidal Science of Flotation, *Anh V. Nguyen and Hans Joachim Schulze*

Surface and Interfacial Tension: Measurement, Theory, and Applications, *edited by Stanley Hartland*

Microporous Media: Synthesis, Properties, and Modeling, *Fredy Romm*

COLLOIDAL BIOMOLECULES, BIOMATERIALS, AND BIOMEDICAL APPLICATIONS

edited by
Abdelhamid Elaissari
CNRS-bioMérieux
Lyon, France

MARCEL DEKKER
Marcel Dekker, Inc. New York • Basel

Although great care has been taken to provide accurate and current information, neither the author(s) nor the publisher, nor anyone else associated with this publication, shall be liable for any loss, damage, or liability directly or indirectly caused or alleged to be caused by this book. The material contained herein is not intended to provide specific advice or recommendations for any specific situation.

Trademark notice: Product or corporate names may be trademarks or registered trademarks and are used only for identification and explanation without intent to infringe.

Library of Congress Cataloging-in-Publication Data
A catalog record for this book is available from the Library of Congress.

ISBN: 0-8247-4779-8

This book is printed on acid-free paper.

Headquarters
Marcel Dekker, Inc., 270 Madison Avenue, New York, NY 10016, U.S.A.
tel: 212-696-9000; fax: 212-685-4540

Distribution and Customer Service
Marcel Dekker, Inc., Cimarron Road, Monticello, New York 12701, U.S.A.
tel: 800-228-1160; fax: 845-796-1772

Eastern Hemisphere Distribution
Marcel Dekker AG, Hutgasse 4, Postfach 812, CH-4001 Basel, Switzerland
tel: 41-61-260-6300; fax: 41-61-260-6333

World Wide Web
http://www.dekker.com

The publisher offers discounts on this book when ordered in bulk quantities. For more information, write to Special Sales/Professional Marketing at the headquarters address above.

Current printing (last digit):

10 9 8 7 6 5 4 3 2 1

PRINTED IN THE UNITED STATES OF AMERICA

Dedication

This book is dedicated to Doctor Christian Pichot for his great scientific contribution in the polymer colloids domain and in honor of his retirement in October 2002.

Christian Pichot completed chemistry studies at Université Pierre et Marie Curie in Paris, France, and then started his research career as a contract laboratory technician at the Faculté des Sciences d'Orsay in a Chemistry-Physics Laboratory directed by Professor M. Magat, one of the founders of Polymer Science in France. After a national service teaching in Ivory Coast, he was welcomed by A. Guyot, Director of Research, under a research contract allowing him to join the CNRS (National Center for Scientific Research) catalysis research institute and to be involved in various topics (physico-chemical characterization of polyphenylsiloxane, degradation studies of PVC models). In October 1972, he got a permanent position at the CNRS and he prepared a doctorate thesis on kinetic anomalies in radical-initiated copolymerization and their explanations, a work that he defended in 1976.

He completed a one-year postdoctoral fellowship (from the National Science Foundation) at the Emulsion Polymers Institute (Lehigh University, Bethlehem, Pennsylvania), under the guidance of Professors J. Vanderhoff and M. S. El-Aasser, giving him the opportunity to be trained in the field of latex synthesis and characterization. When he went back to the Laboratory of Organic Materials, first in Villeurbanne then in Vernaison in 1980, together with J. Guillot he played a significant role in setting up and developing a research team devoted to kinetics, characterization, and properties of radical copolymers produced in dispersed media (mostly in emulsion). He was responsible for training young research scientists who all went on to successful careers in industry. He dealt with a wide range of problems, both applied and fundamental, leading him to publish over 50 high-quality papers. The thesis students and scientists working under his direction particularly appreciated his availability and devotion, along with his friendly manner under all circumstances. He investigated three main topics: first, he tried to establish synthesis–structure–properties relationships in various emulsion (co)-polymers; second, he made a significant contribution to

the knowledge of inverse emulsion polymerization mechanisms, a subject that had been covered only by a restricted number of analyses; finally, he took particular interest in the original behavior of zwitterionic surfactants in emulsion polymerization, especially to produce monodisperse nanosized particles.

In 1988, his scientific reputation helped him to take part in the creation of a joint research unit between the CNRS and bioMérieux, a worldwide company partly located in the Lyon area and dealing with the manufacture of automatic diagnostic analyzers and reagents. This joint project soon resulted in patent registrations, but, above all, Christian Pichot initiated various research subjects related to the preparation of functionalized polymer supports with appropriated properties to interact with biological fluids. For that purpose, he particularly contributed to the design and elaboration either of reactive "linear" copolymers or of functionalized, stimuli-responsive, and magnetic latex particles. He developed a passionate interest in the complexity of these problems, resulting in the establishment of a laboratory considered an exemplary success. He chaired the Chemistry Group of the joint research unit from its creation until 2002, helping this team, now comprising a good balance of researchers with various expertises and coming from the CNRS and bioMérieux, to acquire an internationally acknowledged reputation.

Since 1980, he has been a member of the International Colloid Polymer Group, where he represents all those from the Lyon area working in this field. He is regurlarly invited as a lecturer in various meetings organized by the scientific community, and he has contributed to the creation of a large number of contacts and developed many collaborative research projects. From 1984, he was co-organizer in Lyon of four international symposia dealing mainly with polymers in dispersed media on both academic and applied aspects. He welcomed and trained in the lab many students who have gone on to careers as researchers in France and throughout the world.

Preface

Colloids (latexes and hybrids) provide suitable solid-phase supports as a carrier of various molecules, biomolecules, and active agents. In fact, diverse and varied particles have been developed and explored in numerous biomedical applications.

In biomedical diagnostic, the immobilization (adsorption, covalent grafting, and specific interactions) of biomolecules such as proteins, antibodies, peptides, nucleic acids, bacteria, and viruses onto colloidal particles is of paramount importance. In immunodiagnostic and specific nucleic acids isolation, the captured biomolecules should be adequately immobilized. This is a very critical point since further accessibility and activity of the fixed biomolecules could be affected. One interesting alternative solution to this problem is to take advantage of interactions between a specific domain of a given biomolecule with suitable active chemical functions and including oriented conformations through complexation, dipole–dipole interactions, etc. In this respect, various reactive groups have been recognized to specifically interact with a complementary region of a given biomolecule (or a genetically or chemically modified biomolecule). To control the sensitivity and the specificity of biomedical diagnostic, the covalent coupling of biomolecules onto functionalized particles is envisioned due to the presence of reactive groups onto biological molecules. Whatever the immobilized process, the conformation of the interfacial biomolecules should also be considered in order to exhibit well-defined orientation.

In therapeutic areas, various kinds of colloidal particles have been evaluated as carriers. In fact, in recent years increasing interest has been given to the preparation and utilization of biodegradable particles or stimuli-responsive colloids. All elaborated particles are principally used as a solid support rather than as a micro-container of active molecules. Nowadays, the main objective in therapeutic domain is the elaboration of new colloids—such as smart capsules and well-defined methodologies—in order to enhance the targeting efficiency. One possibility to partially solve this problem is to conduct systematic studies, and to use appropriate stimuli-responsive colloidal particles would be of extreme interest.

This book presents the utilization of colloids in the biomedical field (diagnostic and therapeutic). The chapters present original works, fresh results, new methodologies, and several applications of colloidal particles in biomedicine. The elaboration of biodegradable particles for therapy applications is clearly presented. In addition, interactions (adsorption, desorption, covalent coupling, complexation, specific interactions, etc.) between biomolecules (nucleic acids, proteins, antibodies, peptides, viruses, etc.), colloidal particles, and many fundamental studies are clearly reported.

Abdelhamid Elaissari

Contents

Contributors

Julio Battistoni Facultades de Química y Ciencias, Instituto de Higiene, Montevideo, Uruguay

Michael A. Brook McMaster University, Hamilton, Ontario, Canada

S. Tolga Çamli Hacettepe University, Ankara, Turkey

Jhunu Chatterjee FAMU-FSU, Tallahassee, Florida, U.S.A.

Patrick Couvreur Université de Paris Sud, Chatenay-Malabry, France

Thierry Delair CNRS-bioMérieux, Lyon, France

Frantz S. Deschamps Mainelab S.A., Angers, France

Catherine Dubernet Université de Paris Sud, Chatenay-Malabry, France

David Duracher CNRS-bioMérieux, Lyon, France

Abdelhamid Elaissari CNRS-bioMérieux, Lyon, France

F. Galisteo-Gonzalez University of Granada, Granada, Spain

François Ganachaud Université Pierre et Marie Curie—CNRS, Paris, France

Emmanuelle Imbert-Laurenceau CNRS-bioMérieux, Lyon, France

Haruma Kawaguchi Keio University, Yokohama, Japan

Muxin Liu McMaster University, Hamilton, Ontario, Canada

Bernard Mandrand CNRS-bioMérieux, Lyon, France

Véronique Migonney Institut Galilée, Université Paris-Nord, Villetaneuse, France

J. A. Molina-Bolivar University of Málaga, Málaga, Spain

Lahoussine Ouali CNRS-Institut Charles Sadron, Strasbourg, France

Emile Pefferkorn CNRS-Institut Charles Sadron, Strasbourg, France

Agnès Perrin CNRS-bioMérieux, Lyon, France

Christian Pichot CNRS-bioMérieux, Lyon, France

Amro N. Ragheb McMaster University, Hamilton, Ontario, Canada

Joel Richard Ethypharm S.A., Saint-Cloud, France

Silvina Rossi Facultades de Química y Ciencias, Instituto de Higiene, Montevideo, Uruguay

Serap Şenel Hacettepe University, Ankara, Turkey

S. Slomkowski Center of Molecular and Macromolecular Studies, Lodz, Poland

Serge Stoll Université de Genève, Geneva, Switzerland

Ali Tuncel Hacettepe University, Ankara, Turkey

Ender Ünsal Hacettepe University, Ankara, Turkey

Christine Vauthier Université de Paris Sud, Chatenay-Malabry, France

Raphael Veyret CNRS-bioMérieux, Lyon, France

Paul M. Zelisko McMaster University, Hamilton, Ontario, Canada

COLLOIDAL BIOMOLECULES, BIOMATERIALS, AND BIOMEDICAL APPLICATIONS

1
Biomedical Application for Magnetic Latexes

ABDELHAMID ELAISSARI, RAPHAEL VEYRET, and BERNARD MANDRAND CNRS-bioMérieux, Lyon, France

JHUNU CHATTERJEE FAMU-FSU, Tallahassee, Florida, U.S.A.

I. INTRODUCTION

Magnetic latexes are colloidal composites that combine organic and inorganic materials. The inorganic material may be composed of a metal or iron oxide derivative [1,2] or silica oxide [3,4]. Each of the organic and inorganic components plays a specific role in the properties of the final hybrid material. In inks, cosmetic products, and paints, polymers facilitate compatibility between the pigment and the binder. Thus, the inorganic materials are better dispersed throughout the medium, and there are negligible amounts of aggregated pigments and little aggregation. Consequently, in paints, where aggregate formation is a serious problem, the mechanical properties, durability, and glossiness are improved after the pigment is encapsulated by a polymer layer. The presence of a magnetic material endows the polymer particle with additional properties. For example, iron oxides and silica are used to make conducting polymers [8], to modify the optical properties of polymer films [9], and in ink applications, the paper industry [10,11], high-density recording media [12], and catalyst carriers [13].

Many publications in the biomedical field describe the versatility of magnetic polymer particle applications for magnetic separation of biochemical products [14], cell sorting [15], magnetic particle guidance for specific drug delivery [16], and as a contrast agent in magnetic resonance imaging (MRI) [17,18]. The polymers used as coating agents generally function to protect inorganic material and induce reactive chemical groups capable of fixing biological molecules (via chemical binding), and magnetic iron oxide ensures the migration under an applied magnetic field. To facilitate adaptation of magnetic colloids, a great variety of magnetic composite particles have been developed commercially and are available; they range from a few nanometers to 10 μm and are available as capsules, microgel, and smooth or porous spheres. In this case, the source of the polymer matrix may be from a natural polymer or biopolymer, such as albumin [19] or starch [20,21]; they are eventually biocompatible and biodegradable particles. These colloidal particles may also contain a synthetic polymer derived from glutaraldehyde and cyanoacrylate [22]. Magnetic microspheres with hydrophilic surfaces are generally synthesized using acrylamides, acrylates, and methacrylate monomer derivatives [23]. Hydrophobic magnetic particles are generally constructed from hydrophobic monomers such as styrene [24,25].

Given the variety of processes for synthesizing these latexes, they have a wide range of properties (surface charge density, reactive groups, hydrophilic–hydrophobic surface, particle size, size distribution, surface polarity, and magnetic properties) and can be adapted to many biological applications. The aim of this chapter is to review the various uses of magnetic particles as solid phase or as particle carrier in the biomedical field.

Magnetic particles are used extensively in pharmacy, biology, and medicine to carry biological compounds. Biomolecules fixed to magnetic particles can be transported and can also be separated quickly from complex mediums; thus they can be used for both therapeutic and diagnostic applications. In the therapeutic domain, magnetic latex can guide a drug to and release it at a specific site [26], and can extract tumor cells from the organism in vitro or ex vivo. It is also used to generate sufficient heat in carcinoma cells to inactivate them, i.e., hyperthermia [27].

Diagnostic applications involve the use of magnetic latex for biomolecules extraction, separation, and concentration. Both magnetic and nonmagnetic latex particles have long been recognized as good colloid solid supports for antigen detection by immunological reaction or in agglutination diagnostic tests [28]. Characteristics such as a large developed surface area, ease of use in biomolecule adsorption and chemical grafting, fast separation under magnetic fields, and rapid biomolecule kinetics have made magnetic latex an important solid support in immunoassays. Magnetic particles that bear the target molecules (after capturing them) can be easily and rapidly separated by applying a single magnetic field. This section will briefly detail the method of separation and the requirement for magnetic latex particles.

Many magnetic beads with different characteristics have been described in the literature and used in various applications. The first commercialized monodisperse magnetic beads were polystyrene based (e.g., Dynabeads) [29]. This type of colloidal particle has been used in numerous processes as a solid support for separating biological compounds. Like other magnetic particles, it is synthesized by the precipitation of magnetic ferric oxide fine particles on porous polymer particles of homogeneous size. After the free iron oxide nanoparticles are removed, the composite beads are encapsulated with a polymer layer that fills the pores and brings to the surface the chemical groups needed to immobilize the biomolecules. Particles may also make up a metal oxide core, usually iron oxide, which is surrounded by a polymer shell bearing reactive groups that can be use to immobilize biomolecules [30]. Other magnetic beads are commercially available: Estapor (Merck Eurolab), Serradyn (Seradyn), Magnisort (Dupont) [30], BioMag M4100 and BioMag M4125 (Advanced Magnetic) [31].

Magnetic beads have been used and evaluated in many different diagnostic processes. Their first area of use is separation process. Magnetic particles have obvious advantages for this, whether or not immunological reactions are used. They are also advantageous as a solid phase for immobilizing various biomolecules and labels. Most of these advantageous characteristics do not vary much from one application to the next, but some of them may be dictated by the intended use.

II. MAGNETIC SEPARATION

Magnetic particles have been used as support for the separation, selective isolation, and purification of molecules [32]. For example, in biomedical diagnostics, they can replace the cumbersome steps of centrifugation or filtration [32–34]. The major techniques all involve chemical grafting of biomolecules onto magnetic beads to target specific separation of captured biomolecules or of analytes [29]. Since magnetic supports can be separated from solutions containing other species (e.g., suspended solids, cell fragments, and contaminants), magnetic affinity separation is useful for crude samples [35]. Various magnetic particles can be adapted to these kinds of applications, including large particles (above 1 μm) [29], silica magnetic particles [32], and nanoparticles (below 100 nm).

Therefore, use of a magnetic field to separate composite magnetic particles is, compared with the alternatives, very simple, often cheaper, and above all faster. For example, bacterial control in the food industry requires 10 min of magnetic particle use instead of the 24 h required with the traditional methods of analysis because the target bacteria concentrations are too weak to be characterized and require culturing. Fast magnetic separation results in direct concentration of the bacteria and therefore eliminates this slow step. The need to apply a powerful magnetic field may, however, be quite expensive, depending on the particle properties. Regardless of the use considered, magnetic particle carriers must have the following properties:

- Colloidal and chemical stability in the separation medium
- Nonmagnetic remanence
- Does not release of iron oxide during biomedical applications
- Low sedimentation velocity compared with magnetic separation
- A surface that is biocompatible with the relevant biomolecules
- Allows complete, rapid, and specific separation

Two specific methods are frequently used in biomedical diagnostics for detecting disease with magnetic particles and specific interaction between biological molecules (e.g., antibody/antigen or nucleic acids):

1. *Direct separation.* The biological sample containing the target molecules is mixed with magnetic particles bearing specific antibodies (generally called sensitive particles). After incubation under given conditions (time, buffer composition, temperature) and the subsequent immunological reaction between the immobilized antibody on the particle and the target antigen, the magnetic particles are separated by applying a magnetic field with a single magnet. Such direct capture (Fig. 1) and separation can also be performed with two kinds of antibodies. In this case, the first antibody is fixed on the colloidal particle as a spacer arm for immobilization of the second antibody for specifically capturing the target. The first antibody favors the orientation of the specific antibody immobilized via its Fc part onto the spacer arm–like antibody.
2. *Indirect separation.* The target molecule in a biological sample is first recognized by a specific antibody capable of reacting with the second antibody, which has been chemically grafted onto the surface of magnetic colloidal "carrier" particles. This method requires a thorough knowledge of molecular biology. These indirect methods are often more specific than the direct binding of antibodies [36] (Fig. 2).

On the physical level, the separation of magnetic particles as a function of the magnetic field can be explained and discussed basically on the basis of

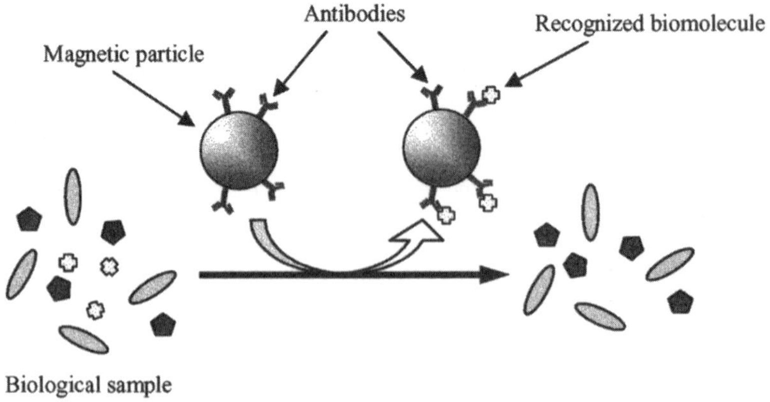

FIG. 1 Schematic illustration of direct specific separation of a target molecule.

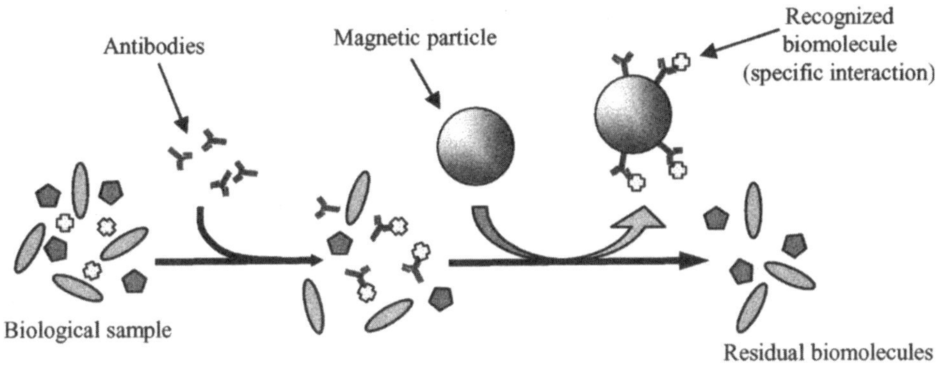

FIG. 2 Schematic illustration of indirect specific separation of a target molecule.

forces acting (magnetic force F_m and electroviscous force F_v) on the magnetic particles placed under the magnetic field (H), as summarized in the following expression (at equilibrium state):

$$F_v + F_m = 0$$

F_v is the friction force (or electroviscous force), which is basically the resistance to the displacement of the particle in a liquid medium; it is expressed as a function of the medium viscosity by Stokes' equation: $-6\,\pi\eta RV$, where R is the hydrodynamic radius of the particle, η is the viscosity of the medium, and V is the separation velocity.

F_m is the magnetic force due to the applied magnetic field (magnetic attraction force) and can be expressed as a function of various parameters:

$$F_m = \mu H \nabla H$$

where H is the magnetic field intensity, ∇H is the gradient of the magnetic field H, μ is the magnetic susceptibility ($\mu = \mu_0 \cdot M_s 4\pi R^3/3$), M_s is the saturated magnetization of the colloidal magnetic particles (principally related to the nature of iron oxide used), and μ_0 is the permittivity of the vacuum.

The separation speed (V) can therefore be expressed from the previous equations and as a function of saturated magnetization:

$$V = \frac{2\mu_0 M_s \cdot R^2}{9\eta} H \nabla H$$

Consequently, the separation speed increases with both the radius of the particle and the saturation of the magnetization. As expressed by the above equation, the saturated magnetization (M_s) is proportional to the content of magnetic material. In a given medium and in the presence of a fixed magnetic field, separation speed (V) therefore depends on the iron content, the magnetic properties of the particles, and the hydrodynamic size of the final microspheres. The above equation can be used to illustrate the parameters that affect the magnetic velocity of the particles under the magnetic field applied.

III. IMMUNOMAGNETIC SEPARATION

Immunoseparation involves the use of antibodies with a high affinity with the biomolecules to be separated, such as antigen or cells. Homogeneous magnetic beads are particularly well fitted for these kinds of applications because of their large, standardized surface of immunological reaction, which enables reliable and rapid biomedical diagnostic tests (such as enzyme-linked immunoassay, ELISA) [29], as various papers [33] and patents have reported. The quantification of immunological reactions (i.e., antibody/antigen/antibody, termed sandwich reaction process) is accomplished by the use of labeled antibodies; then the quantification is performed via fluorescence, chemiluminescence, phosphorescence, and radioactivity analysis [33].

A. Principle of Immunoassay

The term immunoassay refers to any method for measuring the concentration or amount of analyte in solution based on specific interaction between the antibody and the targeted antigen. This method (1) requires that the recognized analyte be physically separated from the residual biomolecules, and (2) employs a labeled antibody using radiolabel, fluorescent dye, enzyme, chemiluminescent molecule, or other label to measure or estimate the bound targets (Fig. 3). This method is termed competitive when the amount of measurable bound label is inversely proportional to the amount of analyte originally in solution. It is termed noncompetitive when the amount of measurable bound label is directly proportional to the amount of analyte originally present in the biological sample.

Large magnetic particles (10–100 μm) have been used in radioimmunoassays (RIAs) to detect and quantify nortriptyline, methotrexate, digoxin, thyroxine, and human placental lactogen. Such large particles must be stirred carefully to avoid sedimentation phenomena. Smaller particles, such as hollow glass or polypropylene or magnetic core (2–10 μm), have also been used and were evaluated in estradiol RIA, with nondrastic problems related to the particle sedimentation [32]. Nowadays various biomedical kits are available that are based on immunodetection and magnetic beads and either enzyme immunoassays (EIAs) or RIAs [34].

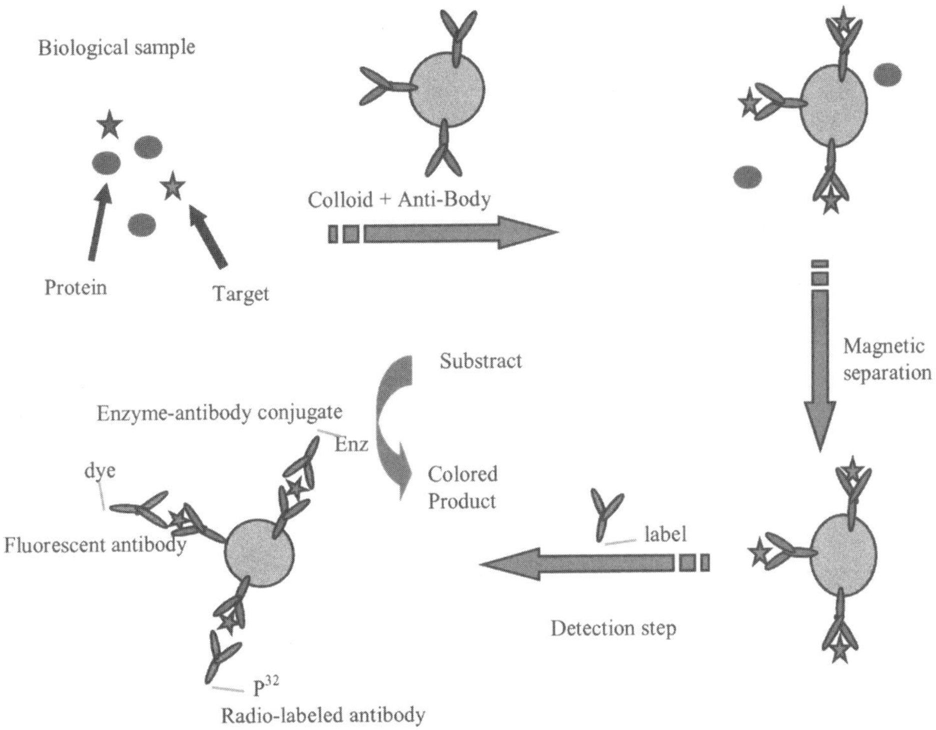

FIG. 3 Schematic illustration of immunoassay using labeled antibody.

B. Immunoaffinity Cell Binding

Magnetic particles are used for negative or positive cell selection. Because hepatocytes require highly specific binding, they are good test cells for developing magnetic particles for cell binding [37] and sorting. Microscopic observation of the interaction between hepatocytes and polystyrene magnetic beads shows that at 20°C the microvillus of the cells enfolds the colloidal particles. The combination of fluorescence microscopy technique analysis and specific immunological interaction between cells and magnetic particles is one of the best methods for visualizing and quantifying cells.

1. Negative Cell Selection

Magnetic particles are utilized for negative cell selection, i.e., in lymphocyte depletion, enrichment of monoclonal antibody–producing cells, and selection of nontumorous bone marrow cells [29]. For example, the latter is important in the autologous bone marrow transplantation used because the treatment for malignant disease often kills the bone marrow cells. The bone marrow cells removed and preserved for transplantation may, however, be contaminated with tumor cells. Magnetic beads seem to be one of the most effective methods reported for removing tumor cells from bone marrow [37].

Allogenic bone marrow transplantation is a treatment for several fatal diseases, including immunodeficiency, leukemia, and aplastic anemia. Graft versus host disease can be avoided by in vitro depletion of T lymphocytes from the donor marrow graft [37]. For example, magnetic beads can be used to remove neuroblastoma cells from bone marrow following isolation of the mononuclear cells by gradient centrifugation. Comparison between porous and compact particles indicates that compact particles are interesting in terms of sensitivity. Indirect coupling of antibody to the particles and use of a cocktail of monoclonal antibodies ensures good binding of the tumor cells. Performing this purging operation twice is reported to be more effective.

Magnetic polymer beads of submicrometer size have also been used to remove B-lymphoma cells and T cells from bone marrow. These procedures use direct or indirect coupling of antibodies to the magnetic particles. Purging B cells twice is again more effective, depleting 10^6 tumor cells compared with 10^4 for a one-time purge. Removal of T cells did not result in damage to stem cells.

2. Positive Cell Selection

In this case, cells are extracted from a simple biological mixture for specific reuse [37]. Magnetic particles are also utilized to separate eukaryotic cells; to perform HLA typing; to count, culture, and separate rare cells; and to detect or study receptor, genetic, precursor, and progenitor cells for transfixion or autografting (e.g., selection of CD34 precursor cells) [29].

Functionally active cells can be selected with direct or indirect binding on magnetic particles. The ratio of beads to target must be low to maintain cell activity. For example, T8 cells and T and B lymphocytes have been bound to polymer magnetic beads without any interference with the effector functions of the bounded cells. The particles released from the cells by overnight culture can be removed by a suitable magnetic separation with a single magnet [37].

Polymer magnetic particles are also used to purify CD8 T cells and HLA II cells, which can then be typed for HLA class I and II antigen respectively. The presence of magnetic beads during serological testing has no effect on the results. Compared to the standard method, the isolation is much faster, and a smaller volume of blood is necessary. The cells were isolated directly from the blood with polymer magnetic particles. After isolation, the cells were lysed, stained, and counted with microscopy techniques [37].

3. Flux Cytometry

This technique allows each kind of cell in a large population to be quantified by processing isolated cells in a detector. In the standard technique, nonmagnetic beads bearing an antibody specific for an antigenic epitope, the antigen, and a labeled antibody (enzyme, fluorescent or radiolabeled compound) are incubated together. Then the excess antibody and beads are washed away. However, the wash step, based on centrifugation, and the redispersion steps are cumbersome. The use of magnetic beads combined with flux cytometry techniques suppresses the drastic problems related to the washing steps. Silica magnetic beads have been explored for this process [32], as have polymer composite latexes.

C. Immunomagnetic Separation of Bacteria

Immunomagnetic separation is based on the specific antigen–antibody reaction common to all of the immunoassays presented above. The immobilization of antibodies on a solid phase spurred the development of "sandwich" tests or ELISA [38]. Selective separation is a function of the specific epitope of the biomolecule under consideration: the efficacy (in term of reactivity) of the immobilized antibodies on the surface of the magnetic particles depends primarily on their orientation on the surface of the particles. They are immobilized either via physical adsorption (electrostatic or hydrophobic interactions), or via a chemical grafting process. The latter method results in more stable immobilized antibodies and is thus used more often in various immunological diagnostic tests [39].

Most bacteria are surrounded by a cellular membrane that protects them from the external environment. They also have a flagella composed of a protein called the fimbria, and the bacteria can be bound to magnetic particles bearing the antifimbria antibody [40].

The bacteria tested for most often in the food industry are *Salmonella*, *Escherichia coli*, and *Listeria*. Numerous detection processes use magnetic particles to give faster results with a detection limit that is the same as or lower than that of conventional techniques [41].

Selective magnetic separation is a means of concentrating bacteria populations without having to go through a culture step to enhance the concentration. After magnetic separation, the target bacteria are visualized by physical methods such as scanning electron microscopy, fluorescence, radioactive labeling, or impedance (or conductivity) measurements of the culture medium. The selection of the method depends on the concentration of the bacteria solution.

After the capture step, the bacteria do not need to be released into any medium; they can develop from the surface of the considered colloidal particles. However, in some cases they are released mechanically by incubation or by magnetic agitation for 10–20 h at 37°C, which takes a long time. Introducing a protein with strong affinity for the magnetic beads or an enzyme capable of destroying the antibody/antigen link permits faster release [42].

Magnetic particle carriers have proven to be a simple means of detecting viruses difficult to obtain in sufficient quantities by artificial culture. Magnetic beads are then used to separate the bacteria for further applications like ELISA, such as spreading on a solid phase, amplification, and impedance measurement [29]. For example, the use of immunomagnetic beads to capture *Listeria* from environmental samples reduces test time and improves sensitivity, compared with the usual methods. The analysis is performed via commercial tests such as Listertest (Vicam, Watertown, MA, USA). Samples are mixed with immunomagnetic beads coated with anti-*Listeria* antibodies. After binding (i.e., biological recognition), the magnetic beads are isolated by applying a magnetic field; they are then planted on medium and incubated overnight. The next day, a replica is made on a thin plastic, and colorimetric detection is performed with anti-*Listeria* antibodies and anti-antibodies linked to alkaline phosphatase (enzyme). Because the method does not use enrichment, the number of *Listeria* colonies is related to the original level of contamination [43].

Using magnetic particles makes it possible to avoid some steps in food (disinfection, heating) that modifies the virus envelope and makes polymerase chain reaction (PCR) amplification unsuitable. This system is used, for example, to detect hepatitis A virus in shellfish. In their review, Olsvik et al. [38] describe examples of parasite detection with magnetic microspheres that reduce analysis time from 6 to at less 1 h and thus permit the diagnosis of a larger number of samples.

D. Collection of a Substance In Vivo

One of the most interesting uses of magnetic particles is for the in vivo capture, detection, concentration, or isolation of the target. This has been reported in various papers but only a few patents. One patent [30] describes a technique to isolate and remove an analyte from a body fluid. A molecule ligand, specific to the target, is immobilized on the surface of a magnetic particle. The particle is then introduced in vivo (in the body fluid) where it is tolerated long enough to enable the ligand to bind biologically to the target. Once the complex (ligand–target) is formed, it is retrieved from the body fluid by the application of a magnetic field.

The method is particularly fitted for detection of substance in gingival crevices for the diagnosis of periodontal disease. IgG against cachectin or interleukin-1 is chemically grafted to carboxyl groups. The magnetic particles are introduced in the cavity with a microdispenser. A magnetic field applied with a device anatomically compatible with the body cavity and adapted for the magnetic dispersion used then collects the magnetic particles.

Magnetic particles for in vivo applications are composed of a magnetic core bearing a hydrophilic-biocompatible biodegradable polymer shell layer or basic silica layer [32]; cells or tissues recognized by the particular bioaffinity adsorbent grafted on the magnetic particle are located, and therapeutic agents immobilized on the particles are delivered by magnetic direction to pathological sites.

IV. SEPARATION AND QUANTIFICATION OF PROTEINS AND ANTIBODIES

The sensitive detection of molecules and biomolecules is one of the most promising research fields. Today various and accessible techniques are used or combined to enhance this sensitivity. To this end, capillary electrophoresis (based on UV or fluorescence analysis), HPLC, fluorescence systems (flux cytometry), PCR, reverse transcriptase (RT) PCR, quantitative PCR, luminescence, phosphorescence, and colorimetric titration have been explored and are widely used. Each technique should be adapted to a given biomolecules, since the detection of nucleic acids (polyelectrolyte-like substances) and proteins (complex copolymers) is totally different. It is then of paramount interest to develop methods based on both specific and nonspecific isolation and concentration of biomolecules. For example, the detection of a protein in a low concentration of a biological sample has been reported and patented [44]. A small magnetic particle is coated with a given protein specific for the target protein; after a capture step leading to protein–protein interaction, the magnetic particles are concentrated in a small volume. Then, after a protein release process, the magnetic particles are separated from the sample and the protein–protein complexes incubated in the presence of a cleaving agent. The bond between the protein conjugates is then cleaved, and the captured protein is ellipsometrically analyzed. Various kinds of magnetic particles can be used for this detection and specific concentration, with particle sizes ranging from 1 to 10 μm. The magnetic part can be derived from iron oxide (i.e., ferrite). The standard illustration involves the use of particles coated with bovine serum albumin (BSA) to detect anti-BSA antibodies [44].

A. Removal of Magnetic Particles

After active cells are selected by immuno affinity magnetic separation, the particles detached from the cells by overnight culture can then be removed by a suitable magnet [37]. This may be due to shedding of the antigen involved in the binding, and this process may not be operative in all cases.

Another method under investigation involves the introduction of a layer of enzymatically degradable polysaccharides between the beads and the cells, or cleavable chemical bonds (S-S), or a Schiff base. Release can also be affected by enzyme digestion with proteinases, such as chymopapain, pronase, or trypsin, or a glycoprotease, to which some antibodies are susceptible, or by the application of ultrasound [29].

One method for detecting a protein involves coating a small magnetic particle with a protein specific for the target protein; the bond between the two proteins is cleaved after separation from the sample, by introducing the magnetic particle into a cleaving agent solution, composed of weak acid or alkaline solutions [44].

Anti-Fab antibodies can also be utilized for soft cell release because they inhibit the binding between antibody and antigen. Larger beads may detach more easily, thus leading to the release of captured cells.

It is worth noting that magnetic beads have been tested in many different immunological applications. These beads, which present a large surface area for binding and are easily separated from any medium, seem particularly well fitted to such applications. In addition to the general characteristics of magnetic beads discussed above, the functional groups on the surface of the particles appear very important for binding active antibodies.

V. NUCLEIC ACID SEPARATIONS

Magnetic separation and isolation of biomolecules always involves interactions between the colloidal particles and the substance to be separated. Interactions in addition to the extremely effective immunological ones have also been explored for separating different biological molecules. Most have been applied to the separation of nucleic acids. Examples of applications include the following [29]:

Purification and concentration of PCR products (or of single-strand DNA fragments) with biotinylated primers and streptavidin-bearing polymer particles, which are used to detect microorganisms or cellular genes, or to label single-strand probes

Specific purification of single-strand DNA or RNA with magnetic particles bearing oligonucleotides (single-strand DNA or RNA fragments of well-defined sequences)

Purification of proteins associated with nucleic acids, such as transcription factors, regulatory genes, and promoters

Purification of mRNA with magnetic beads coated with polythymidylic acid for in vitro translation or gene expression

Nonspecific concentration of nucleic acid followed by PCR or RT-PCR amplification

Specific DNA isolation and extraction to obtain purified nucleic acid without the use of any organic solvent as a precipitating agent is required in biomedical diagnostic tests, for which specificity, rapidity, and ease of handling are incontestably necessary. Some applications have been clearly described and discussed more precisely in the literature.

A. Detection of Mutations

To diagnose leukemia by the detection of mutated mRNA, polymer magnetic beads linked to oligo-dT_{25} are used to purify mRNA after acid guanidium phenol chloroform isolation of crude RNA. Then cDNA extension and PCR are performed in the presence of the magnetic beads.

Mutation detection involves the direct sequencing of the PCR products, obtained with a standard biotinylated primer. They are immobilized by binding onto streptavidin-coated polymer magnetic beads and then by attraction with a permanent magnet during the supernatant removal and washing procedures. Denaturation of the double-stranded PCR product into a single strand is performed in 0.1 M NaOH for 15 min. The single strand, attached to the beads, is then washed and sequenced directly [45].

B. Selection of Differentially Expressed mRNA

Biotinylated oligo-dT is used to synthesize a cDNA library, which is mixed and annealed to the mRNA extracted from the cell of interest. The cDNA/mRNA hybrids are removed with avidin-coated magnetic polymer beads. These two steps are repeated twice, and the remaining mRNA is reverse transcribed and cloned. This process is useful for identifying genes specifically expressed in differentiated cells [45].

Novagen's mRNA isolation system [46] uses superparamagnetic microspheres covalently coated with oligo-dT_{25} to selectively extract mRNA from a variety of sources (Fig. 4). After magnetic separation, the purified mRNA is eluted from the magnetic beads for a second round of purification.

C. Nucleic Acid Extraction and Concentration

The use of cationic colloidal supports (i.e., positively charged particles) in nucleic acid adsorption (i.e., nucleotide multimers) according to size has been explored in numerous studies. The interaction between cationic particles and such negatively charged polyelectrolytes is believed to be based on electrostatic

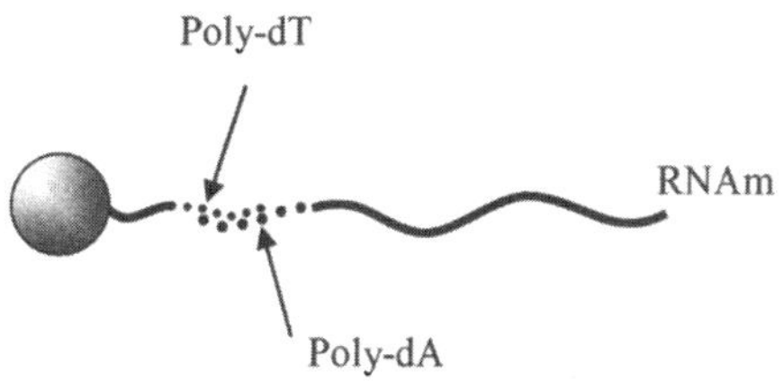

FIG. 4 Selective purification of mRNA with magnetic beads bearing dT oligonucleotide.

attractive forces. After nucleic acid adsorption and magnetic separation of the colloidal particles, conditions such as pH, salinity, and temperature are adjusted to release the nucleic acids from the support for amplification or direct analysis as schematized in Fig. 5.

For the Human Genome Project the Whitehead Institute at MIT developed a method called solid-phase reversible immobilization (SPRI) wherein DNA is captured onto carboxyl-modified encapsulated superparamagnetic microspheres. After the DNA is bound, the beads are washed with ethanol. The DNA is then eluted from the beads in a low ionic strength solution. This method leads to high-quality DNA template purification and can be used with major templates and enzymes responsible for sequencing [47]. It can also be applied to detecting hybridization between complementary nucleic acid sequences. A long polynucleotide that is partially complementary to a short oligonucleotide is chemically grafted onto the magnetic particles. The short probe is used as a spacer arm and does not bind directly to the particles but rather to the first bonded single-stranded nucleic acid. The specific capture of any nucleic acid sequence can then be performed by a hybridization process on the second, short oligonucleotide, as schematically presented in Fig. 6.

The hybridization can also be performed first in solution. The hybrids are separated from the unbounded probes by magnetic particles bearing a short oligonucleotide (with a specific sequence complementary to the target) to capture

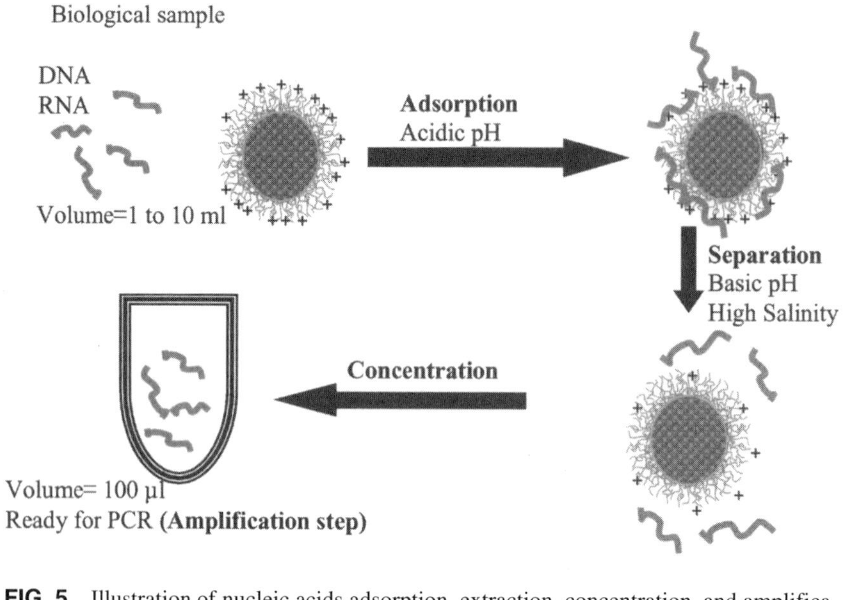

FIG. 5 Illustration of nucleic acids adsorption, extraction, concentration, and amplification.

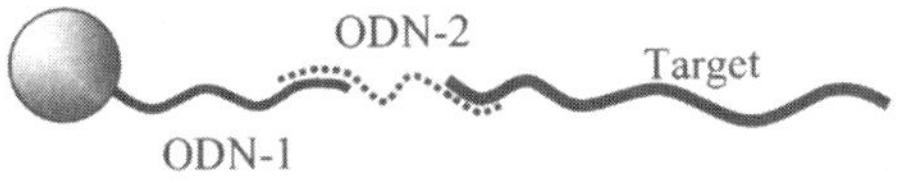

FIG. 6 Schematic illustration of specific capture of nucleic acid fragment by a double-hybridization process onto magnetic polymer particles. ODN-1 oligonucleotide (of a given sequence); ODN-2 oligonucleotides act as a spacer arm and specific to a given part of ODN-1.

the hybrids. The hybridization yield is 20% in solution and only 5% after binding to the beads.

Another method for capturing free nucleic acids in any biological sample uses positively charged magnetic microspheres for immobilization (via electrostatic interaction) of rRNA as well as DNA molecules. The nucleic acids released are not damaged after the capture of rRNA or DNA and can be amplified via a hybridization process with RT-PCR or PCR. The immobilization of rRNA is favored in the presence of urine combined with acetic acid. Efficient rRNA hybridization is obtained with the proper reagents. RNA/DNA hybrids are recovered from the buffer solution and the unhybridized probes left in solution. Captured polynucleotide probes can then be eluted from the beads with the appropriate conditions. Accordingly, nucleic acids can be purified from such biological samples as cell lysate and sputum. For these applications, various cationic magnetic particles are available; these include quaternary ammonium containing magnetic microsphere and poly-D-lysine-functionalized microspheres. In addition, magnetic particles with amine microspheres have been demonstrated to be compatible with a chemiluminescent nonisotopic assay [31] or electrochemiluminescence [48] with streptavidin-bearing magnetic particles.

We see from this description of the use of magnetic beads in molecular biology for various applications that various types of magnetic beads appear to be compatible with many enzyme reactions, even PCR or sequencing. Specific biomolecule separation is the application that has been most widely studied, probably because the major advantage of magnetic particles is their easy removability from any liquid simply, just under a magnetic field, even in an unclear and complex solution.

VI. PROTEIN IMMOBILIZATION ONTO REACTIVE MAGNETIC LATEXES

A. Immobilization of Enzyme

Thermally sensitive magnetic poly(NIPAM/MA/MBA) [*N*-isopropylacrylamide, NIPAM, methacrylic acid (MA), methylenebisacrylamide (MBA)] latex particles are well adapted to immobilize enzyme or drug carriers. Such composite particles contain small amounts of iron oxide material, which induce a low separation rate under a magnetic field. To enhance the magnetic separation velocity, the particles are thermally aggregated [49]. The thermal sensitive property is attributed to the properties of the poly(NIPAM) polymer. Trypsin has been covalently immobilized (via the well-known carbodiimide method) onto carboxylic thermally sensitive magnetic poly(NIPAM/MA/MBA) latex particles. The content of the carboxyl groups is reported to be an important factor in efficient protein binding. Thus, trypsin activity was higher with a high MA content. However, a high MA content enhances the colloidal stability (electrosteric stabilization) when thermal flocculation of the particles is needed. Because the protein (i.e., enzyme and antibody) activity may be sensitive to temperature changes during thermal flocculation (4°–30°C), the stability of the immobilized proteins has been tested. After repeated thermocycles (10 times), 95% of the initial enzymatic activity was retained [49].

B. Immobilization of Antibodies

Different procedures can be used to immobilize a cell-specific antibody to the colloidal particle [37]:

The monoclonal antibody can be chemically grafted directly onto the particles. This is not very efficient, probably because of the antibody orientation, and the need for a spacer arm between the antibody and the particle surface.

A polyclonal antibody can be first immobilized on the beads. After washing, the monoclonal antibody is then bound onto it. The polyclonal antibody in this case is used as a spacer arm and corrects the orientation of the monoclonal antibody far from the solid surface.

In a variant method for cell separation and cell sorting, the monoclonal antibody is first bound to the cells, and then captured onto the particles coated with polyclonal antibodies (indirect method). This technique requires cumbersome washing of the excess antibodies from the cells, leading to cell loss. Washing magnetic particles is much simpler and easy to perform.

The polymer magnetic particles are generally sufficiently hydrophobic to allow a relatively strong physical adsorption of antibodies (or proteinic molecules) via hydrophobic interaction. In addition, the presence of some reactive groups, such as hydroxyl function (−OH), can be used for chemical grafting of antibodies after activation with sulfonyl chlorides. It is also possible to obtain magnetic polymer beads bearing reactive groups such as $-NH_3$, hydrazine, −SH, or −COOH groups on the surface. The covalent coupling of antibody can then be performed via the $-NH_2$, −SH, −COOH, or −CHO (aldehyde) groups of the antibody in question [37].

Recently, thermally sensitive magnetic latexes have been prepared and evaluated as a support in immunoassay. After covalent coupling of anti-α-fetoprotein onto this stimulus-responsive hydrophilic support (via the carbodiimide activation process), performed in optimal coupling (salinity, temperature, and pH) conditions, the sensitive immunomagnetic colloidal particles were tested with ELISA to detect the α-fetoprotein [50].

VII. LABELING PROCESS

A. Accelerated Agglutination

Numerous standard detection methods involve autoagglutination and macroscopic or microscopic visualization. Agglutination techniques are advantageous because they require only one step, with no washes and no addition of reagents. Use of magnetic particles increases sensitivity and rapidity [34]. In this domain, two interesting patents (JP A 5180842 and WO A 8606493) describe a technique that uses a combination of magnetic and nonmagnetic particles. The nonmagnetic particles are colored and coated with the same antigen, which are incubated in two samples, one with and one without the antibody specific for the antigen used. Under a magnetic field, decreased color intensity in the nonagglutinated dispersion of the reaction mix indicates the presence of the antibody [34]. Polymers, possibly including magnetic particles, can be used as coloring agents [51].

Numerous patents describe a method utilizing two kinds of particles—one magnetic and coated with an antigen, the other nonmagnetic and coated with an antiantigen—which are incubated in two samples, one of which does contain and the other of which does not contain the antibody specific for the antigen. After agglutination, the quantity of nonmagnetic and nonagglutinated particles is an indication of the presence of the antibody. This method may not be as sensitive as RIA and EIA methods. The same type of assay has been also patented [34] in a version in which the magnetic and nonmagnetic particles are coated with an antitarget. After incubation of the particles with the biological sample, the magnetic particles are separated under a magnetic field, and the presence of the target is evidenced by observation of the nonagglutinated reaction mix.

For example, the magnetic particle and the nonmagnetic particle can be coated with an antibody against the antigen in question. The evaluation can be done by visual observation, or by measuring the optical density or fluorescence, phosphorescence, luminescence, or chemiluminescence. Nonmagnetic particles can be colored, fluorescent, chemiluminescent, phosphorescent, or luminescent [34]. The technique can identify an antigen in a biological sample (blood, urine, milk, saliva, serum); examples include the degradation product of fibrin, the inhibitor of plasminogen activator, plasminogen activator, C-protein, bacteria, viruses, or antibodies against bacteria, yeast, or viruses. The recommended size of the particles is in between 300 and 800 nm for the magnetic ones, and from 100 to 600 nm for the nonmagnetic.

B. Accelerated Sedimentation Process

Biomedical diagnostics are now targeting the development of basic and easy techniques for disease and bacteria detection. Various investigators [49,50] have described an immunoquantification method that is one illustration of such an application. This method discriminates reacting and nonreacting colloidal particles by whether they sediment or coat the sensitive walls of a special tube. After the biological sample is introduced into a recipient whose walls are coated with a substance (such as antibody) with a specific immunological affinity for the targeted substance, sensitive magnetic particles bearing captured analyte are then added to the sensitive tube with a specific immunological affinity for the captured substance. If the target is captured by the magnetic particles, they will mainly deposit on the walls of the recipient via specific immunological reaction. If the target is not captured, the sensitive colloidal particles will sediment in the tube. Use of magnetic particles enhances sensitivity via the concentration process, rapid immunological reaction on the walls under magnetic field, and finally high sedimentation of the composite particles. In addition, free magnetic particles can be easily removed magnetically [51,52].

C. Contrast Agent in Magnetic Resonance Imaging

The resolution and sharpness of the magnetic resonance image in body scanning depend on spin-spin relaxation time, which is reduced by paramagnetic materials [37,53]. A mono-sized polymer particle carrying magnetic iron oxide, given orally, can produce a negative black contrast in magnetic resonance imaging, thus eliminating the image of the gastrointestinal track and providing a clearer visualization of the other organs in the abdomen. The particle size varies according to the type of test. Small particle sizes should be used (diameter less than 3 μm) for parenteral use.

VIII. CHARACTERISTICS OF THE MAGNETIC MATERIALS FOR DIAGNOSTIC APPLICATIONS

The characteristics of the beads can be modulated according to their intended use. However, for most applications, generally preferred characteristics do not vary.

A. Particle Size and Size Distribution

Both particle size and size distribution are important in the utilization of magnetic particles in the biomedical field. These parameters may affect the efficiency of their utilization in a given application. For example, light intensity can be explored in flux cytometry, but only if particle size is homogeneous (i.e., narrow size distribution), to make multiple analysis and quantification possible

[29]. In any case, monodispersed magnetic particles are particularly appropriate for applications such as immunoseparation because of the large and standardized reaction surface, which makes homogeneous separation reliable and rapid.

B. Specific Surface Area

The separation rate of magnetic particles under a magnetic field is related to the amount of iron oxide in the composite beads and to the particle size [33]. To favor magnetic separation, the colloidal magnetic particles should contain the minimum needed of the magnetic material (less than 12 wt %) and the size should be sufficient [35]. Therefore very small magnetizable nanoparticles (such as ferrofluid dispersions with particle size less than 30 nm) are not easy to separate by applying a magnetic field. In fact, the force induced by the thermal agitation (i.e., Brownian diffusion motion) is higher than the magnetic force induced by the magnet [33]. Hence, magnetic separation of small hybrid particles is difficult.

Large magnetic particles (more than 1 μm in diameter) containing less than 12 wt % iron oxide can respond to a weak magnetic field; however, they generally tend to settle rapidly and have a more limited specific surface area (m^2/g) than smaller particles. Accordingly, the quantity of biomolecules immobilized is low and the sensitivity of the biomedical application reduced [32]. The particles containing less than 40 wt % highly magnetic iron oxide must be large enough (0.1–0.3 μm) to obviate high magnetic field intensity for a long period. Thus, the choice of particle size is difficult. The best magnetic separation is obtained with a large particle size, but sedimentation is higher and sensitivity low due to the low available specific surface for biomolecule immobilization. Thus, the choice of the particle size and the amount of iron oxide may be adapted for different applications.

C. Thermoflocculation of Stimulus-Responsive Magnetic Latexes

Thermally sensitive magnetic latexes have been prepared and evaluated in biomedical domains. Such stimulus-responsive composite particles flocculate as the incubation temperature and salt concentration of the medium increases. The flocculation behavior is related to the reduction of both steric and electrostatic stabilization. Such thermoflocculation is reported to be a reversible process because of the low critical solution temperature of poly[*N*-alkyl(meth)acrylamide] derivative [35,50].

The hydrophilic magnetic and thermosensitive poly(NIPAM/MA/MBA) latex particles have a large surface area and are therefore well fitted to immobilize enzymes or antibodies. Moreover, after thermoflocculation, the particles can be quickly separated by a magnetic field. These thermally sensitive particles can satisfy both a large surface area for biomolecule immobilization and rapid magnetic separation.

Thermally sensitive flocculation (Fig. 7) appears to be a very attractive innovation that resolves the antagonism between rapid magnetic separation and surface area. It is worth noting, however, that this property requires a temperature cycle for flocculation and dispersion, which may not be compatible with all biotechnological devices and applications, especially enzymatic activity.

D. Therapeutic Applications

Magnetic polymer particles must have the appropriate properties for use as a carrier in various in vivo therapeutic applications. The polymer must be biodegradable if the microspheres are to remain inside the organism [55]. The physicochemical and colloidal characteristics of the magnetic carrier and the properties of the continuous phase must be considered. The intensity and distance from the magnetic field applied must be taken into account to check the magnetic particle distribution and displacement before degradation. They must have a biocompatible surface that does not cause an immune system reaction and that generates an efficacious and specific targeting action. Finally, the diameter of the particles is also a crucial parameter. Risks of phagocytosis are limited with particles of diameter greater than 12 μm [26]. On the other hand, small particles (less than 1 μm) can penetrate the capillaries. These act more specifically but do not resist phagocytes.

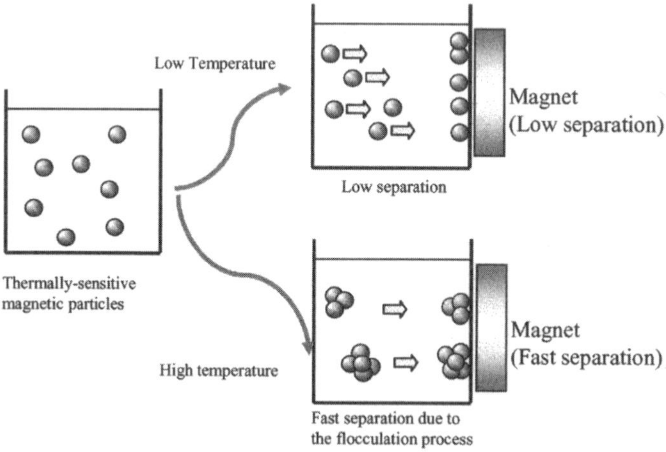

FIG. 7 Schematic illustration of hydrophilic magnetic thermally sensitive latex particles via temperature flocculation process.

The total quantity of medicine introduced into the organism is sometimes limited by toxic effects beyond a certain concentration [26]. By injecting the active ingredient in a polymer capsule or matrix, it is possible to control the diffusion and release of the desired product in the organism. This also avoids repetitive administration of the medicine. The targeting is then well controlled by applying a magnetic field that improves the localization of the chemical product and prevents the particles from being captured by the reticuloendotheliosis system [58]. Because the product is locally concentrated, the total dose administered can be reduced, with a corresponding reduction of toxicity problems. For example, the total dose of doxorubicin, injected freely, is 100 times higher than that introduced with a magnetic carrier for an equivalent concentration at the treatment site.

The active agent is released outside the particles as a function of mechanisms that depend on the properties of the polymer. Adriamycin and doxorubicin in lyophilized albumin magnetic microspheres, conserved at 4°C, diffuse spontaneously at 37°C, in an aqueous solution of 1 g/L NaCl at a rate that depends on the cross-linking of the matrix during synthesis. In the case of certain so-called smart materials, variation of the medium (i.e., pH, temperature, ionic strength) leads to modification of the polymer's properties [59]. For example, certain thermally sensitive polymer-based particles are hydrophilic and expand below their volume phase transition temperature, i.e., poly(NIPAM) particles below 32°C. On the other hand, the particles shrunk above the transition temperature, thus releasing the encapsulated active ingredient, as illustrated in Fig. 8.

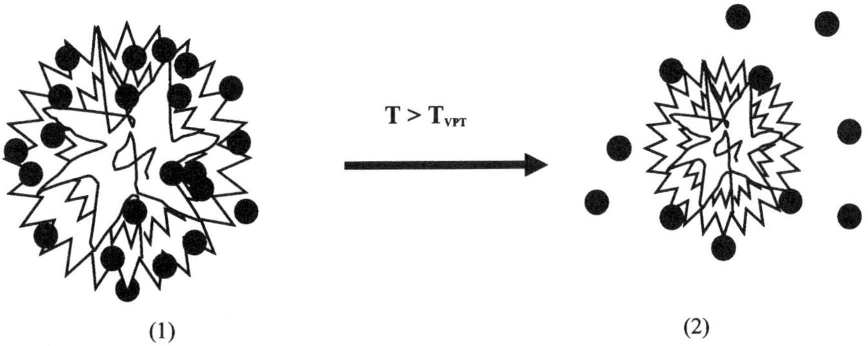

FIG. 8 Release of the active ingredient by controlling incubation temperature. T_{VPT} is the volume phase transition temperature, the T_{VPT} in the case of poly (*N*-isopropylacrylamide) microgel particles is nearly 35°C. (1) Below the T_{VPT} the particles are swollen by the active agent, and (2) above the T_{VPT} the active agent is released.

E. Specific Cell Extraction

Generally, the extraction of diseased cells by magnetic particles is based on the chemical difference between healthy and infected cells. Two steps should be considered. The first involves targeting the microspheres to a specific site by applying a magnetic field. Magnetic guidance makes it possible to reach areas that are difficult to access, and its efficacy usually depends on the properties of the carrier, such as particle size and stability in the environment. Secondly, the infected cells are then recognized and immobilized on sensitized particles. Capture yield varies from patient to patient since recognition depends on the properties of tumor cells and their affinity vis-à-vis the antibodies immobilized on the colloidal particles. Hancok [36], Rembaum [59] and Ugelstad [60] initiated the use of magnetic particles to purify marrow. The method was then extended to other tumors such as the treatment of lymphocytes, leukemia and lung cancer. Using chemotherapy and radiation to cure cancerous tumors can affect neighboring healthy cells. To avoid this problem and reduce secondary effects, treatment is carried out outside the organism (ex vivo) [61]. Specific antigens of tumor cells are fixed on the surface of magnetic particles and make it possible to capture only infected cells [62]. These are then extracted by applying a magnetic field and treated chemically outside the organism (ex vivo) before being reinjected into the patient. Another example of this type of promising application is the study of vascular problems by extracting endothelial cells via coating of magnetic particles with lectin.

IX. CONCLUSION

The advantages of colloidal magnetic particles are, on the one hand, the speed of detection and production of the analysis result and, on the other hand, the simplicity of separation, which makes the system easy to automate. Its specificity and reliability are identical to those already proven for nonmagnetic polymer-based particles (reactive latexes).

In view of the applications in the biomedical field, several properties, such as particle size, size distribution, surface polarity, surface charge density, and amount of magnetic material (i.e., iron oxide content), must be considered carefully for magnetic polymer particles.

The real interest of magnetic particles in biomedical field is clearly evidenced by the extensive literature. These particles have been used in many different applications in biomedical diagnostics and therapeutics. The most common utilization is for rapid specific separation of various biological products, often in association with immunological or nucleic acid interactions. Magnetic particles are also utilized in biomolecule concentration and as a marker in various labeling applications. Most of the criteria required by the magnetic particles are similar in many applications: small particle size to avoid sedimentation, good magnetic response (i.e., high magnetic content in the colloidal particles), high colloidal stability, narrow size distribution, and no residual magnetization when the magnetic field is suppressed (i.e., no remanence). For more literature concerning magnetic particles in medicine and biology, the readers can consult Arshady's *Microspheres, Microcapsules and Liposomes*, Volume 3 [63].

REFERENCES

1. Furusawa, K.; Nagashima, K.; Anzai, C. Synthetic process to control the total size and component distribution of multilayer magnetic composite particles. Colloid Polym. Sci. **1994**, *272*(9), 1104–1110.
2. Chen, T.; Somasundaran, P. Preparation of core-shell nanocomposite particles by controlled polymer bridging. J. Am. Ceram. Soc. **1998**, *81* (1), 140–144.
3. Bamnolker, H.; Nitzar, B.; Gura, S.; Margel, S. New solid and hollow, magnetic and non-magnetic, organic-inorganic monodispersed hybrid microspheres: synthesis and characterization. J. Mater. Sci. Lett. **1997**, *16*, 1412.
4. Caruso, F.; Mohwald, H. Preparation and characterization of ordered nanoparticle and polymer composite multilayers on colloids. Langmuir **1999**, *15*, 8276.
5. Sauzedde, F. Elaboration de latex magnetique hydrophiles fonctionalisés en vue d'applications dans le diagnostic medical. PhD thesis, Claude Bernard University, Lyon, France, 1997.
6. Van Herk, A.M. In *Polymeric Dispersion*: *Principles and Applications*; Asua, J.M., Ed.; NATO ASI Series E: Applied Sciences Vol. 335, 1996: 435–462.
7. Sugimoto, T. In *Fine Particle*: *Synthesis*, *Characterization*, *and Mechanisms of Growth*; Surfactant Science Series Vol. 92. Marcel Dekker: New York, 2000.
8. Neoh, K.G.; Tan, K.K.; Gob, P.L.; Huang, S.W.; Kang, E.T.; Tan, K.L. Electroactive polymer SiO2 nanocomposites for metal uptake. Polymer **1999**, *40*, 887.
9. Sohn, B.H.; Cohen, R.E. Processible optically transparent block copolymer films containing superparamagnetic iron oxide nanoclusters. Chem. Mater. **1997**, *9*, 264.
10. Kommaredi, N.S.; Tata, M.; Vijay, T.J.; McPherson, G.L.; Herman, M.F. Synthesis of superparamagnetic polymer-ferrite composites using surfactant microstructures. Chem. Mater. **1996**, *8*, 801–809.
11. Neveu-Prin, S. Synthèse et caractérisation de nanoparticules magnétiques—elaboration de ferrofluides et des capsules magnétiques. PhD thesis, Pierre–Marie Curie University, 1992.
12. Kwon, O.; Soic, J. New interaction effects with a superparamagnetic latex. J. Magnet. Magnet. Mater. **1986**, *54–57*, 1699–1700.
13. Tamai, H.; Sakura, H.; Hirota, Y.; Nishiyama, F.; Yasuda, H. Preparation and characteristics of ultrafine meta particles immobilized on fine polymer particles. J. Appl. Polym. Sci. **1995**, *56*, 441–449.
14. Patton, W.F.; Kim, J.; Jacobson, B.S. Rapid high-yield purification of cell surface membrane using colloidal magnetite coated with polyvinylamine: sedimentation versus magnetic isolation. Biochim. Biophys. Acta **1985**, *816* (1), 83–92.

15. Kemshead, J.T.; Treleaven, J.G.; Gibson, F.M.; Ugallstad, J.; Rembaum, A.; Philip, T. Removal of malignant cells from marrow using magnetic microsphere and monoclonal antibodies. Prog. Exp. Tumor Res. **1985**, *29*, 249–255.
16. Gupta, P.K.; Hung, C.T. Magnetically controlled targeted micro-carrier system. Life Sci. **1989**, *44*, 175–186.
17. Reynold, C.H.; Anan, N.; Beshah, K.; Huber, J.H.; Shaber, S.H.; Lenkinski, R.E.; Wortman, J.A. Gadolinium-loaded nanoparticles: new contrast agents for magnetic resonance imaging. J. Am. Chem. Soc. **2000**, *122*, 8940.
18. Ogan, M.; Schmiedl, U.; Moseley, M.; Grodd, W.; Paajenen, H.; Brasch, R.C. Albumin labeled with Gd-DTPA: an intravascular contrast enhancing agent for magnetic resonance blood pool imaging: preparation and characterization. Invest. Radiol. **1987**, *22*, 665.
19. Gupta, P.K.; Hung, C.T. Albumin microspheres 1: physico-chemical characteristics. J. Microencapsulation **1989**, *6*, 427–462.
20. Molday, R.S.; Mackenzie, D. Immuno-specific ferromagnetic iron-dextran reagents for labeling and magnetic separation of cells. J. Immunol. Meth. **1982**, *52*, 353–367.
21. Kilocoyne, S.H.; Gorisek, A. Magnetic property of iron dextran. J. Magnet. Magnet. Mater. **1998**, 177–181, 1457–1458.
22. Schütt, W.; Grüttner, C., Häfeli, U.; Zborowski, M.; Teller, J.; Putzar, H.; Schümichen, C. Applications of magnetic targeting in diagnosis and therapy—possibilities and limitations: a mini-review. Hybridoma **1997**, *16* (1), 109–117.
23. Sauzedde, F.; Elaissari, A.; Pichot, C. Hydrophilic magnetic polymer latexes. 1. Adsorption of magnetic iron oxide nanoparticles onto various cationic latexes. Colloid Polym. Sci. **1999**, *277*, 846–855.
24. Charmont, D. Preparation of monodisperse, magnetizable, composite metal/polymer microspheres. Prog. Colloid Polym. Sci. **1989**, *79*, 94–100.
25. Lee, J.; Senna, M. Preparation of monodispersed polystyrene microspheres uniformly coated by magnetite via heterogeneous polymerization. Colloid Polym. Sci. **1995**, *273* (1), 76–82.
26. Langer, R. New methods of drug delivery. Science **1990**, *249* (4976), 1527–1533.
27. Chan, D.C.F.; Kirpotin, D.B.; Bunn, P.A. Synthesis and evaluation of colloidal magnetic iron oxides for the site specific radiofrequency-induced hyperthermia of cancers. J. Magnet. Magnet. Mater. **1993**, *122*, 374.
28. Newman, R.B.; Stevens, R.W.; Gaafar, H.A. Latex agglutination test for the diagnosis of the *Hemophilus influenzae* meningitis. J. Lab. Clin. Med. **1970**, *76*, 107–113.
29. Blondeaux, A.; Caignault, L. La séparation sur bille de taille uniforme. Biofuture **1995**, *147*, 3–8.
30. Rossomando, E.F.; Hadjimichael, J. Method of using magnetic particles for isolating, collecting and assaying diagnostic ligates. US Patent 5,158,871. University of Connecticut.
31. Lyle, J.A.; Nelson, N.C.; Reynold, M.A.; Waldrop, A.A. Polycationic supports for nucleic acid purification, separation and hybridization. EP 0,281,390.
32. Chegnon, M.S.; Groman, E.V.; Josephson, L.; Whitehead, R.A. Magnetic particles for use in separations. EP 0,125,995. Advanced Magnetics Inc.

33. Josephson, L.; Menz, E.; Groman, E. Solvant mediated relaxation assay system. WO 9117428. Advanced Magnetics Inc.
34. Esteve, F.; Amiral, J.; Padula, C.; Solinas, I. Method for assaying an immunological substance using magnetic latex particles and non-magnetic particles. WO 9504279. Societe Diagnostica-Stago.
35. Kondo, A.; Kamura, H.; Higashitani, K. Development and application of thermo-sensitive magnetic immunomicrospheres for antibody purification. Microbiol. Biotechnol. **1994**, 99–105.
36. Hancok, J.P.; Kemshead, J.T. A rapid and highly selective approach to cell separations using an immunomagnetic colloid. J. Immunol. Meth. **1993**, *164*, 51–60.
37. Ugelstad, J.; Berge, A.; Ellingsen, T.; Auno, O.; Kilaas, L.; Nilsen, T.N.; Schmid, R.; Stenstad, P.; Funderud, S.; Kvalheim, G.; Nustad, K.; Lea, T.; Vartdal, F.; Danielsen, H. Monosized magnetic particles and their use in selective cell separation. Macromol. Symp. **1988**, *17*, 177–211.
38. Olsvik, O.; Popovic, T.; Skjerve, E.; Cudjoe, K.S.; Hornes, E.; Ugelstad, J.; Uhlen, M. Magnetic separation technique in diagnostic microbiology. Clin. Microbiol. Rev. **1994**, *7* (1), 43–54.
39. Meza, M. In *Scientific and Clinical Applications of Magnetic Carriers*; Häfeli, U., Schütt, W., Teller, J., Zborowski, M., Eds.; Plenum Press: New York, 1997: 303–309.
40. Safarik, I.; Safarikova, M.; Forsythe, S. The application of magnetic separations in applied microbiology. J. Appl. Bacteriol. **1995**, *78*, 575–585.
41. Ugelstad, J.; Olsvik, O.; Schmid, R.; Berge, A.; Funderud, S.; Nustad, K. Immunoaffinity separation of cells using monosized magnetic polymer beads, in 1993, Vol. 16. 224–244.
42. Ugelstad, J.; Kilaas, L.; Aune, O. In *Advances in Biomagnetic Separation*, Uhlén, M., Homes, E., Olsvik, O., Eds.; Eaton: Oslo, 1993: 1–19.
43. Mitchell, B.A.; Milbury, J.A.; Brookins, A.M.; Jackson, B.J. Use of immunomagnetic capture on beads to recover *Listeria* from environmental samples. J. Food Prot. **1994**, *57*, 743–745.
44. Glaever, I. Diagnostic method and device employing protein-coated magnetic particles. US Patent 4,018,886. General Electric Company.
45. Mizutani, S.; Asada, M.; Wada, H.; Yamada, A.; Kodama, C. Magnetic separation in molecular studies of human leukemia. In *Advances in Biomagnetic Separation*; Uhlen, M., Homes, E., Olsvik, O., Eds.; Eaton: Oslo, 1994: 127–133.
46. McCormick, M.; Hammer, B. Straight A's mRNA isolation system, rapid, high quality poly(A)$^+$ RNA from diverse sources, innovations. 1994, No. 2 (Nov), Novagen Inc.: Madison, WI.
47. Smith, C.; Ekenberg, S.; McCormick, M. The polyATract magnetic mRNA isolation system: optimization and performance. Promega Notes 1990, No. 25, Promega: Madison, WI.
48. Kenten, J.H.; Gudibande, S.; Link, J.; Willey, J.; Curfman, B.; Major, E.; Massey, R.J. Improved electrochemiluminescent label for DNA probe assays: rapid quantitative assays for HIV-1 polymerase chain reaction products. Clin. Chem. **1992**, *38*, 873–879.

49. Kondo, A.; Fukuda, H. Preparation of thermo-sensitive magnetic hydrogel microspheres and application to enzyme immobilization. J. Ferment. Bioeng. **1997**, *84*, 337–341.
50. Sauzedde, F.; Elaissari, A.; Pichot, C. Thermosensitive magnetic particles as solid phase support in an immunoassay. Macromol. Symp. **2000**, *150*, 617–624.
51. Kashlara, A.; Otsuka, C.; Ishikawa, K. Process for preparing particles having monodisperse particle size. EP 0,275,899.
52. Matte, C.; Muller, A. Procédé et Dispositif Magnétique d'Analyse Immunologique sur Phase Solide. EP 0,528,708. Pasteur Sanofi Diagnosics.
53. Matte, C.; Muller, A. Procédé et Dispositif Magnétique d'Analyse Immunologique sur Phase Solide. FR 91 09,242. Diagnostics Pasteurs.
54. Lauterbur, P.C. Magnetic Gels Which Change Volume in Response to Voltage Changes for MRI. US Patent 5,532,006. The Board of Trustees of the University of Illinois.
55. Ibrahim, A.; Couvreur, P.; Roland, M.; Speiser, P. New magnetic drug carrier. J. Pharma. Phamacol. **1982**, *35*, 59–61.
56. Ji, Z.; Pinon, D.I.; Miller, L. Development of magnetic beads for rapid and efficient metal-chelate affinity purifications. J. Anal. Biochem. **1996**, *240*, 197–201.
57. Morimoto, Y.; Okurama, M.; Sugibayashi, K.; Kato, Y. Biomedical applications of magnetic fluids. 2. Preparation and magnetic guidance of magnetic albumin microsphere for site specific drug delivery in vivo. J. Pharmacobio Dynam. **1981**, *4*, 624–631.
58. Hoffman, A.S. Intelligent polymers in medicine and biotechnology. Macromol. Symp. **1995**, *98*, 645–664.
59. Rembaum, A.; Yen, S.P.S.; Molday, R.S. Synthesis and reactions of hydrophilic functional microspheres for immunological studies. J. Macromol. Sci. Phys. **1979**, *A13* (5), 603–632.
60. Treleaven, J.G.; Gibson, F.M.; Ugelstad et al. Removal of neuroblastoma cells from bonemarrow with monoclonal antibodies conjugated to magnetic microspheres. Lancet **1984**, *14*, 70–73.
61. Lea, T.; Vartdal, F.; Nustad, K.; Funderud, S.; Berge, A.; Ellingsen, T.; Schmid, R.; Stenstad, P.; Ugelstad, J. Monosized, magnetic polymer particles: their use in separation of cells and subcellular components and the study of lymphocyte function in vitro. J. Mol. Recog. **1988**, *1* (1), 9–31.
62. Haukanes, B.-I.; Kvam, C. Application of magnetic beads in bioassays. Biotechnology **1993**, *11*, 60–63.
63. Arshady, R. Microspheres, Microcapsules and Liposomes, Volume 3: Radiolabeled and Magnetic Particles in Medicine & Biology. Citus Books: London, 2001.

2

The Agglutination Test

Aggregation of Antibody-Coated Latexes in the Presence of Antigens

SERGE STOLL Université de Genève, Geneva, Switzerland

LAHOUSSINE OUALI and EMILE PEFFERKORN CNRS-Institut Charles Sadron, Strasbourg, France

I. INTRODUCTION

Advances in our knowledge of the mechanism governing aggregation of latex particles in the presence of various destabilizing agents and interest in mechanisms controlling agglutination tests used in biological assays led us to apply the methodology derived from simulation and experimental studies of colloid aggregation to the dynamic characteristics of specific agglutination tests. Polystyrene latex particles have been used as carrier for antigen and antibody reaction in agglutination tests for which the apparition of a visible agglomerate determines the presence of the cross-linking antigen: antibody-coated particles suspended in a buffer solution become unstable when the antigens present in the solution create links between particles [1–3]. The instability of bare polystyrene latex particles may be similarly initiated by a relatively small ligand capable of forming bonds between two single colloids [4,5]. In the latter situation, the ligand concentration determines the kinetics and the mode of aggregation. Actually, on both sides of the optimal flocculation dose leading to fast aggregation the rate of aggregation is strongly decreased. In the absence of antigen, the antibody-supporting particle should be fully stable and the fate of the system should not depend on the ionic strength and/or the presence of other constituents in the suspending phase [6].

This chapter presents two systems differing by the nature of the process leading to agglutination. The first system strictly resembles homocoagulation whereby one ligand binds two similar particles: the antibody-coated particles in the presence of the IgM antigen. The success of the agglutination test depends on the reaction between the gamma globulin molecule IgG adsorbed on the particles and the IgM rheumatoid factor present in the dispersion [7]. The second system (pregnancy test) is more complex since agglutination requires the presence of two chemically different collector systems. Human chorionic gonadotropin (HCG) present in the blood and urine of pregnant women is detected by means of two different monoclonal antibodies. In the commercial assay of Biomerieux (HCG-Slidex), each antibody is covalently linked to a polystyrene latex and the addition of HCG proteins to a mixture of the two sensitized latexes leads to visible agglutination within three minutes. The great selectivity of the test depends on the fact that HCG contains two moieties, each reacting specifically with one type of antibody. In the absence of HCG, no agglutination is observed [8]. Such a process refers to heterocoagulation insofar as adsorbed HCG does not induce aggregation between latex particles sensitized with the same antibody when they are colliding [9–14].

The framework of the chapter concerns the following objectives: (1) determination of the temporal variation of the aggregate mass frequency; (2) establishment of the dynamic scaling laws of the aggregation kinetics; and (3) drawing of conclusions with regard to the mechanism of the aggregation process. The mass frequency curve of the aggregates (usually not represented in this report) was found to satisfactorily characterize the nature of the process, and the intrinsic reactivity of site–ligand interactions could be determined from this characteristic [15–19]. Typical relationships were further established between the exponents of the scaling laws describing the temporal variation of the weight and number average masses of the aggregates. In the first part of the chapter we summarize the fundamental principles required for presentation of the experimental results and the ensuing discussion. The section devoted to antigen characteristics is reduced to a schematic description of the modes of interaction of the different components, a more precise description of the system being irrelevant to the present report, as this investigation was not designed to provide any interpretation of the observed phenomena based on the biochemical nature of the antibodies and the antigens. On the other hand, the section concerning the presentation of typical results of simulation and experiments is more detailed.

Particle counting was implemented to investigate the rate of destabilization of antibody-coated particles in the presence of antigens. Typical results corresponding to aggregation experiments carried out employing bare colloids coated by spherical ligands at different degrees were first presented to illustrate the method and to show the correlation between the aggregate mass characteristics and the mechanism of the aggregation processes.

II. THE AGGREGATION PROCESS

A. Characteristics of Homocoagulation

From numerical studies, irreversible aggregation processes may be considered as being diffusion or reaction limited [16,20]. For aggregation processes resulting from collisions between particles presenting surface site–ligand complexes, the frequency of efficient collisions, which depends on the surface density of active sites as expressed by the relationship of La Mer, has to be taken into account [21]. The occurrence of inefficient collisions leads the aggregation between incompletely coated particles to develop with features of reaction-limited aggregation processes. The kinetics of the aggregation may be expressed in terms of the increase in the weight $S(t)$ and number $N(t)$ average masses of the aggregates with the aggregation time t:

$$S(t) \approx t^{z};\ N(t) \approx t^{w} \tag{1}$$

$$\tau = 2 - (w/z) \tag{2}$$

where the exponents z and w are functions of the surface density of ligand groups. This model ignores the possibility that a particular coating may induce the diffusion-limited aggregation for which $w = z$, as has been determined to occur when rigid ligands are responsible of the colloid aggregation (Fig. 1) [4,5].

The intrinsic reactivity of the site–ligand interaction depends only on the system [4]. This parameter is calculated from the slope of the aggregate mass distribution $c(n,t)$ at small values of the aggregate mass n, according to the usual relationships:

$$c(n,t) \approx n^{-\tau} \tag{3}$$

In Eq. (3), the exponent τ determines the intrinsic reactivity of the aggregates toward further aggregation. For simplicity, when the colloids appear to be poorly reactive τ is equal to 1.5 [22,23]. When the ligand is a rigid polymer, values of 1.5 and 1.64 have been found to correspond to small and high surface coverages, respectively. The slightly higher value of 1.64 indicates the additional difficulty of establishing new site–ligand interactions when the degree of surface occupation is large [4]. For a degree of coverage θ between 0.03 and 0.20, Figure 1 shows z and w values to be close to 1—a figure that usually characterizes the diffusion-limited aggregation process. This results from the systematic establishment of site–ligand interactions when one ligand already fixed to the surface of one particle systematically encounters another free site belonging to a second particle, as has been established on the basis of topological constraints only.

B. Characteristics of Heterocoagulation

Heterocoagulation of latexes of types A and B implies that collisions involving identical particles do not lead to aggregation, successful collisions being those involving A and B particles. If each collision between suitable particles systematically leads to interparticle sticking, the process is diffusion limited. On the contrary, when chemical reactions are required before sticking succeeds, the

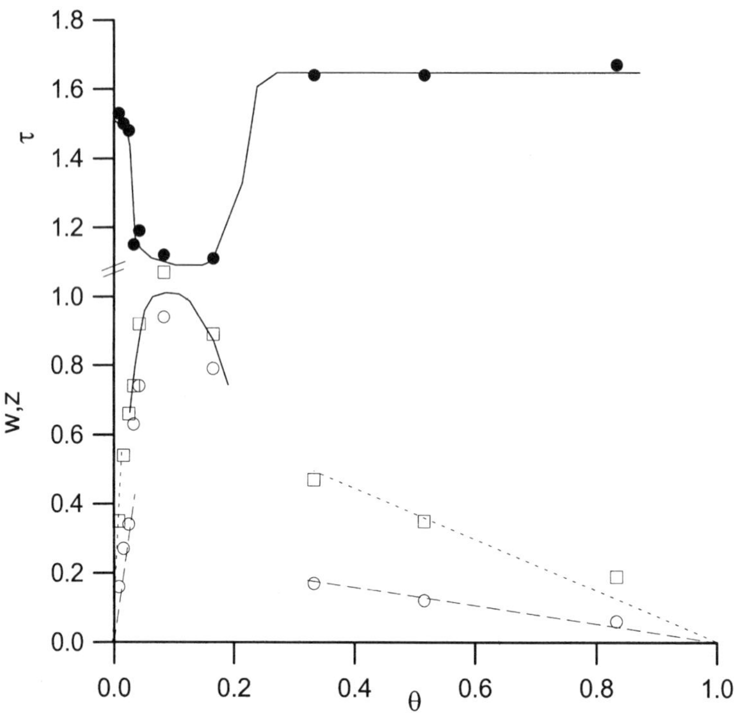

FIG. 1 Representation of the variation of the exponents w (○), z (□) and τ (●) as a function of the degree of surface coverage θ determined using Eqs. (1) and (2) for aggregation of polystyrene latex particles of diameter 900 nm induced by adsorption of a spherical rigid ligand of diameter close to 80 nm. Full surface coverage is expected to occur when 500 ligand molecules are deposited on the latex surface.

process is reaction limited. Studies of the influence of unsuccessful A-A and B-B collisions on the aggregation rate in simulations of heterocoagulation provided the exponents of the dynamic scaling laws as a function of the composition x of the system. The dynamic exponents $z(x)$ and $w(x)$ of the heterocoagulation kinetics expressed by the temporal variations of the weight $S(t)$ and number $N(t)$ average masses were found to be strongly correlated to and independent of the aggregation ability (mobility and reactivity) [11,12]:

$$S(t) \approx t^{z(x)}; \; N(t) \approx t^{w(x)} \tag{4}$$

For reaction-limited heterocoagulation, which seems to control aggregation processes based on site-ligand interactions, the slowing of the kinetics resulting from the existence of inefficient collisions is expressed as follows:

$$z^2(x)/z^2 = (2/3)\ w(x)/w \quad (5)$$

where z and w are the exponents of the temporal variations of $S(t)$ and $N(t)$ derived from simulation studies of homocoagulation. The value of the exponent $z(x)$ of the temporal variation of $S(t)$ is given by:

$$z(x) = zx^k \quad (6)$$

where k, always smaller than 1, depends only on the nature of the process. Therefore, assuming the efficiency of collisions $P(\theta)$ to be correlated with the kinetics of aggregate growth, $P(\theta)$ is expressed by a power law of the surface coverage. Three extreme situations can be envisaged [24,25]:

1. The antigen is initially immunoadsorbed on latex A with a degree of coverage equal to θ' while latex B remains free of antigen. The probability $P(\theta')$ of doublet formation is given by:

   $$P(\theta') \approx (2\theta')^m \quad (7)$$

2. The antigen is rapidly immunoadsorbed on the two latexes A and B with a degree of coverage equal to $\theta = \theta'/2$ on each latex, with the probability of doublet formation being given analogically by:

   $$P(\theta) \approx 2(\theta)^m\ (1 - \theta) \quad (8)$$

   In the case of asymmetrical coverage θ_1 and θ_2, with $\theta_1 + \theta_2 = \theta'$, Eq. (8) is modified to:

   $$P(\theta_1,\theta_2) \approx (\theta_1)^m + (\theta_2)^m - [\theta_1(\theta_2)^m + \theta_2(\theta_1)^m] \quad (9)$$

   Asymmetrical situations were studied by Meakin and Djordjevic, who determined the reaction rate as a function of the composition of the system and the functionality of the latices [10]. These authors concluded that S(t) and N(t) always followed a power law behavior after an initial non-scaling growth regime.
3. The kinetics of the immunoadsorption limits the aggregation rate. In this situation, the probability of doublet formation increases with antigen concentration and the aggregation is delayed.

III. GENERAL ASPECTS OF ANTIBODY–ANTIGEN INTERACTIONS

The Y-shaped IgG antibody takes up approximately the volume of a lens-shaped spheroid with a diameter of 14.08 nm and a thickness of 2.46 nm. Using results of Ref. 26, we calculated that a latex particle of 900 nm diameter adsorbs approximately 7500 gamma globulin molecules at pH 8.6. The immunogenicity of IgG molecules is a function of the surface coverage, and apparently the unsaturated layer has the same immunogenetical characteristics as the heat-denatured molecule [27]. The normal behavior of the physically adsorbed IgG thus requires full surface coverage by IgG molecules.

The HCG glycoprotein (30,000 Da) consists of two subunits α and β. Subunit α presents some structural similarities with other pituitary glycoprotein hormones, while subunit β contains a carboxy terminal peptide specific to HCG [28–31]. Monoclonal anti-αHCG and anti-βHCG antibodies were prepared and covalently bound to polystyrene latexes. The HCG-mediated link established between the two sensitized latexes may be schematized as shown in Fig. 2. Obviously, three-dimensional alternate combinations of the latexes are expected to emerge during aggregate growth.

IV. MATERIALS AND METHODS

A. The System IgG-IgM

The bare polystyrene latex particles were prepared under emulsifier-free conditions by the two-step procedure [32]. Particles of 900 nm diameter developed in 0.03 mol/L NaCl at pH 7.3 a surface potential of −13.5 mV. The density was 1.045 g/mL. Fully IgG-coated latex particles were prepared by using the conventional procedure. The antigen IgM was provided by Biomerieux at a concentration of 30 IU/mL. One liter of the aqueous phase at pH 7.3 employed to disperse the IgG-coated particles contained 7.2 g Glycocoll, 1.6 g NaCl, and 1.5 g trishydroxyaminomethane. This preparation constitutes the suspending medium of the slide test.

Figure 3 shows the electrophoretic mobility of the system IgG-IgM as a function of the amount of IgM added per gram of IgG-coated latex particles

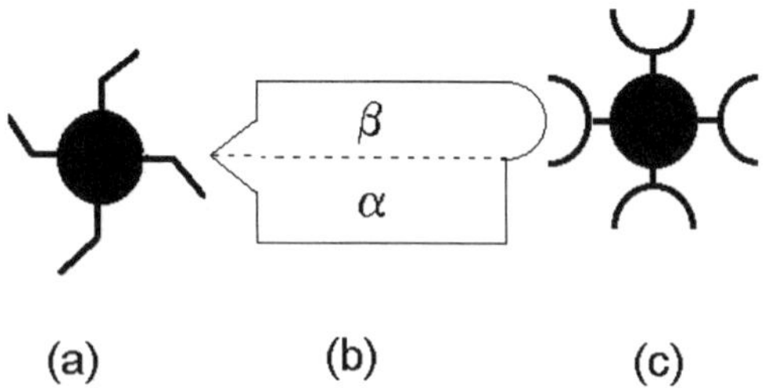

FIG. 2 Schematic representation of the elementary structure of the link formed by the HCG molecules (b) between two latex particles, one bearing the monoclonal anti-αHCG determinant (a) and the other bearing the monoclonal anti-βHCG determinant (c).

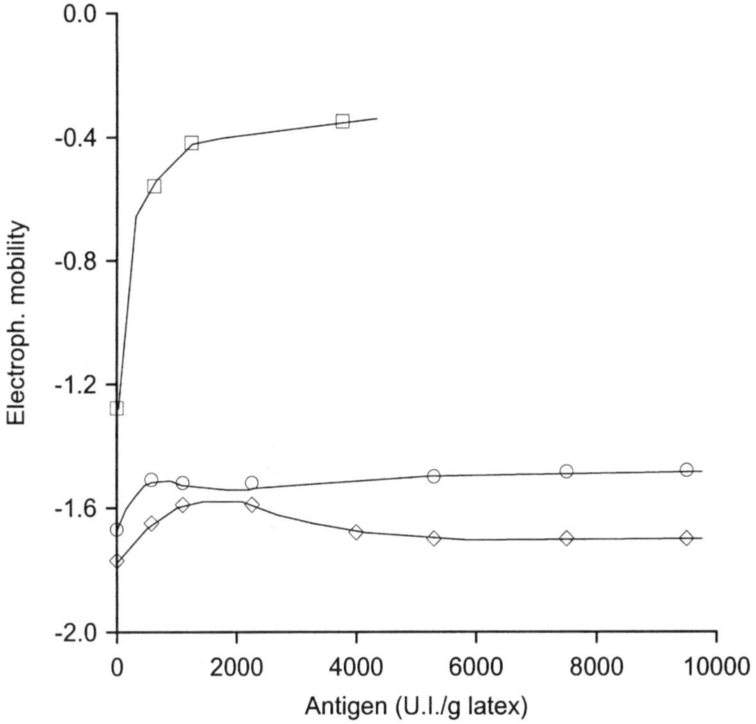

FIG. 3 Representation of the electrophoretic mobility (μm s^{-1} V^{-1} cm) of antibody-coated latex particles as function of the concentration of antigen in the solution for the different systems: IgG-IgM in 0.03 mol/L NaCl at pH 7.3 (□), anti-βHCG-HCG (○) and anti-αHCG-HCG (◇) in 0.05 mol/L NaCl at pH 7.6.

dispersed in the standard solution. Increased surface coverage with IgM molecules strongly decreases the electrophoretic mobility of the coated particles.

B. Pregnancy Test

The bare latex was a polystyrene latex from Rhône-Poulenc having a diameter of 0.8 μm and bearing both sulfate and carboxylate surface groups. The density as determined from titration was 4.1 μC/cm^2 for the sulfate groups and 6.73 μC/cm^2 for the carboxyl groups. The carboxyl groups of the latex particles were treated with the water-soluble activating agents *N*-hydroxysulfosuccinimide and 1-(3-dimethylaminopropyl)-3-ethylcarbodiimide [33,34]. The particles were then centrifuged and washed with the buffer Tris 0.1 M at pH 8.5 to eliminate excess activating agent.

The activated particles were dispersed in the anti-αHCG or anti-βHCG antibody solution. On completion of adsorption, the sensitized latexes were centrifuged and washed using the same Tris buffer. The amount of anti-αHCG or anti-βHCG antibody covalently coupled was determined by the difference in antibody concentration before and after reaction (depletion procedure) and found to be equal to 2.37 mg/m^2 for the two proteins. The two types of sensitized latexes were dispersed in a specific buffer of ionic strength 0.15.

HCG protein was obtained from Sigma Chemical Company at a concentration of 1000 IU/mL in 0.01 M sodium phosphate buffer at pH 7.2. The protein was used without further purification, and dilutions were carried out with the buffer used to disperse the latex suspensions.

Figure 3 shows the electrophoretic mobility of the two latex systems coated with the anti-αHCG and anti-βHCG molecules as a function of the amount of HCG molecules added per gram of coated latex particles dispersed in 0.05 M NaCl at pH 7.6. Adsorption of HCG molecules only slightly affects the electrophoretic mobility of the sensitized latex particles.

C. Aggregation Experiments

All experiments were performed at 18°C. Experiments with the system IgG-IgM were carried out using a particle concentration of 4.5 g/L. Experiments with the HCG protein were performed using latex suspensions at a concentration of 10 g/L. In the latter case, two different procedures were employed. In a first series of experiments, the two sensitized latexes were mixed simultaneously with HCG proteins, homogenization of the system being effected by gentle tumbling performed twice before the onset of perikinetic aggregation. In a second series of experiments, one sensitized latex was mixed with a given amount of HCG protein and the mixture was left at rest for 24 h, at which time the electrophoretic mobility of the protein–latex complex was determined to be constant. The complex was then added to the other sensitized latex to initiate aggregation. To avoid particle and aggregate sedimentation in the resting dispersions, experiments were carried out in a mixture of salt, deuterium oxide, and water with a density of 1.045 g/L. Actually, to determine the aggregation characteristics, investigation were conducted for one week, whereas in the agglutination test the systems working at high concentrations produce aggregates under slow agitation within a few minutes [35].

D. Particle Counting

The aggregate mass distribution was determined with the Coulter counter (Coultronics) [36,37], and the particular method of analysis being implemented to calculate the mass distribution $c(n,t)$ was reported in [38]. Precautions were taken to perform nondestructive mass distribution analyses. Samples of the IgG-

IgM systems were diluted up to 100 times in aqueous solutions of identical pH and electrolyte concentrations as the latex dispersion. For the pregnancy test, samples were diluted in Isoton (Coultronics), which is an electrolyte solution of ionic strength close to 0.15. The aggregate mass distribution $c(n,t)$ vs. n (the aggregate mass being defined as the number of particles composing the aggregate) is periodically determined from the histogram, and the average masses $S(t)$ and $N(t)$ are defined by:

$$S(t) = \frac{\Sigma_n n^2 c(n,t)}{\Sigma_n nc(n,t)} \quad N(t) = \frac{\Sigma_n nc(n,t)}{\Sigma_n c(n,t)} \tag{10}$$

The reduced concentration $c(n,t)S^2(t)/N_0$ is usually expressed as a function of the reduced mass $n/S(t)$ by:

$$c(n,t)\ S^2(t)/N_0 = F[n/S(t)] \tag{11}$$

Equation (11) allows portrayal of the mass frequency determined at different moments by a single function where N_0 represents the initial concentration of latex particles in the suspension and $c(n,t)$ the concentration of aggregates composed of n particles at time t.

V. AGGREGATION CHARACTERISTICS OF THE IgG-IgM SYSTEM

The agglutination characteristics were determined for three systems differing by the IgM concentration (IU/g latex).

A. Aggregation in the Presence of 3.77×10^3 IU of IgM per Gram of IgG-Coated Latex

The weight and number average masses of aggregates are represented as a function of aggregation time in Fig. 4; the vertical dashed line indicates the transition between two rates of aggregation. In the first period, the aggregate masses grow like $S(t) \approx N(t) \approx t^{0.66}$ whereas in the second period $S(t) \approx N(t) \approx t^{0.33}$ is valid. The variations of $S(t)$ and $N(t)$ are concomitant and the values of $z = w$ being smaller than 1 determine the diffusion-limited process to be reversible [39,40]. Collisions between latex particles systematically lead to aggregate formation, but some links are unstable and break up, reverting to the situation that existed prior to collision. This mechanism that preserves the aggregate mass polydispersity, $S(t)/N(t) = 2$, is different from the reaction-limited process for which some collisions fail to induce aggregation. The bell-shaped mass frequency curves (not represented) corresponding to the establishment of the asymptotic domain ($z = w = 0.33$) indicates that the distribution of 3770 IU IgM molecules on the

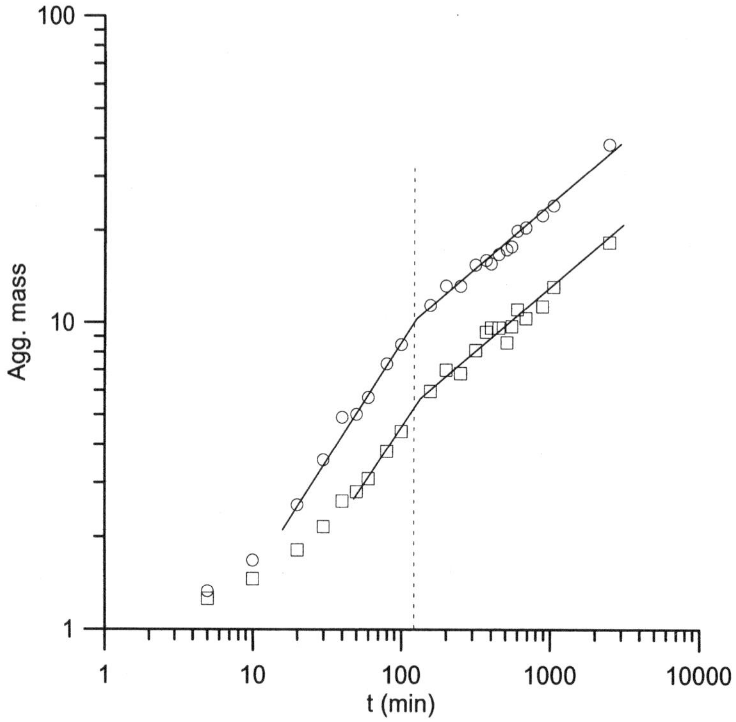

FIG. 4 Concentration of IgM = 3770 IU/g IgG-latex: representation of the number $N(t)$ (□) and $S(t)$ (○) average masses of the aggregates as a function of the aggregation time (min) (log-log scale).

IgG-coated latex allows each collision between coated latex particles to be efficient toward aggregation (τ being close to 1 as shown in Fig. 1). However, the rate of aggregate growth indicates that some links are not definitely established and break up [41]. The retardation effect due to the partial reversibility in aggregation on $S(t)$ and $N(t)$ may be expressed as follows:

$$S(t) \approx N(t) \approx t/t^{f} \tag{12}$$

with $f = 0.66$.

B. Aggregation in the Presence of 1.24 × 10³ IU of IgM per Gram of IgG-Coated Latex

As previously, the aggregation rate expressed by the increase in $S(t)$ and $N(t)$ with time is described by scaling laws that are different for periods I to IV, as shown in Fig. 5. Four domains of growth of the aggregate average masses $S(t)$ and $N(t)$ can be determined to occur with increasing periods of aggregation:

Domain	Period (min)	$S(t) \approx$	$N(t) \approx$	$S(t)/N(t) =$
I	20–100	$t^{0.90}$	$t^{0.23}$	$t^{0.67}$
II	100–500	$t^{0.90}$	$t^{0.90}$	3.0
III	500–1000	$t^{0.46}$	$t^{0.90}$	$t^{-0.44}$
IV	1000–4000	$t^{0.46}$	$t^{0.46}$	2.5

The following comments may be made:

1. In domain I, the aggregation is reversible but the fragmentation does not revert to the preexistent situation since $N(t)$ increases very slowly while $S(t)$ increases at a faster rate. This figure reveals break-up of a given aggregate to release two aggregates of very different masses. For example, when the collision between two doublets gives rise to quadruplets and the subsequent fragmentation leaves a triplet and a single particle, $N(t)$ remains constant whereas $S(t)$ increases from 2 to 2.5. This process leads to an increasing mass polydispersity as shown in Fig. 6 for the corresponding period I.
2. In domain III, the existence of a bell-shaped mass distribution demonstrates that aggregates of a given mass are preponderant and that the suspension does not contain isolated particles. Since $S(t) \approx N^{1/2}(t)$, one may conjecture that aggregate break up does not restore the preexisting situation but produces fragments of nearly equivalent masses.
3. In domains II and IV, the situation corresponding to paragraph 1 is reproduced and the increase of aggregate mass described by $z = 1$ is limited by a concomitant fragmentation. Deviation form irreversible diffusion-limited aggregation is expressed by Eq. (12) with $f = 0.10$ and 0.54 in domains I and IV, respectively. Actually, fragments of small mass resist break up at a greater extent than do aggregates of higher masses. When 1240 IU of IgM molecules is distributed on the surface of 1 g IgG-coated latex, the aggregate stability is higher than for a coating of 3770 IU, as observed in paragraph 1.

C. Aggregation in the Presence of 0.62×10^3 IU of IgM per Gram of IgG-Coated Latex

The temporal variations of the average masses $S(t)$ and $N(t)$ represented in Fig. 7 show the aggregation process to develop extremely slowly up to doublet formation and to really start after 600 min. In domain I, the aggregation develops with features of the reaction-limited aggregation process: $S(t)$ and $N(t)$ increase

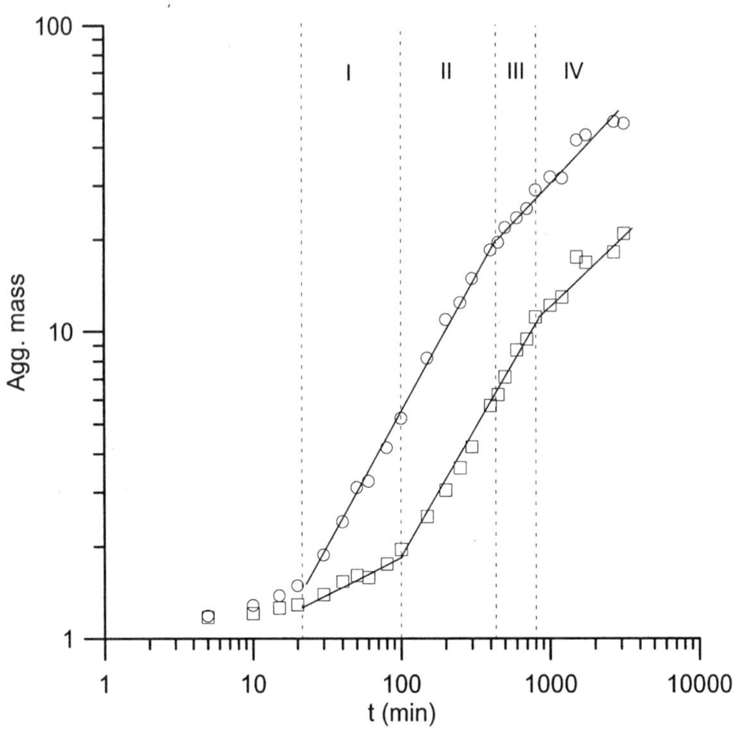

FIG. 5 Concentration of IgM = 1240 IU/g IgG-latex: representation of the number $N(t)$ (□) and $S(t)$ (○) average masses of the aggregates as a function of the aggregation time (min) (log-log scale).

as $\exp(t)$ (not represented) [7]. This behavior later develops in domain II according to power laws as usually observed in the reaction-limited aggregation process:

$$S(t) \approx t^{2.18}, \quad N(t) \approx t^{1.27} \tag{13}$$

and the value $\tau = -1.42$ of slope of the reduced mass distribution agrees with that calculated using Eq. (3). In domain III, the mass distribution is represented by a bell-shaped distribution, which indicates that for such large aggregates the reaction-limited process tends to a situation where all collisions succeed insofar as a very large number of attempts are possible when one aggregate is exploring the total external envelope of the nearest neighbor. Once more, this behavior is usually observed in the long term in reaction-limited aggregation.

When the IgG-coated latex is only covered at 620 IU of IgM molecules per

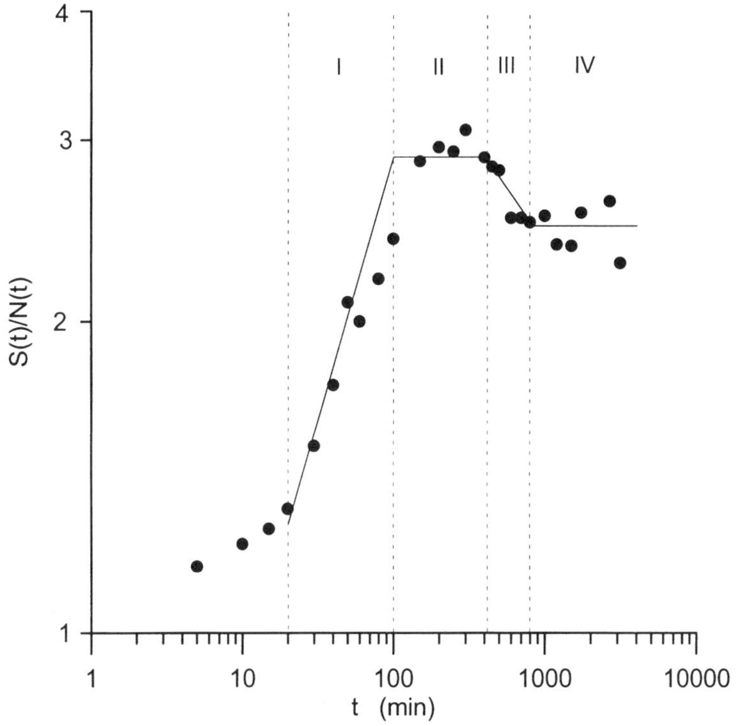

FIG. 6 Concentration of IgM = 1240 IU/g IgG-latex: representation of the aggregate mass polydispersity $S(t)/N(t)$ as a function of the aggregation time.

latex, the aggregation process is reaction limited insofar as the process develops with features that were evidenced in numerical simulation of the reaction-limited process, and establishment of doublets resulting from the sticking of two single particles is a very time-consuming process [22]. However, when the doublet concentration has reached a typical value, collisions between doublets and further between aggregates of greater masses lead to efficient sticking and the resulting aggregates appear to be extremely stable. Conversely to situations described in paragraphs 1 and 2, in the present case, the aggregation is not reversible, and the initially observed delay cannot be related to fragmentation but must be attributed to the fact that the low density of the IgM molecules on the IgG-coated latex does not favor efficient collisions.

When the system displays bell-shaped mass frequencies, which means that aggregates of a given mass are preponderant in the system, it should be noted that the aggregate mass present at the greatest concentration depends on the IgM

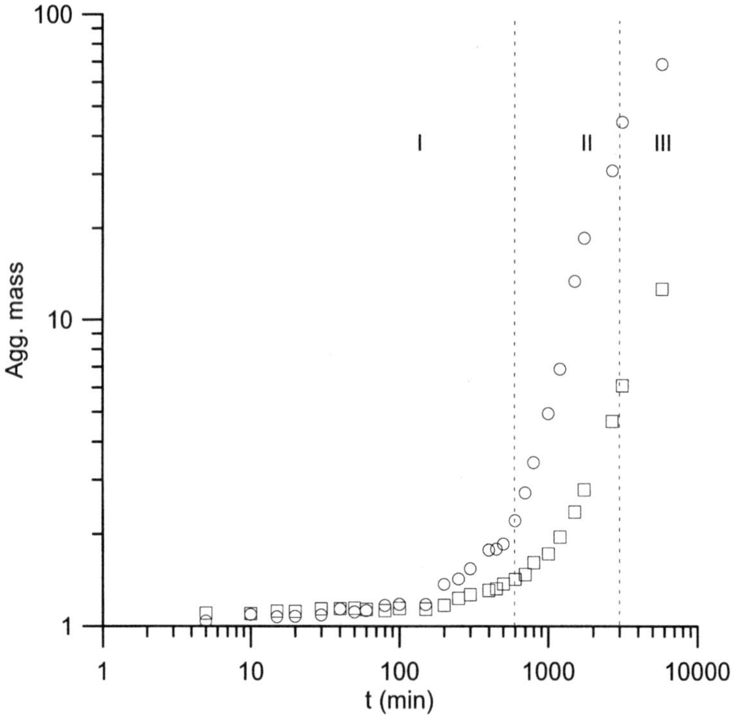

FIG. 7 Concentration of IgM = 620 IU/g IgG-latex: representation of the number $N(t)$ (□) and $S(t)$ (○) average masses of the aggregates as a function of the aggregation time (min) (log-log scale).

concentration as represented in Fig. 8 using the reduced concentration and mass. When the concentration of IgM molecules passes form 620 to 1240 and 3770 IU/g latex, the mass of aggregates present at the greatest concentration goes from 0.18 to 0.28 and $0.44 \times S(t)$, respectively. For an unknown system, the position of the concentration peak may serve to estimate the concentration of IgM molecules as shown in Fig. 9, without reference to additional tests.

VI. AGGREGATION CHARACTERISTICS OF THE ANTIBODY-SENSITIZED LATEX–HCG SYSTEM

The agglutination characteristics were determined for three systems differing by the HCG concentration (IU/g latex).

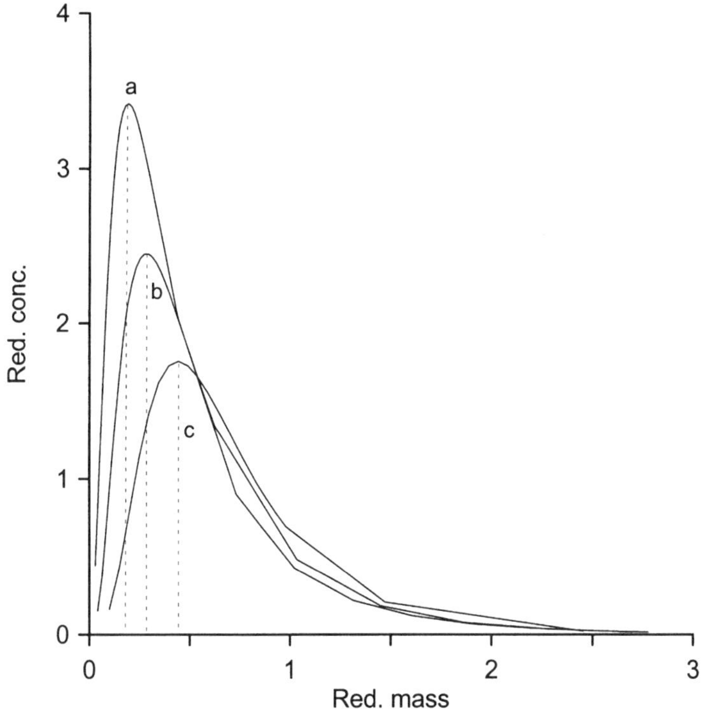

FIG. 8 Representation of the reduced concentration of aggregates as a function of their reduced mass for experiments carried out in the presence of (a) 620, (b) 1240, and (c) 3770 IU/g IgG-coated latex.

A. Aggregation as a Function of the Concentration of HCG per Gram of Antibody-Sensitized Latex

The three constituents—latex-anti-αHCG, latex-anti-βHCG, and HCG molecules—were mixed simultaneously. The parameter of interest was the HCG concentration, which was set to 16.6, 33, 166, and 833 IU/g latex. Figure 10 represents the average masses $S(t)$ and $N(t)$ to increase as a function of aggregation time for aggregation experiments carried out in the presence of 33 IU/g antibody-coated latex particles. Two domains I and II of variation of $S(t)$ and $N(t)$ are observed, and the dashed line indicates the transition toward a slowing down of the aggregation rate after an initial period of 200 min. The global aggregation rate increases strongly with the HCG concentration because the time corresponding to the transition from domain I to domain II decreases from

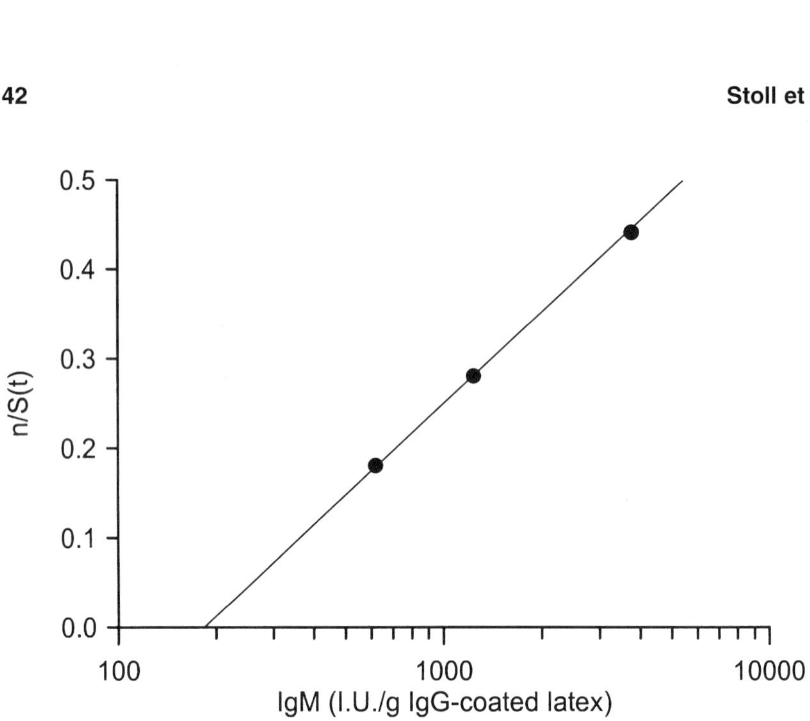

FIG. 9 Representation of the reduced mass $n/S(t)$ of aggregates present at the greatest concentration as a function of the concentration of IgM molecules expressed by IU/g IgG-coated latex.

2000 to 100 min when the HCG concentration increases from 16.6 to 833 IU/g latex.

The determination of the initial slope τ of the aggregate mass distribution [Eq. (3)] provides direct information on the intrinsic reactivity of the colliding particles and aggregates. Results of Fig. 11 shows a strong decrease in τ with the square root of the HCG concentration, and the value of τ may serve to determine the HCG concentration without doing additional reference tests, as noted for the IgG-IgM system.

The decrease of the slope of τ as a function of $[\mathrm{HCG}]^{0.5}$ leads to estimate $m = 0.5$ in Eq. (8) and $\theta \ll 1$ since the plateau of the electrophoretic mobility of the systems [latex-antiαHCG + HCG] and [latex-antiβHCG + HCG] is only obtained for HCG concentrations greater than 5000 IU/g latex.

B. Heterocoagulation Involving One Latex Bearing the Complex [Antibody-Antigen] and the Second Latex Bearing Only the Second Antibody (33 IU/g Latex)

The complexes [antibody specific for the αHCG determinant + HCG (complex A) and [antibody specific for the βHCG determinant + HCG (complex B)] were

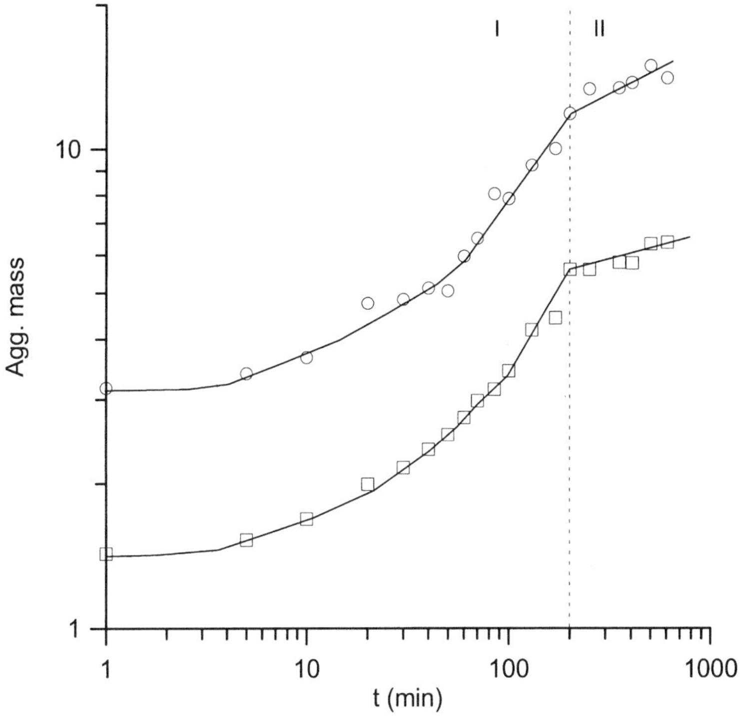

FIG. 10 Representation of the weight $S(t)$ (○) and number $N(t)$ (□) average masses of the aggregate as a function of time for the system aggregating in the presence of 33 IU HCG/g antibody-coated latex.

prepared by mixing the two sensitized latexes separately with HCG protein at the concentration of 33 IU/g latex and leaving the suspensions at rest for 24 h. Two types of aggregation experiments were carried out concomitantly. Complex A was added to the latex sensitized with the antibody specific for the βHCG determinant; likewise, complex B was added to the latex sensitized with the antibody specific for the αHCG determinant. Figure 12 represents the corresponding variations of $S(t)$ (curve a) and $N(t)$ (curve b) as a function of time (log-log scale). In both cases the initial fast regime abruptly changed to a slow regime. The complete similarity of the aggregation rates led us to conclude that the intrinsic reactivities of the latex bearing the antibody–HCG complex and the latex bearing only the antibody are similar for the two determinants. However, heterocoagulation in systems that contain the complex A (or B) and the other sensitized latex but no free HCG molecules produces aggregates of lower masses than does the experiment carried out with the system containing free

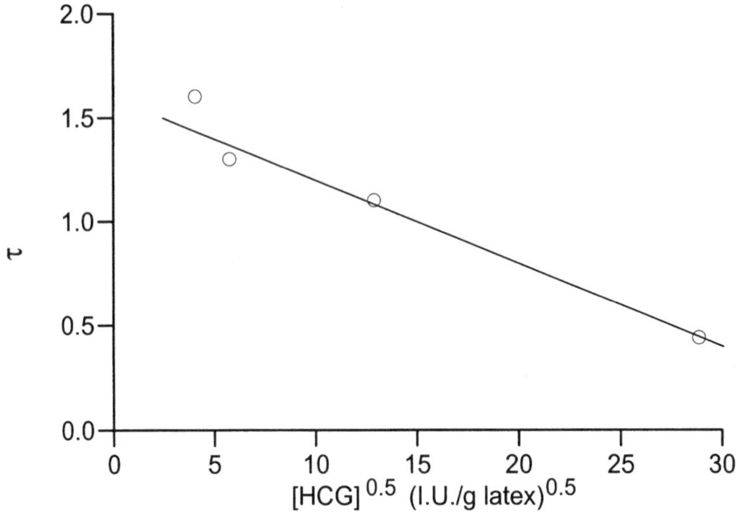

FIG. 11 Representation of the slope τ [Eq. (11)] of the reduced mass distribution in the domain I (see Fig. 10) of the aggregation process as a function of the square root of the HCG concentration expressed in IU/g antibody-coated latex.

HCG molecules and the two types of sensitized latexes. The transition from domains I to II occurs at the same time in all of these experiments.

C. Coagulation Involving One Latex Bearing the Antibody Specific for the αHCG Determinant, a Second Latex Bearing the Antibody Specific for the βHCG Determinant, and the Solubilized HCG Protein at the Concentration of 33 IU/g Latex

The three constituents were mixed simultaneously and left at rest to allow the onset of perikinetic aggregation. Experiments were carried out using different proportions (20%, 50%, and 80%) of a given sensitized latex, and the corresponding variations of $S(t)$ and $N(t)$ as a function of time (log-log scale) are shown in Fig. 13. Under these conditions, the kinetics and extent of aggregation were determined to be functions of the composition of the system [10–13]. When the two latexes are present in equal proportions, the average masses $S(t)$ (curve c) and $N(t)$ (curve d) increase at their most rapid rate to values of $S(t) = 13$ and $N(t) = 5$, after which aggregation abruptly slows. One notes that the masses increase faster than in the previous experiments of heterocoagulation [curves a and b in Fig. 13, for the system complex A (or B) and the other

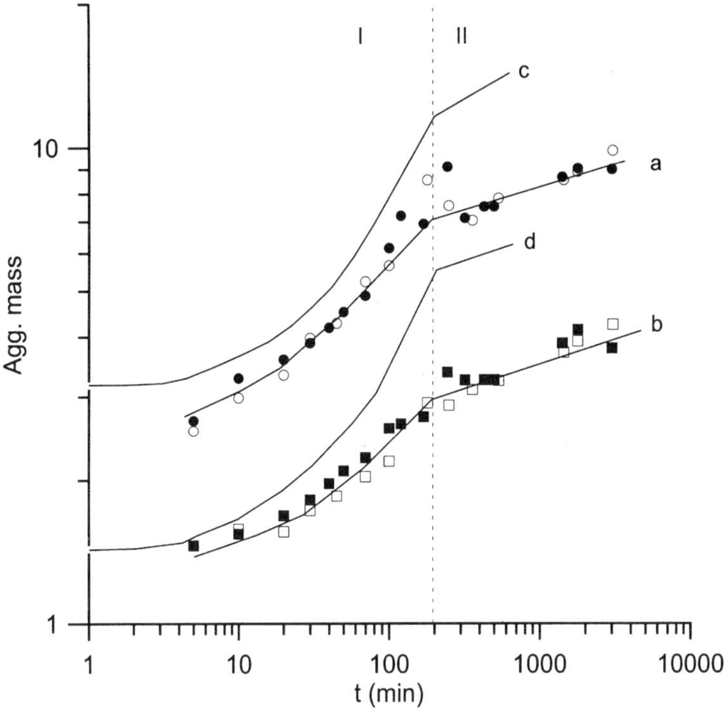

FIG. 12 Representation of the average masses $S(t)$ (curves a and c) and $N(t)$ (curves b and d) as a function of aggregation time (log-log scale). The HCG concentration is 33 IU/g latex. Curves a and b: Heterocoagulation in a mixture containing equal proportions of one latex bearing the complex [antibody–antigen] and a second latex bearing only the antibody: (●, ■), complex [anti-αHCG + HCG] and the [anti-βHCG]; (○, □), complex [anti-βHCG + HCG] and the [anti-αHCG]; curves c and d correspond to experiments containing initially equal proportions of latex bearing the [anti-βHCG] and the [anti-αHCG] and HCG at the concentration of 33 IU/g latex (see Fig. 10).

sensitized latex present in equal proportions]. As noted further, the aggregation rate did not depend on the nature of the complex A or B present in the system.

For the dissymmetrical systems, the aggregation rate is expected to not depend on the nature of the [latex + anti-αHCG] or [latex + anti-βHCG] present in great excess in the system insofar as it was assumed that the HCG molecules were distributed over the surface of the two sensitized latex particles in proportion to the developed areas. This assumption is not valid since the system 20% anti-αHCG + 80% anti-βHCG aggregates faster than does the system 80% anti-αHCG + 20% anti-βHCG. Moreover, the transition from domain I to II at 200

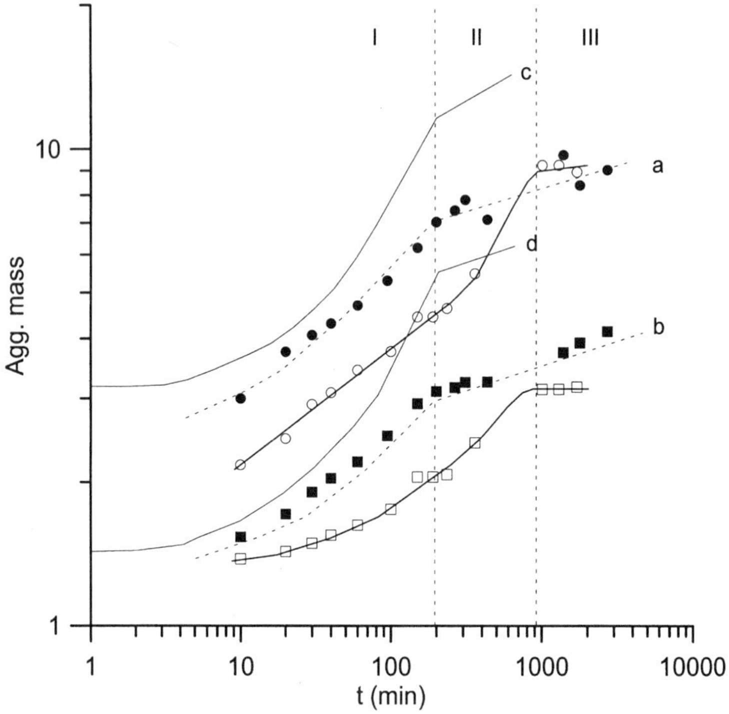

FIG. 13 Aggregation in a mixture containing the two types of sensitized latex particles and the solubilized HCG protein at a concentration of 33 IU/g latex. Representation of the weight $S(t)$ (circle) and number $N(t)$ average masses (square) as a function of time for experiment carried out in the presence of 33 IU HCG/g latex and different mixtures of sensitized latex particles: (○, □) [80% anti-αHCG + 20% anti-βHCG]; (●, ■) [20% anti-αHCG + 80% anti-βHCG]; curves c [$S(t)$] and d [$N(t)$] correspond to [50% anti-αHCG + 50% anti-βHCG]. Curves a and b refer to aggregation of complexes A or B (no free HCG molecules) in the presence of the second sensitized latex particles at the same latex concentration.

min appears to be delayed to 900 min (transition from domain II to III) for the latter system. Therefore, the anti-βHCG-coated latex, which controls the extent of aggregation insofar as it is present at the low concentration of 20% in one experiment, gives rise to the complex B, which has longer activity toward aggregation. Conversely, the anti-αHCG-coated latex, which controls the aggregation in the second experiment, gives rise to the complex A, which is more efficient toward aggregation because the average aggregate masses increase faster. Since we have no theoretical or numerical support to interpret the prolonged activity of the complex B, the higher efficiency of complex A may be analyzed taking into account results of the numerical study of Meakin and Djordjevic addressed to cluster–cluster aggregation in two-monomer systems of different functionalities [10]. This study led us to conclude the following: when two sensitized latex particles are suspended in equal portions in a solution containing a small amount of HCG proteins (full coating of the sensitized latex particles is not allowed), the solute HCG molecules do not adsorb evenly on the two sensitized latexes but preferentially adsorb on the anti-αHCG sites. The former analysis based on the electrophoretic mobility of the complexes A and B (Fig. 3) erroneously led us to the inverse conclusion in Ref. 8.

This result deserves a supplementary comment. The uneven distribution of HCG molecules between the two sensitized latexes cannot result from a different covering of the bare latex particles with anti-αHCG or anti-βHCG molecules since the surface density of the two antibodies was determined to be identical and equal to 2.37 mg/m^2. On the other hand, the different aggregation rates observed for the dissymmetrical systems cannot result from differences in the affinity of the two determinants for the α and β moieties of the HCG molecules since the reactivity of complex A (or B) toward the other sensitized latex was determined to be measured by the same value of the exponent τ. Therefore, our conclusion initially derived from the extension of the La Mer model [21] and finally based on the studies of Meakin and Djordjevic [10] of the existence of an uneven adsorption of HCG molecules on the two sensitized latexes is thus confirmed.

Finally, the effect of the slow rotation that is applied in the slide test has been ignored in the present study, but we imagine that the process may increase the absolute aggregation rate and slightly affect the fragmentation mechanism of the aggregate [42]. One may imagine that slight agitation will disrupt aggregates of low internal cohesion and preserve aggregates comprising highly linked particles. This effect has been investigated recently to interpret the internal cohesion of aggregates resulting from orthokinetic or perikinetic processes [43].

VII. SUMMARY

The bases of the antibody–antigen reaction that serves in agglutination tests were examined by implementing usual perikinetic aggregation experiments and analyzed in the light of results of numerical simulations of cluster formation by diffusion- or reaction-limited processes. In the reversible diffusion-limited aggregation process, the collision between two colloids systematically gives rise to a single aggregate of larger mass. This newly formed aggregate does not systematically constitute a stable structure, and Brownian motion may disrupt the aggregate either by reverting the system to the prior situation or by producing fragments that are different from the previously colliding colloids. In reaction-limited processes, a given fraction of interparticular collisions is inefficient toward sticking, and the success of the collision may depend on the aggregate mass. In the present chapter, the theme of antibody-coated latex particles aggregating in the presence of antigens has been revisited and some conclusions of the previous work were interpreted in the light of more recent investigations. The major idea is that, in some cases, reaction-limited aggregation processes may establish dispersion structures that are common to those induced by diffusion-limited processes, in spite of the fact that aggregation obviously requires site–ligand interactions. This comment applied to the IgG-IgM system.

For the pregnancy test, it has been demonstrated that the anti-αHCG-sensitized latex particles interact with the HCG protein at a greater extent than do the anti-βHCG-sensitized ones. Therefore, when the anti-αHCG- and the anti-βHCG-sensitized latex particles are evenly present in the system, aggregation developed with features of the heterocoagulation of particles characterized by different coordination numbers. In all cases and as initially expected from the success of agglutination tests, the aggregation process that has been implemented in this study may serve to directly determine the concentration of IgM or choriogonadotropin molecules from the values of the exponents of the aggregation kinetics and mass frequency functions. Since high accuracy in the determination of the HCG concentration is not a major challenge in the normal-pregnancy situation, precise determination of the concentration of HCG molecules and degradation products may serve to detect anomalous situations and more serious diseases [44–48].

ACKNOWLEDGMENTS

The company Biomerieux is acknowledged for providing the latexes and some reagents as well as for financial support. The authors thank A. Theretz, C. Pichot, A. Elaissari, and B. Mandrand for helpful discussions.

REFERENCES

1. Singer, J.M.; Chang, M.; Daniel, J.C. In *Future Directions in Polymer Colloids*; NATO ASI Series E Applied Sciences 138, Position Paper 2; 1987; 315.
2. von Schulthess, G.K.; Benedek, G.B.; De Blois, R.W. Measurements of the cluster size distributions for high functionality antigens cross-linked by antibody. Macromolecules **1980**, *13*, 939–945.
3. Kondo, A.; Kawano, T.; Itoh, F.; Higashitani, K. Immunological agglutination kinetics of latex particles with physically adsorbed antigens. J. Immunol. Meth. **1990**, *135*, 111–119.
4. Elaissari, A.; Pefferkorn E. Aggregation mode of colloids in the presence of block copolymer micelles. J. Colloid Interface Sci. **1991**, *143*, 343–355.

5. Csoban, K.; Pefferkorn, E. Perikinetic aggregation induced by chromium hydrolytic polymer and sol. J. Colloid Interface Sci. **1998**, *205*, 516–527.
6. Ortega-Vinuesa, J.L.; Molina-Bolivar, J.A.; Hidalgo-Alvarez, R. Particle enhanced immunoaggregation of $F(ab')_2$ molecules. J. Immunol. Meth. **1996**, *190*, 29–38.
7. Stoll, S.; Lanet, V.; Pefferkorn, E. Kinetics and modes of destabilization of antibody-coated polystyrene latices in the presence of antigen: reactivity of the system IgG-IgM. J. Colloid Interface Sci. **1993**, *157*, 302–311.
8. Ouali, L.; Pefferkorn, E.; Elaissari, A.; Pichot, C.; Mandrand, B. Heterocoagulation of sensitized latexes in the presence of HCG protein: the pregnancy test. J. Colloid Interface Sci. **1995**, *171*, 276–282.
9. Healy, T.W.; Wiese, G.R.; Yates, D.E.; Kavanagh, B.V. J. Colloid Interface Sci. **1972**, *42*, 647–649.
10. Meakin, P.; Djordjevic, Z.B. Cluster-cluster aggregation in two-monomer systems. J. Phys. A Math. Gen. **1986**, *19*, 2137–2153.
11. Stoll, S.; Pefferkorn, E. Monte Carlo simulation of controlled colloid growth by homo- and heterocoagulation in two dimensions. J. Colloid Interface Sci. **1996**, *177*, 192–197.
12. Stoll, S.; Pefferkorn, E. Kinetics of heterocoagulation. J. Colloid Interface Sci. **1993**, *160*, 149–157.
13. Aisunaidi, A.; Lach-hab, M.; Gonzalez, A.E.; Blaisten-Barojas, E. Cluster-cluster aggregation in binary mixtures. Phys. Rev. E **2000**, *61*, 550–556.
14. Furusawa, K.; Anzai, C. Heterocoagulation behaviour of polymer latices with spherical silica. Colloid Surf. **1992**, *63*, 103–111.
15. Vicsek, T.; Family, F. Critical dynamics in cluster-cluster aggregation. Family, F.; Landau, D.P., eds. In *Kinetics of Aggregation and Gelation*; North-Holland: Amsterdam, 1984; 111–115.
16. Jullien, R.; Botet, R. *Aggregation and Fractal Aggregates*; World Scientific: Singapore, 1987.
17. Meakin, P. Formation of fractal clusters and networks by irreversible diffusion-limited aggregation. Phys. Rev. Lett. **1983**, *51*, 1119–1122.
18. Meakin, P.; Vicsek, T.; Family, F. Dynamic cluster-size distribution in cluster-cluster aggregation: effects of cluster diffusivity. Phys. Rev. B **1985**, *31*, 564–569.
19. Pefferkorn, E. The role of polyelectrolytes in the stabilization and destabilization of colloids. Adv. Colloid Interface Sci. **1997**, *73*, 127–200.
20. Meakin, P. Fractal aggregates. Adv. Colloid Interface Sci. **1988**, *28*, 249–331.
21. La Mer, V.K. Filtration of colloidal dispersions flocculated by anionic and cationic polyelectrolytes. Discuss. Faraday Soc. **1966**, *42*, 448–453.
22. Ball, R.C.; Weitz, D.A.; Witten, T.A.; Leyvraz, F. Universal kinetics in reaction-limited aggregation. Phys. Rev. Lett. **1987**, *58*, 274–277.
23. Lin, M.Y.; Lindsay, H.M.; Weitz, D.A.; Ball, R.C.; Klein, R.; Meakin, P. Universal reaction-limited colloid aggregation. Phys. Rev. A **1990**, *41*, 2005–2020.
24. Hogg, R. Collision efficiency factors for polymer flocculation. J. Colloid Interface Sci. **1984**, *102*, 232–236.
25. Molski, A. On the collision efficiency approach to flocculation. Colloid Polym. Sci. **1989**, *267*, 371–375.

26. Waldmann-Meyer, H.; Knippel, E. A surface charge density model for structure and orientation of polymer-bound proteins. J. Colloid Interface Sci. **1992**, *148*, 508–516.
27. Platsoucas, C.D.; Wilkins, T.A.; Hansen, F.N.; Jolley, M.E.; Nustad, K.; Papamichail, M.; Ugelsad, J.; Wang, C.J. Biomedical applications of polymer colloids: future directions. In El-Aasser, M.S.; Fitch, R.M.; eds. *Future Directions in Polymer Colloids*; NATO ASI Series E Applied Sciences 138, Position Paper 1; Kluwer Acad. Dordrecht Pub., 1987; 307.
28. Birhen, S.; Armstrong, E.G.; Kolks, M.A.G.; Cole, L.A.; Agosto, G.M.; Krischevsky, A.; Canfield, R.E. The structure of the human chorionic gonadotropin-β subunit core fragment from pregnancy urine. Endocrinology **1988**, *123*, 572–583.
29. Bohler, H.; Cole, L.A. HCG and related molecules in pregnancy serum and urine. Assisted Reprod. Rev. **1993**, *3*, 48–48.
30. Cole, L.A. HCG and related molecules. Diagn. Endocrinol. Metab. **1994**, *12*, 207–224.
31. Kardana, E.M.; Lustbader, J.; Cole, L.A. Carbohydrate and peptide structure of the α- and β-subunits of HCG from normal and aberrant pregnancy and choriocarcinoma. Endocrinology **1997**, *7*, 15–32.
32. Goodwin, J.W.; Hearn, J.; Ho, C.C.; Ottewill, R.H. Studies on the preparation and characterization of monodisperse polystyrene lattices. III. Preparation without added surface active agents. Colloid Polym. Sci. **1974**, *252*, 464–471.
33. *Ultrogel, Magnogel and Trisacryl. Practical Guide for Use in Affinity Chromatography and Related Techniques*. Réacrifs IBF: France, 1983.
34. Pelton, R. Chemical reactions at the latex-solution interface. In Candau, F.; Ottewill, R.H.; eds. *Scientific Methods for the Study of Polymer Colloids and Their Applications*; NATO ASI Series C Mathematical and Physical Sciences 303; Kluwer Acad. Pub., Dordrecht 1990; 493.
35. BioMérieux SA. *Practical Guide to HCG Slidex.*
36. Matthews, B.A.; Rhodes, C.T. Studies of the coagulation kinetics of mixed suspensions. J. Colloid Interface Sci. **1970**, *32*, 332–338.
37. Matthews, B.A.; Rhodes, C.T. Some observations on the use of the Coulter counter model B in coagulation studies. J. Colloid Interface Sci. **1970**, *32*, 339–348.
38. Pefferkorn, E.; Varoqui, R. Dynamics of latex aggregation. Modes of cluster growth. J. Chem. Phys. **1989**, *91*, 5679–5686.
39. Stoll, S.; Pefferkorn, E. Modes of spontaneous and provoked cluster fragmentation. I. During diffusion-limited aggregation. J. Colloid Interface Sci. **1992**, *152*, 247–256.
40. Stoll, S.; Pefferkorn, E. Modes of spontaneous and provoked cluster fragmentation. I. During reaction-limited aggregation. J. Colloid Interface Sci. **1992**, *152*, 257–264.
41. Cohen, R.D. The self similar cluster size distribution in random coagulation and breakup. J. Colloid Interface Sci. **1992**, *149*, 261–270.
42. Le Berre, F.; Chauveteau, G.; Pefferkorn, E. Shear induced aggregation/ fragmentation of hydrated colloids. J. Colloid Interface Sci. **1998**, *199*, 13–21.
43. Tatek, Y.; Pefferkorn, E. Structural and stability characteristics of agglomerated clusters. Colloid Polym. Sci. **2001**, *279*, 1183–1191.

44. Kardana, E.M.; Lustbader, A.; Cole, L.A. Carbohydrate and peptide structure of the α- and β-subunits of HCG from normal and aberrant pregnancy and choriocarcinoma. Endocrinology **1997**, *7*, 15–32.
45. Cole, L.A.; Kohorn, E.; Kim, G. Detecting and monitoring trophoblast disease. New perspectives in measuring HCG levels. J. Reprod. Med. **1994**, *39*, 193–200
46. Knight, G.J.; Palomaki, G.E.; Neveux, L.M.; Fodor, K.K.; Haddow, J.E. HCG and the free β-subunit as screening test for Down syndrome. Prenat. Diagn. **1998**, *18*, 235–245.
47. Towner, D.R.; Shaffer, L.G.; Yang, S.P.; Walgenbach, D.D. Confined placental mosaicism for trisomy 14 and maternal uniparental disomy in association with elevated second trimester maternal serum human chorionic gonadotropin and third trimester fetal growth restriction. Prenat. Diagn. **2001**, *21*, 395–398.
48. Maymon, R.; Cuckle, H.; Sehmi, I.K.; Herman, A.; Sherman, D. Maternal serum human chorionic gonadotropin levels in systemic lupus erythematosus and antiphospholipid syndrome. Prenat. Diagn. **2001**, *21*, 143–145.

3
Latex Immunoagglutination Assays

J. A. MOLINA-BOLIVAR University of Málaga, Málaga, Spain
F. GALISTEO-GONZALEZ University of Granada, Granada, Spain

I. INTRODUCTION

Recent years have heralded an increase in the use of clinical diagnostic methods involving immunological procedures because they are specific and have high sensitivity. Of the many heterogeneous and homogeneous immunological assay methods available, those based on the agglutination of latex particles continue to be widely used in biology and medicine for the detection of small quantities of an antibody or antigen in a fluid test sample. Some advantages of these assays are that the procedures are simple, widely applicable, and nonhazardous, and test results are obtained in a very short time. The agglutination reaction involves in vitro aggregation of microscopic carrier particles (usually of polymeric nature, referred to as latex). This aggregation is mediated by the specific reaction between antibodies and antigens, one of which is immobilized on the surface of the latex particles to enhance the sensitivity and extend the point of equivalence. In one format, a fluid containing the ligand of interest is introduced into a suspension of the sensitized carrier particles, and the presence of agglutination is noted as indicative of the ligand. The degree of agglutination plotted as a function of the agglutinant concentration follows a bell-shaped curve similar to that for precipitin. The agglutination reaction may be used in several different modes to detect an antigen or antibody (the ligand of interest), and each has its own limitations and applications:

1. *A direct latex agglutination test for the detection of the presence of an antigen or hapten in a biological sample.* The biological sample is mixed with a suspension containing antibodies against that antigen bound to latex particles (Fig. 1). If antigen is present in the sample it will react with the antibodies to form an aggregate. If no antigen is present in the sample the mixture will keep its appearance as a smooth suspension. This method is

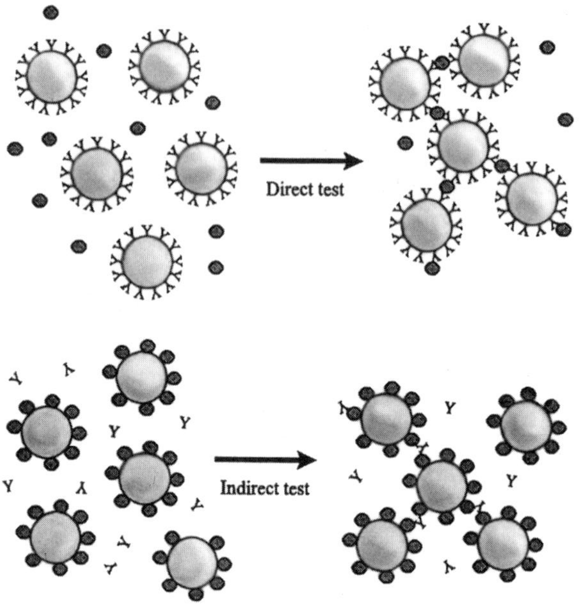

FIG. 1 Latex immunoagglutination assay: antibody-coated particles agglutinated by antigen molecules (direct test); antigen-coated particles agglutinated by antibody molecules (indirect test).

applicable to the detection of polyvalent antigens, e.g., proteins and microorganisms.

2. *An indirect latex agglutination test for the detection of an antibody in a biological sample.* This works based on similar principles whereby antigens of the antibody corresponding are bound to latex particles (Fig. 1). This approach is applicable to mono- and polyvalent antigens, e.g., drugs, steroid hormones, and proteins.
3. *An agglutination inhibition mode using antigen immobilized particles.* A fixed quantity of antibody is mixed with a dilution of the test sample containing the ligand of interest. This reaction mixture is then combined with the antigen immobilized carrier particles. The degree to which the ligand of interest (antigen) in the test sample inhibits the aggregation of the carrier particles that would otherwise have occurred, indicates the concentration of ligand present in the sample.
4. *An agglutination inhibition mode with antibody immobilized particles.* A fixed quantity of antigen is mixed with a dilution of the test sample containing the ligand of interest (a specific antibody) which inactivates a portion

of the antigen. This reaction mixture is then combined with the antibody-immobilized carrier particles. The degree to which the ligand (antibody) present in the test sample inhibits the aggregation of carrier particles, in comparison to that which would otherwise have occurred, indicates the concentration of antibody present.

Latex immunoagglutination assay was first described in 1956 by Singer and Plotz [1] and applied to rheumatoid factor. One can realize the importance of this kind of assays when perusing the specialized literature. In the last decade alone more than 400 publications in medicine and veterinary journals reported the use of latex immunoagglutination assays as analysis or research tools. The popularity of this diagnostic technology is illustrated by the fact that in 1992 there were over 200 commercial reagents available employing this approach to detect infectious diseases from "strep throat" to AIDS [2]. These include bacterial, fungal, parasitic, rickettsial, and viral diseases. The tests are also useful for cancer detection and for identification of many other substances (hormones, drugs, serum proteins, etc). The most familiar application of latex immunoassays is the pregnancy determination. In this procedure, a suspension of latex particles covered by human chorionic gonadotropin (HCG) is mixed with a drop of urine. New latex applications and technologies are still being devised and applied to new analytes.

Immunoglobulins are bifunctional molecules that not only bind to antigens but also initiate a number of other biological phenomena such as complement activation and histamine release by mast cells (activities in which the antibody acts as a directing agent). These two kinds of functional activities are localized to different portions of the molecule: the antigen binding activity to the Fab and the biological activities to the Fc portion of the molecule. Structurally they have a tetrameric arrangement of pairs of identical light and heavy polypeptide chains held together by non-covalent forces and usually by interchain disulfide bridges. Each chain consists of a number of loops or domains of more or less constant size. The N-terminal domain of each chain has greater variation in amino acid sequence than the other regions, and it is this factor that imparts the specificity to the molecule. There are five types of heavy chains, which distinguish the class of immunoglobulins IgM, IgG, IgD, IgA, and IgE, and two types of light chains. Of these, immunoglobulin G (IgG) is the most abundant and its structural characteristics are better understood.

In the main immunoagglutination assays, the particles employed to adsorb the antibody or the antigen are latex particles rather than another kind of solid support (e.g., sheep or human erythrocytes, metal sols, etc.). This is due to the following factors: (1) Uniform size particles can be synthesized with diameter in the range 50–10,000 nm. The monodispersity is an important property for detection of immunoagglutination by light scattering techniques. (2) A wide selection of functional groups can be incorporated onto the latex surface to bind proteins covalently or to achieve colloidal stability. (3) Biological molecules adsorb strongly to the hydrophobic surface of latex particles.

In the recent past, numerous researchers have done extensive work in an attempt to optimize the multiple variables affecting the reproducibility, detection limit, analytical range, sensitivity, and reliability of latex immunoassays. Some of the parameters that must be taken into account are the size of the particles used to immobilized the biological molecule, the surface charge density and hydrophilic–hydrophobic nature of the particle surface, the means of attachment of the biological molecule to the latex, the experimental conditions for immobilization and agglutination, and the optical method for detecting the extent of immunoaggregate formation [3].

The attachment of molecules to latex particles can be achieved through physical adsorption or covalent coupling. Polymer engineering has facilitated the synthesis of latex particles with surface reactive groups that enable covalent coupling of protein molecules to the particle. In addition, spacer groups may be introduced between the particle surface and the immunoproteins. The spacer groups are thought to permit a degree of freedom to the reagent moiety separating it from the particle surface, thereby lending enhanced specificity.

Apart from some visual methods for detecting qualitatively the agglutination of sensitized particles, there exist optical techniques to quantify the agglutination. The most important ones are turbidimetry, nephelometry, angular anisotropy, and photon correlation spectroscopy (light scattering measurements). The require particle size is different for qualitative (bigger particles) than for quantitative methods (smaller particles). In general, for visual slide agglutination the particle diameter range is 0.2–0.9 μm, whereas for light scattering immunoassays the diameter range is 0.01–0.3 μm.

The major problem of particle-based assays, which require careful attention, is the nonspecific agglutination. The presence of this nonspecific agglutination has been one of the main reasons why, for a long time, latex immunoagglutination tests were considered to be semiquantitative at best [4]. Nonspecific agglutination can be caused by a variety of factors:

(1) Many body fluids, such as serum, often contain other undefined substances in addition to the particular analyte of interest. Such substances can cause or inhibit agglutination. The mechanisms by which they interfere are poorly understood, and no particular causative agent or set of conditions is responsible for these effects. Moreover, interferences of these types cannot be corrected by comparison of the assay results with a similar assay using a sample not containing the analyte in question as a blank sample because the blank may not be truly representative of the serum under test. As a result, much time and effort has been expended in the search of eliminating nonspecific interferences. Some methods of reducing nonspecific interferences in latex immunoagglutina-

tion assays are as follows: massive dilution of the test sample; addition of detergents; covering of the bare surface of the sensitized particles with inactive proteins; rigorous pretreatment of the test sample including heat treatment for 30 minutes at 56°C; and enzymatic treatment with proteases reaction. These procedures are time consuming and can carry with them the undesirable effect of drastically reducing the potential sensitivity and accuracy of the immunoassay as a result of the required manipulations.

(2) Sometimes nonspecific agglutination occurs by a bridging mechanism. This mechanism assumes that the biomolecule attached to the particle has chains or loops extending to the dispersion medium sufficiently far to encounter another particle, provoking the unspecific linking of the two particles. The agglutination by bridging phenomena is important at low degrees of coverage. This process can be eliminated using an inactive protein to cover the free surface of antibody-coated particle.

(3) After the protein coating procedure, the latex particles show low colloidal stability and the aggregation occurs at pH and ionic strength values reproducing the physiological conditions. This self-aggregation process is undesirable. The difficulty in keeping the protein-coated particle system colloidally stable is the main reason that half of all latex immunoagglutination testing is unsuccessful.

II. PHYSICAL ADSORPTION

Under most conditions globular proteins, such as enzymes and immunoglobulins, show a strong tendency to adsorb at interfaces. This surface-active behavior of proteins is utilized in various biomedical and biotechnological applications. In many of those applications the sorbent material is supplied as a dispersion to reach a large surface area to volume ratio and, hence, to accommodate sufficiently large amounts of adsorbed protein in a given volume. Examples are the immobilization of enzymes on solid matrices in biocatalysis [5] and of immunoglobulins in clinical diagnostics [6]. Possible advantages of immobilization are, among others, reusability and apparent stabilization of the protein, as well as visual amplification of the antigen–antibody reaction in the case of immunodiagnostic tests.

Effective control of any of these processes requires an understanding of the driving force(s) for adsorption, which is a complex process. Investigations on simple "model" systems, consisting of a well-characterized protein, a well-characterized sorbent, and an aqueous solvent containing only nonbuffering ions, have provided the most reliable and meaningful data on the process of protein adsorption [7–10]. For instance, protein and sorbent hydrophobicity, charge distribution, protein structural stability, solution pH, and ionic strength are known to influence the affinity of a protein for a given interface. These findings form the basis of a qualitative theory, originally proposed by Norde and Lyklema

[11–16], which indicates that four effects—namely, structural rearrangements in the protein molecule, dehydration of the sorbent surface, redistribution of charged groups in the interfacial layer, and protein–surface polarity—usually make the primary contributions to the overall adsorption behavior. However, many important questions remain unanswered, and a unified, predictive theory is not in sight. Thus, protein adsorption research is in need of novel theoretical and experimental approaches that complement and expand our knowledge of the adsorption mechanism.

The mechanism of adsorption can be studied by systematically changing the physical properties of the protein, the sorbent surface, and the solution. In this way information is obtained about the nature of the interactions responsible for the adsorption process. A detailed understanding of the influence of the various interactions on the adsorbed state of the proteins is required to optimize the functioning of the immobilized proteins in their applications.

A. Adsorption Isotherms

When protein adsorption occurs it takes a certain time before the adsorbed amount and the protein concentration in solution reach their equilibrium values. When the equilibrium adsorption values are determined as a function of the equilibrium protein concentration in solution, an adsorption isotherm is obtained. Measurement of an adsorption isotherm is the starting point of most protein adsorption studies. By measuring adsorption isotherms under several experimental conditions one tries to determine which of the above-mentioned interactions plays an important role in the adsorption process. It depends on the experimental conditions, such as pH, ionic strength, charge of the adsorbent surface, and nature of the protein, and gives information on the affinity of the protein for the surface. In an adsorption isotherm two parts can be distinguished. The rising initial part at low concentrations indicates the affinity of the protein for the surface. When the isotherm rises very steeply it is called a "high-affinity" isotherm. This means that almost all protein present in the system is at the surface and the amount remaining in solution is minimal. The slope of the isotherm at low surface coverage is a measure of the first three of the four interactions discussed previously. In general, the adsorption increases up to an amount where the total surface is saturated with protein. This amount is called the "plateau value" of the isotherm, and in general it depends on the experimental conditions, the conformation of the adsorbed molecules, the affinity of the proteins for the surface, and the surface coverage–dependent lateral interaction.

The shape of a typical protein adsorption isotherm is very similar to the well-known Langmuir gas adsorption isotherm, and in some publications the protein adsorption is interpreted accordingly. However, the typical irreversibility of protein adsorption against dilution [17–19] shows that one of the essential conditions for a Langmuir isotherm is not fulfilled in the case of adsorption of most proteins. Many studies demonstrate that protein–surface interaction includes several stages, two of which are always specified: (1) adsorption of protein onto the surface and (2) conformational changes or reorientations of adsorbed protein. For the other proposed models of three-stage and multistage adsorption [20,21], quantitative descriptions have not been developed.

The protein–surface interaction has often been shown to be quite heterogeneous. Beginning by solid surfaces, some polymers consist of crystal and amorphous phases, and the existence of hydrophilic and hydrophobic domains on the surface of multiblock copolymers [22] can lead to this heterogeneity of protein–surface interactions. Electrostatic interactions between protein and strong charge sites of the surface can lead to dispersion of the entropy of the adsorption when the distance between the sites on the surface is as large as a protein molecular size [23]. Regarding proteins, they are known to have anisotropic properties and, thus, the activation energy, as well as other characteristics of adsorption, will depend on which sites of the adsorbing molecule (hydrophobic or hydrophilic, charged or uncharged) directly interacts with the surface. That is why one should consider the heterogeneity of the protein–surface interaction rather than the heterogeneity of the protein and surface separately.

It should be noted also that protein–protein interactions might lead to additional heterogeneity in the adsorption rate constant. Thus, multilayer adsorption can be characterized by a set of adsorption rate constants, which correspond to different mechanisms of formation of each protein layer. Several proteins with different adsorption or desorption rate constants, lateral dynamics, heat adsorption, and so on have been described [24,25]. These observations have suggested that there are weakly and strongly bound proteins on different surfaces. However, the processes underlying such evident heterogeneity are not clear. Many of theses models take into consideration the heterogeneity of the protein–surface interaction, which can affect the kinetics of protein adsorption even more than the electrostatic repulsion [26].

B. Immunoglobulin Adsorption

The literature concerning the study of immunoglobulin G adsorption at solid–liquid interfaces has a long and confusing history [20,27,28]. We note specifically that (1) experimental adsorption isotherms performed in different laboratories on quite similar systems often conflict; and (2) minor changes in experimental conditions (pH, ionic strength, temperature) may result in major differences in the measured adsorption. These studies are difficult due to the complex interactions involved, and they suggest that immunoglobulin adsorption on solid surfaces takes place with a rather low experimental reproducibility.

Perhaps one of the most striking features that crystallographic studies have revealed is that of molecular flexibility. This kind of flexibility is expected to facilitate the formation of antibody–antigen complexes. The Fab and Fc fragments are relatively compact; however, the whole IgG molecule is not compact (its scattering curves are anomalous and the radii of gyration of the whole molecule are larger than expected for overall close packing of regions). This segmental flexibility could explain why the dimensions of immunoglobulin G vary, and why the distance between binding sites of an antibody on an elongated molecule is 12 nm (crystalline state) but molecules can expand to reach 25 nm (end-to-end solution distance) [29–33]. This segmental flexibility might explain the poor agreement between the IgG adsorption data obtained by different authors. The IgG_1, IgG_2, IgG_3, and IgG_4 subclasses of human IgG contain two, four, five, and two disulfide bridges, respectively, between heavy chains, whereas mouse IgG_1, IgG_{2a}, and IgG_{2b} contains three bridges each and guinea pig IgG_2 also contains three [33]. Hence, the flexibility of these IgG molecules (Y- or T-shaped molecules) would be different as would their dimensions. Also, the area per molecule depends on the configuration of the IgG at the solid–liquid interface: the projected area in an end-on configuration is 20 nm^2, whereas side-on is 103 nm^2 [34]. A monolayer of side-on IgG is reported to correspond to an adsorbed amount of about 3 mg m^{-2}, while a monolayer of end-on IgG corresponds to approximately 15 mg m^{-2} [35].

IgG adsorption is usually an irreversible process; there is practically no desorption of antibodies by dilution of IgG-coated polymer particles when they are diluted at pH 7, as can be seen in Fig. 2 [36]. Thus, although adsorption isotherms from solution appear to be of the Langmuir type, it is not possible to determine equilibrium thermodynamics binding constants from this kind of experiment. Adsorption isotherms of IgG on polymer supports usually developed well-defined plateaus that were in the range of those calculated for a close-packed monolayer of IgG molecules [28,36–43]. The results obtained with the adsorption of a monoclonal antibody (MAb) (IgG_1 isotope directed against hepatitis B antigen, HBsAg) on cationic and anionic polystyrene latex particles are shown in Fig. 3. Even when the protein has the same charge sign as its adsorbent, adsorption occurs spontaneously. These results constitute an example of well-defined plateaus [37]. It should be noted, however, that step-like adsorption isotherms [27] and others without a clear plateau value [44] have also been reported. These discrepancies stress the necessity for proper characterization of both IgG and polymer supports used in IgG adsorption studies.

The conformational stability of a protein is mainly determined by intramolecular factors and solvent interactions (hydration of interfacial groups). Nevertheless, solubility is determined primarily by intermolecular effects (protein–protein interactions), but protein molecules are solvated, so that hydration effects are also involved in changes in solubility. The energy of the hydration interaction will depend on the groups placed in the interfacial zone of the protein, and then solubility and conformational stability are closely related. Solubility is a good index of denaturation and undergoes a minimum in the neighbor-

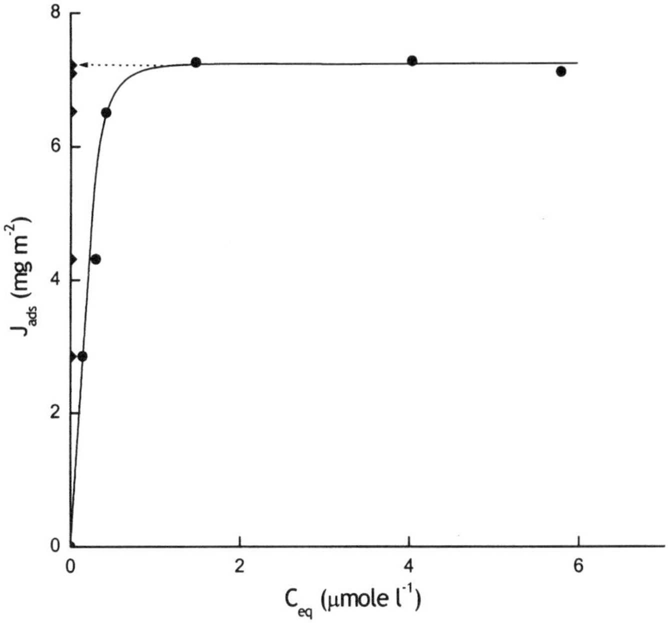

FIG. 2 Desorption of IgG from cationic polystyrene (PS) latex by dilution at pH 7.2, 2 mM ionic strength and (20 ± 1)°C. Adsorption values (●) and final values (◆).

hood of the isoelectric pH. From the adsorption point of view the solubility of a protein is of major importance, as the method for determining adsorbed protein amount is based on the difference between the initial and the supernatant concentration. If protein molecules denature in the process, they form aggregates and precipitate in the centrifugation step. This amount of protein should be quantified as adsorbed, and it could be a cause of error [37].

Several investigators [28,36–38,40,42,43] have observed a maximum in the amount of IgG adsorbed with pH, and indicate that it is due to the decrease in conformational stability of the IgGs with increasing net charge on the molecule. This results in a greater tendency for structural rearrangements of the adsorbing molecules that create a larger surface area per molecule and cause a small amount of IgG to be adsorbed. Furthermore, at pH values away from the isoelectric point of the IgG, there is an increased electrostatic repulsion between adsorbed molecules that leads to a smaller amount of adsorbed IgG. Maximal protein adsorption around the isoelectric point (IEP) has been reported for IgG [28,37,38]. Figure 4 shows the adsorption plateau values of the above MAb on cationic and anionic polystyrene latex beads as a function of pH, and we can

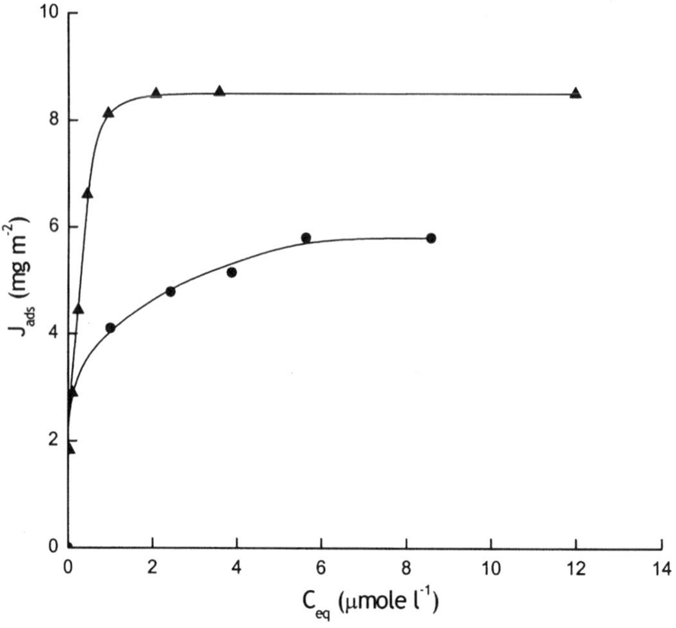

FIG. 3 Adsorption isotherms of IgG on cationic (▲) and anionic (●) PS latex at pH 5.5, ionic strength 2 mM and (20 ± 1)°C.

see that maximal values occur in the neighborhood of the IEP of the protein [36]. Nevertheless, some authors have shown that the maximum appears in the IEP of the immunoglobulin–carrier complex [38].

C. Factors That Influence Adsorption

The major types of interactions that are relevant in immunoglobulin adsorption from aqueous solution are (1) hydrophobic interaction, (2) Coulomb interaction, and (3) hydrogen bonding. The effects of electrostatic charge and potential (which can be controlled by varying the pH and ionic strength in the system), the hydrophobicities of the protein and the sorbent surface, and the chemical compositions of the sorbent and the medium on the rate of adsorption and on the adsorbed amount at equilibrium provide insight into the relative importance of the above-mentioned interactions. Other factors that may also influence immunoglobulin adsorption onto surfaces include intermolecular forces between adsorbed molecules, solvent–solvent interactions, strength of functional group bonds, chemistry of solid surface, topology, and morphology. Adsorption of IgG on hydrophobic surfaces is usually an irreversible process and occurs rapidly. It

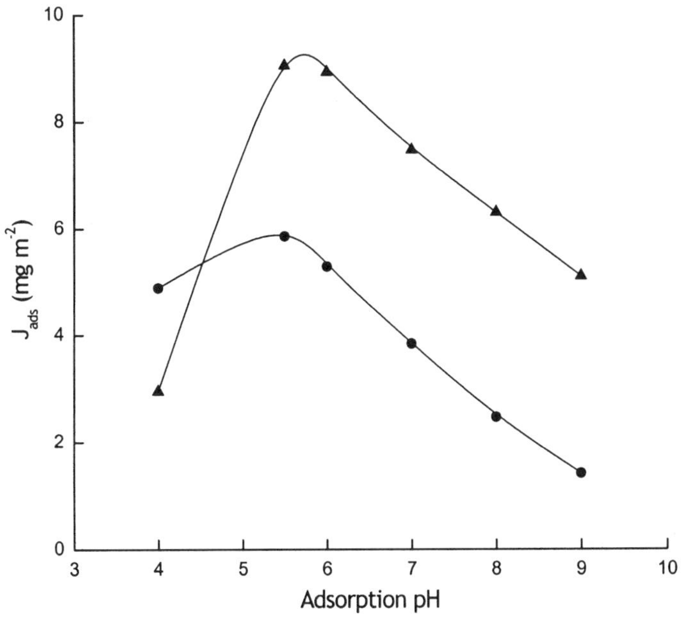

FIG. 4 Maximum adsorption of monoclonal IgG on cationic (▲) and anionic (●) PS latex as a function of pH at 2 mM ionic strength and (20 ± 1)°C.

should be noted that the characteristics of $F(ab')_2$ and IgG adsorption are very similar [45,46].

The major driving force for protein adsorption onto polymer surfaces is the dehydration of hydrophobic side groups [11,47], which is almost completely due to the entropy increase in water that is released from contact with hydrophobic components, and the surface dehydration also favors the protein adsorption. It seems reasonable to assume that antibody adsorption on hydrophobic surfaces is driven entropically as well. However, electrostatic forces at low ionic strength can play a certain role in IgG adsorption even on hydrophobic surfaces. This role has been shown by the adsorption of monoclonal antibodies onto surfaces with different signs of surface charge [36,38,43,48]. The initial slopes in the adsorption isotherms give information about the affinity between IgG and the adsorbent surface. With this aim, several authors have studied adsorption isotherms on systems that vary the possible electrostatic interactions between the components. Figure 5 shows the adsorption of a monoclonal antibody (IEP 5.5) at neutral adsorption pH on positively and negatively charged surfaces. We can see the differences in affinity between the IgG molecules and the polymer surfaces when the electrostatic forces influence the adsorption process. Effectively,

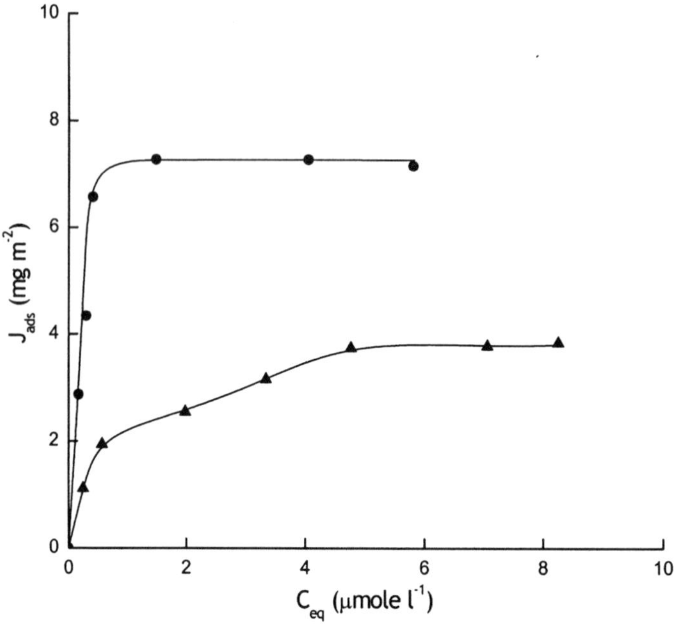

FIG. 5 Adsorption isotherms of monoclonal IgG on cationic (●) and anionic (▲) PS latex at pH 7, 2 mM ionic strength and (20 ± 1)°C.

the initial adsorption values for the anionic surface do not coincide with the total adsorption line, showing that electrostatic repulsion between negative charges makes the approach of IgG molecules to the surface difficult.

The electrostatic forces can give rise to fractionation in the adsorption of polyclonal antibodies on charged surfaces. Since polyclonal antibodies are in fact mixtures of IgG molecules with different physical properties, preferential adsorption of any fraction can take place at low ionic strength. To check this possibility, some authors [43] have analyzed by isoelectrofocusing (IEF) the supernatants after IgG adsorption onto charged surfaces. These authors have demonstrated that, at pH 7 and 9, preferential adsorption is partly determined by electrostatic factors; the IgG molecules with the highest IEP are preferentially adsorbed on negatively charged surfaces, whereas at pH 5 no preferential adsorption is observed. Most single-component IgG adsorption from buffer studies simulates physiological conditions, implying that the ionic strength is relatively high. Under those experimental conditions the electrostatic forces between protein and adsorbent are negligible. An exception is the serum competition, where IgG adsorption from a multicomponent protein solution is a phenomenon completely distinct from single-component IgG adsorption from buffer.

Antibody adsorption to, and desorption from, adsorbent surfaces is a function of the nature of both antibody and the surface, and can be dependent on time, temperature, ionic strength, pH, protein concentration, and surface tension [39]. IgG molecules adsorbed onto a surface are in a dynamic state. Although adsorbed IgG molecules generally do not desorb as a result of simple dilution, they can be displaced by an increase in ionic strength. Certainly, ionic strength exerts a pronounced effect on the adsorption of IgG molecules on charged surfaces. As ionic strength increases, the electrostatic forces between the IgG molecules and the adsorbent decreases. Under these experimental conditions, hydrophobic interactions are predominant in the adsorption mechanism of IgG molecules on a surface. Figure 6 shows the plateau values of adsorption as a function of pH at increasing ionic strength. In this case, the adsorbed IgG maximum is less dependent on pH at high ionic strength. Nevertheless, the effect of ionic strength on the adsorbed amount is different at adsorption pH 4, 7, and 10, as can be seen in Fig. 7. At neutral pH an increase in the ionic strength

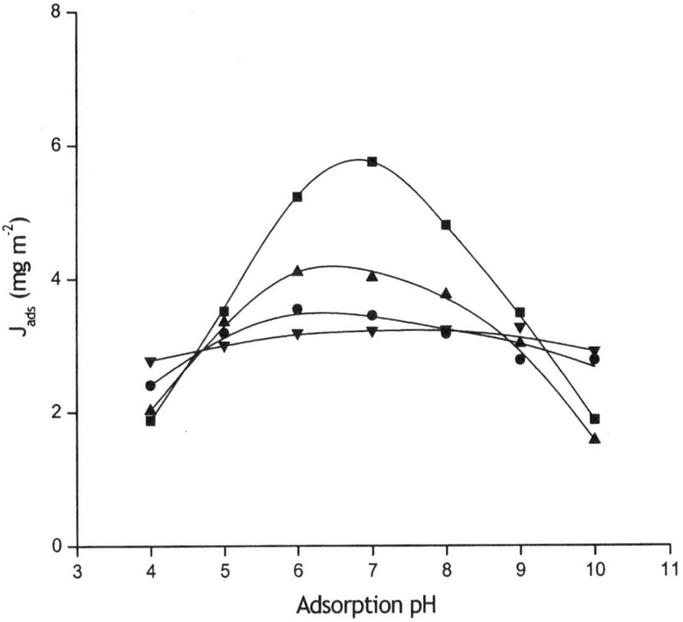

FIG. 6 Maximum adsorption as a function of pH at different NaCl ionic strengths for rabbit IgG: 2 mM (■), 20 mM (▲), 50 mM (●), 100 mM (▼).

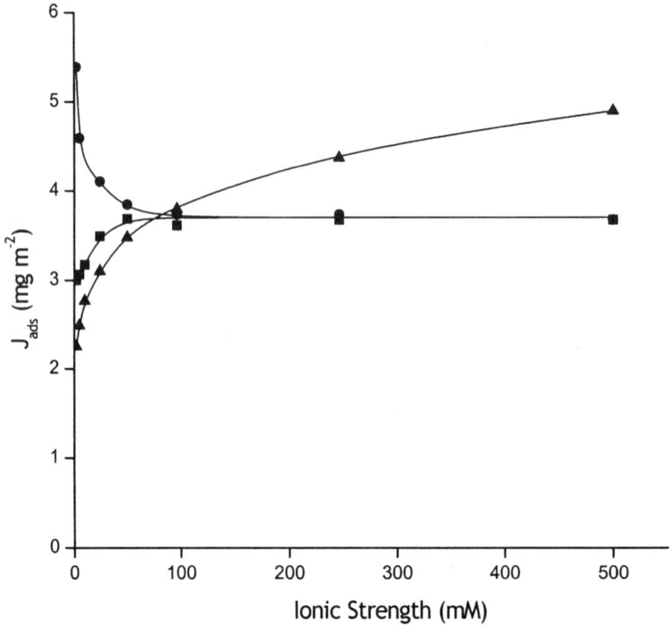

FIG. 7 Maximum adsorption as a function of NaCl ionic strength at different adsorption pH for rabbit IgG: pH 4 (▲), pH 7 (●), pH 9 (■).

implies a decrease in the plateau value, whereas at pH 4 and 10 this value increases. This trend seems to indicate that the structural stability of polyclonal IgG molecules decreases as ionic strength increases at neutral pH, whereas when the polypeptide chains are highly charged (pH 4 and 10) an increase in electrolyte concentration provokes a larger screening of the net charge on the IgG molecules and, thus, an increasing conformational stability of the IgG.

D. Electrokinetics

To determine the role of electrostatic interactions in IgG adsorption, several authors [36–38,40,42,43] have studied the electrokinetic behavior of IgG-coated surfaces. Furthermore, the electrophoretic mobility of IgG-coated surfaces can be suitable to predict the colloidal stability of these systems. Figure 8 shows the electrophoretic mobility of IgG-coated polystyrene beads (IgG-PS) as a function of the amount of adsorbed IgG (Γ). With increasing Γ the absolute value of mobility decreases to reach a plateau value. This decrease is dependent on the pH, i.e., on the charge of the IgG molecules. In Fig. 9 mobilities at complete coverage of the IgG-PS complexes are given as a function of resuspension pH

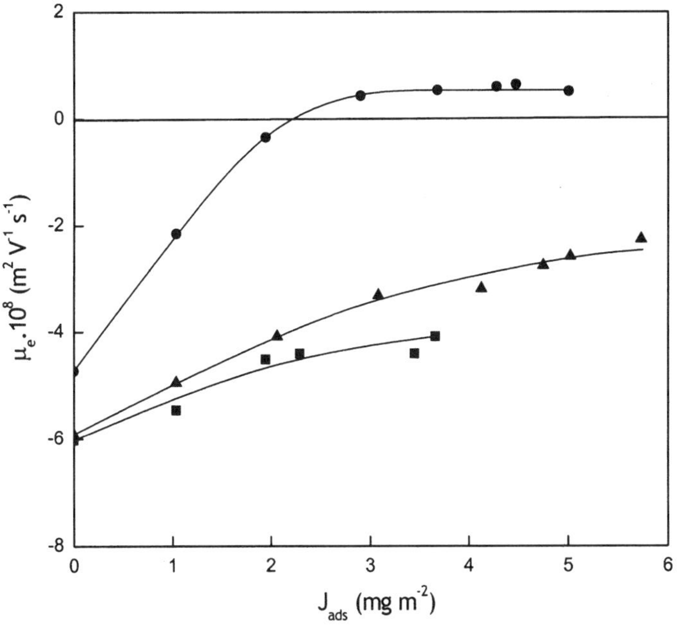

FIG. 8 Electrophoretic mobility vs. adsorbed amount of rabbit IgG at 2 mM NaCl and three different adsorption-resuspension pH values: pH 5 (●), pH 7 (▲), pH 9 (■).

for both types of charged surfaces. There is a significant difference between IgG adsorbed on negatively and positively charged surfaces. This difference can be related to electrostatic interactions between IgG and the charged surface. The IEPs of the polyclonal IgG-PS complexes are 6 and 8 for the anionic and cationic PS beads, respectively. This difference indicates that the surface charge must compensate, at least partly, the charge of the IgG molecules. This effect has also been seen with monoclonal antibodies on positively and negatively charged surfaces [36,38]. Also, it should be noted that, in both cases, the mobility of the IgG-PS complexes is decreased in comparison with the mobility of the bare PS beads, which could explain the extremely low colloidal stability of the polyclonal IgG-coated surfaces. The structure of the electrical double layer (EDL) of polyclonal IgG-coated polystyrene beads has been studied by Galisteo et al. [49]. The main conclusions drawn from these studies are (1) that ions in the electrical double layer surrounding the IgG–polymer surface (especially those under the hydrodynamic slipping plane) have a greater ionic mobility when the electric charge in the protein molecule has the same sign as the electric

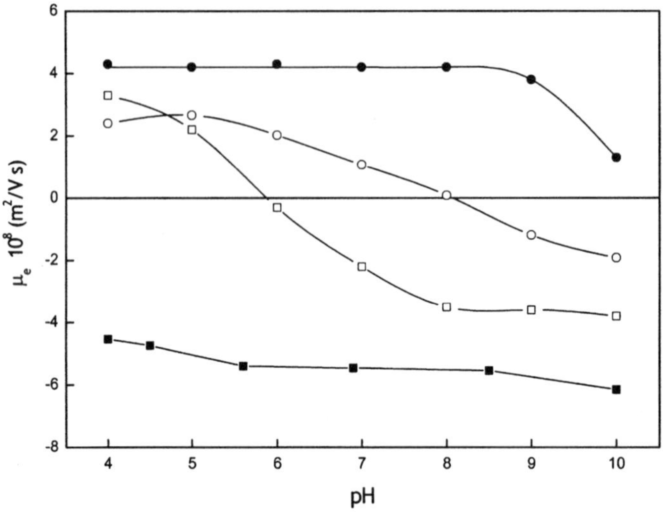

FIG. 9 Electrophoretic mobility vs. resuspension pH: bare cationic PS beads (●), saturated cationic PS beads (○), bare anionic PS beads (■), saturated anionic PS beads (□).

groups in the particle surface; and (2) that the anomalous surface conduction mechanism is more pronounced in this case in the surface charge region.

A different approach to the electrophoretic mobility of antibody-carrying latex particles has been used by Nakamura et al. [50,51]. According to these authors, the ζ potential loses its meaning for the IgG-latex particles, since the electrophoretic mobility is insensitive to the precise position of the slipping plane. Upon analysis, they conclude that the depth of the bound IgG layer varied from 3.5 to 8.5 nm, and it decreased with increasing ionic strength, suggesting that conformation of the bound IgG becomes more compact following addition of electrolyte. It is considered that the addition of excessive electrolyte ions reduces the intramolecular and intermolecular electrostatic interaction of the IgG molecules bound to the surface of latex particles. This new approach to the interpretation of the electrophoretic mobility even provides conformational information about the IgG that is fixed on polymer surfaces, which is its major advantage against the classical electrical double-layer theory.

III. COVALENT COUPLING

A. Why Attach Covalently?

One of the several advantages for the selection of latex particles as solid support for immunoagglutination tests is that latex may be "custom synthesized" to fit the needs of a given application. A wide variety of monomer combinations may be chosen to produce latex with desired surface characteristics. In this way, reactive groups can be incorporated into microspheres by an emulsion copolymerization process in which, for example, one monomer is styrene and the other is methyl methacrylate, methacrylic acid, chloromethylstyrene, acrylamide, etc. A small amount (<5%) of the functionalized monomer is generally used in the copolymerization reaction. Other functional groups can be obtained by using different initiators. These functional groups can be employed as reaction sites to covalently bind different ligands as proteins. On the other hand, the nature of the particle surface may be modified by introducing a given degree of hydrophobicity or hydrophilicity by monomer selection. In fact, it is considered advantageous for immunomicrospheres to be hydrophilic in order to avoid nonspecific interactions. The selection of latex particles as solid support for medical diagnostic tests offers a great deal of flexibility in the design of the reactive.

The ideal polymer support for latex immunoagglutination should allow the attachment of proteins in a controlled manner, resulting in a colloidally stable system with the required surface concentration of the immobilized proteins, and retaining a maximum of their biological activity. This immobilization could be achieved not only by physical adsorption, the most conventional approach, but also by covalent coupling. The latter might have, in principle, some advantages from the point of view of its application in the development of new immunodiagnostic tests:

1. The functioning of latex immunoassays depends on the capability of the immobilized antibodies to bind antigens. It is important that antibodies adsorb retaining a maximum of their biological activity. However, the antibody molecules immobilized onto polystyrene by physical adsorption retained only a small fraction (≤10%) or even lost completely their binding activity due to the protein denaturation on the surface [52–54]. This adsorption-induced reduction in the effective immunological activity can be caused by changes in the protein structure. Furthermore, surface activity could be enhanced if the antibodies are covalently coupled, by reducing the rearrangement of the protein molecules during and after adsorption [17,55].
2. Covalent attachment is permanent, leaving no unbound material after clean-up. It may prevent elution of bound protein during storage, thus increasing shelf life.
3. The hydrophobic surface of polystyrene beads may contribute to nonspecific binding of other ligands, producing false positives in the immunoagglutination tests. Chemical binding of the antibodies would aid in the elimination of false results.
4. Covalent coupling is a uniform coating procedure. As a consequence of having reactive groups over the entire surface of the latex, it is possible to

completely cover the surface with protein. On the other hand, the protein coverage is more easily controlled by covalent coupling, especially when the desired quantity of adsorbed protein is low.

5. Achieving the correct spatial orientation for the bound protein can be difficult via physical adsorption. Covalent attachment, on the other hand, can orient the molecule properly, if the correct coupling chemistry is chosen, improving the activity of the bound proteins and resulting in lower reagent consumption.
6. In a standard technique of immunoassays, washing buffers containing surfactants are used to remove loosely bound proteins from the device surface and to avoid the nonspecific adsorption of proteins such as the first and the second antibodies on the surface. Since there is a possibility that adsorbed proteins on the latex may be desorbed by surfactants, it is advisable to use covalent coupling of the proteins.
7. Smaller molecules, such as antigens or $F(ab')_2$ fragments, do not adsorb strongly to the hydrophobic surface of polystyrene beads. These smaller ligands are candidates for covalent coupling to the particle surface. Protein A or protein G covalently attached can be used to capture a variety of immunoglobulins.
8. In order to reduce nonspecific reactions and self-aggregation of the particles, a large variety of blockers (bovine serum albumin, casein, nonionic surfactants, polyethylene glycol, etc.) are added to the storage buffer to block the exposed hydrophobic surface of the polymeric microspheres. This is especially useful with antibody-coated particles because the IgG molecule has a low charge density and presents low colloidal stability. Also, a separate incubation in a higher concentration of blocker is recommended before storage in order to saturate the latex surface. The additives could displace the antibodies physically adsorbed [56,57]. For a lot of antigen–antibody systems, blocking agents of proteinic nature cannot be used because it is possible to find cross-reaction and alteration of the antigen–antibody reactivity that falsify the result of the immunodiagnostic test [58].
9. Some evidence indicates that one can attach 10–40% more protein via covalency than for physical adsorption [59,60].
10. There are latex applications that require thermocycling. In these cases, reactives with protein covalently bonded are desired because the covalent bond is more thermally stable.

B. Functionalized Particles

Many kinds of functionalized latexes with reactive surface groups suitable for covalent protein immobilization have been described. Some of the more common choices include the following:

- Carboxyl

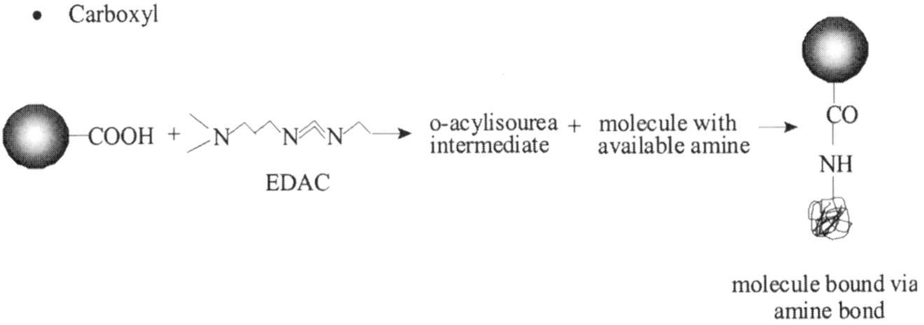

- Amino

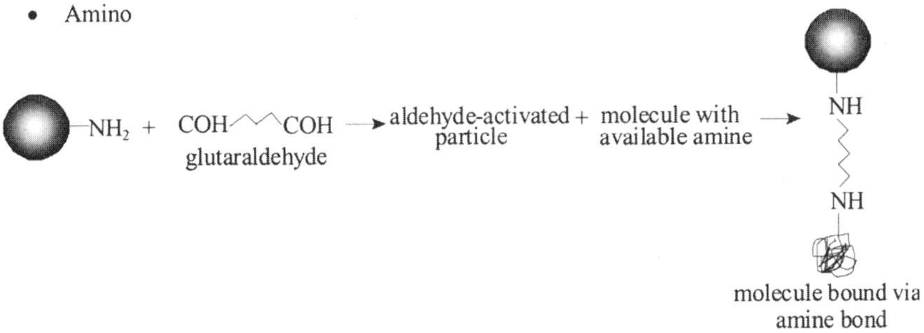

The most commonly used surfaces for attachment of ligands are carboxyl and amino groups, for the following reasons:

1. These groups have proven to be very stable over time.
2. The chemistries involved in attaching ligands to either of these groups have been widely explored, and several options exist for each.
3. The existence of terminal amino and carboxyl groups on proteins is universal, ensuring their availability for complementary attachment to one or the other functional group on the surface of the microspheres.

A number of special linkers can be used to convert one surface functional group on a microsphere to another. For example, amino microspheres can be converted to carboxylic particles by reacting with succinic anhydride [61]. Conversely, carboxylic particles can be converted to amino microspheres through water-soluble carbodiimide-mediated attachment of a diamine [62]. Also, sulfhydryl particles can be made by reacting amino particles with iminothiolane [63].

The functionalized groups may be used as sites for the attachment of spacer arm molecules. These spacer arms have functional groups at the distal end for the flexible bonding of proteins. In this way, antibody molecules extend away from the latex surface into the aqueous medium. This approach may minimize

- Hydroxyl

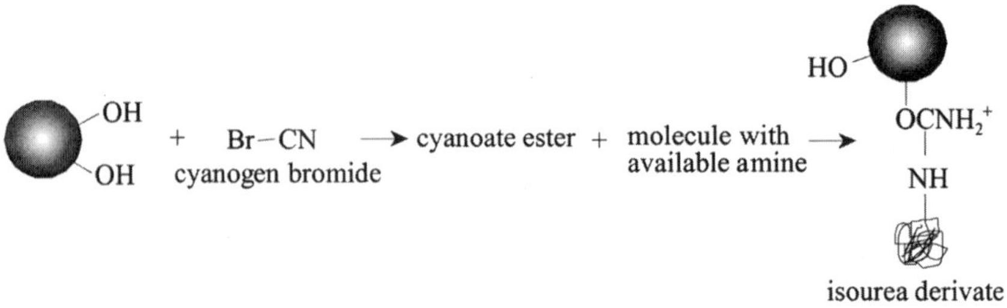

- Chloromethyl

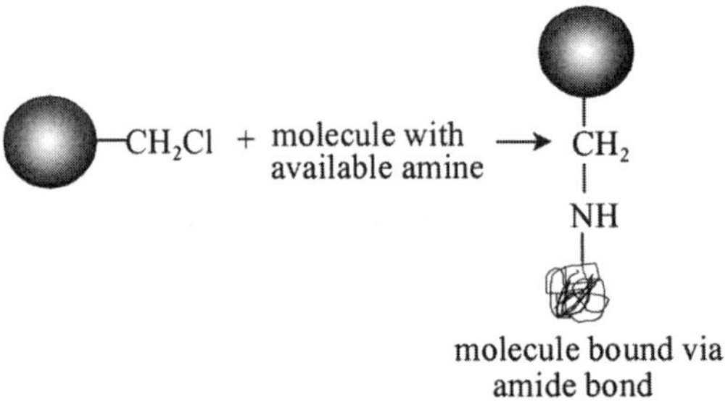

- Aldehyde

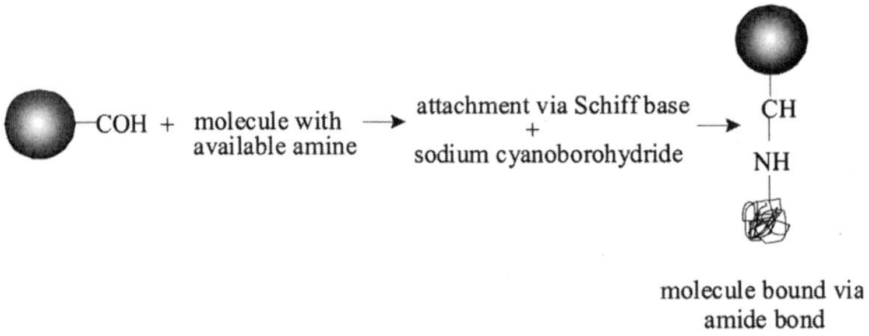

protein denaturation and the antigen–antibody recognition could be easier because the coupled antibodies are set off from the surface [64,65]. For example, it is possible to attach amine spacer arm molecules to activated chlorine groups at the latex surface.

Some functional groups borne by the latex are unreactive as such and need to be activated prior to protein immobilization: cyanogen bromide is used to bind hydroxyl groups from the latex surface to amine groups in the protein at alkaline pH, and glutaraldehyde is used to link amino groups present on the latex to amino groups on the antibody molecule. Most of the work on the covalent binding of proteins has been conducted on carboxylated latexes [66,67]. The methodology required for coupling proteins to unactivated latexes is tedious (involving several steps before the activated groups react with free groups of the protein), expensive, and more time consuming than one-step coupling. The carboxyl groups have to be activated by the 1-ethyl-3-(3-dimethylaminopropyl)carbodiimide (EDAC). The intermediate obtained on reaction of a carboxylic acid group with a carbodiimide is fairly unstable, especially in water, and has to be quickly mixed with the protein to be immobilized. This is the major drawback of any immobilization protocol based on the use of highly reactive and unstable intermediates. The balance between covalent coupling and unproductive side reactions depends to some extent on activation reaction conditions. The ionic composition of the reaction, the pH, and the buffer type can greatly enhance or inhibit the total binding of protein to particles. A number of variations exist for EDAC-mediated coupling. The degree of coupling is dependent on the density of the reactive groups, i.e., carboxyl groups on the polymer. Sufficient amounts of functional groups should be present to provide adequate coupling of antibody. The covalent attachment of the IgG molecule to carboxylated particles improves the immunoreactivity of antibodies when compared with physical adsorption, maintaining its immunoreactivity after long periods of storage [68].

Some authors have described the preparation of diazotized polystyrene for use in the separation and purification of antibodies [69,70] and for latex agglutination testing [71]. This functionalized latex with active diazo groups can be coupled to phenol and imidazole groups of antibodies via a covalent bond linkage. This immunopolystyrene diazonium latex reagent showed a positive agglutination reaction of 78–91% when mixed with serum from patients with leptospirosis.

Different authors have indicated that the use of aldehyde groups could simplify the covalent bond of the protein due to the direct reaction between the aldehyde groups of the latex and the primary amino groups in the protein molecules [72,73] by forming an imine derivate with concomitant water elimination. Rembaum et al. [74] described polymerization of acrolein to prepare microspheres that could be used as substrates for immobilization of proteins, but the microspheres were porous. Bale et al. [75] proposed a method for providing beads nonporous since reaction between an immobilized ligand and other reactant is expected to be faster on nonporous surfaces due to diffusional considerations. Alternatively, preformed polymeric latex could be modified to contain aldehyde groups [76].

In general, hydrophilic surfaces may have a lower level of nonspecific interactions than hydrophobic surfaces [77,78]. In this sense, Koning et al. [79]

proposed the synthesis of core-shell particles containing a hydrophilic polymeric shell with aldehyde groups. These particles have been used to detect human chorionic gonadotropin (HCG) in urine and serum. The results were compared with hydrophobic latex particles with the same antibodies physically adsorbed. The results showed that the functionalized latex presents a less nonspecific interaction and a higher detection limit. Hydrophilic particles with functional aldehyde groups can be produced by polymerization of glutaraldehyde and of acrolein at high pH [80]. New approaches can be accomplished by producing particles with a uniform distribution of functional group areas separated by hydrophilic areas [81,82]. The former can be used for attaching proteins, the latter for inhibiting nonspecific effects. Ideally, particles should be available with different percentages of the two types of areas in order to optimize assay concentration ranges.

The aldehyde groups tend to decompose with time, losing the capacity to bind the proteins. As suggested by Kapmeyer et al. [83,84], a possibility is to produce latex particles with acetal groups on the surface. These groups can be transformed to aldehyde groups at the moment to produce the covalent coupling of the proteins, by moving the medium to acid pH. Peula et al. [85–87] prepared acetal latexes, which permitted the covalent coupling of IgG anti-C-reactive protein (anti-CRP) in a simple way by changing the pH of the suspension to pH 2. The latex–protein complexes showed a good immunological response that was not disturbed by the presence of a nonionic surfactant in the reaction medium and was stable with time.

Preactivated microparticles have been developed with surface groups, which are sufficiently reactive to directly couple with proteins. There is no requirement for a separate preliminary activating step. This convenience allows fewer handling and transfer steps. An example of preactivated groups is vinylbenzyl chloride [88–90], where the reaction occurs by nucleophilic displacement of the chloride atom of chloromethylstyrene groups. Such latexes have limited shelf life due to the inevitable hydrolysis or oxidation of the reactive groups in aqueous media. The stability of chloromethyl function is strongly dependent on temperature. The hydrolysis rate increases with increasing temperature. At the storage temperature of 4°C some hydrolysis occurs, and after a long period of time (1 year) approximately 20% of the surface chloromethyl groups disappear [91]. Nevertheless a significant proportion of the reactive groups are retained. A chloromethylstyrene monomer can be polymerized onto a polystyrene core in any proportion to other nonactivated containing monomer, to produce a shell having from 5% to 100% chloromethylstyrene monomers [92].

Sarobe et al. [93] have studied the covalent immobilization of lysozyme on chloromethyl latexes. As can be seen in Fig. 10, the initial steps of the adsorption isotherm indicate that all the adsorbed protein becomes covalently bound. This result is general if a certain surface density of functional groups exists on

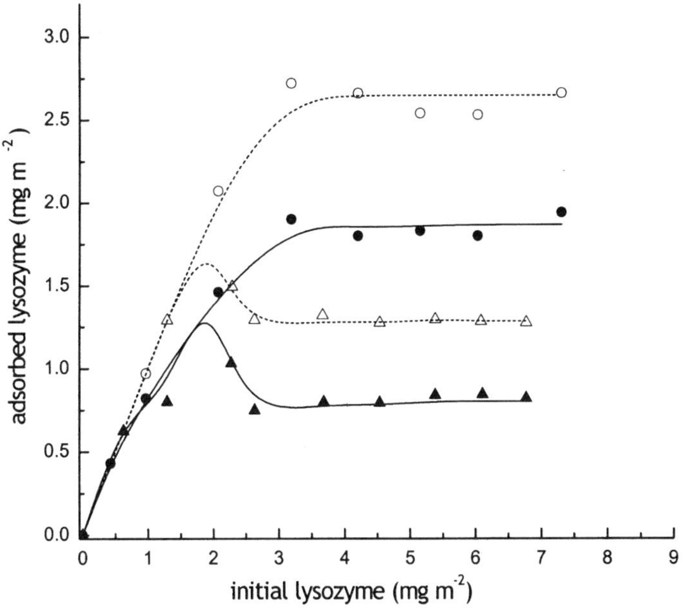

FIG. 10 Adsorption isotherm of lysozyme on chloro-activated latex at two pH: (▲) pH 7; (●) pH 11. Total adsorption, open symbols; covalent binding, closed symbols.

the surface (>21 μmol/g polymer). As physical contact occurs prior to chemical linking, a sufficiently high number of chloromethyl groups are needed to ensure that covalent binding can take place. With more adsorbed protein at the surface, the covalent extent decreases to a more or less constant value between 60% and 70% independent of the number of chloromethyl groups and pH.

A method to show the existence of covalency between protein and functionalized latex is treatment of protein–latex system with surfactants under appropriate conditions able to recover all the physically adsorbed protein from the particle surface. In the case of chloromethyl latexes, the determination of free chloride ions after protein adsorption and comparison with a blank could show the covalent attachment (Fig. 11). The kinetic of the aminochloromethyl reaction at the interface is slow. Although physical adsorption achieves saturation after some minutes of sensitization [94], covalent binding is quite slower, needing several hours to be completed (5–6 h). This means that the first contact between the protein and surface is always physical, while chemical linking develops later. Nustad et al. [95] demonstrated that adsorptive binding to core-shell particles occurred rapidly followed by slow covalent coupling. It has been claimed that adsorption is a necessary prerequisite to covalent coupling and that

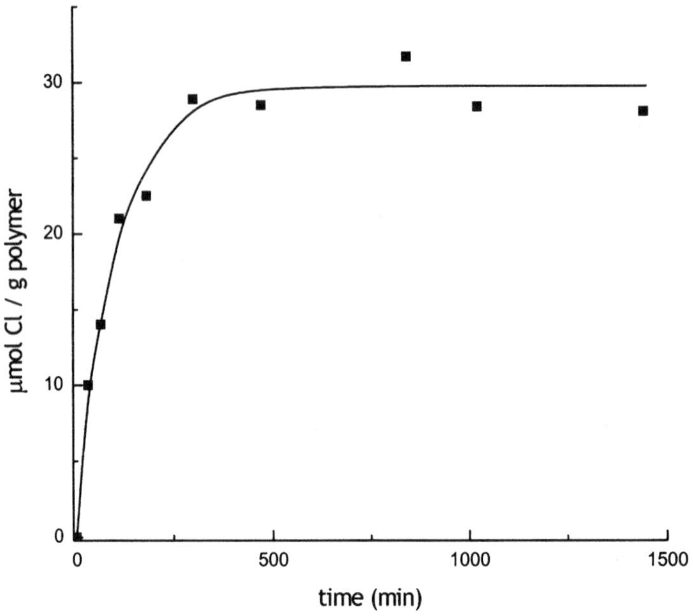

FIG. 11 Evolution of chloride ion release as a function of sensitization time for protein-saturated conditions.

an excessively high charge density on the latex surface will actually decrease the yield of covalent coupling [96].

In practice, with preactivated microparticles a blocking step in the process of preparing antibody-coated latex has to be included to eliminate the unreacted functional groups (in the case of chloromethyl groups with inert amines). Figure 12 shows a functionalization scheme of chloro-activated latexes to obtain functional groups other than chloromethyl groups through a linked spacer arm and without a linked spacer arm [97].

The immunoreactivity of IgG or $F(ab')_2$ antibody molecules covalently bound to the surface of chloro-activated latex has been compared to passive physical adsorption to a conventional polystyrene latex [98,99] (Fig. 13). For both antibodies an improvement in the immunoresponse is observed for the covalent union to latex particles. The desorption of physically attached protein from the surface with time reduces the period for which latex agglutination tests may be stored. Molina et al. [99] indicated that the storage period for IgG and $F(ab')_2$ antibodies covalently attached to chloro-activated surface is higher than for physical adsorption.

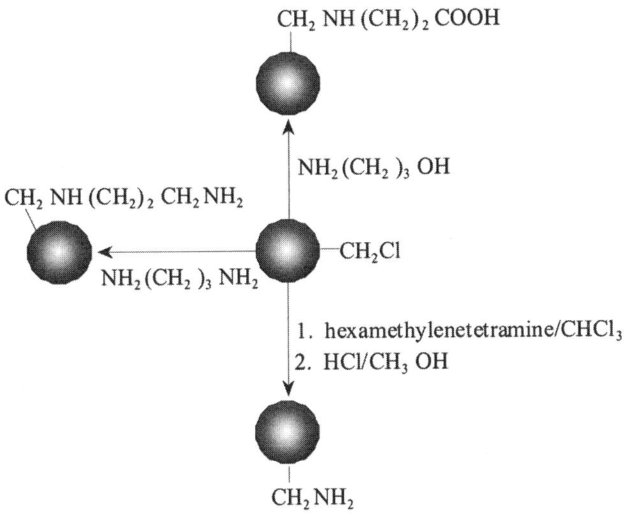

FIG. 12 Functionalization of chloromethyl-activated particles.

In covalent coupling of proteins to functionalized latexes a variety of conditions need to be tested for each case. In many of the coupling procedures extensive multipoint binding takes place during covalent attachment of the proteins and latex particle because there is a very large number of functional groups on the particle surface. This multiple binding may render either the antigen or the antibody inactive. The binding procedures have to be adapted to minimize the protein denaturation.

IV. DETECTING IMMUNOAGGLUTINATION

A. Visual Observation

The light scattered from a monodisperse suspension of particles makes the latex look milky in appearance. If the suspension of particles is aggregated by some process then the microspheres clump together to produce a coarse granular suspension that resembles curdled milk in appearance. Many of the latex agglutination tests developed are performed manually and the agglutination is detected by visual observation. In these tests larger particles of several hundred nanometers have to be used, the most common size being 0.8 μm. It has been established that about 100 clumps must be seen to determine agglutination and that these clumps must be about 50 μm to be seen by eye [100]. For a particle size

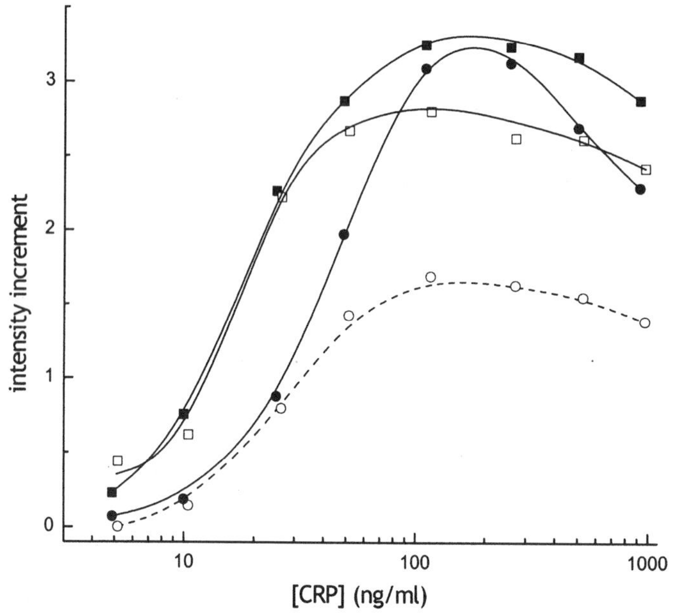

FIG. 13 Immunoreactivity of IgG anti-CRP (■) and F(ab′)$_2$ anti-CRP (●) molecules physically (open symbol) and covalently (closed symbol) adsorbed on latex particles. The immunoreactivity is measured by nephelometric monitoring of the immunoaggregation reaction with human CRP for 10 min. Both IgG-latex conjugates have the same protein coverage. Also, both F(ab′)$_2$-latex conjugates have the same protein coverage.

of 0.8 μm, about 10^5 latex particles is required to make one visible aggregate, and about 10^7 particles is needed to determine agglutination in a given test. Based on these calculations and assuming that about 10 bonds are required per particle to hold them together, Bangs has evaluated the sensitivity of such a manual test to be in the order of picograms [100]. Although quite useful in the laboratory and cheap due to the absence of equipment needs, the major limitation of these manual assays is that the only quantitation possible is to use serial dilutions of the test fluid and to look for the disappearance of immunoagglutination.

B. Light Scattering Techniques

Latex looks like milky white because the submicrometer particles scatter visible light. Developments in instrumentation have permitted the automatically full quantification of the immunological agglutination by light scattering, thus avoiding the subjectiveness of manual detection. These instrumental methods are far more sensitive than visual detection of the aggregates. Light scattering can be measured with rather inexpensive and simple instruments that are commercially available. The sensitivity, reproducibility, and detection limits of latex immunoagglutination tests depend on the technique used to detect the aggregated product.

Monitoring of the antigen–antibody reaction by measurement of light scattering has been known for the past 50 years, although the broader concept of agglutination had been explored in the 1920s for microbiological assays [101]. Light scattering methods can be divided into two major categories: methods that measure time-average scattering (static light scattering) and methods that observe the scattering fluctuation as a function of time (dynamic light scattering). Both methods can give information on the agglutination state of the protein-coated particles. During the immunoagglutination process particles aggregate with increasing diameter, resulting in the appearance of very large particles. The scattering particle size can vary from nanometers to millimeters. This change in the particle size of containing suspension provokes a dramatically increase in the scattered light. Light scattering studies applied to particle immunoassays have been published and reviewed [102–104]. Sensitivity, reproducibility, detection limit, reaction time, amount of particles needed, and availability of the required detection device are some characteristics that depend on the chosen technique [105]. There are different variables that must be optimized to obtain the best results. Some of them are specific for each technique, whereas others can be applied to all of them [106]. All of these techniques allow an increase in sensitivity and improve standardization, and the procedures involved may be automated. The difficulty for users, if they have not already purchased an instrument, is in how to choose the most appropriate system for their needs. In most cases this has probably been determined by the ease of use of the instrument itself.

1. Light Scattering Theory

When the light impinges on a particle, its electrons are subjected to a force in one direction and its nuclei to a force in the opposite direction, causing the electrons about the particle to oscillate in synchrony with the electric field of the incident light. Thus, an oscillating dipole is induced in the particle by the incident light. This oscillating dipole becomes a source of electromagnetic radiation, reradiating light at the same wavelength of the incident light and in all directions. This radiation from the particle is called scattered light. The theories of light scattering can be divided into three different regimes, depending on the relation between the particle size and the wavelength of the incident light (λ). In 1871 Rayleigh developed a theory for light scattering by a very small dielectric sphere [107,108]. When the dimensions of the particle are much smaller than the wavelength of the incident light (diameter $< 0.1\lambda$), then the entire particle is subjected to the same electric field strength at the same time. The intensity of the scattered light (I) at an angle θ between the incident and the scattered beam is given by the expression:

$$\frac{I}{I_0} = N\frac{8\pi^2\alpha^2}{r^2\lambda^4}(1 + \cos^2\theta) \tag{1}$$

where N is the number of nonabsorbing particles per unit volume, I_0 is the intensity of the incident beam, r is the distance from the particle to the detector face, and α is the polarizability of the spherical particles given by:

$$\alpha = a^3\frac{n^2 - 1}{n^2 + 2} \tag{2}$$

where a is the radius of the particle, and $n = n_p/n_m$ is the ratio of refractive index of the particle, n_p, to that of the surrounding medium, n_m. As can be seen, the intensity of scattered light is proportional to the square of the particle volume and to $1/\lambda^4$. Hence, the scattering from larger particles may dominate the scattering from smaller particles, and a decrease in the wavelength will substantially increase the scattering intensity. The ratio of light scattered forward to light scattered backward at any pair of supplementary angles centered on 90° is known as the dissymmetry ratio. The scattering pattern for Rayleigh scatterer is symmetrical about the line corresponding to the 90° scattering angle.

When the particles are larger, the particle cannot be considered as a point source and some destructive interference between light originating from different sites within the particle will occur. The Rayleigh theory is no longer valid and must be modified. The physical basis of the modification, known as the Rayleigh-Gans-Debye theory, is that a particle of arbitrary shape is subdivided into volume elements [109,110]. Each element is treated as a Rayleigh scatterer excited by the incident beam, which is assumed to be unperturbed by the presence of the rest of the particle. The Rayleigh-Gans-Debye theory is valid in the region $\lambda/20 < 2a < \lambda$, where the radiation envelope will become asymmetrical with more light being scattered forward ($\theta \leq 90°$) than backward ($90° \leq \theta \leq 180°$). The light scattering intensity at 90° is much less than the intensity at the forward (0°) angle due to destructive interferences. The relatively enhanced forward scattering with increasing particle size can be used as an index to predict the particle size. The evaluation of the dissymmetry ratio during the immunoagglutination process give valuable information, especially at various times after initiating the reaction.

For particles larger than the wavelength of incident light ($2a \geq 10\lambda$) the Mie theory is used [111]. For this size particle region the scattered light progressively decreases with increasing θ, and eventually minima and maxima may appear in the radiation diagram (Fig 14). The number and position of minima

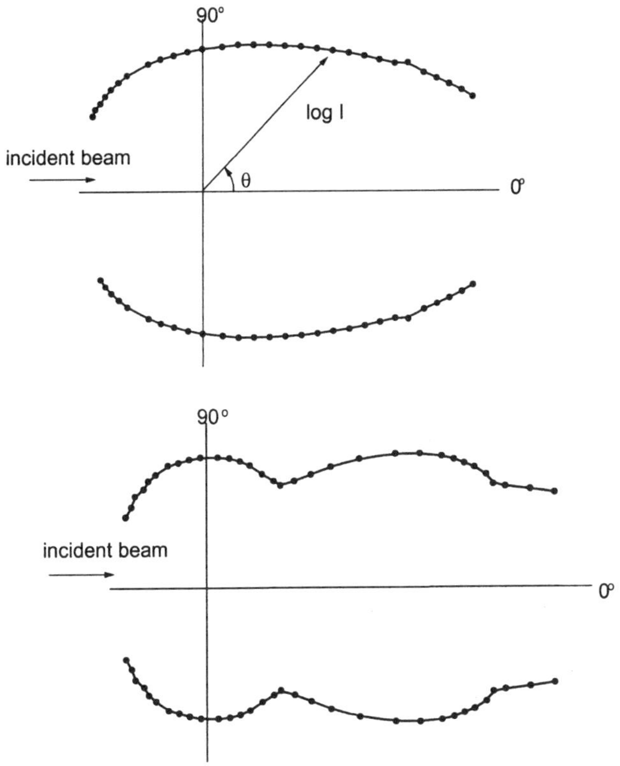

FIG. 14 Light scattering intensity distribution for two particle diameters: (top) 186 nm, (bottom) 665 nm.

and maxima which appear depend on the size parameter ($\pi 2a/\lambda$) and the polarizability of the particle. For all previous theoretical considerations the dispersion is assumed dilute; light is scattered by single particles independent of other particles present.

There are a number of techniques based in light scattering phenomena to detect latex particles' immunoagglutination: turbidimetry, nephelometry, angular anisotropy, and photon correlation spectroscopy. Now we will discuss these techniques in regard to their relative merits for latex immunoagglutination assays.

2. Turbidimety

Turbidimetry involves measurement of the intensity of the incident beam as it passes through the sample. The light beam may pass through a suspension or be absorbed, reflected, or scattered by the particles. As a consequence, the intensity of light decreases as it is transmitted through the suspension. For nonabsorbing particles the decrease in light intensity due to scattering is expressed as turbidity, τ:

$$\tau = \frac{1}{l} \ln \frac{I_0}{I} \tag{3}$$

where I_0 and I are the intensities of the initial and transmitted beams, respectively, and l is the length of the light path, usually the sample thickness. As can be seen, the turbidity is a measure of light attenuation caused by scattering. The spectrophotometer measures increased turbidity (i.e., the reduction in the intensity transmitted light), which is due to the increasing particle size resulting from the immunoagglutination reaction. This increased turbidity is a direct measure of the immunoagglutination caused by the analyte or an indirect measure of the immunoagglutination inhibition caused by the analyte. For dispersions with aggregating particles, turbidity measurements at two wavelengths may also be used to follow the aggregation process.

This technique is rapid and easy to use. With turbidity no special equipment is required other than a spectrophotometer, which is generally available in clinical laboratories. There are fully automatic spectrophotometers that not only measure transmitted light automatically at a desired time but also dilute, pipette, and transfer to the cuvette the convenient volumes of reagents buffers and samples, incubate at a programmed temperature and make the necessary calculations using the selected algorithms and calibration curves [112]. The possibility of running latex agglutination tests into these automatic analyzers allows the processing of hundreds of samples in a short time without investment in new instrumentation or personnel.

To optimize the turbitidy change, which occurs during immunoagglutination, it is important to select the appropriate particle size. The number of antigen–antibody bridges between pairs of particles during the immunoagglutination is about 2–10. With larger particles, the shear forces across these bridges may result in disruption of agglutinates when pumped at high speed in automatic machines. Thus, particles of smaller diameter may yield more robust assays. For particles to agglutinate they must first collide so that antigen–antibody bridges can form. For molecules and small particles diffusion is sufficiently rapid to produce the initial collisions necessary for aggregate formation. If the particles are larger diffusion is reduced (i.e., the agglutination kinetics) because the diffusion coefficient is inversely proportional to particle size. Small particles are desirable because of the requirement for increasing the collision frequency between particles or aggregates to enhance the rate of immunoaggregate production.

For the turbidimetric detection of the particle size change it is imperative that the particle size and the incident light wavelength be chosen with care since the turbidity reaches a maximum with time. This maximum occurs when the signal change exceeds the optical limits of the measuring system. It has been observed by photon correlation spectroscopy that changes in aggregates size continue beyond the plateau observed in turbidimetric assays [113]. The optimal performance may be a function of the ratio of the particle diameter to the illumination wavelength, and the refractive index of particle. Thus, the selection of particle material, particle size, and wavelength of detection of the immunoagglutination reaction are all important factors in optimizing assay sensitivity. For particles that are small in comparison with the wavelength of light, the scattering increases with the inverse fourth power of the wavelength. Shorter wavelengths, such as 340 nm, give larger signal differences during immunoagglutination than longer wavelengths, such as 450 nm. On the other hand, the higher the refractive index of the particles at the wavelength of choice, the higher the light scattering signal. In general, the refractive index of a material is greater at shorter wavelengths. Particles with a polyvinylnaphthalene core have been proposed to enhance sensitivity of latex immunoagglutination assays [114]. Galvin et al. claimed that for the lowest detection limits particles should be in the size range 40–70 nm, with a high refractive index but low absorbance at the wavelength of light used [115].

Heller and Pangonis gave some information about how to optimize the particle size to wavelength ratio in turbidimetric assays [116]. Sharp absorbance changes during the agglutination process can be obtained if the value of the term $2\pi a/\lambda$ is in the range 1–2. This theoretical prediction, obtained from Mie's theory, has often been corroborated experimentally. Different authors provided some evidence that minimal detection limits are achieved if the light wavelength used was in the order of 340 nm for particles in the size range 40–70 nm [117–119]. This recommendation has been frequently followed by Price et al. [120–123]. These authors even claimed that minimal detection limits are achieved using a wavelength of 340 nm independent of the particle size [101], which is a questionable statement. If latex particles are bigger (i.e., diameters from 100 to 400 nm) the optimal wavelength would be in the 450- to 700-nm range [124–127], as shown in Fig. 15. If larger particles are used instead, the infrared region on the spectrum should be employed [128]. Finally, small particles of high refractive index with short wavelength detection are preferred for high sensitivity in the turbidimetric assays. There is a practical limit in the ultraviolet region for measurement of sample in serum because of light absorption by proteins and other components. Thus, convenient wavelengths are those in excess of approximately 320 nm. Turbidimetry measurements require a higher particle number than the other light scattering techniques.

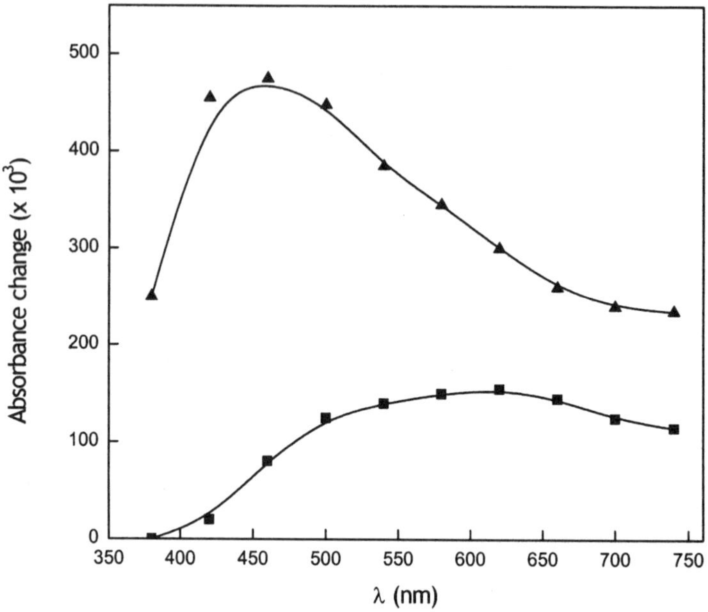

FIG. 15 Turbidity change for agglutination of latex-F(ab′)$_2$-aCRP in presence of a constant concentration of CRP as a function of wavelength: (■) particle with 180 nm diameter; (▲) particle with 340 nm diameter.

3. Nephelometry

Nephelometry refers to the measurement of the light scattered at an angle θ from the incident beam when the incident beam is passed through the sample. The scattering theories show the importance of choosing a forward scattering angle for the study of particles with size approaching the wavelength of the incident light (Fig. 14). Common nephelometers measure scattered light at right angles to the incident light. The ideal nephelometric instrument would be free of stray light; neither light scatter nor any other signal would be seen by the detector when no particles are present in solution in front of the detector. However, due to stray light-generating components in the optics path as well as in the sample cuvette or sample itself, a truly dark-field situation is difficult to obtain when making nephelometric measurements. The sensitivity of nephelometric measurements clearly depends on this background signal [129,130]. On the other hand, the sensitivity of this technique also depends on the intensity of the light source with the highest sensitivity being achieved with a laser light source [131]. Some nephelometers are designed to measure scattered light at an angle lower than 90° in order to take advantage of the increased forward scatter intensity caused by light scattering from larger particles (immunoaggregates). Unfortunately, forward scattering optical systems are harder to construct but several manufacturers now have forward scattering nephelometers [132]. A good example of nephelometers that have been specially designed to operate with latex immunoagglutination is the Behring Nephelometer Analyzer with the following main features: (1) the light source is a red diode with a wavelength of 850 nm; (2) the detector is a photodiode that measure the scattered light in the forward direction at small angles (13–24°); (3) the detection limit is as low as 10 ng/mL for some measuring systems. Dedicated instrumentation is thus required for nephelometry, whereas turbidimetry is more broadly applicable [102].

In nephelometry the change in the intensity of the scattered light after a time is measured because the scattering species rapidly increase size. The scattered light is proportional to the initial antigen concentrations when measured in the presence of a fixed antibody–latex complex. Calibration curves can therefore be generated by plotting the intensity increment values against antigen concentrations. The concentration of the same antigen in an unknown sample can then be determined by measuring the intensity increment value under identical conditions and extrapolating on the calibration curve. Figure 16a and b show the importance of the angle in nephelometric observations. These figures correspond to the agglutination kinetics of latex particles sensitized with two different $F(ab')_2$ anti-CRP coverages for various low angles of measurement of the light scattering intensity (5°, 10°, and 20°). The intensity of light scattered by small clusters is weak and linear (low antibody coverage or beginning of the process), and is better monitored at higher angles (Rayleigh's scattering). Nevertheless, in the case of high coverage, after some time aggregates grow and light scattering amplification increases, preferably at lower angles (Mie's scattering).

Figure 17a and b show the immunoreactivity (scattered light intensity increment) as a function of CRP concentration at three different angles after 10 min of reaction for two latex-$F(ab')_2$ anti-CRP complexes. As can be seen, the main response features (intensity increments, detection limit, and sensitivity) are dependent on the scattering angle and antibody coverage. For the lowest coverage (Fig. 17a) the intensity increments increase with increasing the light scattering angle, and the shape of these curves coincides with that of the precipitin curve proposed by Heidelberger and Kendall [133]. Such response can be explained considering that an antigen molecule acts as a bridge to coagulate two sensitized particles. It is easy to understand why before reaching the maximum the immunological response increases as CRP concentration does. At higher antigen concentrations the system seems to lose reactivity. It may be due to the blocking of the antibody active sites by antigens; thus, the bridging process is unfavored. Nevertheless, in the case of a higher protein coverage (Fig. 17b) the change in the scattered intensity for the 20° angle does not show the typical bell curve of

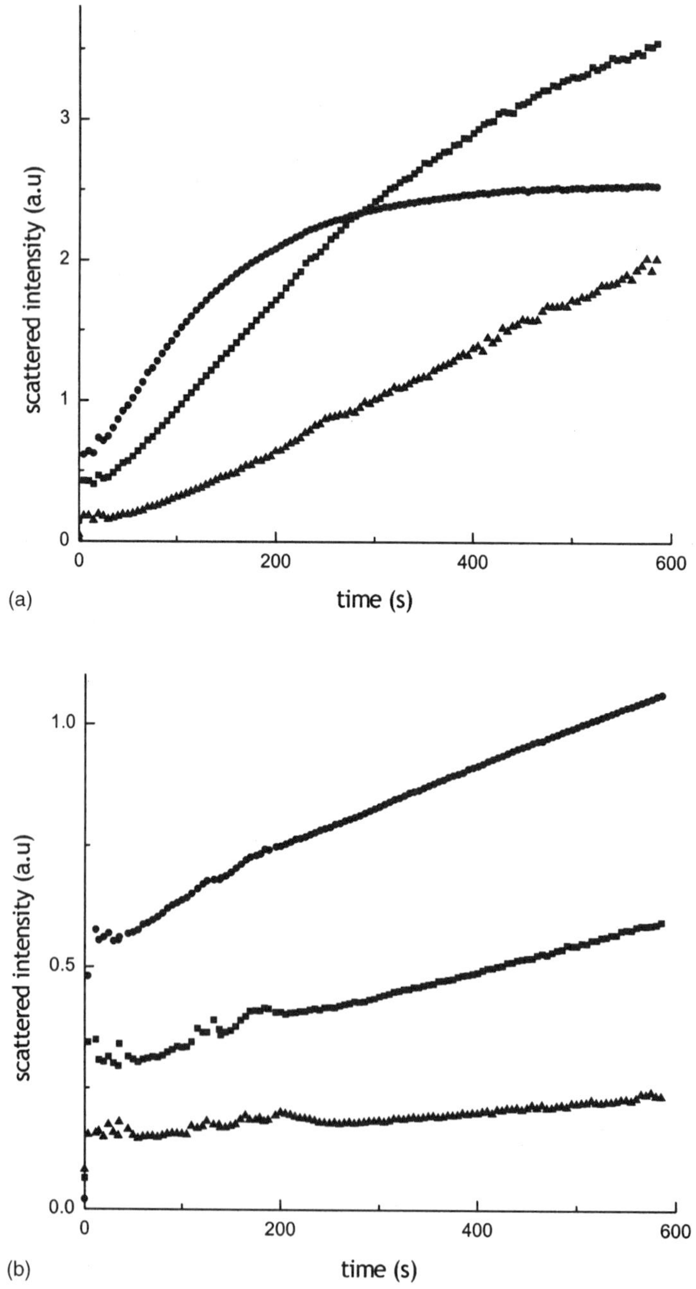

FIG. 16 Scattered light intensity at various angles for the immunoagglutination of particles covered with $F(ab')_2$ anti-CRP. (▲) 5°; (■) 10°; (●) 20°: (a) 0.9 mg/m^2; (b) 2.1 mg/m^2.

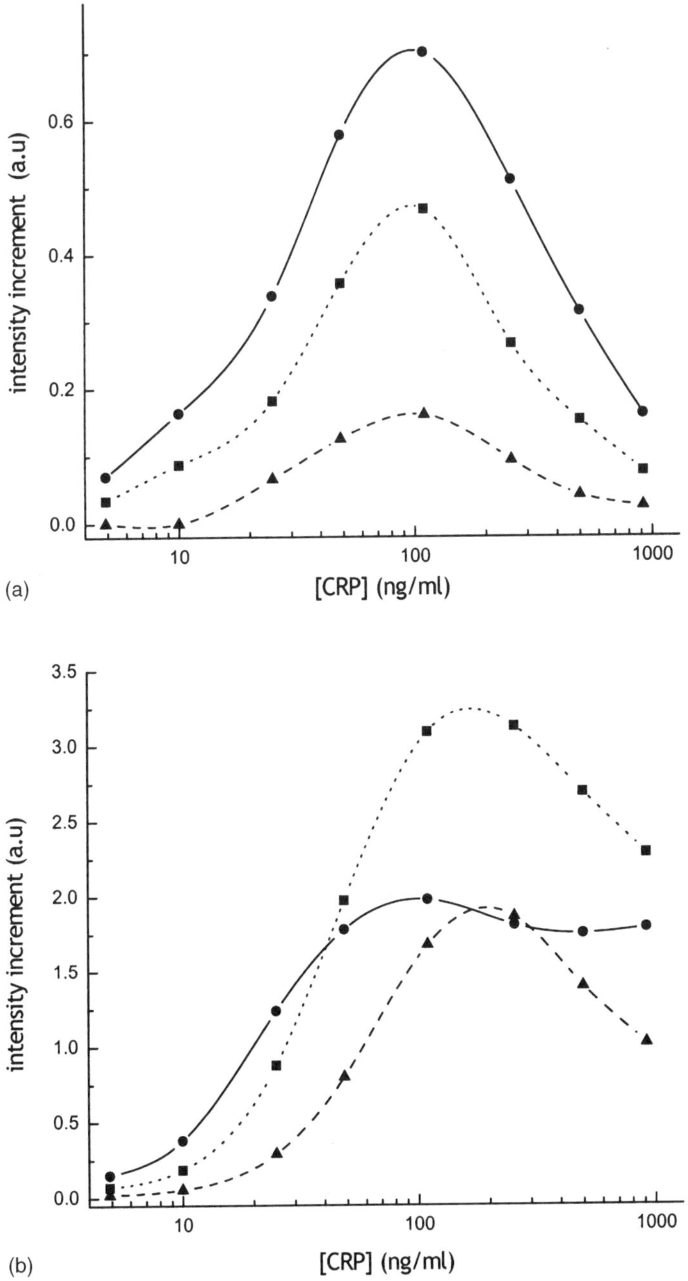

FIG. 17 Immunoreactivity as a function of CRP concentration at three different angles (▲) 5°; (■) 10°; (●) 20° for two latex-F(ab′)$_2$ anti-CRP complexes: (a) 0.9 mg/m^2; (b) 2.1 mg/m^2.

the immunoprecipitin reaction, and gives an apparent plateau when the CRP concentration is above 50 ng/mL. This plateau represents a limitation of the nephelometric technique for high concentrations of antigen at 20°. It is important to emphasize that equilibrium and equivalence points in these assays are illusory. They are produced not only by the classical Heidelberger-Kendall immune aggregation phenomena but also optically [102]. When monitoring the light scattered by an immunoaggregate, further optical considerations may influence the apparent kinetics.

The preceding figures demonstrate that it is advisable to choose a 20° angle for agglutination processes of short-duration, low-antigen concentration or complexes with low antibody coverage. In these cases, the measurements made at 20° are more sensitive than at 10° or 5° angles because the size of the aggregates is small enough. Nevertheless, if the size of the aggregates (or the initial single particles) is high, a smaller angle is preferred. Montagne et al. has followed the immunoagglutination of latex particles at different scattering angles (from 9.8° to 40.2°), concluding that the best light scattering amplification is obtained at small angles (9.8° and 12.2°) [134].

Although some authors have claimed that turbidimetric responses are more reproducible and much simpler to reach than those obtained by nephelometry [135,136], the latter has been successfully employed by many over the last 20 years [128,137–139]. Ortega et al. has demonstrated that both techniques provide similar detection limit, although turbidimetry is slightly more reproducible [105]. No longer reaction times are used with turbidimetric and nephelometric detection systems. On the other hand, nephelometry is best performed in dilute dispersion (in which background signal is reduced).

4. Angular Anisotropy

Angular anisotropy is a technique in which the ratio of the intensity of light scattering at two different angles is measured (dissymmetry ratio), usually one above and one below 90°. If the nephelometer measures the angular distribution of the scattered intensity a calibration curve can be obtained where the dissymmetry ratio is shown as a function of antigen concentration. The intensity of the light scattered at small angles is directly proportional to the square volume of the aggregates, whereas for higher angles this dependence on volume is drastically reduced. This feature was studied by Von Schulthess et al. who proposed a new strategy to analyze latex agglutination immunoassays [140]. These authors applied angular anisotropy to the detection of human chorionic gonadotropin (HCG), measuring the scattered light at 10° and 90° to maximize sensitivity and to skip source light fluctuations. This method yields high sensitivity, at least theoretically [140], provided that the two angles and the carrier particle size are properly chosen. The above authors demonstrated that the particles acting as antibody carriers should have a radius (a) in the range $\lambda/4 < a < \lambda$. They also showed that the angular anisotropy technique could be very useful if one angle is lower than or equal to 15°, and the other is in the 60–100° range. Ortega et al. has shown that angular anisotropy is a very sensitive technique detecting 1 ng/mL of CRP, after longer reaction time [105].

5. Photon Correlation Spectroscopy

Photon correlation spectroscopy (PCS) is based on the fact that the intensity of light scattered from a latex suspension, when it is illuminated with a coherent light, fluctuates with time, depending on Brownian movement and therefore on the average diffusion coefficient, which could be correlated to the particle size [112]. A photon correlation spectroscopy instrument is essentially a multiangle laser nephelometer. The PCS-based immunoassays generally have greater sensitivity than nephelometric and turbidimetric detection systems, although to achieve this longer reaction times are used due to the reduced particle numbers present in the dispersion. The device for PCS experiments is relatively expensive, as it is necessary to use a correlator. Furthermore it is difficult to find clinical laboratories with the sophisticated and extremely delicate equipment for carrying out dynamic light scattering measurements. This technique was first applied to latex immunoagglutination test for the detection of anti-bovine serum albumin by Cohen and Benedek in 1975 [141] and they claimed a sensitivity of 20 ng/ml in a highly reproducible way. Different authors have indicated that photon correlation spectroscopy offers lower detection limits, and use little reagent, but have longer assay times than the classical optical techniques of turbidimetry and nephelometry [105,142].

V. THE PROBLEM OF LOW COLLOIDAL STABILITY

A. Colloidal Stability

One of the basic requirements for a protein–latex complex to be applicable to clinical diagnostics is colloidal stability under immunological conditions. However, as indicated in the introduction, the most serious problem in latex immunoagglutination assays is that the system can lose its colloidal stability after antibody adsorption step. This low colloidal stability of latex–antibody complexes in the reaction medium may provoke the nonspecific agglutination of particles. The isoelectric point of most polyclonal IgG molecules used in latex immunoassays tests is in the range 6.5–8.5; in addition, they present a low charge density. Therefore, when the particles are covered by IgG the nonspecific agglutination process takes places under physiological conditions (pH 7.4 and ionic strength 150 mM), since there is almost no electrostatic repulsion between them. Antibody-coated particles must be completely stable in the absence of the antigen. That is, agglutination must only be triggered by the presence of the specific antigen rather than by the experimental conditions of the test. But, what is colloidal stability? When a cube with 1 cm of edge inmersed in a fluid medium is divided into many small colloidal cubes with 10 nm of edge, the surface of the system increases from 6 cm^2 to 600 m^2 (Fig. 18). This increasing area process is accompanied by a change in the free energy given by the expression $dG = \gamma_{SL} dA$, where γ_{SL} is the solid–liquid interfacial surface tension expressed in J/m^2. If the interfacial surface tension is positive, the colloidal dispersion is thermodynamically unstable ($\Delta G > 0$) and the particles tend to assemble to reduce the interfacial area (aggregation phenomena). These colloids are generally called lyophobic. On the other hand, if the interfacial surface tension is negative, the colloid is said to be lyophilic, the free energy of the system is negative, and the particles are thermodynamically stable. An example of lyophobic colloid is latex particles, whereas a typical lyophilic system is microgel. The term colloidal stability is refer to the ability of a suspension to resist aggregation. The colloidal stability may be either thermodynamic or kinetic. Lyophilic colloids are systems thermodynamically stable whereas lyophobic colloids are kinetically stabilized. The kinetic stability is a consequence of an energy barrier opposing collisions between the particles and possible aggregation subsequently. The stability control of suspensions warrants detailed attention because development of different applications of these systems to biophisics, pharmacy, agriculture, medicine, and modern technologies is dependent to a large extent on a better understanding and manipulation of the colloidal stability.

The tendency to aggregation of lyophobic colloids is attributable to the universal attractive van der Waals forces. In some cases, this attractive force between particle and medium is stronger than that between particles, with the result that the colloidal state is preferred, i.e., the system is lyophilic [143]. For lyophobic colloids only when the attractive van der Waals force is counteracted by a repulsive force can some degree of stability be obtained. When the particles have charges on the surface the colloid may be electrostatically stabilized. In some cases, a suitable polymer can be adsorb on the particle surfaces. As the two surfaces are brought closer together the concentration of polymer units increases in the overlap region with a resulting increase in the osmotic pressure.

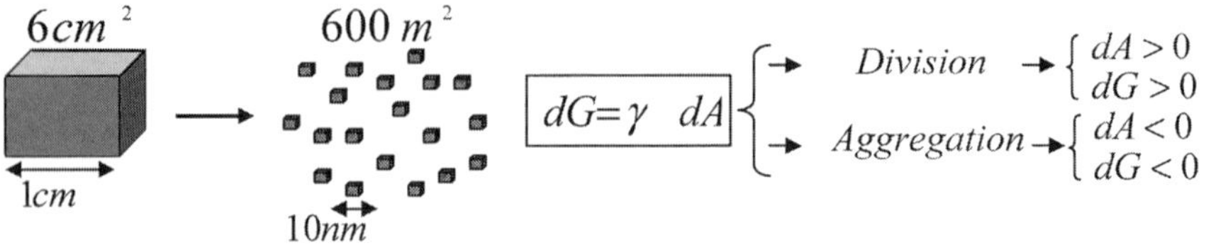

FIG. 18 Fragmentation process of a cube.

This tends to bring in solvent from the surrounding medium, with a consequent repulsive force to separate the particles. This polymer-induced stability is referred to as steric stability.

The Derjaguin-Landau-Verwey-Overbeek (DLVO) theory of colloidal stability occupies a central position in colloid science. According to this theory, the stability is predicated on the notion that two independent types of forces govern the interaction between similar colloidal particles immersed in polar (especially aqueous) solutions: attractive van der Waals forces and repulsive electrostatic forces due to the net charge of the particle [144,145]. Electrostatic repulsion decays approximately exponentially with the separation distance H between two particles, whereas the van der Waals forces are proportional to H^{-1}. As a consequence, the interaction–energy distance curve is characterized by the presence of a shallow, secondary minimum at longer separation distances, an interaction barrier closer to the surface, and a deep primary minimum at short separation. The maximal potential represents the energy barrier opposing aggregation. If particles approach each other with sufficient kinetic energies to overcome this energy potential, aggregation occurs and the suspension is destabilized. Adding salt to a dispersion initiates aggregation by suppressing the electrostatic repulsion between particles. The energy barrier decreases with increasing electrolyte concentration and disappears above certain salt concentration called critical coagulation concentration (CCC). The study of this concentration is a practical way to determine colloidal stability.

The DLVO theory has been extensively tested and reviewed [146–149], and it stands today as the only quantitative theory of the colloidal and biocolloidal sciences. However, experimental investigations of the aggregation properties of a wide range of colloidal dispersions suggest that not all systems can be explained using the DLVO theory. When water is the dielectric medium in which colloidal particles are suspended, the theory generally fails to predict the stabilities of very hydrophobic or very hydrophilic particle suspensions. For example, the colloidal stability of silica at its isoelectric point [150], prevention of bubble coalescence at high ionic strength [151], deposition of PS latex on glass surfaces [152], swelling of clays [153], and many hydrophilic colloidal particles, and most biological surfaces and macromolecules, remain separated in aqueous solution even in high salt or in the absence of any net surface charge [154,155]. Reported results with amphoteric charged latex [156] also points to a deviation in the behavior respect to the classical DLVO theory. Many studies based on atomic force microscopy have thrown light on the limitations of the classical DLVO theory. Direct investigations of the interaction potential between silica surfaces [157–159] and mica surfaces [160] in aqueous electrolyte solutions have revealed agreement with DLVO at separations above a few nanometers, but at smaller separations a short-range repulsive force appears, often termed a "hydration" interaction.

B. Improvement of the Colloidal Stability of Antibody-Covered Particles

Different strategies have been proposed in the literature to solve the problem of the false diagnoses resulting from the nonspecific agglutination of the immunolatex for low colloidal stability. These strategies have been proposed to increase the stability of the antibody–latex system. The most relevant are the following:

1. Posttreatment to cover nonoccupied parts of the latex surface of the sensitized microspheres with a second inactive protein acting as a stabilizer [161, 162]. On many occasions the biomolecule used is bovine serum albumin (BSA), a globular protein that is able to emphasize the colloidal stability of antibody-covered particles. The BSA molecule is a highly-charged protein at physiological pH. It supplies electrostatic stabilization to antibody-covered particles [163, 164]. Moreover, BSA is easily obtainable in significant amounts, which is why it is commercially available at reasonably low prices. There are two different methods for preparing latex–antibody complexes with coadsorbed BSA: (1) sequential adsorption where in a first step the antibody adsorption is carried out and after centrifugation the complex is resuspended in a solution of BSA at constant concentration [165–167]; (2) competitive coadsorption whereby the adsorption of both proteins occurs in a single step [166,168]. For competitive coadsorption the adsorption phenomena become more complicated because two types of proteins are in the medium. Peula et al. observed that the colloidal stabilization of IgG-covered particles appears when the coverage of coadsorbed BSA is high and at pH 7 and 9, and not stable at pH 5 (isoelectric point of albumin) [169]. These authors also indicated that it is necessary to find the adequate equilibrium between the amount of IgG that produces a good immunological response and the amount of BSA responsible for the colloidal stability. These ways of stabilizing antibody–latex particles may have some disadvantages, e.g., only antibody–latex complexes with low antibody coverage can be stabilized or coadsorption with inactive proteins may involve partial or complete displacement of the preadsorbed molecules of antibody.

Other proteins coadsorbed together with the antibody molecules on the particle surface are casein or rabbit serum albumin [170].

2. The use of a detergent as stabilizer molecules can also have powerful effects on nonspecific agglutination. They mainly stabilize antibody-coated particles by means of steric forces, although electrostatic repulsions can also be generated by ionic surfactants. The surfactant concentration requires careful optimization because excess surfactant inhibits the antibody–antigen reaction. The aim is to balance producing a stable antibody-coated particle and not producing inhibition of the immunological reaction. However, this strategy to preserve the colloidal stability of particles presents some disadvantages, since such molecules can desorb the previously adsorbed antibody [171,172]. Only when covalent

coupling was used to attach the antibody to particle surface is the use of detergent advisable. But, even in this case, some surfactants are capable of unfolding polypeptide structures, making the antibody lose its own immunoreactivity [172, 173].

3. Coadsorption of lipids and antibody on the polymer surface. The goal of this strategy is to obtain interfacial structures that are similar to biological membranes [174–176]. This stabilization strategy has not been greatly investigated, probably because the lipids usually show a high insolubility in aqueous media, which is a considerable drawback [177].

4. Use of monoclonal antibodies with isoelectric point far from the physiological pH. The stability of such systems improves considerably [178], but the use of monoclonal antibodies would increase the cost of test.

5. An IgG molecule consists of three domains; two of them, the so-called Fab parts, contain the antigen binding site which is located on the top, whereas the third domain is the Fc region. The $F(ab')_2$ fragment can be obtained from IgG molecule by pepsin digestion. It has been demonstrated that the use of the $F(ab')_2$ antibody fragment is useful in the development of latex immunoagglutination tests [179–183] because the colloidal stability of $F(ab')_2$–latex complexes is higher than that of the IgG–latex complexes. As the $F(ab')_2$ fragment shows an isoelectric point range more acidic than the whole IgG molecule (4.6–6.0) at physiological pH, the $F(ab')_2$-covered particles are electrostatically stabilized. On the other hand, as the divalent $F(ab')_2$ fragment has a similar avidity for antigen as the intact IgG molecule it can be used perfectly in the development of latex assays. Moreover, the use of $F(ab')_2$ instead of the whole IgG eliminates false positives in diagnostic tests due to the presence of the rheumatoid factor [184].

6. Covalent coupling of IgG on hydrophilic particles instead of hydrophobic surfaces [185]. As previously indicated, the hydrophilic particles present highest colloidal stability but the physical adsorption of proteins on hydrophilic surfaces is energetically unfavored (nonspontaneous).

7. Colloidal particles covered with IgG or $F(ab')_2$ fragment present an "anomalous" colloidal stability (not explained by the DLVO theory) at high electrolyte concentration [186,187]. As can be seen in Fig. 19, the stability diagram presents two regions: a DLVO region and a non-DLVO region. In the DLVO region the colloidal stability proceeds as expected, decreasing with increasing the electrolyte concentration until a minimum is reached at the so-called critical coagulation concentration (CCC) where every interparticle collision is effective. At the CCC the repulsive part of the total interaction potential is completely shielded. Nevertheless, above a certain electrolyte concentration, known as critical stabilization concentration (CSC), a change in the trend is observed and the coagulation diminishes with increasing ionic strength (non-DLVO region). This phenomenon is, of course, quite contrary to the DLVO theory, since addition of

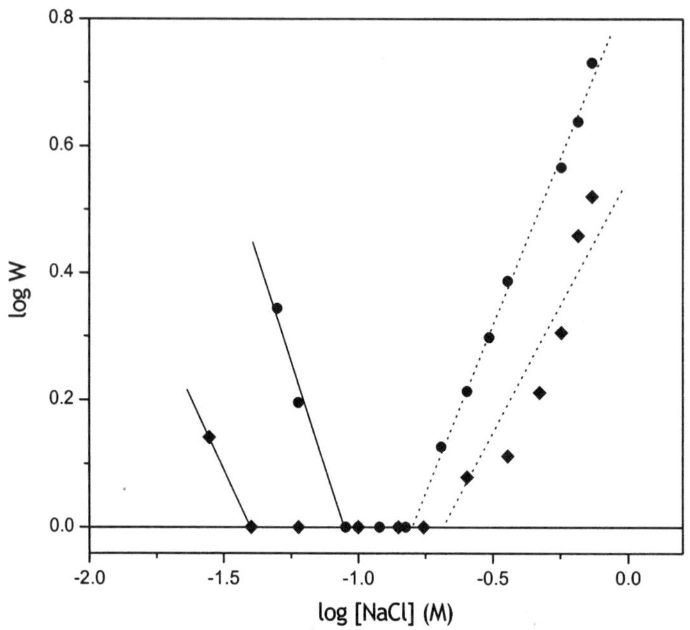

FIG. 19 Dependence of the logarithm of stability factor (*W*) on the logarithm of NaCl concentration for IgG-latex conjugate (●) and F(ab′)$_2$-latex conjugate (◆).

electrolyte is generally expected to cause coagulation. This anomalous stability at high electrolyte concentrations has been explained by the so-called hydration forces [126,186–188]. It is well established that water molecules and hydrated cations strongly bind to the hydrophilic patches of the latex–protein complex. A simplified explanation of the hydration force is that it represents the opposition exerted by water molecules surrounding ions, charges, or polar groups on the complex surface to be removed, thus imposing an energy barrier to a close approximation between particles [189]. The repulsive hydration force strongly depends on the type and concentration of electrolytes present in solution, and becomes stronger for increasing hydration energy of the cations involved [190, 191].

It is clear, as shown in Fig. 19, that particles coated by antibody become stable at high salt concentrations. Now the question is, will the presence of the specific antigen trigger the agglutination of such a stable system? The answer is affirmative, as demonstrated by Molina et al. [99,138]. Stable antibody–latex conjugates in 170 mM NaCl plus 300 mM $Mg(NO_3)_2$ aggregate when the antigen is added to the sample. Therefore, this colloidal stabilization strategy based on adding salts to the reaction medium aims to be totally successful in the development of latex immunoagglutination tests. Moreover, it is cheap and requires minimal handling.

In a latex immunoagglutination test the aim is to selectively allow agglutination to occur with an analyte. The initial latex reagent test system must be sensitive, specific, stable, and reproducible in behavior. In order to accomplish this, a delicate balance between the attractive and repulsive forces must be achieved by manipulation of pH and ionic strength of the reaction medium, as well as by the extent of particle surface coverage by both nonspecific and specific molecules such that the repulsive forces maintain particle stability at short range (less than the bridge length of an antibody molecule) while at the same time keeping the repulsive energy less than the antigen–antibody bonding energy.

REFERENCES

1. Singer, J.M.; Plotz, C.M. Am. J. Med. **1956**, *21*, 888.
2. Bangs, L.B. Amer. Clin. Lab. News **1988**, *7* (4A), 20.
3. Ortega-Vinuesa J.L., Molina-Bolivar J.A., Hidalgo-Alvarez R. J. Immunol. Meth. **1996**, *190*(1), 29.
4. Newman, D.J.; Henneberry, H.; Price, C.P. Ann. Clin. Biochem **1992,** *29*, 22.
5. Malcata, F.X.; Reyes, H.R.; García, H.S.; Hill, C.G., Jr.; Amundson, C.H. J. Am. Oil. Chem. Soc. **1990**, *67*, 870.
6. Karlsson, R.; Michaelsson, M.; Matson, L. J. Immunol. Meth. **1991**, *145*, 229.
7. Haynes, C.A.; Norde, W. J. Colloid Interface Sci. **1995**, *169*, 313.
8. Norde, W.; Galisteo-González, F.; Haynes, C.A. Polym. Adv. Tech. **1995**, *6*, 518.
9. Arai, T.; Norde, W. Colloids Surf. **1990**, *51*, 1.
10. Andrade, J. In *Surface and Interfacial Aspects of Biomedical Polymers*; Andrade, J.D., Ed.; Plenum Publishers: New York, 1985; Vol. 2, 1.
11. Norde, W.; Lyklema, J. J. Colloid Interface Sci. **1978**, *66*, 257.
12. Norde, W.; Lyklema, J. J. Colloid Interface Sci. **1978**, *66*, 266.
13. Norde, W.; Lyklema, J. J. Colloid Interface Sci. **1978**, *66*, 277.
14. Norde, W.; Lyklema, J. J. Colloid Interface Sci. **1978**, *66*, 285.
15. Norde, W.; Lyklema, J. J. Colloid Interface Sci. **1978**, *66*, 295.
16. Norde, W.; Lyklema, J. J. Colloid Interface Sci. **1979**, *71*, 350.
17. Norde, W. Adv. Colloid Interface Sci. **1986**, *25*, 267.
18. Jönsson, U.; Ivarsson, B.; Lundström, Y.; Berghem, L. J. Colloid Interface Sci. **1982**, *90*, 148.
19. van Enckevort, H.J.; Dass, D.V.; Langdon, A.G. J. Colloid Interface Sci. **1984**, *98*, 138.
20. Soderquist, M.E.; Walton, A.G. J. Colloid Interface Sci. **1980**, *75*, 386.
21. Andrade, J.D.; Herron, J.; Hlady, V.; Horsley, D. Croat. Chem. Acta **1987**, *60* (3), 495.

22. Grainger, D.; Okano, T.; Kim, S.C. In *Advances in Biomedical Polymers*; Gebelein, C.G., Ed.; Plenum Publishers: New York, 1987; 229.
23. Lyklema, J. In *Surface and Interface Aspects of Biomedical Polymers*; Andrade, J.D., Ed.; Plenum Publishers: New York, 1985; Vol. 1.
24. Burghardt, T.P.; Axelrod, D. Biophys. J. **1981**, *33*, 455.
25. Bohnert, J.L.; Horbett, T.A. J. Colloid Interface Sci. **1986**, *111*, 363.
26. Sevastianov, V.I.; Kulik, E.A.; Kalinin, I.D. J. Colloid Interface Sci. **1991**, *145*, 191.
27. Fair, B.D.; Jamieson, A.M. J. Colloid Interface Sci. **1980**, *77*, 525.
28. Bagchi, P.; Birnbaum, S.M. J. Colloid Interface Sci. **1981**, *83* (2), 460.
29. Silverton, E.W.; Navia, M.A.; Davies, D.R. Proc. Natl. Acad. Sci. USA **1977**, *74* (11), 5140.
30. Sarma, V.R.; Silverton, E.W.; Davies, D.R.; Terry, W.D. J. Biol. Chem. **1971**, *246* (11), 3753.
31. Pilz, I.; Kratky, O.; Licht, A.; Sela, M. Biochemistry **1973**, *12* (24), 4998.
32. Pilz, I.; Kratky, O.; Karush, F. Eur. J. Biochem. **1974**, *41* 91.
33. Yguerabide, J.; Epstein, H.F.; Stryer, L. J. Mol. Biol. **1970**, *51*, 573.
34. Brash, J.L.; Lyman, D.J. J. Biomed. Mater. Res. **1969**, *3*, 175.
35. Merrill, E.W.; Salzman, E.W.; Dennison, K.A.; Tay, S.-W.; Pekala, R.W. In *Progress in Artificial Organs 1985*; Nosé, Y., Kjellstrand, C., Ivanovich, P., Eds.; ISAO Press: Cleveland, 1986.
36. Martin A., Puig J., Galisteo F,. Serra J., Hidalgo-Alvarez R. J. Dispersion Sci. Tech. **1992**, *13*(4), 399.
37. Serra, J.; Puig, J.; Martín, J.; Galisteo, F.; Gálvez, M.J.; Hidalgo-Álvarez, R. Colloid Polym. Sci. **1992**, *270*, 574.
38. Elgersma, A.V.; Zsom, R.L.J.; Norde, W.; Lyklema, J. Colloids Surf. **1991**, *54*, 89.
39. Bale, M.D.; Danielson, S.J.; Daiss, J.L.; Gopper, K.E.; Sutton, R.C. J. Colloid Interface Sci. **1989**, *132* (1), 176.
40. Kondo, A.; Oku, S.; Higashitani, K. Biotechnol. Bioeng. **1991**, *37*, 537.
41. Bale, M.D.; Mosher, D.F.; Wolfarth, L.; Sutton, R.C. J. Colloid Interface Sci. **1988**, *125* (2), 516.
42. Martín, A.; Puig, J.; Galisteo, F.; Serra, J.; Hidalgo-Alvarez, R. J. Dispersion Sci. Technol. **1992**, *13* (4), 399.
43. Galisteo, F.; Puig, J.; Martín, A.; Serra, J.; Hidalgo-Alvarez, R. Colloids Surf. B Biointerfaces **1994**, *2*, 435.
44. Ronner, J.A.; Lensen, H.G.W.; Olthuis, F.M.F.G.; Smolders, C.A. J. Feijen. Biomater. **1984**, *5*, 241.
45. Ortega, J.L.; Hidalgo-Alvarez, R. Colloids Surf. B Biointerfaces **1993**, *1*, 365.
46. Ortega, J.L.; Hidalgo-Alvarez, R. J. Biomater. Sci. Polym. Ed. **1994**, *6* (3), 269.
47. Norde, W. Adv. Colloid Interface Sci. **1986**, *25*, 267.
48. Kato, K.; Sano, S.; Ikada, Y. Colloids Surf. B Biointerfaces **1995**, *4*, 221.
49. Galisteo, F.; Moleón, J.A.; Hidalgo-Alvarez, R. J. Biomater. Sci. Polym. Ed. **1993**, *4* (6), 631.
50. Nakamura, M.; Ohshima, O.; Kondo, T. J. Colloid Interface Sci. **1992**, *149* (1), 241.

51. Nakamura, M.; Ohshima, O.; Kondo, T. J. Colloid Interface Sci. **1992**, *154* (2), 393.
52. Butler, J.E.; Ni, L.; Nessler, R.; Joshi, K.S.; Suter, M.; Rosenber, B.; Chang, J.; Brown, W.R.; Cantarero, L.A. J. Immunol. Meth. **1992**, *150*, 77.
53. Molina-Bolívar, J.A.; Galisteo-González, F.; Quesada-Pérez, M.; Hidalgo-Álvarez, R. Colloid Polym. Sci. **1998**, *276*, 1117.
54. Quesada, M.; Puig, J.; Delgado, J.M.; Peula, J.M.; Molina, J.A.; Hidalgo-Álvarez, R. Colloids Surf. B Biointerfaces **1997**, *8*, 303.
55. Limet, J.N.; Moussebois, C.H.; Cambiaso, C.L.; Vaerman, J.P.; Mason, P.L. J. Immunol. Meth. **1979**, *28*, 25.
56. Tan, J.S.; Martic, P.A. J. Colloid Interface Sci. **1990**, *136*, 415.
57. Rapoza, R.J.; Horbett, T.A. J. Colloid Interface Sci. **1990**, *136*, 480.
58. Wahlgren, M.C.; Nygren, H. J. Colloid Interface Sci. **1991**, *142*, 503.
59. Lutanie, E.; Voegel, J.C.; Scaf, P.; Freund, M.; Cazenave, J.P.; Schmitt, A. Proc. Natl. Acad. Sci. USA **1992**, *89*, 9890.
60. Douglas, A.S.; Monteith, C.A. Clin. Chem. **1994**, *40*, 1833.
61. Gounaris, A.D.; Perlman, G.E. J. Biol. Chem. **1967**, *242*, 2739.
62. Kurzer, F.; Douraghi-Zadeh, K. Chem. Rev. **1967**, *67*, 107.
63. Kenny, J.W.; Sommer, A.; Traut, R.R. J. Biol. Chem. **1975**, *250*, 9434.
64. *Particle Technology Microparticle Immunoassays Techniques*. Particle Technology Divison. Seradyn Inc. Procedures; 1988.
65. Kitano, H.; Iwai, S.; Okubo, T.; Ise, N. J. Am. Chem. Soc. **1987**, *109*, 7608.
66. Borque, L.; Bellod, L.; Rus, A.; Seco, M.L.; Galisteo-González, F. Clin. Chem. **2000**, *46*, 1839.
67. Kondo, A.; Kawano, T.; Higashitani, K. J. Ferment. Bioeng. **1992**, *73*, 435.
68. Ortega-Vinuesa, J.L.; Bastos-González, D.; Hidalgo-Álvarez, R. J. Colloid Interface Sci. **1996**, *184*, 331.
69. Gyenes, L.; Sehon, A.H. J. Biochem. Physiol. **1960**, *38*, 1235.
70. Filippusson, H.; Hornby, W.E. Biochem. J. **1970**, *120*, 215.
71. Liu, Y.; Dong, G.; Zhao, Y. J. Immunol. Meth. **1989**, *124*, 159.
72. Charleux, B.; Fanget, P.; Pichot, C. Makromol. Chem. **1992**, *193*, 205.
73. Bale, M.D.; Danielson, S.J.; Daiss, J.L.; Goppert, K.E.; Sutton, R.C. J. Colloid Interface Sci. **1989**, *132*, 176.
74. Rembaum, A.; Chang, M.; Richards, G.; Li, M. J. Polym. Sci. Polym. Chem. Ed. **1984**, *22*, 609.
75. Oenick M.D.B., Warshawsky A. Colloid Polym. Sci. **1991**, *269*(2), 139.
76. Rembaum, A.; Yen, R.C.K.; Kempner, D.H.; Ugelstad, J. J. Immunol. Meth. **1982**, *52*, 341.
77. Elwing, H.; Nilsson, B.; Svensson, K.E.; Askendahl, A.; Nilsson, U.R.; Lundstrom, I. J. Colloid Interface Sci. **1988**, *125*, 139.
78. Wahlgren, M.; Arnebrant, T. Trends Biotechnol. **1991**, *9*(6), 201.
79. Koning, B.L.J.C.; Pelssers, E.G.M.; Verhoeven, A.J.C.M.; Kamps, K.M.P. Colloids Surf. B Biointerfaces **1993**, *1*, 69.
80. Kumakura, M.; Suzuki, M.; Adachi, S.; Kaetsu, I. J. Immunol. Meth. **1983**, *63*, 115.

81. Saethre, B.; Mork, P.C.; Ugelstad, J. J. Polym. Sci A Polym. Chem. **1995**, *17*, 2951.
82. Ugelstad, J.; Mork, P.C.; Schmid, R.; Ellingsen, T.; Berge, A. Polym. Int. **1993**, *30*, 157.
83. Kapmeyer, W.H.; Pauly, H.E.; Tuengler, P. J. Clin. Lab. Anal. **1988**, *2*, 76.
84. Kapmeyer, W.H. Pure Appl. Chem. **1991**, *63*, 1135.
85. Peula, J.M.; Hidalgo-Álvarez, R.; Santos, R.; Forcada, J.; de las Nieves, F.J. J. Mater. Sci. Mater. Med. **1995**, *6*, 779.
86. Peula, J.M.; Hidalgo-Álvarez, R.; de las Nieves, F.J. J. Colloid Interface Sci. **1998**, *201*, 132.
87. Peula, J.M.; Hidalgo-Álvarez, R.; de las Nieves, F.J. J. Colloid Interface Sci. **1998**, *201*, 139.
88. Suen, C.H.; Morawetz, H. Makromol. Chem. **1995**, *186*, 255.
89. Miraballes-Martínez, I.; Martín-Rodríguez, A.; Hidalgo-Álvarez, R. J. Biomater. Sci. Polym. Ed. **1997**, *8*, 765.
90. Bon, S.A.F.; van Beek, H.; Piet, P.; German, A.L. J. Appl. Polym. Sci. **1995**, *58*, 19.
91. Verrier-Charleux, B.; Graillat, C.; Chevalier, Y.; Pichot, C.; Revillon, A. Colloid Polym. Sci. **1991**, *269*, 398.
92. Sarobe, J.; Forcada, J. Colloids Surf. A Physicochem. Eng. Asp. **1998**, *135*, 293.
93. Sarobe, J.; Molina-Bolívar, J.A.; Forcada, J.; Galisteo, F.; Hidalgo-Álvarez, R. Macromolecules **1998**, *31*, 4282.
94. Galisteo, F. PhD thesis. University of Granada, Spain, 1992.
95. Nustad, K.; Johasen, L.; Ugelstad, J.; Ellingse, T.; Berge, A. Eur. Surg. Res. **1984**, *16*, 80.
96. Seaman, G.V.F.; Goodwin, J.W. Am. Clin. Prod. Rev. **1986**, June, 25.
97. Marge, S.; Nov, E.; Fisher, I. J. Polym. Sci A Polym. Chem. **1991**, *29*, 347.
98. Molina-Bolívar, J.A.; Galisteo-González, F.; Hidalgo-Álvarez, R. J. Biomater. Sci. Polym. Ed. **1998**, *9*, 1089.
99. Molina-Bolívar, J.A.; Galisteo-González, F.; Hidalgo-Álvarez, R. J. Biomater. Sci. Polym. Ed. **1998**, *9*, 1103.
100. Bangs, L.B. J. Clin. Immunoassay **1990**, *13*(3), 127.
101. Price, C.P.; Newman, D.J. *Principles and Practice of Immunoassay*; Stockton Press: New York, 1991.
102. Newman, D.J.; Henneberry, H.; Price, C. Ann. Clin. Biochem. **1992**, *29*, 22.
103. Whicher, J.T.; Blow, C. Ann. Clin. Biochem. **1980**, *17*, 170.
104. Price, C.P.; Spencer, K.; Whicher, J. Ann. Clin. Biochem. **1983**, *20*, 1.
105. Ortega-Vinuesa, J.L.; Molina-Bolívar, J.A.; Peula, J.M.; Hidalgo-Álvarez, R. J. Immunol. Meth. **1997**, *205*, 151.
106. Kimura H. J. Immunol. Meth. **1980**, *38*(3–4), 353.
107. Strutt, J.W. Phil. Mag. **1871**, *41*, 107.
108. Kerker, M. *The Scattering of Light and Other Electromagnetic Radiation*; Academic Press: New York, 1969.
109. Debye, P. Ann. Physik. **1915**, *46*, 809.
110. Gans, R. Ann. Physik. **1925**, *76*, 29.

111. Mie, G. Ann. Physik. **1908**, *25*, 377.
112. Javier Gella, F.; Serra, J.; Gener, J. Pure Appl. Chem. **1991**, *63*, 1131.
113. Deverill, I.; Lock, R.J. Ann. Clin. Biochem. **1983**, *20*, 224.
114. Litchfield, W.J.; Craig, A.R.; Frey, W.A. Clin. Chem. **1984**, *30*, 1489.
115. Galvin, J.P.; Looney, C.E.; Leflar, C.C.; Luddy, M.A.; Litchfield, W.J.; Freytag, J.W.; Miller, W.K. In *Clinical Laboratory Assays: New Technology and Future Directions*; Nakamura, R.M., Ditto, W.R., Tucker, E.S. III, Eds.; Masson: New York, 1983; 73–95.
116. Heller, W.; Pangonis, W. J. Chem. Phys. **1957**, *26*, 498.
117. Montagne P., Laroche P., Cuilliere M.L., Varcin P., Pau B., Duheille J. J. Clin. Lab. Anal. **1992**, *6*(1), 24.
118. Medcalf, E.A.; Newman, D.J.; Gilboa, A.; Gorman, E.G.; Price, C.P. J. Immunol. Meth. **1990**, *129*, 97.
119. Medcalf, E.A., Newman, D.J., Gorman, E.G., Price, C.P. Clin. Chem. **1990**, *36*, 446.
120. Newman, D.J.; Kassai, M.; Craig, A.R.; Gorman, E.G.; Price, C.P. Eur. J. Clin. Chem. Clin. Biochem. **1996**, *34*, 861.
121. Thakkar, H.; Newman, D.J.; Holownia, P.; Davey, C.L.; Wang, C.C.; Lloyd, J.; Craig, A.R.; Price, C.P. Clin. Chem. **1997**, *43*, 109.
122. Thomson, J.C.; Craig, A.R.; Davey, C.L.; Newman, D.J.; Lonsdale, M.L.; Bucher, W.J.; Nagle, P.D.; Price, C.P. Clin. Chem. **1997**, *43*, 2384.
123. Thakkar, H.; Cornelius, J.; Dronfield, D.M.; Medcalf, E.A.; Newman, D.J.; Price, C.P. Ann. Clin. Biochem. **1991**, *28*, 407.
124. Kondo, A.; Uchimura, S.; Higashitani, K. J. Ferment. Bioeng. **1994**, *78*, 164.
125. Ortega-Vinuesa, J.L.; Gálvez-Ruiz, M.J.; Hidalgo-Álvarez, R. Prog. Colloid Polym. Sci. **1995**, *98*, 233.
126. Molina-Bolívar, J.A.; Ortega-Vinuesa, J.L. Langmuir **1999**, *15*, 2644.
127. Ortega-Vinuesa, J.L.; Bastos-González, D.; Hidalgo-Álvarez, R. J. Colloid Interface Sci. **1996**, *184*, 331.
128. Sorin, T.; Ifuku, Y.; Sato, L.; Kohno, H.; Mizokami, M.; Tokuda, H.; Nakamura, M.; Yakamoto, M. Clin. Chem. **1989**, *35*, 1206.
129. Grange J., Roch A.M., Quash G.A. J. Immunol. Meth. **1977**, *18*(3–4), 365.
130. Buffone G.J., Savory J., Cross R.E., Hammond J.E. Clin Chem. **1975**, *21*(12), 1731.
131. Kusnetz, J.; Mansberg, H.P. In *Automated Immunoanalysis*; Ritchie, R.F., Ed.; Marcel Dekker: New York, 1978; Vol. 1, 1–43.
132. Anderson, R.J.; Sternberg, J.C. In *Automated Immunoanalysis Part 2*; Ritchie, R.F., Ed.; Marcel Dekker: New York, 1978; 409–469.
133. Heidelberger, M.; Kendall, F.W. J. Exp. Med. **1935**, *62*, 467.
134. Montagne, P.; Varcin, P.; Cuilliere, M.L.; Duheille, J. Bioconj. Chem. **1992**, *3*, 187.
135. Skoug, J.W.; Pardue, H.L. Clin. Chem. **1988**, *34*, 300.
136. Skoug, J.W.; Pardue, H.L. Clin. Chem. **1988**, *34*, 309.
137. Kapmeyer, W.H.; Pauly, H.E.; Tuengler, P. J. Clin. Lab. Anal. **1988**, *2*, 76.
138. Molina-Bolívar, J.A.; Galisteo-González, F.; Ortega-Vinuesa, J.L.; Schmitt, A.; Hidalgo-Álvarez, R. J. Biomater. Sci. Polym. Ed. **1999**, *10*, 1093.

139. Marchand, J.; Varcin, P.; Riochet, D.; Montagne, P.; Cuilliere, M.L.; Duheille, J.; Pau, B. Biopolymer **1992**, *32*, 971.
140. Von Schultness, G.K.; Giglio, M.; Cannell, D.S.; Benedek, G.B. Mol. Immunol. **1980**, *17*, 81.
141. Cohen, J.R.; Benedek, G.B. Immunochemistry **1975**, *12*, 239.
142. Von Schultness, G.K.; Cohen, R.J.; Benedek, G.B. Immunochemistry **1976**, *13*, 963.
143. Fitch, R.M. *Polymer Colloids: A Comprehensive Introduction*; Academic Press: San Diego, 1997.
144. Derjaguin, B.V.; Landau, L. Acta Physicochim. USSR. **1941**, *14*, 633.
145. Verwey, E.J.W.; Overbeek, J.T.G. *Theory of the Stability of Lyophobic Colloids*; Elsevier: Amsterdam, 1952; Vols. 1 and 2.
146. Lyklema, J. In *Molecular Forces*; Pontif. Acad. Sci. Scripta Varia, North Holland, Amsterdam; Wiley-Interscience, New York, 1967; Vol. 31, 181, 221.
147. Napper, D.H.; Hunter, R.J. In *Hydrosols*; Kerker, M., Ed.; MTP Int. Rev. Sci., Surface Chem. and Colloids, Butterworth: London, Series 1, 1972, 7, 241; Series 2, 1975, 7, 161.
148. Tsaur, S.L.; Fitch, R.M. J. Colloid Interface Sci. **1987**, *115*, 463.
149. Overbeek, J.T.G. Adv. Colloid Interface Sci. **1982**, *16*, 17.
150. Grabbe, A.; Horn, R.G. J. Colloid Interface Sci. **1993**, *157*, 375.
151. Lessard, R.R.; Zieminski, S.A. Ind. Eng. Chem. Fundam. **1971**, *10*, 260.
152. Elimelech, M. J. Chem. Soc. Faraday Trans. **1990**, *86*, 1623.
153. Israelachvili J.N., Adams G.E. J. Chem. Soc. Farad. Trans. **1978**, *74*, 975.
154. LeNeveu, D.M.; Rand, R.P.; Parsegian, V.A. Nature **1976**, *259*, 601.
155. Broide, M.L.; Tominc, T.M.; Saxowsky, M.D. Phys. Rev. E **1996**, *53*, 6325.
156. Healy, T.W.; Homola, A.; James, R.O.; Hunter, R.J. Faraday Discuss. Chem. Soc. **1978**, *65*, 156.
157. Ducker, W.A.; Senden, T.J.; Pashley, R.M. Nature **1991**, *353*, 2239.
158. Chapel, J.P. J. Colloid Interface Sci. **1994**, *162*, 517.
159. Vigil, G.; Xu, Z.; Steinberg, S.; Israelachvili, J. J. Colloid Interface Sci. **1994**, *165*, 367.
160. Pashley, R.M.; Israelachvili, J. J. Colloid Interface Sci. **1984**, *97*, 446.
161. Okubo, M.; Yamamoto, Y.; Uno, M.; Kamei, S.; Matsumoto, T. Colloid Polym. Sci. **1987**, *265*, 1061.
162. Tamai, H.; Hasegawa, M.; Suzawa, T. Colloid Surf. **1990**, *51*, 271.
163. Peula, J.M.; Callejas, J.; de las Nieves, F.J. In *Surface Properties of Biomaterials*; West, R., Batts, G., Eds.; Butterworth-Heinemann: Oxford, 1994.
164. Peula, J.M.; de las Nieves, F.J. Colloids Surf. A **1994**, *90*, 55.
165. Puig, J.; Fernández-Barbero, A.; Bastos-González, D.; Serra-Domenech, J.; Hidalgo-Álvarez, R. In *Surface Properties of Biomaterials*; West, R.,Batts, G., Eds.; Butterworth-Heinemann: Oxford, 1994.
166. Peula, J.M.; Hidalgo-Álvarez, R.; de las Nieves, F.J. J. Biomater. Sci. Polym. Ed. **1995**, *7*, 231.
167. Ortega-Vinuesa, J.L.; Hidalgo-Álvarez, R. Biotech. Bioeng. **1995**, *47*, 633.
168. Peula, J.M.; Puig, J.; Serra, J.; de las Nieves, F.J.; Hidalgo-Álvarez, R. Colloids Surf. A **1994**, *92*, 127.

169. Peula-García, J.M.; Hidalgo-Álvarez, R.; de las Nieves, F.J. Colloid Polym. Sci. **1997**, *275*, 198.
170. Miraballes-Martínez, I.; Martín-Rodríguez, A.; Hidalgo-Álvarez, R. J. Dispersion Sci. Technol. **1996**, *17*, 321.
171. Feng, M.H.; Morales, A.B.; Poot, A.; Beugeling, T.; Bantjes, A. J. Biomater. Sci. Polym. Ed. **1995**, *7*, 415.
172. Poot A., Beugeling T., Vanaken W.G., Bantjes A. J. Biomed. Mat. Res. **1990**, *24*(8), 1021.
173. Vermeer, A.W.P.; Norde, W. Colloids Surf. A **2000**, *161*, 139.
174. Nakashima, K.; Fujimoto, Y.; Kido, N. Photochem. Photobiol. **1995**, *62*, 671.
175. Arai, T.; Yasuda, N.; Kitamura, H. In *Communication in the 7th ICSCS Congress.* Compiegne, France, 1990.
176. Ortega-Vinuesa, J.L.; Galvez-Ruiz, M.J.; Hidalgo-Álvarez, R. J. Mater. Sci. Mater. Med. **1995**, *6*, 754.
177. Ortega-Vinuesa, J.L.; Bastos-González, D. J. Biomater. Sci. Polym. Ed. **2001**, *12*, 379.
178. Elgersma, A.V. In *Competitive Adsorption of Albumin and Monoclonal γ-Globulin Molecules on Polystyrene Surfaces.* PhD dissertation, University of Wageningen, The Netherlands, 1990.
179. Kawaguchi, H.; Sakamoto, K.; Ohtsuka, Y.; Ohtake, T.; Sekiguchi, H.; Iri, H. Biomaterials **1989**, *10*, 225.
180. Lievens, M.M.; Woestyn, S.; De Nayer, P.; Collet-Cassart, D. Eur. J. Clin. Chem. Clin. Biochem. **1991**, *29*, 401.
181. Buijs, J.; Lichtenbelt, J.W.T.; Norde, W.; Lyklema, J. Colloids Surf. B **1995**, *5*, 11.
182. Ortega-Vinuesa, J.L.; Galvez-Ruiz, M.J.; Hidalgo-Álvarez, R. Prog. Colloid Polym. Sci. **1995**, *98*, 233.
183. Borque, L.; Rus, A.; Bellod, L.; Seco, M.L. Clin. Chem. Lab. Med. **1999**, *37*, 899.
184. Pereira, A.B.; Theofilopoulos, A.N.; Dixon, F.J. J. Immunol. **1980**, *125*, 763.
185. Montagne, P.; Varcin, P.; Cuilliere, M.L.; Duheille, J. Bioconj. Chem. **1992**, *3*, 187.
186. Molina-Bolívar, J.A.; Galisteo-González, F.; Hidalgo-Álvarez, R. Colloids Surf. B 1996, *8*, 73.
187. Molina-Bolívar, J.A.; Galisteo-González, F.; Hidalgo-Álvarez, R. Phys. Rev. E **1997**, *55*, 4522.
188. Delgado-Calvo-Flores, J.M.; Peula-García, J.M.; Martínez-García, R.; Callejas-Fernández, J. J. Colloid Interface Sci. **1997**, *189*, 58.
189. Leikin, S.; Kornyshev, A.A. Phys. Rev. A **1991**, *44*, 2.
190. Pashley, R.M.; Quirk, J.P. Colloids Surf. **1984**, *9*, 1.
191. Pashley, R.M. Adv. Colloid Interface Sci. **1982**, *16*, 57.

4

Capture and Detection of Biomolecules Using Dual Colloid Particles

AGNÈS PERRIN CNRS-bioMérieux, Lyon, France

I. INTRODUCTION

Colloidal microparticles have for decades found wide applications for in vitro diagnosis (IVD) because they are cheap, versatile, and compatible with automation. Usually organopolymeric nanospheres are used; however, other materials, such as metals [1] or minerals like silica [2], are also described. Besides diversity of materials and of physical properties, a large scale of size is available, from a few nanometers for metal microspheres to several dozen micrometers for silica clusters. Under appropriate reaction conditions, it is possible to produce monodisperse particles leading to reliable and reproducible results.

In the frame of IVD applications, particles are often functionalized with biomolecules (nucleic probes, antibodies, cells, etc.) that are specific to searched analytes. Theses ligands can be directly adsorbed onto particle surface or chemically grafted through a large variety of functional chemical groups (amine, aldehyde, carboxyl, thiol, epoxy, etc.) obtained during polymerization or afterward. So-called conjugated particles are used as solid phase for the capture of the targeted molecule or as labels for its detection. A recent review of Kawaguchi [3] describes in details most significant applications of functional polymer microspheres in fields such as physicochemical or specific bioseparation of cells, bacteria, proteins, DNA or RNA, immunoassays, and DNA diagnosis. Also, nanoparticles are widely used in histology and cellular labeling [4].

After a brief and nonexhaustive description of the state of art on the application of nanospheres in diagnosis as solid phase and/or as detection label, the aim of this chapter is to describe an original homogeneous assay method based on a duplex colloid particle format, taking advantage of microspheres' properties to create a sensitive, versatile, and robust assay.

II. SUBMICROMETER COLLOIDAL PARTICLES IN DIAGNOSIS

A. Particles as Solid Phase

The main advantage of microparticles as solid phase is to offer a much larger specific area than that of a flat solid support. For example, the surface developed by 10-mg, 200-nm-diameter, latex particle dispersion is more that 70 times that of one well of a standard 96-well microtitration plate. Therefore, the probability for one analyte to be captured by one immobilized ligand is improved. Also, diffusion limitation, which is a crucial problem in the case of liquid–solid interfacial reactions where no efficient stirring is possible, is greatly lowered in an homogeneous phase assay [5]. A liquid-like reaction is performed that results in improved kinetic coefficients, and permits a quicker and more reproducible reaction.

Washing steps to eliminate excess reagents are surely technically less easy than those performed on a plane surface. Above 0.5 μm diameter and depending on their density, particles can be separated by sedimentation. But below that size microspheres remain in suspension in the solution. They can be extracted from the liquid medium by filtration or centrifugation or by magnetic attraction in the case of magnetic particles.

Applications, especially those implying magnetic particles and reviewed in depth by Ugelstadt et al. [6], are very wide. Most often, microparticles are a key reagent for purification of biomolecules or for sandwich-like assays: the analyte—RNA or DNA [7,8], immunoreagents [9], bacteria [10], virus [11]—is captured on the conjugated particle through a specific interaction and detected in a second step by a labeled molecule (fluorophore [12,13], enzyme [14]) or by electrochemistry [15].

Not only academic researchers, but also the biggest diagnosis companies understood the potentiality of microspheres as solid phase support and currently exploit them in automated systems such as Centaur (Bayer, Leverkusen, Germany) [16], LCx (Abbott, North Chicago, IL, USA) [17], Access (Beckman Coulter) [18], and Cobas Amplicor (Roche, Basel, Switzerland) [19]. Microbeads already proved their interest in growing field of lab-on-a-chip and microarrays, as demonstrated by Bioarray Solutions Ltd (Piscataway, NJ, USA) and Illumina (San Diego, CA, USA) that propose custom bead arrays for protein profiling and molecular typing.

B. Particles as Markers

Using individually detectable microparticles for assaying analytes might be considered in a first approach as an easy way to reach the ultimate sensitivity, i.e., detecting the unique molecule, assuming that this single molecule is bound to a single particle. But diffusion limitation, crucial in extremely diluted medium at the liquid-plane support interface, is still enhanced in the case of particles compared to molecules. Such sensitivity is only reached in small analytical volumes for the detection of molecular labels with technically advanced methods such as fluorescence correlation spectroscopy [20] or total internal reflection fluorescence microscopy [21].

Nevertheless, particulate labels hold a huge interest for lots of applications since their intrinsic signal is stronger and easier to detect than that of a single molecule, which finally results in an improved sensitivity. For example, fluorescent particles are brighter and most stable against photobleaching than the usual organic fluorophores. They were used for sandwich-type immunoassay [22] or for probing specific sequences on DNA molecules [23]. Using extremely powerful instruments such as atomic force microscopes, almost all types of particles might be individually counted to elaborate sensitive immunoassays [24,25]. Also, optical chips are greatly improved while using nanoparticles as mass labels [26], as well as techniques based on refractometry using high refractive index gold beads [27]. The same gold labels greatly facilitate cells quantification by electron or optical microscopy [4]. More specific applications took advantages of magnetic particles as markers for superconducting quantum interface device (SQUID)-based remanence measurements [28], magnetic permeability [29], magnetoresistance [30], or as perturbators of magnetic field frequency [31].

C. Particles as Capture Supports and Markers

1. Agglutination Assays

Agglutination assays could be defined as tests in which the same sort of micrometric particles are used both as solid phase to capture targeted molecules and as detection labels because cross-reactions induce the formation of large particles clusters that become detectable macroscopically. The principle has been known for decades [32] and exploited in several commercial tests [33] for the rapid and simple identification of infectious agents such as bacteria [34,35] or for following metabolic parameters or infections in patients [36–38]. The same principle can also be applied to the detection of polynucleotides [39]. Commonly latex particles, dyed or not, are involved in these tests, but some authors proposed to use magnetic particles [40] or to take advantage of optical properties of gold nanoparticles [39] whose color changes from red to blue under aggregation.

Agglutinations assays are appreciated for their low cost and simplicity of use. Methods of detection can be very basic such as the observation of particle aggregates by the naked eye [41]. In this case sensitivity is limited by the number of microspheres required to make one visible clump (typically 10^{-12} M for an IgG with 1-μm-diameter particles). Turbidity measurement, particle number counting, and dispersion scattering are more sensitive methods. The Copalis Multiplex technology developed by DiaSorin [42] readily discriminates weak variations in latex diameter and permits multiple simultaneous assays by the use of beads mixture. Quartz crystal microbalance (QCM), sensitive to mass variation, is also a sensitive tool in immunoassays [43].

2. Other Types of Tests

More sophisticated, particularly elegant, and more sensitive than agglutination tests where the presence of the analyte is detected by the aggregation of microspheres, single-particle assays are available. They are based on the analysis of each single particle that is, on the one hand, the support of the capture reaction and that, on the other hand, affords informations on the analyte's nature and concentration.

Flow cytometry is one method of choice for these applications because it permits one to count particles and to reveal their identity via a suitable label. This idea was notably that of Fortin et al. [44] who describes a cell assay where the same latex particle is used as capture support and as a detection label. The principle is currently being commercially exploited by Luminex (Austin, TX, USA) [45], which developed an original fluorescent bead immunoassay perfectly adapted for the simultaneous detection of dozen of analytes for applications in DNA hybridization, immunoassays, and enzyme/substrate activity. Individual microspheres are color-coded using an internal blend of various proportion of two dyes, giving each microsphere a distinct spectral address. Detection is based on a sandwich system including a fluorophore-labeled molecule. The complex identity—internal dye and fluorophore on the surface—is identified trough a channel illuminated with laser beams. Another clever multiplexed system, the nano bar codes particles test proposed by Nicewarner et al. [46], contains multistrips of Au and Ag that are optically detectable. Varying the length of these stripes permits one to obtain distinguishable particles onto which several kinds of ligands can be bound.

D. Dual Particle–Based Assays or Dumbbell Assays

After the description of systems in which the same sort of particle is used as solid support, as detection label, or both, another category of tests includes those in which two different kinds of particles have the role of solid support and of detection label, respectively. The interest in such assays is based on combining advantages of both previously described methods. To resume, use of microparticles as solid support enhances reaction yield, and use of microparticles as detection labels offers improved sensitivity. The choice of two different particles allows the possibility to optimize each of them independently for the considered application. A few examples of such systems—later called dumbbells—that may include magnetic or nonmagnetic particles are described in the literature as shown below.

1. Nonmagnetic Dumbbells

Bains [47] proposed a concept whereby two latex particles of different density are covalently coupled with two different DNA probes that do not hybridize to each other but both hybridize to a target. The probes are mixed with the sample and, in the presence of DNA targets, specific aggregates of the two particles types are formed, having a third, intermediate density. A suitable density gradient allows the separation of this intermediate density class.

The conjugated use of latex beads onto which are captured lymphocytes and of gold or silver metal beads to form a sandwich assay permitted Siiman [1] to simultaneously count in blood several types of cell populations with the help of a flow cytometer.

An example of a commercial system based on two different particles is proposed by Packard under the name Alphascreen. Two small beads are brought into proximity by specific biological interactions. Upon laser excitation, a chemical signal is generated by the donor bead that remains undetected without the close proximity of the acceptor bead.

2. Magnetic Dumbbells

Magnetic particles greatly facilitate the manipulation of dumbbells by the opportunity they offer to separate them from the excess reagents and to drive them at a desired location. In the Laser Magnetic Immunoassay of Mizutani [48], viruses captured on microbeads are let to react with antibodies labelled with magnetic particles. Magnetically labelled antigens are attracted and concentrated at one point of the surface, onto which is focused a laser beam, making an interference fringe whose intensity indicates the amount of virus.

Magnetic properties of ferromagnetic labels are also exploited in the system of Kriz et al. [29], whose principle is based on the adsorption of the target analyte onto silica carrier particles detected by functionalized magnetomarkers. The magnetic permeability variation in the presence of labelled silica permitted to assay a model protein.

The team of Lim [49,50] proposes one assay based on the simultaneous antigen binding on two types of reagent particles. Both an indicator latex particle coated with a monoclonal antibody specific to a determinant of the considered antigen and a magnetic bead modified with another monoclonal antibody specific of another determinant are involved. After reaction, magnetic beads are separated with a magnet, and the resultant supernatant color or turbidity is read.

III. A NEW CONCEPT OF MAGNETIC DUMBBELLS: LATMAG ASSAY

A. LATMAG Assay

The common characteristic of all dumbbell assays described above is that they are intrinsically limited in sensitivity by the fact that a significant number of labeled particles is needed to reach a detectable signal. None of the assays allows the detection of the unique particle because detection systems are not sensitive or specific enough.

A new magnetic dumbbell system or LATMAG (latex and magnetic particle) assay tries to bring a solution to this challenge. The principle is as follows [51]: the analyte to be assayed (antigen, DNA, bacteria, etc.) is captured between a magnetic bead and a latex particle, these two colloids being functionalized by ligands specific to two different sites of the analyte (epitope or nucleotide sequence). Dumbbells are separated from reaction medium under a magnetic field to discard the excess of free latex, resuspended, and assayed to determine the amount of latex particles (Fig. 1).

The main originality of LATMAG is that magnetic particles are chosen so that they do not interfere during the detection. In the work of Lim [49], large Dynabeads magnetic particles (3 μm diameter) are used. The turbidimetric detection is processed on the supernatant after magnetic sedimentation, and a decrease of the optical density as a function of antigen concentration is observed, while the amount of latex particles forming a complex with magnetic beads increases. Generally, competition assays are known to be less sensitive and to display a shorter dynamic range than direct assays. The choice of small magnetic particles that cannot be detected permits one to carry a direct assay where the amount of latex involved in the dumbbell complex increases as function of the analyte concentration.

On the other hand, latex particles are chosen so that they can be individually counted under a microscope. Thus, the ultimate potential sensitivity of the assay is the detection of the unique particle, i.e., the unique molecule. Nevertheless, parameters such as capture yields between particles and analytes and nonspecific interactions between particles are likely awaited to corrupt the sensitivity.

LATMAG offers other potential interests. First, this assay is a one-step reaction wherein all relatively cheap reagents are incubated together, followed by a magnetic separation that can be easily processed without sophisticated instrumentation. The detection requires a basic microscope for counting latex when the utmost sensitivity is required. In other cases, the naked eye is sufficient to detect the presence of turbid latex particles after resuspension of the reaction medium. Second, LATMAG is a rapid test since the target capture occurs in a stirred homogeneous suspension of colloid particles, and it was shown above that capture on beads is much more favorable than that on flat support. Finally, displaying a large range of applications, LATMAG can be applied to the detection of nucleic acid targets, antigens, bacteria, cells, and, more generally, any analyte possessing at least two different specific sites that can be recognized by a ligand bound on a particle.

B. Magnetic Capture Particle

The first quality of a magnetic particle candidate for LATMAG utilization is the absence of interference during the detection of latex, thus offering two main possibilities:

The diameter should be significantly different from that of the latex particle for an easy distinction under the microscope. Small magnetic particles whose turbidity in the absence of latex is low are preferred in a view to enhance sensitivity in the case of a naked-eye detection. Furthermore, diffusion coefficient and, consequently, particle reactivity is higher for smaller particles.

The colloidal stability of the magnetic particle should remain excellent after binding of ligand, especially during magnetic separation steps. Formation of aggregates would reduce sensitivity by hiding reactive sites and lowering reaction yields. Also, the principle of a magnetic particle significantly smaller than the detection particle has not been applied so far and brings originality to the assay.

The magnetic particle should be chosen so that its tendency to adsorb nonspecifically on latex particle is as low as possible. Indeed, the formation of such a complex cannot be distinguished from those issued from the specific reaction. Detected even in the absence of the analyte, they are the main limitation for assay sensitivity.

Particle functionality, i.e., the density of active groups at the bead surface, is targeted to be as high as possible, so that the minimal number of particles can be added per trial to limit as far as possible nonspecific interactions between particles.

Last but not least, an accurate compromise must be found between particle size, which should be as small as possible as shown earlier, and the magnetic content of particles, which should be as high as possible for an efficient attraction of dumbbells under magnetic field. Ideally, one magnetic particle should be able to carry in a minimum of time—a few minutes—one latex particle. If several magnetic particles are needed to displace one latex bead, intrinsic sensitivity is dramatically degraded since at the final stage of detection one detected bead corresponds to several analytes captured on the surface.

TABLE 1 Some Commercially Available Magnetic Particles

Manufacturer	Particle diam. (nm)	No. reactive sites/μm^2	Magnetic sedimentation speed (min)
Miltenyi [52]	50	Unknown	> 60
Immunicon [55]	145	0.27	10
Seradyn [53]	750	0.02	3
Dynal [54]	2800	0.07	1

Several commercially available magnetic particles, whose main characteristics are summed up in Table 1, are evaluated. Colloidal stability of all these particles established by size measurement by light scattering is acceptable.

Miltenyi particles [52] are constituted by a ferric oxide core whose diameter is around 10 nm, surrounded by a thick layer of polysaccharide for a final diameter of 50 nm. Because of their low magnetic content, these particles cannot be separated by a common magnet. A specially designed column creating a strong magnetic gradient is available for this purpose. Although Miltenyi microspheres give positive results in the LATMAG process, their cumbersome and expensive separation step is an obstacle to their further exploitation.

Seradyn [53] and Dynal [54] display a very interesting magnetic separation speed and a satisfying surface functionality. But experimentally, it is observed that the dumbbell formation is limited by the particle diameter probably too large to lead to an efficient reaction rate. Finally, Immunicon [55] magnetic microspheres are chosen as favorite candidates for LATMAG.

FACING PAGE

FIG. 1 General LATMAG protocol. 1. *Reagents for LATMAG.* Magnetic particles coated with the first ligand are incubated with latex particles modified with the second ligand, in the presence of the target which bear receptors for both types of ligands. 2. *Formation of dumbbells.* Particles and targets react together to form magnetic complexes or dumbbells. 3. *Separation of dumbbells from unreacted latex particles by magnetic sedimentation.* In the presence of a magnet, free magnetic particles and dumbells are attracted against the test tube wall. Free latex particles are discarded with the supernatant. 4. *Counting of latex particles after resuspension of magnetic dumbbells.* Preparations are deposited onto a glass slide and analyzed under the microscope to estimate the number of latex particles, i.e., the number of dumbbells.

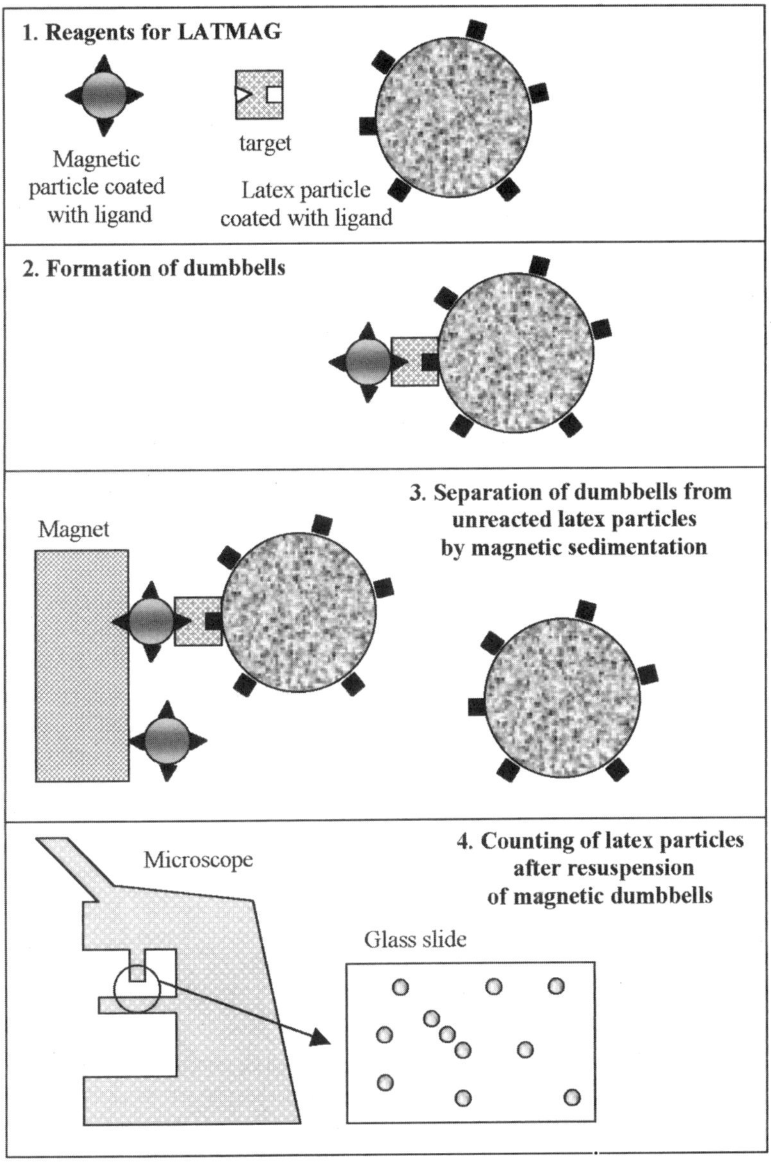
1. Reagents for LATMAG
Magnetic particle coated with ligand
target
Latex particle coated with ligand
2. Formation of dumbbells
Magnet
3. Separation of dumbbells from unreacted latex particles by magnetic sedimentation
Microscope
4. Counting of latex particles after resuspension of magnetic dumbbells
Glass slide

C. Marker Particles

Recommendations about the diameter of latex particles depend on the detection method:

For a naked-eye or a spectrophotometric detection, the sensitivity, i.e., the minimal number of particles that can be detected per trial, increases as function of particle diameter. For the same number of particles, the optical density measured at a given wavelength (380 nm) increases as function of particle diameter (Fig. 2).

Nonfluorescent particles with a diameter above 300 nm are readily detected under microscope. Using larger particles allows to work with an objective of lower magnification and to improve sample speed analysis and/or accuracy since a larger surface can be analyzed.

Much smaller fluorescent particles (<100 nm) are detectable with a fluorescent microscope since it is not necessary to resolve them optically for individually counting them. Sensitivity is linked to the density, intensity, and stability of fluorophores entrapped inside the particle or chemically bound to the surface.

Requirements concerning colloidal stability and surface functionality are the same than for magnetic particles, i.e., an excellent colloidal stability after ligand binding and a high density of reactive sites. Experimentally, the density of reactive groups can be estimated with ^{35}S-radiolabeled molecules.

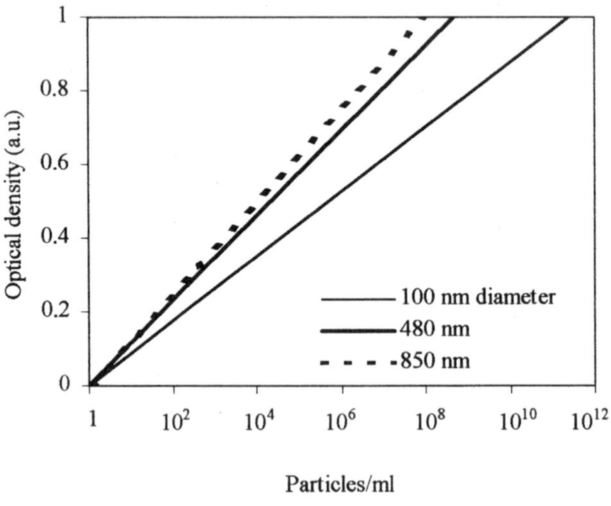

FIG. 2 Variation of optical density measured at 380 nm as function of polystyrene latex particle concentration, for different particle diameters.

IV. CAPTURE AND DETECTION OF BIOMOLECULES BY LATMAG

A. Generality

Several applications are described below to demonstrate the wide possibilities of LATMAG that can be relevant to the quantitative detection of almost all biomolecules. Protocols are almost the same whatever the analyte to be assayed. Incubation of magnetic and latex biosensitized particles occurs in the presence of targeted molecules, in an appropriate buffer containing salts and nonionic detergents. Afterward, the test tube is placed in the strong magnetic field generated by a rare earth alloy. After magnetic sedimentation, the supernatant is removed. Dumbbells and free magnetic particles that cannot be physically separated at this stage are dispersed in the desired buffer, and observed by eye, analyzed by spectrophotometry to measure the absorbance, or deposited onto a glass slide that is placed under a microscope to determine the number of latex particles. Reference samples without target are always needed to evaluate the amount of nonspecific adsorption of magnetic particles onto latex microspheres. Finally, after subtraction of this background, the detection signal can be reported as a function of searched analyte concentration. The sensitivity threshold is defined as the minimal target molecule concentration enabling a detection signal higher than that obtained for the background noise measured on 10 samples plus three times the standard deviation.

B. Application to Immunology

1. Antibody Detection

LATMAG can be applied for the detection of antibodies (Fig. 3) whose presence is researched, among numerous examples, in viral infections (anti-P24 antibodies) or for detecting pregnancy (anti-HCG antibodies). The strategy is to capture biotinylated IgG [56] between magnetic streptavidin particles and latex microspheres modified to exhibit affinity toward the targets. In Fig. 4 are reported images realized from microscope glass slides onto which are deposited preparations for different target IgG concentration. Only latex particles can be detected in these conditions because magnetic particles are too small to be optically resolved. An increase of the latex particle number per unit area as a function of antibody concentration is observed, proving that dumbbell complexes, bridged by IgG, are formed between latex and magnetic particles and that they can be separated from the medium by an appropriate magnetic field.

Dumbbells can also be studied more precisely by transmission electron microscopy (TEM) (Fig. 5). The commercial magnetic particles displays a highly heterogeneous structure, which makes them look like "corn flakes" [57]. On the other hand, size difference between magnetic and latex particle is very impres-

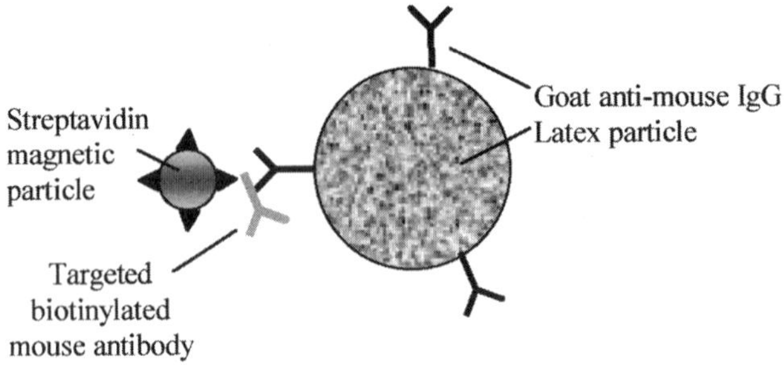

FIG. 3 LATMAG protocol for the detection of antibodies. Goat anti-mouse IgG latex particles are incubated with streptavidin magnetic particles and biotinylated mouse antibodies, to form dumbbells those number is a direct function of the antibody concentration.

sive, and it is difficult to imagine that such a small particle can carry the latex sphere load. Actually, one question remains in suspense: is it possible to magnetically attract one latex particle when only one magnetic bead is bound to its surface? TEM does not bring the answer because magnetic spheres hidden below latex are not imaged. If the answer is negative, the assay sensitivity is intrinsically limited by the fact that more than one target molecule should bind

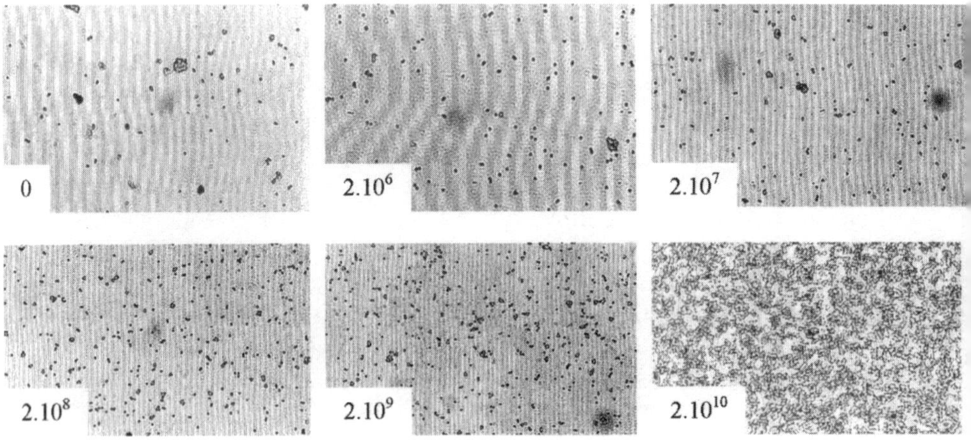

FIG. 4 Images of microscopic analysis of dumbbells for increasing concentration of anti-TSH targeted antibodies (from 0 to 2.10^{10} molecules/mL). The objective is $\times 40$. Latex particles (1.1 μm) are easily counted while magnetic particles (0.15 μm) are not optically resolved.

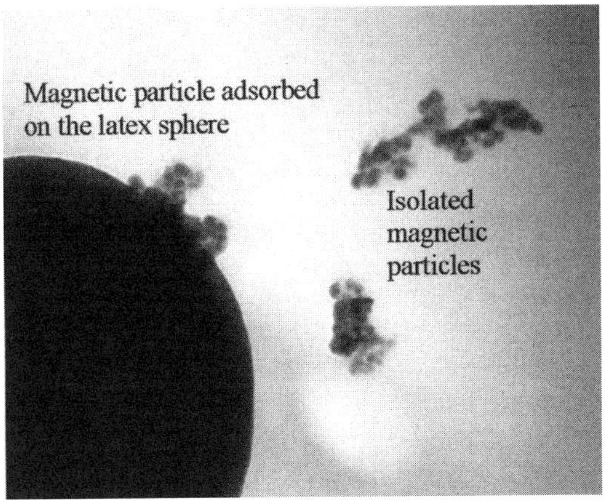

FIG. 5 TEM picture of a close-up view of the surface of a latex particle (1.1 μm diameter) bound to a magnetic particle. Isolated magnetic particles that did not participate to the reaction are also present around the dumbbell. Their shape is that of irregular and polydisperse clusters whose mean diameter is about 150 nm.

a latex for having a positive signal, i.e., it is not possible to detect a unique molecule.

Even in the absence of targeted antibodies, a few latex spheres are found in the sample that can be free latex or magnetic complexes formed by the nonspecific adsorption of magnetic particle onto latex spheres. At high antibody concentration, chains of particles are observed with optical as well as with electron microscopy, due to the reaction of several antibodies on the same latex particle, each antibody being able to bind one magnetic particle. Thus, the separation of such complexes is much quicker thanks to their consequent size and the magnetic content (Fig. 6).

Dumbbells can be quantified per unit area (Fig. 7) after image treatment. While it is not possible to count all latex particles present on the microscope slide, surfaces analyzed are sufficiently large to be statistically representative of the sample. The dynamic range of the assay spread on 6 $\log_{10}$ units; the sensitivity threshold is estimated at 2×10^6 molecules/mL, equivalent to a molar concentration of 3×10^{-15} mol/L or a weight concentration of 5×10^{-10} g/L, which is sensitive enough for many assays. For example, the threshold for pregnancy is around 10 mUI/mL of HCG, approximately equivalent to 10^{-6} g/L. Also, LATMAG sensitivity is interesting compared to other dual-particle-based tests. As a matter of fact, Lim et al. [49] obtained a sensitivity of 1.6×10^{-2} g/L for the detection of anti-LPS antibodies by assaying the supernatant after magnetic sedimentation. In the case of the LATMAG assay, positive latex particles are

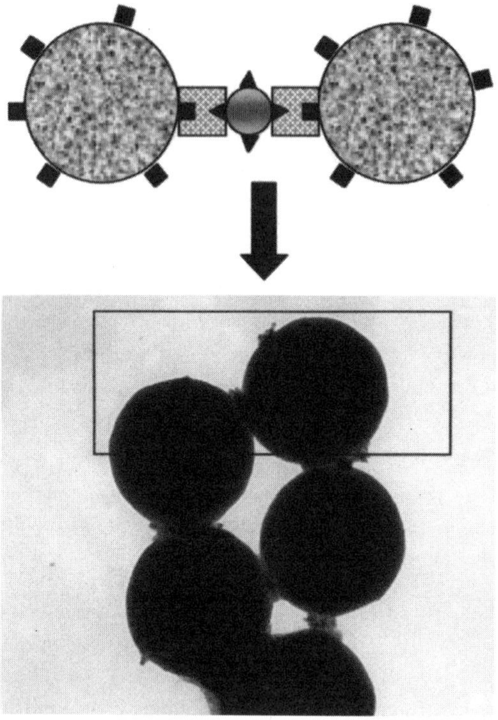

FIG. 6 Formation of chains of dumbbells for high analyte concentration. Several targets specifically bound on the same latex particle permit the formation of several bridges between both types of particles, leading to clusters whose size can reach 10 μm.

directly counted at the issue of magnetic separation, while Lim measured the spectrocolorimetric signal of negative latex particles.

2. Antigen Detection

Among numerous antigens whose presence and concentration are often assayed by biological sampling, a physiological parameter is chosen as a model illustrating the immunological detection of proteins by LATMAG (Fig. 8). Thyroid-stimulating hormon (TSH) is a major regulating factor of hormone synthesis and secretion, and sensitivity to that parameter is needed for the diagnosis of hyperthyroidism [58].

To demonstrate that several detection methods are compatible with LATMAG, dumbbell suspensions are assayed by spectrophotocolorimetry to measure the turbidity brought by the presence of latex particles (Fig. 9). At very low antigen

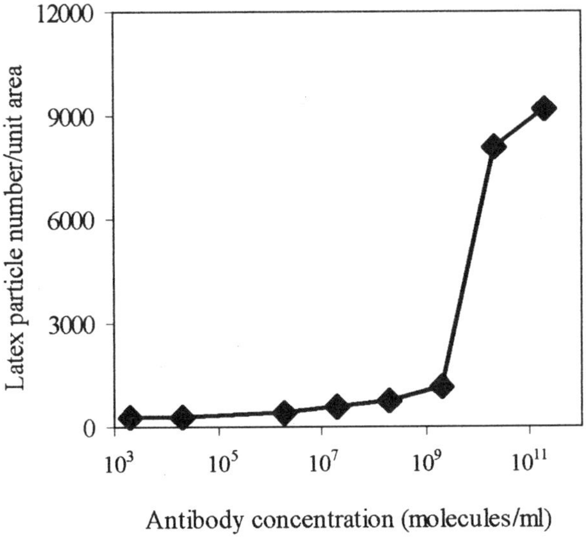

FIG. 7 Variation of particle number/unit area evaluated from images realized by optical microscopy, as a function of antibody concentration.

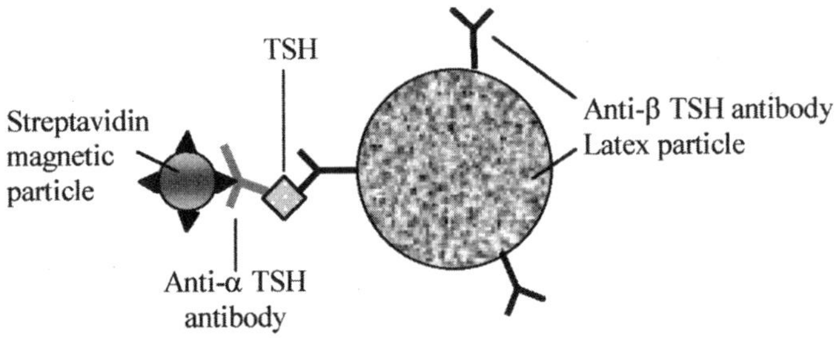

FIG. 8 LATMAG protocol for the detection of TSH molecules. Polystyrene latex particles were modified by grafting anti β-TSH antibodies. On the other hand, biotinylated anti α-TSH antibodies were captured onto streptavidin magnetic particles. In the presence of TSH, the formation of dumbbells was evidenced by measuring the optical density or turbidity of the magnetic suspension after the elimination of free latex particles.

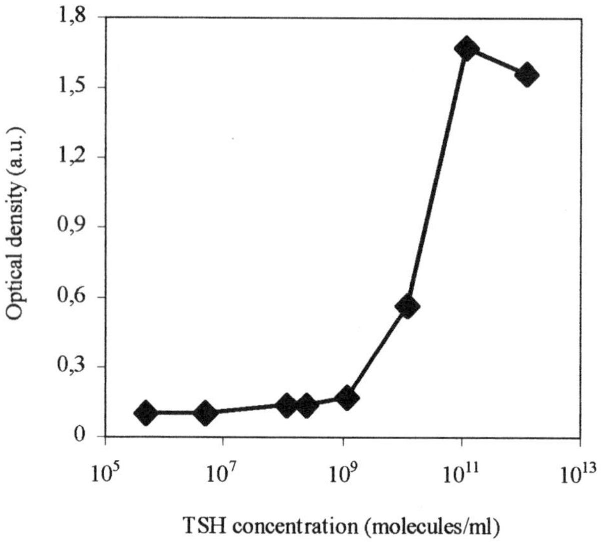

FIG. 9 Variation of optical density of dumbbell solution after resuspension as function of TSH concentration in solution.

concentration, the weak absorbance measured is that of magnetic particles, which give an optical density signal even in the absence of specific reaction. This is not observed when detection is carried out with a microscope because magnetic particles are too small to be optically resolved. This background is added to that of false dumbbells, making the method less sensitive although quicker to perform.

For higher TSH concentration, the optical density increases as function of TSH concentration to reach a saturation level where all latex particles are involved in dumbbells and where the supernatant at the issue of magnetic separation is completely clear. Detection limit is estimated at about 6×10^{-8} g/L of TSH, or 2×10^{-12} mol/L which proves that an interesting sensitivity could be achieved with LATMAG even without using a microscope for the detection. A spectrophotometer is used for accurate quantification, but it should be underlined that exactly the same sensitivity is obtained with the naked eye, making this method particulary suitable under conditions in which technical equipment is poor and only a qualitative or semiquantitative assay is needed.

C. Application to DNA Hybridization

DNA or RNA hybridization reactions are widespread, and their applications become more and more diversified with the take-off of DNA chips. Compared to immunological tests based on antibody or antigen assay, they often allow an earlier diagnosis of infection since nucleic acids, which indicate the presence of the infectious agent, can be evidenced before the immune response induces antibody production.

The demonstration of DNA assay by LATMAG is carried out with a nucleic target from a *pol* region of the HIV-1 genome [57]. Two oligonucleotides complementary of two different regions of the target, modified so that they both could be immobilized on particles (Fig. 10), are used as capture probes to hybridize the DNA target between the two particles. One of the nucleic probes is modified with biotin to bind streptavidin magnetic particles. The other one is bound to an enzyme (horseradish peroxidase) and its capture on latex microsphere modified with anti-HRP antibodies is awaited. For both types of particles, the number of each specific nucleic probes added in the reaction mixture is carefully optimized so that it is less than the number of available capture sites. Therefore, no free hybrids should remain in solution to damage assay sensitivity. In the same view, to enhance the reaction, and unlike previously described applications, the capture of the target is carried out in two steps. First, nucleic capture probes and DNA targets are hybridized in stirred solution. Then magnetic and latex microspheres are added to the mixture to capture hybrids and form dumbbells. As a matter of fact, it is well known that hybridization of

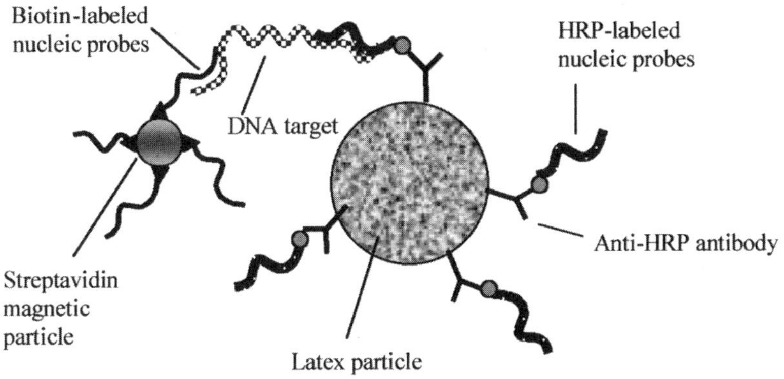

FIG. 10 LATMAG protocol for the detection of HIV DNA targets (72 mers). AntiHRP antibodies were covalently grafted onto polystyrene latex microspheres. In a first step, two nucleic probes (29 and 32 mers) respectively modified at their 5′ end by biotin residues or by the horseradish peroxidase, and specific of two distinct regions of the HIV target, were hybridized with the target. Then streptavidin magnetic particles and anti-HRP latex microspheres were added to form the dumbbell structure. After the elimination of free latexes, complexes were deposited onto a Mallassez cell, covered with a coverslip, and their volume concentration was estimated.

nucleic probes and DNA targets is more efficient if probes are free in solution rather than immobilized onto particles [59].

A precise quantification of dumbbells can be completed here with the help of tools usually useful in microbiology such as the Mallassez cell. This etched glass slide comprises different types of microcavities those volume is perfectly defined. Counting the number of particles in one or several chambers allows one to evaluate the total number of dumbbells, knowing the final volume of resuspended magnetic complexes.

This approach makes it possible to estimate the reaction yield, i.e., the ratio [(number of dumbbells)/(number of targets)] for each target concentration. As shown in Fig. 11, the best reaction yield is obtained for low target concentrations and does not exceed 5%. Several parameters play a rule in this yield. First, the reaction is limited by the affinity constant between anti-HRP antibodies and HRP-labeled oligonucleotides, which is much lower than that of biotin/streptavidin [60]. Furthermore, some immobilized antibodies may have lost part of their activity [61]. Second, nonspecific adsorption of one particle on the other induces a background signal that should be subtracted from that of dumbbells for the estimation of reaction yield. Finally, an unknown number of dumbbells are surely lost during separation steps because they are too weakly magnetic.

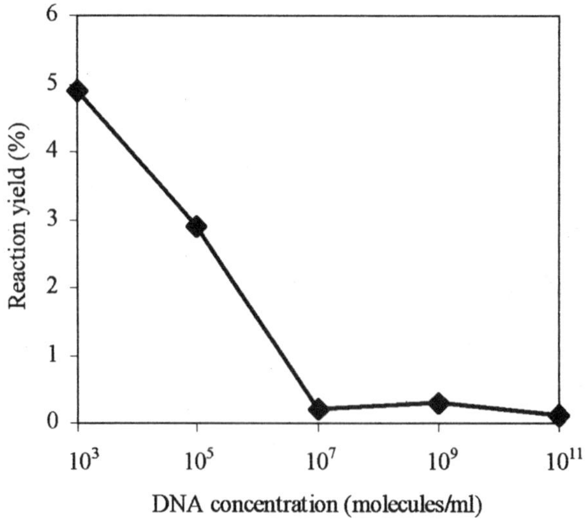

FIG. 11 Variation of reaction yield, i.e., the ratio between the number of dumbbells at the issue of the reaction and the initial number of DNA targets introduced in the reaction medium, as a function of DNA concentration. (From Ref. 57. Copyright 2001 Bioconjugate Chem.)

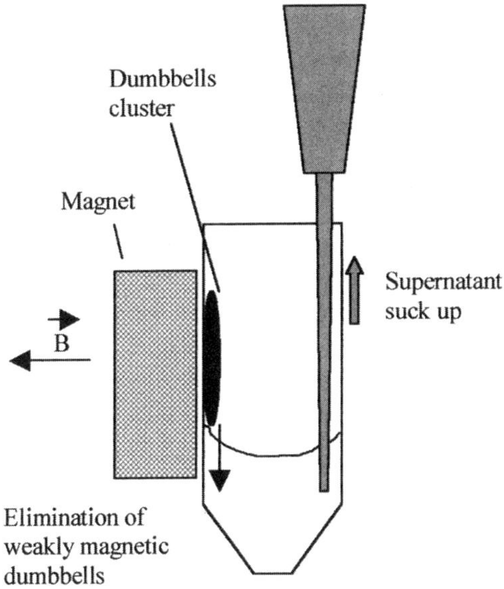

FIG. 12 Washing process by magnetic sedimentation. The test tube is placed against the magnet to permit the separation of dumbbells. The supernatant is sucked up by a pipette. Most dumbbells stay on place, but the weakly magnetic complexes are carry away because of capillary forces exerted by the liquid meniscus.

As a matter of fact, interfacial strengths applied by the liquid meniscus on the dumbbells cluster is believed to drag them down (Fig. 12). Beyond a certain target concentration, there is statistically more than one target per latex. As depicted in the case of the antibody assay in Fig. 6, chains of particles are formed leading to a drop of the reaction yield since the number of dumbbells increases more slowly than the number of targets (Fig. 13). From this curve, the sensitivity threshold can be estimated at 10^6 molecules/mL or 4×10^{-11} g/L or 2×10^{-12} mol/L [57].

D. A Liquid Cell Microcomponent to Improve LATMAG Sensitivity

With multiple objectives (1) to retain weakly magnetic dumbbells and avoid the loss of specific signal, (2) to increase binding kinetics, and (3) to reduce the number of nonspecific dumbbells, a liquid microcell designed and produced by the LETI-CEA at Grenoble is used [62]. This microcomponent issued from silicon etching technology is depicted in Fig. 14. The component itself is made of silicon, recovered by a sealed transparent glass coverslip, allowing the cell

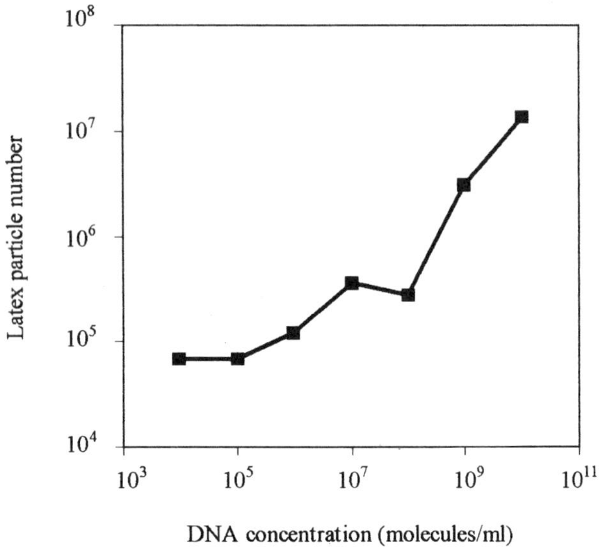

FIG. 13 Variation of the total number of latex particles at the issue of the reaction, as function of the initial DNA concentration in solution. (From Ref. 57. Copyright 2001 Bioconjugate Chem.)

to be studied by optical microscopy. One operture permits the introduction of reagents in the reaction microchamber. The lower wall of the cell is thin (<2 mm) so that magnetic particles injected in the chamber are under the influence of a magnet placed below the component.

The protocol for LATMAG should be somewhat modified to fit and to take profit of the chip. It is known that one of the main factors limiting sensitivity is the formation of unspecific complexes between magnetic particles and latex, leading to the presence of dumbbells even in the absence of DNA target. One solution to limit this adsorption is to reduce the number of particles. On the other hand, reduction of the volumic concentration of reagents is awaited to lower reaction yield and, thus, sensitivity. Magnetic properties of microspheres are exploited to concentrate "half-dumbbells," i.e., targets specifically captured on magnetic particles (without latex particles) from the initial volume of several hundred microliters until a final volume of less than a few microliters that is injected in the cell. Latex particles are added to this solution just before injection so that their final concentration is identical to that of standard conditions. In this case the number of latex spheres is divided proportionally to the volume reduction.

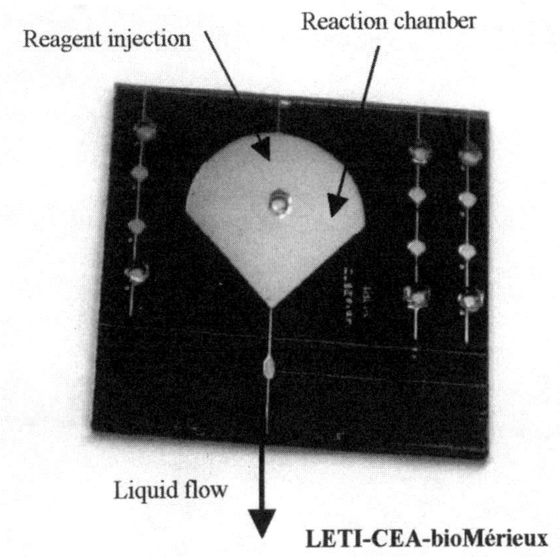

FIG. 14 The liquid flow cell for the optimization of LATMAG assay. The reaction volume (less than 10 μL) comprising fluorescent latex particles and hybrids captured on magnetic particles is introduced in the silicon microcomponent. Formation of dumbbells occurs quickly. Particles are brought together with a magnet, surmounted by a cone to focus line field, sweeping the lower surface of the cell. The magnetic cluster is analyzed under the microscope to determine its average level of fluorescence.

The second idea is the use of fluorescent latex instead of unlabeled white particles. These commercial particles are strongly fluorescent and are almost insensitive to bleaching [63]. Bearing carboxylic groups, they can be modified as well, and they greatly facilitate the detection of dumbbells in the component whose surface is not as clean as that of a microscope slide. The risk that dusts might be confused with dumbbells and degrade test performance is almost absent.

Molecules and particles meeting and binding are enhanced by the microscopic dimensions of the chip whose height does not exceed 100 μm. Therefore, a short incubation period of a few minutes is sufficient to reach the saturation against 1 h for all other tests previously described. After the reaction, dumbbells are brought together in a few seconds with the help of a magnet sweeping the bottom surface of the reaction chamber and topped by a conical iron piece to focus particles on a restricted area of a few mm^2 (Fig. 15). No washing is carried out at the issue of magnetic sedimentation; fluorescence intensity of the magnetic cluster is measured under the microscope. Therefore, besides gain of

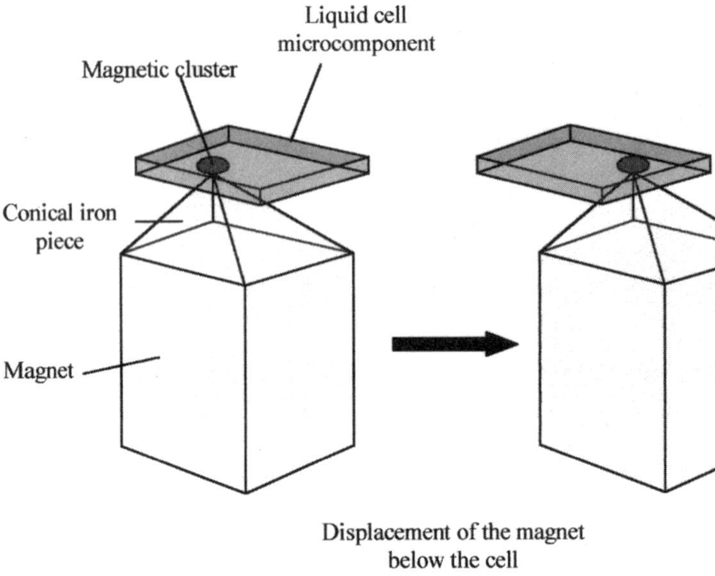

FIG. 15 Magnetic separation of dumbbells into the liquid cell component with the help of a magnet tapped by a conical iron piece, sweeping the bottom surface of the cell to concentrate dumbbells in a spot.

time, the loss of weakly magnetic dumbbells is clearly avoided. A direct microscope analysis of the magnetic cluster is possible in the microcomponent whose upper surface is planar and transparent.

Fluorescent magnetic clusters of dumbbells are obtained for high target concentration (Fig. 16). A strong density of fluorescent particles is observed just above the magnet top that concentrates them on a restrained area, surrounded by isolated free latex microspheres. The quantification can be carried out by measuring the intensity at different places of the spot. The interest with regard to previously described microscopic investigations in which only a fraction of the sample is studied is that the entire signal can be analyzed in one image, leading to a more precise and sensitive quantification.

The variation of fluorescence intensity as a function of DNA concentration permits estimation of the detection threshold at 3×10^5 molecules/mL (Fig. 17), which is three times lower than the same assay carried outside the chip. Apart from its benefit on sensitivity, LATMAG in the microcell is easier and quicker to perform in terms of reduction of incubation and of magnetic separation duration, washing steps uselessness, and simplification of the detection process. The chip can be reused several times without noticeable degradation of performance.

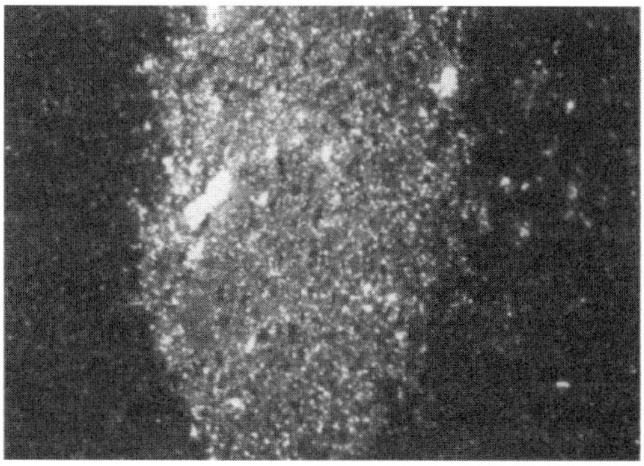

FIG. 16 Image obtained by fluorescent microscopy of the magnetic cluster of dumbbells in the microcomponent at the issue of the process. The mean intensity is estimated and related to the concentration of DNA targets. The cluster is surrounded by free fluorescent particles that are not involved in dumbbells but do not interact with the specific signal.

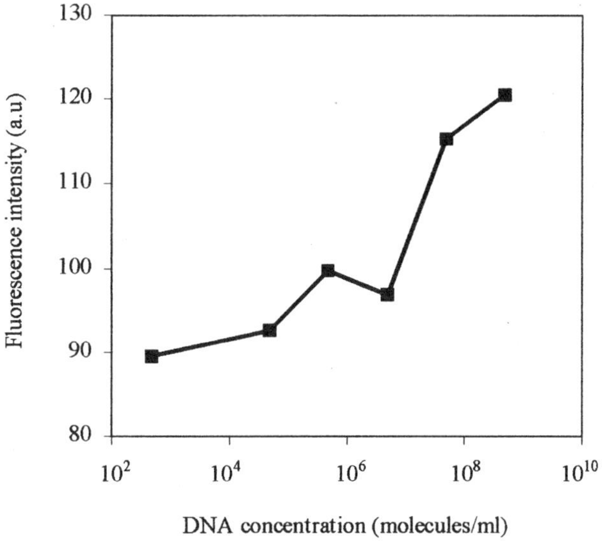

FIG. 17 Variation of the fluorescence intensity of the magnetic cluster as a function of DNA targets concentration. The dynamic range of the camera that permits to quantify the signal spread from 0 to 255 grey levels.

V. CONCLUSION

This chapter provides information pointing to the huge potentiality of latex particles in the field of diagnosis. Polymeric nanospheres are described as a very interesting alternative to plane capture solid phase since they bring a higher specific surface and they improve reaction yields. They have been employed for decades in agglutination tests that permit the very simple and rapid determination of a pathogen in a sample. Magnetic particles also offer the possibility of a convenient and rapid separation, and most large diagnosis companies proposed automated systems using such microspheres. Other nanospheres such as fluorescent latex or gold beads also proved their strong interest as sensitive and original markers for the detection of biological reactions. From these achievements was developed a generic assay based on the combination of two types of particle, one playing the role of solid phase and the other of label: LATMAG is born. Few constrains are imposed because it has been demonstrated that the assay can be applied to the detection of all types of analytes bearing at least two different receptors such as cells, bacteria, tissue, proteins, and nucleic acids. Nature and size of suitable particles are very diversified as function of customer specifications in terms of cost, rapidity, sensitivity, and assay complexity. On one hand, for basic applications LATMAG can be carried out without any associated instrument since only a magnet and a transparent tube allowing the naked-eye detection are necessary, making it interesting for point-of-care applications. On the other hand, when the utmost sensitivity is targeted, a specific liquid cell microcomponent issued from silicon technology allows one to obtain a detection threshold of 5×10^{-16} mol/L. Thus, LATMAG is a promising tool in fields where high sensitivity is required, such as early infectious diagnosis or bacterial identification in food.

Beyond diagnosis, applications of polymeric nanospheres, already very diversified and convincing, are still very likely to have a great future in academic research as well as in industry. The imagination of researchers could be the only barrier limiting their applications in biological, chemical, and physical sciences.

REFERENCES

1. Siiman, O.; Gordon, K.; Burshteyn, A.; Maples, J.A.; Whitesell, J.K. Immunophenotyping using gold or silver nanoparticle-polystyrene bead conjugates with multiple light scatter. Cytometry **2000**, *41*, 298–307.
2. Thomas, N.E.; Sobanski, M.A.; Coakley, W.T. Ultrasonic enhancement of coated particle agglutination immunoassays: influence of particle density and compressibility. Ultrasound Med. Biol. **1999**, *25*, 443–450.
3. Kawaguchi, H. Functional polymer microspheres. Prog. Polym. Sci. **2000**, *25*, 1171–1210.
4. de Harven, E.; Soligo, D. Immunogold labeling of human leukocytes for scanning

electron microscopy and light microscopy: quantitative aspects of the methodology. Scanning Microsc. **1987**, *1*, 713–718.

5. Henry, R. Real-time measurements of DNA hybridization on microparticles with fluorescence resonance energy transfer. Anal. Biochem. **1999**, *276*, 204–214.
6. Ugelstad, J.; Stenstad, P.; Kilaas, L.; Prestvik, W.S.; Herje, R.; Berge, A.; Hornes, E. Monodisperse magnetic polymer particles. Blood Purif. **1993**, *11*, 349–369.
7. Matsunaga, T.; Nakayama, H.; Okochi, M.; Takeyama, H. Fluorescent detection of cyanobacterial DNA using bacterial magnetic particles on a mag-microarray. Biotechnol. Bioeng. **2001**, *73*, 400–405.
8. Hakala, H.; Maki, E.; Lonnberg, H. Detection of oligonucleotide hybridization on a single microparticle by time-resolved fluorometry: quantitation and optimization of a sandwich type assay. Bioconj. Chem. **1998**, *9*, 316–321.
9. Lukin, Y.V.; Pavlova, I.S.; Generalova, A.N.; Zubov, V.P.; Zhorov, O.V.; Martsev, S.P. Immunoreagents based on polymer dispersions for immunochemical assays. J. Mol. Recog. **1998**, *11*, 185–187.
10. Romalde, J.R.; Magarinos, B.; Lores, F.; Osorio, C.R.; Toranzo, A.E. Assessment of a magnetic bead-EIA based kit for rapid diagnosis of fish pasteurellosis. J. Microbiol. Meth. **1999**, *38*, 147–154.
11. Samoylova, T.I.; Smith, B.F. Flow microsphere immunoassay-based method of virus quantitation. Biotechniques **1999**, *27*, 356–361.
12. Huang, S.C.; Swerdlow, H.; Caldwell, K.D. Binding of biotinylated DNA to streptavidin-coated polystyrene latex. Anal. Biochem. **1994**, *222*, 441–449.
13. Yu, H.; Ahmed, H.; Vasta, G.R. Development of a magnetic microplate chemifluorimmunoassay for rapid detection of bacteria and toxin in blood. Anal. Biochem. **1998**, *261*, 1–7.
14. Stark, M.; Reizenstein, E.; Uhlen, M.; Lundeberg, J. Immunomagnetic separation and solid-phase detection of *Bordetella pertussis*. J. Clin. Microbiol. **1996**, *34*, 778–784.
15. Perez, F.G.; Mascini, M.; Tothill, I.E.; Turner, A.P. Immunomagnetic separation wih mediated flow injection analysis amperometric detection of viable *E. coli* O157. Anal. Chem. **1998**, *70*, 2380–2386.
16. Hall, J.; Jilg, W.; Hottentrager, B. Performance of a chemiluminescent immunoassay for HbsAg on the new high-throughput and fully automated ACS:Centaur™ system. Clin. Lab. **1998**, *44*, 349–354.
17. Gilchrist, A.; Solomon, N.; Erickson, D.; Sikand, A.; Bauer, K.A.; Kruskall, M.S.; Kocher, O. Automated detection of the G20210A prothrombin mutation using the LCx microparticle enzyme immunoassay. Clin. Chem. Acta **2001**, *314*, 249–254.
18. Liu, X.; Turner, B.P.; Peyton, C.E.; Reisner, B.S.; Okorodudu, A.O.; Mohammad, A.A.; Hankins, G.D.V.; Weissfeld, A.S.; Petersen, J.R. Prospective study of IgM to *Toxoplasma gondii* on Beckman Coulter's Access™ immunoassay system and comparison with Zeus ELISA and gull IFA assays. Diagn. Microbiol. Infect. Dis. **2000**, *36*, 237–239.
19. Poljak, M.; Marin, I.J.; Seme, K.; Brinovec, V.; Maticic, M.; Meglic-Volkar, J.; Lesnic, G.; Vince, A. Second-generation hybrid capture test and Amplicor monitor test generate highly correlated hepatitis B virus DNA levels. J. Virological Meth. **2001**, *97*, 165–169.

20. Kinjo, M.; Rigler, R. Ultrasensitive hybridization analysis using fluorescence correlation spectroscopy. Nucleic Acids Res. **1995**, *23*, 1795–1799.
21. Harada, Y.; Funatsu, T.; Murakami, K.; Nonoyama, Y.; Ishihama, A.; Yanagida, T. Single-molecule imaging of rna polymerase-DNA interactions in real time. Biophys. J. **1999**, *76*, 709–715.
22. Hall, M.; Kazakova, I.; Yao, Y.M. High sensitivity immunoassays using particulate fluorescent labels. Anal. Biochem. **1999**, *272*, 165–170.
23. Taylor, J.R.; Fang, M.M.; Nie, S. Probing specific sequences on single DNA molecules with bioconjugated fluorescent nanoparticles. Anal. Chem. **2000**, *72*, 1979–1986.
24. Perrin, A.; Theretz, A.; Lanet, V.; Vialle, S.; Mandrand, B. Immunomagnetic concentration of antigens and detection based on a scanning force microscopic immunoassay. J. Immunol. Meth. **1999**, *224*, 77–87.
25. Perrin, A.; Theretz, A.; Mandrand, B. Thyroid stimulating hormone assays based on the detection of gold conjugates by atomic force microscopy. Anal. Biochem. **1998**, *256*, 200–206.
26. Kubitschko, S.; Spinke, J.; Bruckner, T.; Pohl, S.; Oranth, N. Sensitivity enhancement of optical immunosensors with nanoparticles. Anal. Biochem. **1997**, *253*, 112–122.
27. Schneider, B.H.; Dickinson, E.L.; Vach, M.D.; Hoijer, J.V.; Howard, L.V. Highly sensitive optical chip immunoassays in human serum. Biosensors Bioelectronics **2000**, *15*, 13–22.
28. Kötitz, R.; Trahms, L.; Koch, H.; Weitschies, W.; Rheinländer, T.; Semmler, W.; Bunte, T. SQUID based remanence measurements for immunoassays. IEEE Trans. Appl. Superconductivity **1997**, *7*, 3678–3681.
29. Kriz, K.; Gehrke, J.; Kriz, D. Advancements toward magneto immunoassays. Biosensors Bioelectronics **1998**, *13*, 817–823.
30. Baselt, D.R.; Lee, G.U.; Natesan, M.; Metzger, S.W.; Sheehan, P.E.; Colton, R.J. A biosensor based on magnetoresistance technology. Biosensors Bioelectronics **1998**, *13*, 731–739.
31. Richardson, J.; Hawkins, P.; Luxton, R. The use of coated paramagnetic particles as a physical label in a magneto-immunoassay. Biosensors Bioelectronics **2001**, *16*, 989–993.
32. Singer, J.M.; Plotz, C.M. The latex fixation test. I. Application to the serologic diagnosis of the rheumatoid arthritis. Am. J. Med. **1956**, *21*, 888.
33. Van Griethuysen, A.; Bes, M.; Etienne, J.; Zbinden, R.; Kluytmans, J. International multicenter evaluation of latex agglutination tests for identification of *Staphylococcus aureus*. J. Clin. Microbiol. **2001**, *39*, 86–89.
34. Mayayoshi, T.; Hirai, Y.; Kanemasa, Y. A latex agglutination test for the detection of *Mycoplasma pneumoniae* in respiratory exudates: a comparative study with a commercially available DNA-probe test. Microbiol. Immunol. **1992**, *36*, 149–160.
35. Fach, P.; Popoff, M.R. Detection of enterotoxigenic clostridium prefringens. I. Food and fecal samples with a duplex PCR and the slide latex agglutination test. Appl. Environ. Microbiol. **1997**, *63*, 4232–4236.
36. Newman, D.J.; Kassai, M.; Craig, A.R.; Price, C.P. Validation of a particle en-

hanced immunoturbidimetric assay for serum beta 2-microglobulin on the Dade aca. Eur. J. Clin. Chem. Clin. Biochem. **1996**, *34*, 861–865.
37. Sohma, K.; Kinoshita, Y.; Fujino, R.; Saito, T. A novel magnetic particle agglutination in microtiter plates for rapid detection of human T-lymphotropic virus type 1. J. Clin. Lab. Anal. **1995**, *9*, 59–62.
38. Harchali, A.A.; Montagne, P.; Ruf, J.; Cuilliere, M.L.; Bene, M.C.; Faure, G.; Duheille, J. Microparticle-enhanced nephelometric immunoassay of anti-thyroid peroxidase autoantibodies in thyroid disorders. J. Clin. Chem. **1997**, *40*, 442–447.
39. Elghanian, R.; Storhoff, J.J.; Mucic, R.C.; Letsinger, R.L.; Mirkin, C.A. Selective colorimetric detection of polynucleotides based on the distance-dependent optical properties of gold nanoparticles. Science **1997**, *277*, 1078–1081.
40. Johne, B.; Jarp, J.A. A rapid assay for proteins A in *Staph. aureus* strains, using immunomagnetic monosized polymer particles. Acta Pathol. Microbiol. Immunol. Scand. **1988**, *96*, 43–49.
41. Sohma, K.; Kinoshita, Y.; Fujino, R.; Saito, T. A novel magnetic particle agglutination in microtiter plates for rapid detection of human T-lymphotropic virus type 1. J. Clin. Lab. Anal. **1995**, *9*, 59–62.
42. Benecky, M.J.; Post, D.R.; Schmitt, S.M. Detection of hepatitis B surface antigen in whole blood by coupled particle light scattering (Copalis™). Clin. Chem. **1997**, *43*, 1764–1770.
43. Aizawa, H.; Kurosawa, S.; Tanaka, M.; Wakida, S.; Talib, Z.A.; Park, J.W.; Yoshimono, M.; Marutsugu, M.; Hilborn, J.; Miyake, J.; Tanaka, H. Conventional diagnosis of Treponema pallidum in serum using latex piezoelectric immunoassay. Mater. Sci. Eng. C **2001**, *17*, 127–132.
44. Fortin, M.; Hugo, P. Surface antigen detection with non-fluorescent, antibody-coated microbeads: an alternative method compatible with conventional fluorochrome-based labeling. Cytometry **1999**, *36*, 27–35.
45. Vignali, D.A.A. Multiplexed particle-based flow cytometrix assays. J. Immunol. Meth. **2000**, *243*, 243–255.
46. Nicewarner-Pena, S.R.; Freeman, R.G.; Reiss, B.; He,L.; Pena, D.J.; Walton, I.D.; Cromer, R.; Keating, C.D.; Natan, M.J. Submicrometer metallic barcodes. Science **2001**, *294*, 137–141.
47. Bains, W. A simple latex agglutination format for DNA probe-based tests. Anal. Biochem. **1998**, *260*, 252–255.
48. Mizutani, H.; Suzuki, M.; Mizutani, H.; Futjiwara, K.; Shibata, S.; Arishima, K.; Hoshino, M.; Ushijima, H.; Honma, H.; Kitamura, T. Sensitive detection of viral antigens with a new method, "Laser Magnet Immunoassay." Microbiol. Immunol. **1991**, *35*, 717–727.
49. Lim, P.L. A one-step two-particle latex immunoassay for the detection of Salmonella typhi endotoxin. J. Immunol. Meth. **1990**, *135*, 257–261.
50. Lim, P.L.; Tam, F.C.H.; Cheong, Y.M.; Jegathesan, M. One step 2 minute test to detect typhoid specific antibodies based on particle separation in tubes. J. Clin. Microbiol. **1998**, *36*, 2271–2278.
51. Perrin, A.; Thertz, A.; Mandrand, B. Method for Detecting an Analyte Using Two Types of Particles. International Patent WO0114880, 2001.

52. Miltenyi, S.; Müller, W.; Weichel, W.; Radbruch, A. A high gradient magnetic cell separation with MACS. Cytometry J. Soc. Anal. Cytol. **1990**, *11*, 231–238.
53. Dorgan, L.; Magnotti, R.; Hou, J.; Engle, T.; Ruley, K.; Shull, B. Methods to determine biotin-binding capacity of streptavidin-coated magnetic particles. J. Magnet. Magn. Mater. **1999**, *194*, 69–75.
54. Gilmartin, K. Magnetic applications in sphere molecular biology. Am. Biotechnol. Lab. **1991**, *9*, 38–40.
55. Fletcher, R.C.; Piccoli, S.P. Capacity of a colloidal goat anti-mouse IgG ferrofluid to negatively deplete cells. Prog. Clin. Biol. Res. **1994**, *389*, 79–87.
56. Kim, M.J. Homogeneous assays for riboflavin mediated by the interaction between enzyme-biotin and avidin-riboflavin conjugates. Anal. Biochem. **1995**, *231*, 400–406.
57. Perrin, A.; Martin, T.; Theretz, A. A dumbbell-like complex formation for DNA target assay. Bioconj. Chem. **2001**, *12*, 678–683.
58. Hueston, W.J. Treatment of hypothyroidism. Am. Fam. Phys. **2001**, *64*, 1717–1724.
59. Soderlund, H. DNA hybridization: comparison of liquid and solid phase formats. Ann. Biol. Clin. **1990**, *48*, 489–491.
60. Livnah, O.; Bayer, E.A.; Wilchek, M.; Sussman, J.L. Three-dimensional structures of avidin and avidin-biotin complex. Proc. Natl. Acad. Sci. USA **1993**, *90*, 5076–5080.
61. Butler, J.E.; Ni, L.; Nessler, R.; Joshi, K.S.; Suter, M.; Rosenberg, B.; Chang, J.; Brouwn, W.R.; Cantarero, L.A. The physical and functional behavior of capture antibodies adsorbed on polystyrene. J. Immunol. Meth. **1992**, *150*, 77–90.
62. Puget, P.; Pouteau, P.; Ginot, F.; Caillat, P. Procédé et dispositif de transport et concentration d'un analyte présent dans un échantillon. French Patent FR2817343, 2002.
63. Burgess, S.C.; Davison, T.F. Counting absolute numbers of specific leukocyte subpopulations in avian whole blood using a single-step flow cytometric technique: comparison of two inbred lines of chickens. J. Immunol. Meth. **1999**, *227*, 169–176.

5

Polymer Particles and Viruses

Biospecific Interactions and Biomedical Applications

EMMANUELLE IMBERT-LAURENCEAU CNRS-bioMérieux, Lyon, France

VÉRONIQUE MIGONNEY Institut Galilée, Université Paris-Nord, Villetaneuse, France

I. INTRODUCTION

Viruses are pathogenic agents requiring cell infection to replicate. The simplest viral form is constituted of a protein-made capsid containing a nucleic acid (DNA or RNA), called the nucleocapsid. The nucleocapsid of enveloped viruses is surrounded by an envelope membrane containing glycoproteins. In the latter case, cell infection may occur via specific receptor on the cell membrane. Viruses are present in various environments such as water, biological fluid, air, and earth. In order to prevent these environments from viral infection, techniques allowing the detection or capture and/or concentration, as well as removal of viral particles, are of great importance.

Virological control of drinking water is of constant main interest. Most parts of enteroviruses from human feces remain present in wastewater even after treatment, thus, they represent a risk for public health. In addition, the detection of enterovirus in water is problematic because of the low amount of viral particles in large volumes of water. The World Health Organization (WHO) recommends the analysis of at least 100–1000 L of water, which requires a concentration step before analysis. Moreover, when present in wastewater, viruses are rarely monitored because reference methods for their detection involve cell culture, which is time consuming, expensive, and can only detect a limited number of viral types.

Classical methods used to concentrate virus are based on precipitation with chemical compounds such as ammonium sulfate or aluminum sulfate, filtration or adsorption followed by elution step on membrane or fiber, and adsorption on systems such as glass powder. Many efforts have been made in order to optimize detection techniques of low amounts of virus in large volumes of water sample. This can be reach by modifying the pH between adsorption and elution process of viral particles. Indeed, from a 50 L water sample containing 10^4 particle-forming units (PFU) of poliovirus, it was shown that at pH 3.5 viruses (positively charged) adsorbed onto glass beads (negatively charged). Elution performed at pH 11.5 led to recover 55% of viral particles with a concentration factor of 2000 allowing detection of 1 PFU/L of water [1]. However, this kind of process requires water acidification to pH 3.5, which could induced viral inactivation. One solution was to use positively charged filters in order to prevent pH variation. Vilagines et al. have shown that sodo-calcic glass wool adsorbed poliovirus type 1 from pH 3 to pH 9 via electrostatic and hydrophobic interactions, without modifications of the analyzed sample [2]. From 100 L of water contaminated with poliovirus type 1 0.5×10^1 to 1.4×10^6 PFU/L, 70% recovering were obtained independently on viral charge and pH (7.2–8.0). But only 48% and 57% of viral recovering were obtained from, respectively, 5 L of wastewater and 10 L of treated wastewater, both contaminated with 10^7–10^8 PFU of poliovirus type 1 [2]. These results indicated that virus concentration and recovery were dependent on many factors such as viral type, presence of other components (organic matter, bacteria), and sample origin. Recovering rates remain low probably because of the nonspecificity of the involved techniques.

On the other hand, a great deal of research has been carried out to develop techniques for virus removal or inactivation in protein bioproduct. Indeed, therapeutic agents deriving from human and animal tissues have the potential of containing viral contaminants. Removal or inactivation of microorganisms from liquids by filtration is of major industrial importance in manufacturing processes and in quality control. In the blood products industry, a number of well-established techniques, including heat treatment, solvent detergent treatment, and pasteurization, were set up. However, the threat of new and more resistant viruses contaminating these products has to be taken into account. This was highlighted in 1990/1991 by an outbreak of hepatitis A virus (HAV) among 52 hemophiliacs in Italy [3].

Membrane filters remove virus particles by two main processes: (1) the entrapment or size exclusion of virus particles as they pass through the membrane; (2) the adsorption of virus particles to the membrane. The important factors for the adsorption of virus onto membrane filter are electrostatic and hydrophobic interactions, chemical composition of the membrane, and ratio of the membrane pore diameter to the virus diameter. The most efficient adsorption occurs when membrane and virus charges are opposite or of low intensity. But application problems occur because of the rapid clogging of the membrane pores leading to a decrease in the filtration flow. Hou et al. observed that viruses (poliovirus, influenza virus, bacteriophage MS-2) removal by filtration on various membranes was mostly dependent on the adsorption process when virus diameters are lower than membrane pore diameter, and that electropositive filters were more efficient than electronegative ones [4]. In the same way, by using poly(vinylidene fluoride) (PVDF) microporous membrane, Oshima et al. have shown that it was possible to remove 10^4–10^6 of viral particles greater than 50 nm in diameter, such as hepatitis B virus and simian virus 40, independently of the type of fluid, whereas when viral agent was smaller than 50 nm, such as hepatitis C virus, removal (10–10^2) depended on the type of fluid and on the viral particle size [5]. However, the porous skin layer of the filtration membrane can exhibit large pore mechanical defects that allow the permeation of viral particles and thus reduce efficiency of the membrane.

In order to reach complete removing or inactivation of possible HIV infection from transfused plasma or blood products, research tends to develop anti-HIV membrane able to decrease HIV infectivity, thus increasing the potential supply of safe plasma product. To reach this aim, Owada et al. have studied the efficiency of a polypropylene-polyethyleneimine (PP-PEI) membrane (pore size = 450 nm, 3.8 cm^2 surface area) toward HIV-1 particles from seropositive plasma (in vivo strain) or HIV-1 strain (in vitro strain) obtained from peripheral blood mononuclear cells (PBMCs) [6]. Results were different because both HIV-1 strains displayed differences in their electrostatic behavior. Indeed, a complete reduction of the infectious titer was observed in the case of in vivo HIV-1 strain whereas only a decrease was observed for the in vitro strain. In vivo HIV-1 strain is probably surrounded by antibodies leading to aggregates of viruses that become larger than the pore size of the membrane. Thus, PP-PEI membrane should only decrease infectious titer optimally from 10 to 10^2, which is not competitive when compared to the other methods using photosensitization or solvent detergent leading to decrease infectious titer from 10^6.

At least, immunological or virological experiments often require highly concentrated retrovirus suspensions. But retroviruses are very sensitive to inactivation by chemical or physical agents. In addition to contaminant removal, filtration media have application in the concentration and analysis of microorganisms.

Makino et al. have developed a rapid method (1 h) to concentrate (×30) live retrovirus such as HIV or MuLv (c-type murine leukemia viruses) [7]. This method was based on filtration through cuprammonium-regenerated cellulose hollow fiber (BMM) developing a surface area of 0.03 m^2 and leading to minimal loss of viral activity. 5.89×10^5 PFU from 1.20×10^6 PFU of MuLv was recovered on the BMM 75 filter (mean pore size = 75 nm). This kind of filter did not adsorb viruses but entrapped them in the hollow fibers; recovery was performed by dynamic washing of the fibers in the opposite direction of the filtration.

All techniques described above are nonspecific filtration processes based on the use of particulate (glass powder), fiber (glass wool, epoxy fiber glass, cellulose hollow fiber) and membrane (nitrocellulose, charge-modified cellulose, PVDF, PP-PEI) systems. For concentration application, high concentration factors can be obtained but at the same time the recovery rate of infectious viral particles remains low because of virus inactivation during the process. For virus removal application, the obtained reduction titers are not sufficient to guarantee safe bioproducts. Indeed, the nonspecificity of these methods introduces uncontrolled parameters such as nonspecific adsorption of nonviral components—organic matter, proteins, bacteria—which can compete with viruses onto the system. Furthermore, membrane filtration processes mostly lead to rapid clogging of the membrane pore inducing a decrease in the efficiency of the method. Anyway, after concentration/capture or removal, detection and identification of the viral type requires further techniques such as cell culture, electron microscopy, enzyme-linked immunosorbent assay (ELISA) or radioimmunoassay (RIA), reverse transcriptase–polymerase chain reaction (PCR). The latter should probably be the most sensitive method to detect virus in aqueous solution; nevertheless, it involves a concentration step that allows the removal of inhibiting agents of RT-PCR.

To improve the efficiency and sensitivity of concentration/capture, removal, and detection of viral particles, it becomes necessary to develop specific systems. Specificity or biospecificity can be described by the "lock and key" system; recognition between two biological species, such as a ligand and its receptor, is a result of interactions between complementary chemical groups on both components. These interactions are highly specific because the best "key" will interact with the best "lock." In order to allow specific interactions between a material surface (natural or artificial) with a biological entity such as a virus, bioactive sites have to be created. However, the question is, do we want direct or indirect interactions between material surface and viral particle? The answer will depend on the application as well as on the viral type, enveloped or not. Indirect interactions can be established through virus-adsorbed specific molecules (antibodies, matrix proteins, ions, polysaccharides, etc.) on the material surface. The adsorption process depends on the physicochemical properties of the material and can be achieved by electrostatic and/or hydrophobic interactions. Direct interactions between viral particle and material surface involve the presence of biospecific recognition sites on surface. These sites can be created by functionalizing the surface with chemical groups or through chemical linkage of virus-specific molecules to the material surface.

The aim of this chapter is to present indirect and direct biospecific interactions between polystyrene beads and viral particles. Indeed, material has to be chosen in order to be suitable for various applications. Material such as polystyrene beads has the advantage of being (1) commercially available, (2) easily functionalized, (3) able to develop large surface area, (4) already in use for diagnostic and chromatographic processes. In the following, biomedical applications will be discussed further.

II. INDIRECT BIOSPECIFIC INTERACTIONS OF VIRAL PARTICLES WITH LATEX

Stable latex particle reagent can be used to diagnose a variety of infectious diseases, such as rotavirus, adenovirus, and canine parvovirus (CPV) infections. Indeed, polystyrene beads (latex particles) are agglutinated by antigen when coated with antibodies. The principle of the latex agglutination (LA) test is shown in Fig. 1. Briefly, latex microspheres coated with specific antibodies react in the presence of the antigen. A positive reaction is characterized by macroscopically visible agglutination of the latex particles.

During winter, 50% of children younger than 2 years suffer from severe diarrhea caused by rotavirus infection. Among the several methods for detecting rotavirus (an enveloped virus), electron microscopy, RNA-polyacrylamide gel electrophoresis, ELISA, and EIA are the most popular. The inconvenience of all these techniques is the required time (several days) to obtain results. Thus it is necessary to get rapid, reproducible, sensitive and simple diagnostic test for rotavirus infection. Agglutination is a rapid method for the detection of rotavirus antigen in stool specimens. Amongst all the detection methods available, LA is the simplest one. Indeed, as large amounts of antigens are excreted during rotavirus gastroenteritis, it should be possible to diagnose this disease by latex agglutination reaction. The latex agglutination method for the rotavirus detection was first described by Sanekata et al. in 1981 [8]. Then several LA tests were developed with various latex sizes and different antibody preparations. The most important part of them will be presented and discussed in the following.

Hughes et al. have sensitized commercially latex beads (0.797 μm in diameter, Sigma Chemical Co.) with antirotavirus antibodies from an antiserum preparation or from IgG preparation obtained from the antiserum [9]. The LA test was performed by incubating antibody-coated latex beads with clarified stool suspensions for 15 min at 36°C. Results showed that latex beads were more sensitive for rotavirus detection when coated with IgG than the latex beads coated with whole immune serum. Indeed, the antirotavirus antibody concentration was greater in the IgG preparation than in the whole immune serum. Moreover, in addition to IgG, the whole immune serum contains several proteins that can adsorb onto the latex beads and thus decreased the number of available sites for IgG adsorption. The latex reaction was found to be two- to eightfold less sensitive than the EIA technique (Rotazyme, Abott).

Commercially available LA kits (Slidex Rotakit from bioMérieux, Rotalex from Orion) have been compared with classical techniques [electron micros-

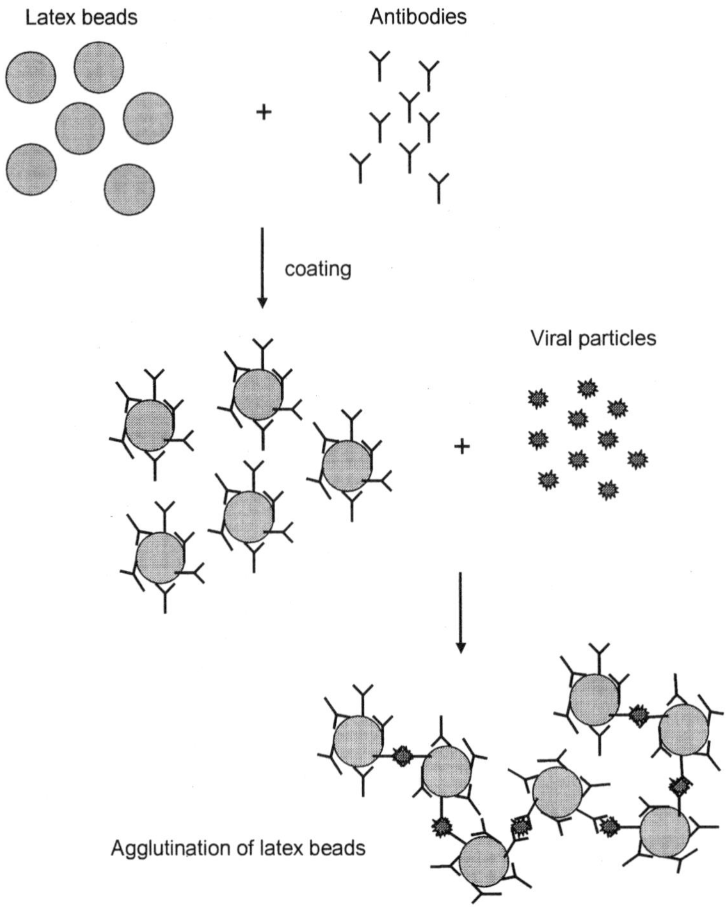

FIG. 1 Latex agglutination test for the detection of viral particles.

copy, Rotazyme (EIA) from Abott] used to detect rotavirus [10] or with their home-made LA test [11]. Results were obtained faster with LA kit (within 25 min) than with classical methods and gave 100% specificity. However, the percent agreement of LA test with classical methods dramatically decreased as the virion count decreased, indicating a lower sensitivity for the LA kit at low viral titer [10]. Treatment of latex beads with 1% bovine serum albumin (BSA) before the coating with antirotavirus IgG fraction led to sensitivity and specificity equal to those of EIA test kit or electron microscopy analysis [11]. Indeed, BSA is a nonspecific protein that adsorbed nonspecifically onto most polymer surfaces. So it was used as precoating of material in order to decrease nonspecific adsorption of proteins [12].

Canine parvovirus (CPV), a nonenveloped virus of 22 nm diameter, causes hemorrhagic enteritis and myocarditis in young dogs and puppies. After symptom onset virus is excreted in the feces for several days. The virus can be infectious for a very long time because of extreme resistance of parvoviruses to environmental inactivation. Thus, there is a real need for a rapid and simple test that can be used by veterinarians to diagnose CPV infections.

Bodeus et al. have compared the sensitivity of the following detection methods: classical slide test with particle counting, and LA test involving polystyrene beads (0.8 μm in diameter) sensitized with purified monoclonal antibodies against CPV antigen [13]. Results were rapidly obtained in 5 min (slide test) to 30 min (particle counting), indicating that the antigen concentration detected by slide test is about 100 ng/mL whereas the particle counter led to 4 ng/mL of detected viral antigen, which is the same sensitivity as RIA technique.

In other respects, Esfandiari et al. have described one-step immunochromatographic test (Rapid Parvovirus Test) for the detection of CPV in canine feces samples [14]. In this test, latex particles have 0.3 μm diameter, are dyed blue, and are commercially available as polystyrene beads. The binding of parvovirus particles present in the fecal sample to CPV-mAb_1-conjugated latex suspension led to a positive blue band in the T zone within 3–5 min (Fig. 2). This test showed about the same sensitivity and specificity as available ELISA tests.

Nakamura et al. have perfected a latex agglutination test for the detection of infectious bursal disease virus (IBDV). The test is commonly performed using either an agar gel precipitin (AGP) test or ELISA [15]. The minimal titer of virus that agglutinated anti-IBDV monoclonal antibody–bound latex microspheres (0.01 mL of 3.98×10^4 PFU strain IQ IBDV-sensitive) was 10–40 times higher than the observed titer value in the AGP test. Therefore, the sensitivity of this LA test is sufficient to detect IBDV antigen in infected bursal tissue within 10 min instead of 48 h.

To conclude on the various latex agglutination tests for the detection of virus particles in a sample, the results presented above are reported in Table 1. It appears that all LA tests described herein have similar sensitivity (about 10^6 viral particles, or 100 ng/mL) except the LA test using flow cytometry, which displays a higher sensitivity. The simplicity, high specificity, and speed of LA kits make them useful for rapid screening of symptomatic patients. Moreover, the LA test is very simple to perform because it does not require expensive equipment and is sensitive enough for routine diagnostic requirements. Several authors [9,14,15] also indicated the good conservation of biological activity of the antibody-coated latex particles with time. However, even if monoclonal antibodies were used to coat latex microspheres, sensitivity of the LA technique is not optimal. Indeed, antibody can adsorb onto the latex particle surface through

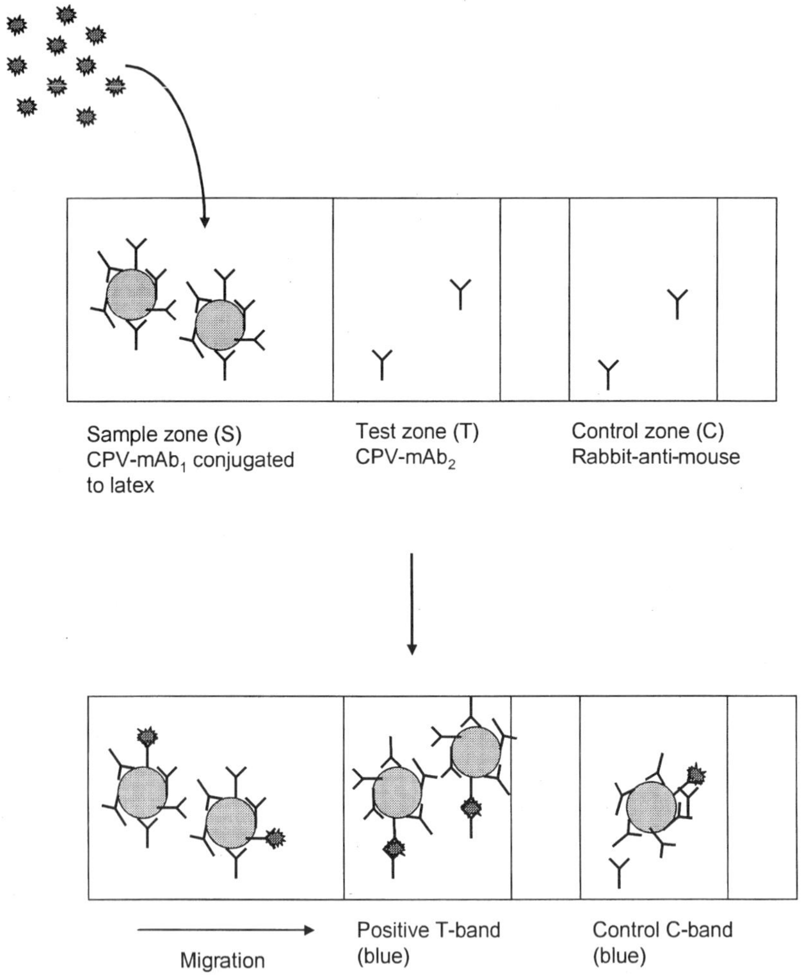

FIG. 2 Positive result in immunochromatographic test for the detection of canine parvovirus (CPV) in fecal samples. After migration, bound CPV to CPV-mAb$_1$ conjugated to latex was detected in the T zone by reaction with the CPV-mAb$_2$ leading to a blue band.

TABLE 1 Comparison of Latex Agglutination Tests

LA test	Ref. 9	Ref. 10	Ref. 13 (flow)	Ref. 13 (slide)	Ref. 14	Ref. 15
Latex	62.5 μg	500 μg	2.5 mg	2 mg	833 μg	250 μg
Antibody	0.2 mg	3 mg	0.2 mg	0.2 mg	0.3 mg	dil. 1:400 (ascite)
Sample volume	50 μL	100 μL	25 μL	20 μL	170 μL	10 μL à 50 %
Incubation time	15 min, 360°C	3 min, R.T.	30 min, R.T.	5 min, R.T.	3–5 min, R.T.	10 min, R.T.
Sensitivity	9.0 10^5 viruses	3.10^7 virions/mL	4 ng/mL	100 ng/mL	dil. 1:500	4.10^4 PFU

R.T., room temperature.

its Fc fragment and/or via its Fab fragment (Fig. 3), leading to a decrease in the efficiency of the binding with the antigen. This kind of coating, which involves nonspecific interactions between antibodies and polymer surface, is the same as the ELISA one. Therefore, it is not surprising that the sensitivity of both techniques, i.e., LA test and ELISA, is in the same range.

III. DIRECT BIOSPECIFIC INTERACTIONS OF VIRUSES WITH LATEX PARTICLES

A. Bound Antibody Latex

In order to increase the efficiency of antigen binding onto modified polymer and the sensitivity of virus detection, oriented binding of antibodies onto polymer surface appeared to be of great interest. To reach this aim, virus-specific antibodies were covalently bound onto latex particles by using carbodiimide reagent techniques [16,17]. Flow cytometry was used to detect and to quantify virus-bound latex particles by fluorescence. This technique requires the microspheres to be coated with a unique specific antibody.

Iannelli et al. have developed several methods involving flow cytometry to detect and identify simultaneously one to six types of virus [16]. Three viruses causing severe losses in many cultivated plant species were studied: cucumber mosaic virus (CMV), tomato mosaic virus (ToMV), and potato virus Y (PVY). Virus-specific antibodies were first covalently bound to latex microspheres (3 μm, 6 μm, or 10 μm in diameter) by using carbodiimide reagent. Then, to detect one single virus, such as CMV, anti-CMV latex particles (3 μm in diameter) were incubated first with leaf extract, second with primary rabbit anti-CMV

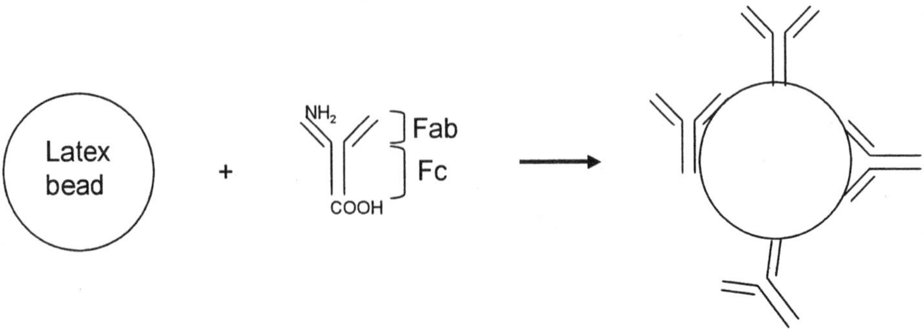

FIG. 3 Nonspecific adsorption of antibodies onto latex particle. Adsorption of the antibody can be carried out, for example, by the Fc fragment (via the C-terminus end), by one or the two Fab moieties (via the N-terminus end), or by the Fc and Fab moieties.

antibodies, and third with secondary anti-rabbit FITC antibodies (Fig. 4a). The limit of detection was 7 pg for CMV and PVY, and 4 pg for ToMV. In order to detect three viruses, two procedures were performed as follows:

1. Anti-CMV latex particles (3 μm in diameter) mixed with anti-CPY latex particles (3 μm in diameter) were incubated first with leaf extract, second with primary rabbit anti-CMV and mouse anti-CPY antibodies. In parallel, anti-ToMV latex particles (6 μm in diameter) were incubated first with leaf extract, second with primary rabbit anti-ToMV antibodies. Then, all latex particles were mixed and incubated with secondary anti-rabbit phycoerythrin (PE) and anti-mouse FITC antibodies (Fig. 4b). CMV and CPY were identified and detected on the basis of different fluorescence (FITC and PE), whereas ToMV was identified and detected on the basis of different M_r of latex particles (6 μm vs. 3 μm in diameter).
2. Anti-CMV latex particles (3 μm in diameter) mixed with anti-ToMV latex particles (6 μm in diameter) and with anti-PVY latex particles (10 μm in diameter) were incubated first with leaf extract, second with an anti-CMV FITC, anti-PVY FITC, and anti-ToMV FITC antibodies mixture (Fig. 4c). The three viruses were identified and detected on the basis of different M_r of latex particles.

Authors indicated that the latex diameter did not significantly influence the sensitivity of the test, which was equal for the three described above [16]. The critical point was the optimal antibody amount bound to the latex particles. Indeed, more than 1 μg antibody/3 $\times 10^6$ particles led to aggregation of latex microspheres, whereas less than 1 μg antibody/3 $\times 10^6$ particles decreased the sensitivity of the test. Furthermore, Iannelli et al. suggested that these methods could be useful to concentrate viruses when present at too low concentration or if interfering substances are present in the sample.

Likewise, Samoylova et al. used flow cytometry to quantify adenovirus (a nonenveloped virus of 70–80 nm in diameter) bound to polystyrene-based microspheres [17]. To accomplish this, latex containing carboxylic acid functions were carbodiimide activated for covalent binding of anti-adenovirus monoclonal antibodies. Bound viruses were detected and quantified by total fluorescence measurement using the nucleic acid dye TOTO-1. Maximal adsorption of viruses onto latex was reached within 30 min. Moreover, virus capture is efficient when viral particles are sufficiently closed to microspheres. Indeed, three processes may occur simultaneously: (1) Brownian motion of the virus, (2) decay of the virus, and (3) virus capture by the microsphere. Saturation of virus-specific sites present on the surface of latex particles was reached for 2×10^5 PFU/microsphere, and the limit of detection was 500 PFU/microsphere.

At least, latex particles were used as immunospecific markers for viral detection in electron microscopy technique due to their uniform size, shape, and

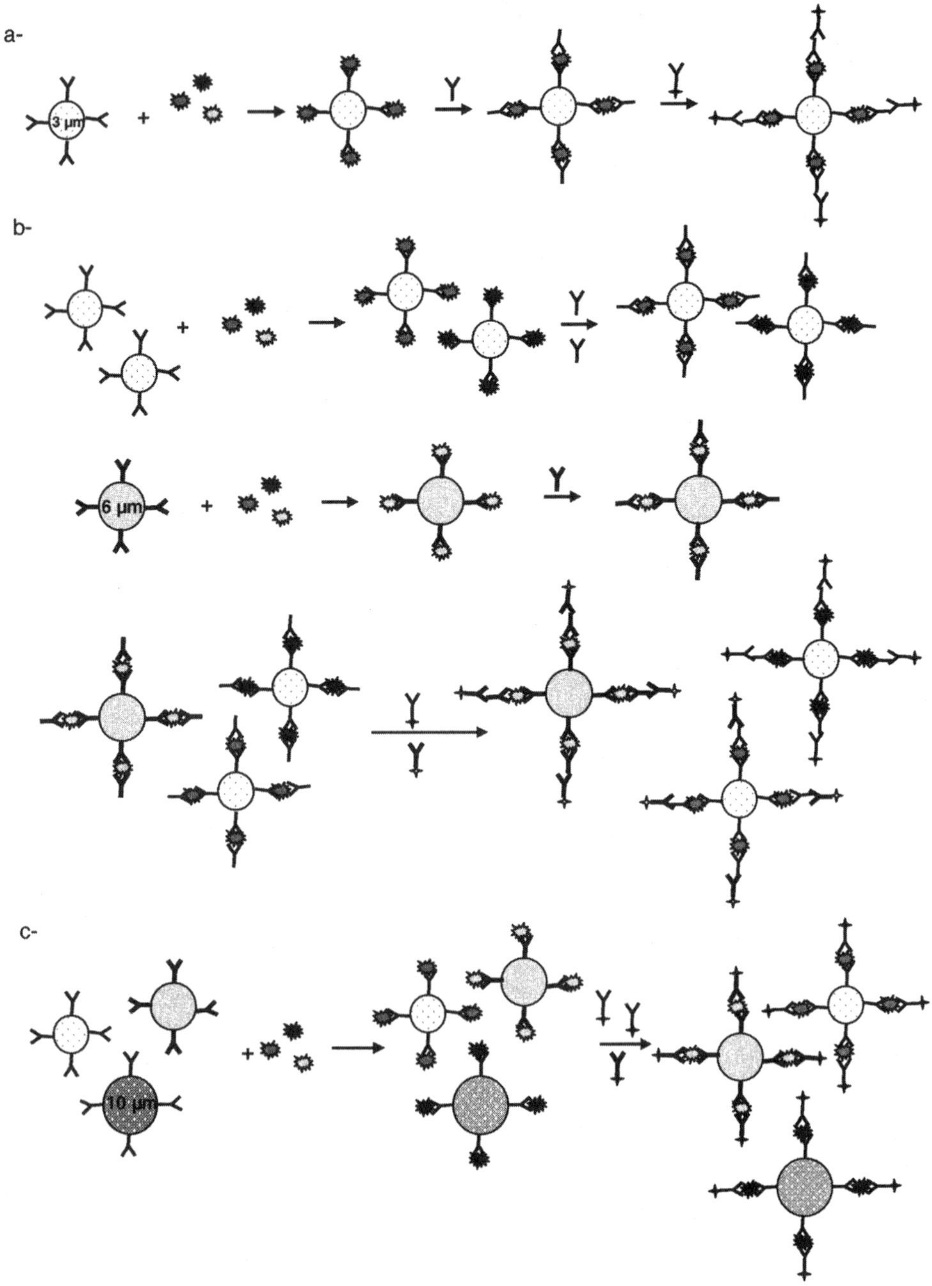
a-
3 µm
b-
6 µm
c-
10 µm

electron density. Gonda et al. have covalently grafted antibodies to latex particles (200 nm in diameter) functionalized by aromatic amine groups for the surface labeling of Rauscher murine leukemia–enveloped virus (RLV) [18].

The covalent binding of antibodies to latex particles was probably the first step toward obtaining biospecific polymers retaining biological activity. Nevertheless, this is not sufficient because, depending on the chemistry, a part of bound antibodies could be inactived. Indeed, bound antibodies should keep available their antigen-binding Fab fragment to increase the sensitivity of virus detection. Carbodiimide activation technique used by Iannelli et al. [16] and Samoylova et al. [17] does not induce "specific," or regioselective, linkage of the antibody. Carbodiimide activates carboxylic acid functions, grafted or present on the macromolecular chains of the polymer, to react with primary amine groups. However, primary amine are present at the amino terminal part (N-terminal part) of the antibody and in side chain of lysine residues. So, if binding involves the N-terminal part of the antibody, its antigen recognition activity will be completely inactivated. Moreover, lysine residues are distributed anywhere in the antibody sequence, even in its biological active part—Fab fragment. Bound reaction with the primary amine functions located in the side chains of lysine residues could induce well-oriented bound antibodies keeping their biological activity as well as few or no active bound antibodies depending on the site of the

---

FACING PAGE

FIG. 4 Identification of cucumber mosaic virus (CMV), tomato mosaic virus (ToMV), and potato virus Y (PVY) by various cytofluorometric tests [16]. (a) Single-parameter test; (b) multiple parameter test 1; (c) multiple parameter test 2.

- cucumber mosaic virus (CMV)
- potato virus Y (PVY)
- tomato mosaic virus (ToMV)
- anti-CMV antiboby
- anti-PVY antiboby
- anti-ToMV antibody
- anti-mouse FITC antibody
- anti-rabbit PE antibody

linkage. Antibody binding to diazotized latex particles is no more regioselective because reaction occurs at histidine and tyrosine residues, which could also be involved in the antigen binding activity. Thus, it appears very difficult to covalently bind antibodies onto polymer surface getting them well oriented and keeping their antigen recognition properties.

In conclusion to this section, sensitive detection, identification, concentration, and/or purification of viruses require the development of biospecific interactions between viral particles and latex through other virus-specific ligand than antibodies.

B. Bound Lectin Latex

The enveloped glycoprotein gp120 of human immunodeficiency virus type 1 (HIV-1) is a mannose-rich glycoprotein that strongly interacts with some lectins such as Con A. Akashi et al. [19] and Hayakawa et al. [20] described covalent immobilization of Con A and other mannose-specific lectins onto modified polystyrene nanospheres (360 nm in diameter) in order to capture HIV-1 virion. As shown in Fig. 5, polystyrene beads were functionalized by carboxylic acid groups brought by poly(*tert*-butyl methacrylate), which allow covalent binding of lectins by reaction with their amino groups using carbodiimide activation. Interaction between Con A–immobilized latex particles and HIV-1 virions led to aggregation of the nanospheres allowing their separation from noncaptured virions by centrifugation. 0.3 μg of Con A/cm^2 [19] and 0.91 μg of Con A/cm^2 [20] were assessed. Results indicated that HIV-1 capture activity of Con A–immobilized nanospheres was dose dependent. A significant reduction of the viral infectious activity—up to 2×10^3—was reached at 2 mg/mL concentration [20]. Furthermore, identical activity in HIV-1 capture was obtained with the other tested mannose-specific lectins whereas no HIV-1 capture activity was detectable with galactose-specific lectin [19]. Authors concluded that both selection of immobilized lectins and diameter of nanospheres are determining parameters for the capture activity and selectivity. Indeed, interaction between Con A and gp 120 is not as specific as that with anti-gp 120 antibodies, since lectins recognize the oligosaccharide chain of gp 120.

The efficiency of lectin-immobilized latex particles in biomedical applications is not clear because of the low specificity of the interaction with HIV-1 virion, which could be canceled under in vivo conditions. Indeed, other biological components, including oligosaccharide moieties, should present higher affinities for lectins than HIV-1 virion. Moreover, Con A is known to exhibit various toxicities in vivo. Nevertheless, the Con A–latex nanospheres could be used in ex vivo experiments or for basic research applications such as a chromatographic matrix to purify or concentrate enveloped viruses.

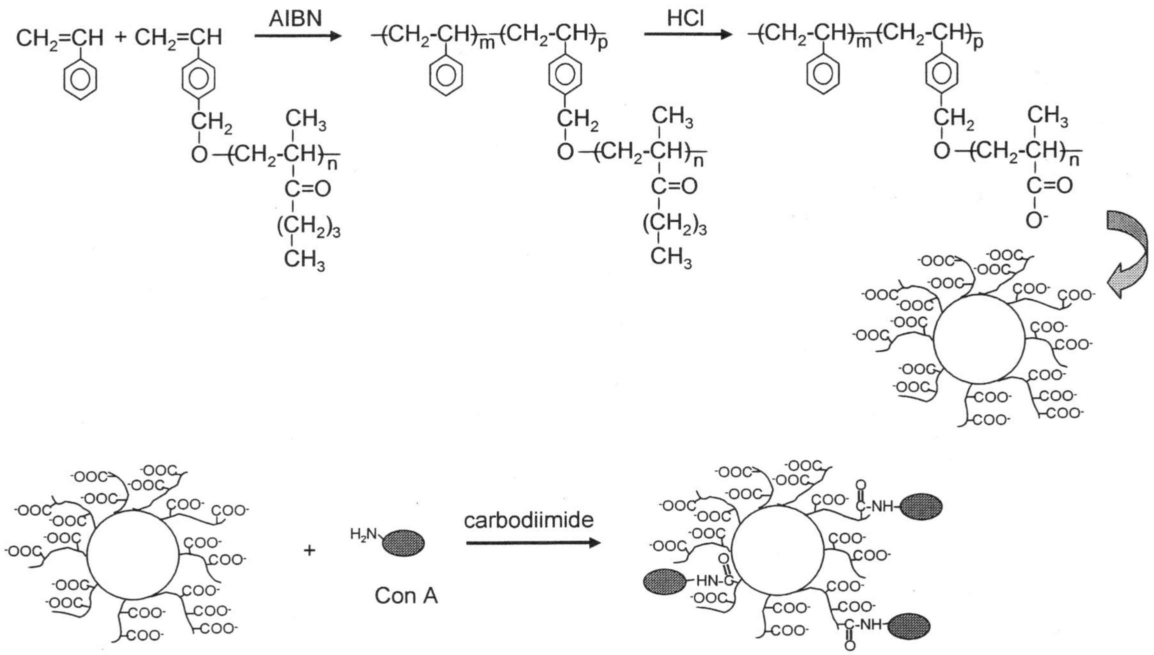

FIG. 5 Functionalization of polystyrene beads by poly(*tert*-butyl methacrylate) and covalent binding of concanavalin A (Con A) via activation with a water-soluble carbodiimide.

IV. BIOMIMETIC POLYSTYRENE BEADS

Human arbovirus outbreaks spread throughout the world with important centers of infection located in intertropical zones. Infections by these viruses cause three kinds of pathologies: fever, meningitis or encephalitis, hemorragic fever. Among them, the most well-known infections are caused by yellow fever, dengue, and Ebola viruses from the Flaviviridae family, and by Sindbis, eastern equine encephalitis, and western equine encephalitis viruses from the Togaviridae family. Rapid detection, isolation, and identification of these enveloped viruses are necessary steps to diagnosis and to observe epidemics. Techniques involving cell culture, ELISA, or polymerase chain reaction (PCR) are commonly used to detect these viruses, but they remain time consuming and/or expensive.

Another approach to obtain direct biospecific interactions between viral particles and polystyrene beads consists of functionalizing the surface of polystyrene microspheres with appropriate chemical groups. Indeed, we previously showed that random copolymerization of suitable functionalized monomers or random functionalization of macromolecular chains allow creation of active sites that can interact specifically with biological entities such as proteins or cellular receptors [21–23]. The probability of appearance of these active sites depends on the nature, number, and distribution of the chemical groups on macromolecular chains and on the overall chemical composition of polymers. Furthermore, several studies have demonstrated that the probability of appearance of active sites shows a maximal value for a precise chemical composition of the polymer and that composition corresponds to the maximal value of the observed biological activity. The composition of these "synthetic active sites" is defined to mimic the biological site involved in the "natural" interaction. As an example, polystyrene derivatives substituted with leucine groups were shown to specifically interact with epidermal growth factor (EGF) receptor [24] or with CR1 receptor [25], which are two glycoproteins of the cell membrane. According to this method, we developed functionalized polystyrene beads and studied their interactions with two viruses belonging to the arbovirus group.

A. Viruses

Kedougou virus (50–70 nm in diameter), a flavivirus from the Flaviviridae family, and Babanki virus (45–50 nm in diameter), an alphavirus from the Togaviridae family, were kindly provided by Dr. A. Jouan (Unité de Virologie du Centre de Recherche du Service de Santé des Armées "Emile Pradé," La Tronche, France). Babanki virus is a derivatized form of Sindbis virus that prevails in northern Europe as well as in Africa [26,27]. These two viruses were chosen because they display different envelope glycoproteins [28,29].

Kedougou and Babanki viruses were multiplied in Vero cells (from kidney monkey). They induced cytopathic effects allowing the determination of the infectious titer which corresponds to the 50% tissue culture infective doses per mL ($TCID_{50}$/mL) by the Reed and Munch statistical method.

For the analysis of interaction between viral particles and modified polystyrene beads, viruses were radiolabeled with ^{3}[H]uridine and purified by ultrafiltration on Biomax-100 membrane (Millipore) and by ultracentrifugation on a linear saccharose gradient (10–40%). Saccharose was removed by ultrafiltration on Biomax-100 membrane with culture medium.

B. Synthesis and Characterization of Functionalized Polystyrene Beads

Polystyrene microspheres SX3 (35–75 μm in diameter, Biorad) were modified in three steps (Fig. 6). First, 10 mol of monochlorosulfonic acid (HSO_3CI) (excess) was added to polystyrene in dichloromethane (CH_2Cl_2). The mixture was stirred at room temperature for 1 h, then the suspension was filtered and washed with CH_2C1_2. This nucleophile substitution led to poly(parachlorosulfonyl)styrene—a reactive intermediate—immediately used in the second step of the functionalization process. Poly(parachlorosulfonyl)styrene was condensed with dimethyl ester of amino acids in CH_2Cl_2 at room temperature in the presence of triethylamine (Et_3N). Resin was extensively and successively washed with ethanol (100%) and aqueous solutions of sodium hydroxide (from 10^{-2} M to 2 M, and back to 10^{-2} M) in order to gently hydrolyze ester groups. Five amino acids (AAs) were substituted: arginine (R), aspartic acid (D), phenylalanine (F), asparagine (N), and serine (S). Functionalized polymers were characterized by elemental analysis and their chemical compositions are reported in Table 2. Results showed that 77% ± 10% of styrene units were substituted by sodium sulfonate groups (PS-SO_3Na) and by amino acid sulfamide groups (PS-SO_2AA); the substitution rate is directly related to the amount of introduced dimethyl ester amino acid in the reaction mixture. In addition to its amino acid substitution percentage, each polymer was characterized by R values with R = [COO^-/ (COO^- + SO_3^-)]. This parameter gives the molar ratio of carboxylate upon the total anionic groups present along the macromolecular chains of the polymers. The different values of R reported in Table 2 allowed comparison of the different functionalized polymers.

In aqueous medium, functionalized polystyrene beads displayed swelling properties varying with their chemical composition. In order to evaluate and normalize the surface area really accessible to biological molecules such as protein and virus, human serum albumin (HSA) adsorption experiments were achieved on all the synthesized polystyrene derivatives. It is well known that HSA adsorbs most of the polymer surfaces in a nonspecific way [30–32]. Previous studies showed that the amount of adsorbed HSA at saturation on calibrated nonporous silica beads equaled approximately 0.12 μg/cm^2 of available surface

Step 1 : Reaction of monochlorosulfonic acid

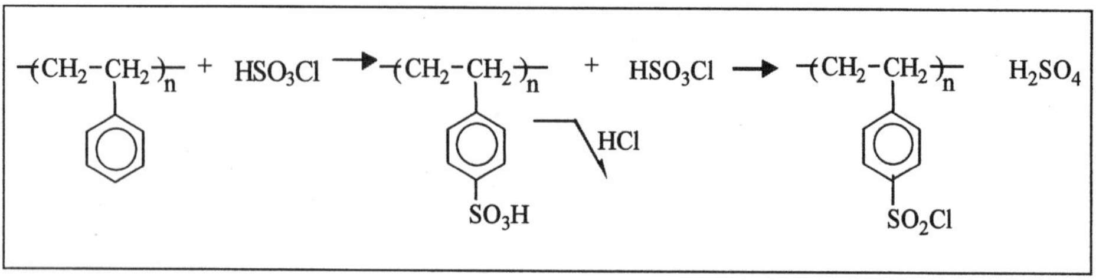

Step 2: Condensation of dimethyl ester amino acid

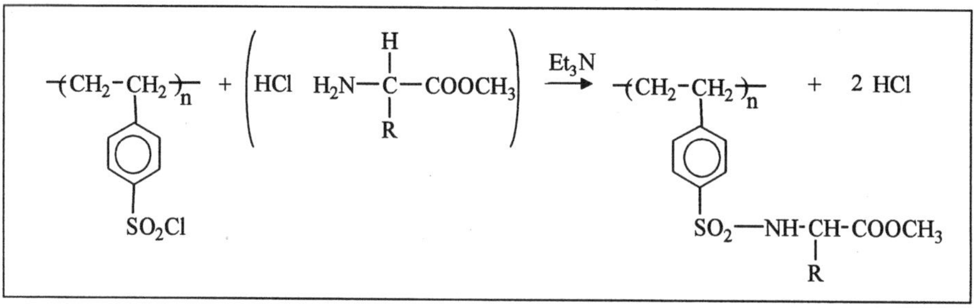

Step 3: Hydrolysis of methyl ester group

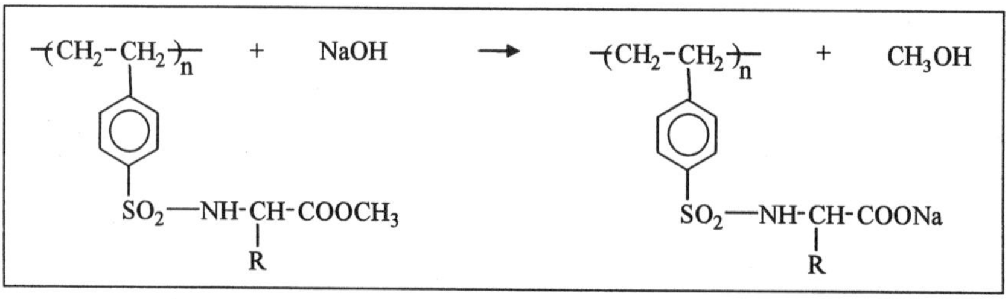

[30]. Therefore, adsorption experiments of radiolabeled HSA (^{125}I-HSA) onto polystyrene derivatives were performed; the analysis of the adsorption isotherms showed that the adsorption followed a one-site Langmuir law and allowed determination of maximal adsorption capacity of the polymers (B_{max}) and the apparent affinity constant K_a. The available surface area for portein or virus was determined for each polymer as S_A (cm^2/mg) = $B_{max}/0.12$ (Table 2). Results showed that S_A varied with the chemical composition of the polymer. Indeed, except for phenylalanine derivatives, S_A values of the polymers decreased vs. the amino acid substitution rate to reach a plateau value, at which all the polymers displayed the same swelling behavior. Phenylalanine is a hydrophobic amino acid

FACING PAGE

FIG. 6 Functionalization of polystyrene beads SX3 (Biorad) by various amino acids.

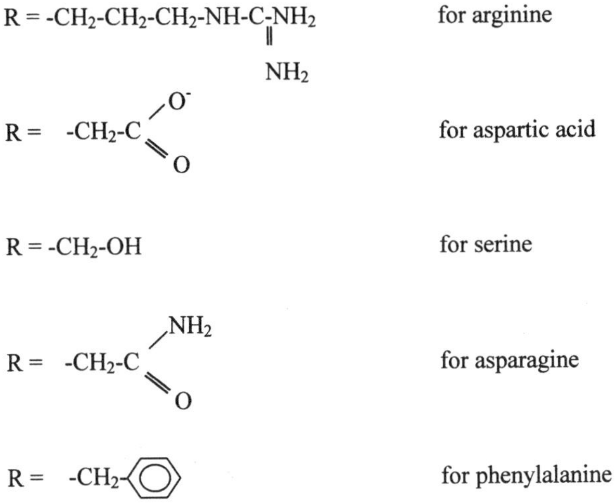

that can induce the development of several interactions with the hydrophobic domains of albumin, this may explain the different behavior of this polymer as compared to the other polystyrene derivatives. Therefore, the calculated mean value of K_a was about 2×10^5 M^{-1} and corresponded to a nonspecific adsorption value. Taking this result into account, surface of the different polymers was precoated with BSA before the adsorption of viruses onto polymers in order to screen nonspecific sites. Thus, any observed variation in viruses adsorption will be due to the chemical composition of the polymer and not to its physical properties.

C. Interactions of Viruses with Functionalized Polystyrene Beads

Adsorption experiments of viruses Babanki and Kedougou on polystyrene derivatives were performed by incubating purified radiolabeled viruses (various concentrations) with polymers, for 10 min at room temperature, under gentle stirring. Indeed, kinetic studies have shown that the adsorption rate of viruses onto polymers reached a plateau value within 10 min of incubation time and stayed constant until 60 min where a little decrease was observed. After several washings with culture medium, polymer suspension was β counted. The bound radioactivity was then directly related to the infectious titer. Bound virus particles onto PS-SO_2D vs. free virus are shown in Fig. 7. Results showed that it was

TABLE 2 Chemical Composition of Modified Polystyrene Beads Obtained from Elementary Analysis and Surface Area Calculated from Bovine Serum Albumin Adsorption

Polymer	PS (%)	PS-SO_3Na (%)	PS-SO_2AA (%)	Molar ratio $R = COO^-/(COO^- + SO_3^-)$	Surface S_A (cm^2/mg)
PS-SO_2R					
PS-SO_2R_4	33	63	4	0.06	0.44
PS-SO_2R_{13}	32	55	13	0.19	0.17
PS-SO_2R_{20}	28	52	20	0.29	0.10
PS-SO_2R_{38}	26	36	38	0.52	0.18
PS-SO_2D					
PS-SO_2D_8	16	76	8	0.18	3
PS-SO_2D_{15}	21	64	15	0.32	2.4
PS-SO_2D_{29}	24	47	29	0.55	1.1
PS-SO_2D_{42}	30	28	42	0.75	0.4
PS-SO_2D_{48}	32	20	48	0.83	1.8
PS-SO_2D_{50}	19	31	50	0.76	1.3
PS-SO_2D_{55}	23	22	55	0.84	0.6
PS-SO_2F					
PS-SO_2F_{10}	27	63	10	0.14	0.8
PS-SO_2F_{21}	34	45	21	0.32	4.1
PS-SO_2F_{30}	30	40	30	0.43	3.4
PS-SO_2F_{46}	24	31	46	0.60	4.2
PS-SO_2F_{63}	24	13	63	0.83	3.6
PS-SO_2S					
PS-SO_2S_7	21	72	7	0.09	2.6
PS-SO_2S_{16}	20	64	16	0.2	0.9
PS-SO_2S_{24}	28	48	24	0.33	1
PS-SO_2S_{44}	18	38	44	0.54	0.6
PS-SO_2S_{53}	19	28	53	0.65	0.6
PS-SO_2S_{62}	14	24	62	0.72	1.1
PS-SO_2S_{72}	3	25	72	0.74	0.9
PS-SO_2N					
PS-SO_2N_{15}	25	60	15	0.2	2.8
PS-SO_2N_{27}	24	49	27	0.36	3
PS-SO_2N_{35}	17	48	35	0.42	1.4
PS-SO_2N_{36}	23	41	36	0.47	1.7
PS-SO_2N_{40}	25	35	40	0.54	1.6
PS-SO_2N_{42}	19	39	42	0.52	1.7

PS, styrene unit; PS-SO_3NA, sodium sulfonate styrene unit; PS-SO_2AA, styrene unit substituted with sulfamide amino acid—R (arginine), D (aspartic acid), F, (phenylalanine), S (serine), N (asparagine).

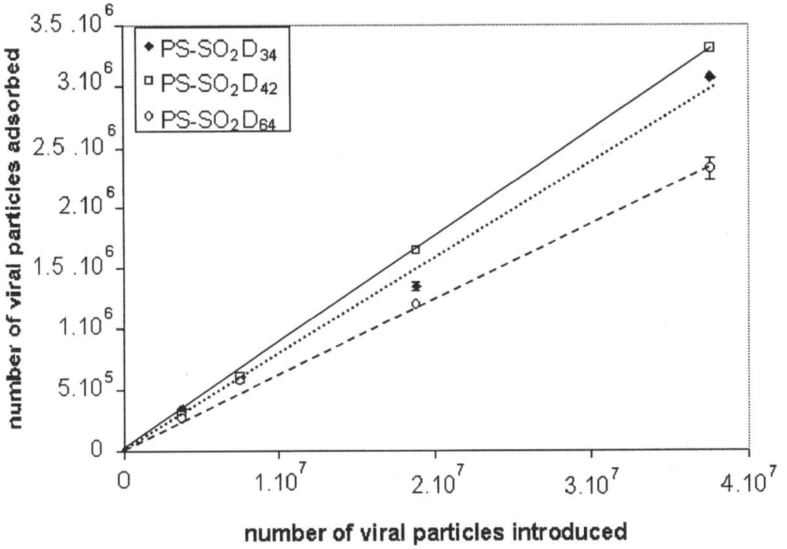

FIG. 7 Adsorption of Babanki virus onto polystyrene beads substituted with aspartic acid. Each point is the mean value of triplicate data.

impossible to reach and determine polymer's maximal adsorption capacities (B_{max}) for viruses as well as affinity constant. Indeed, the observed straight line (Fig. 7) corresponded to the tangent to origin of isotherm and did not present any further plateau value corresponding to the saturation of the surface. The equation of the straight line can be derived from the isotherm equation:

$$B/[(B_{max} - B)*F] = K_a \qquad (1)$$

where B represents the concentration of bound viruses onto polymer surface and F the concentration of free viruses in solution; assuming that $B << B_{max}$; therefore $B = (K_a^* B_{max}) * F$.

In order to compare the virus adsorption level of the different functionalized polymers, we defined the polymer adsorption rate of virus (expressed as % per cm^2) "equaling $(K_a^* B_{max}) * 100$". Results presented in Fig. 8a and b indicated that the adsorption curves was similar for both viruses and that the adsorption rate was dependent on the chemical composition of the polymer. Indeed, two different behaviors of polystyrene derivatives were observed:

1. The adsorption rate of viruses varies with the chemical composition of the polymer. This is the case for arginine- and aspartic acid–substituted poly-

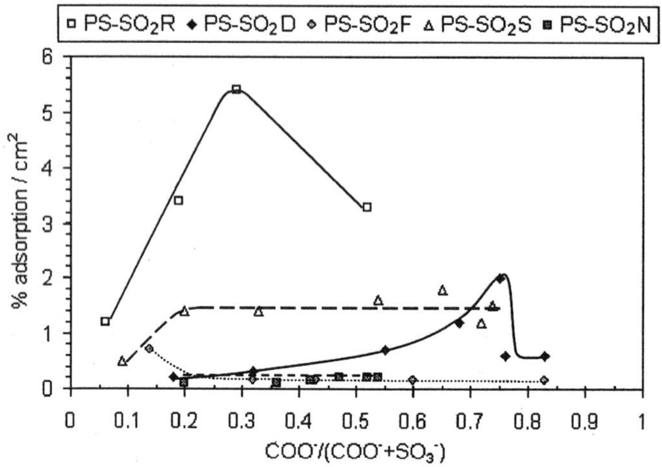

(a)

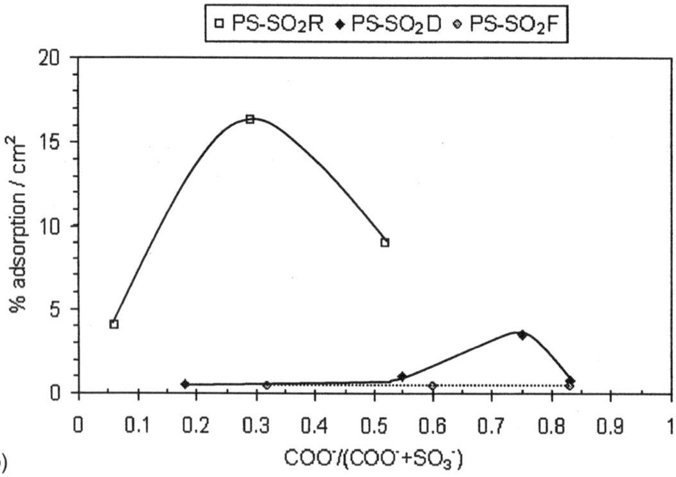

(b)

FIG. 8 Adsorption rate in percent per square centimeter of viral particles vs. the chemical composition of the functionalized polystyrene beads. (a) Adsorption rate of Babanki virus; (b) adsorption rate of Kedougou virus.

styrene derivatives, $PS\text{-}SO_2R$ and $PS\text{-}SO_2D$. Moreover, these two polymers presented a maximal adsorption rate for a given composition of the polymer. The maximal rates were obtained for $PS\text{-}SO2R_{20}$ which correspond to 20% of arginine substitution and R value = 0.29, and for $PS\text{-}SO_2D_{42}$ which correspond to 42% of aspartic acid subtitution with R value = 0.75. These results suggest that for these precise chemical compositions of the polymer, specific interactions are developed between viral particles and bioactive sites present on the surface of the modified polystyrene beads.

2. The adsorption rate of viruses is almost constant whatever the chemical composition of the polymer suggesting no specificity toward viral particles. It is the case for phenylalanine-, asparagine-, and serine-substituted polystyrene derivatives $PS\text{-}SO_2F$, $PS\text{-}SO_2N$, and $PS\text{-}SO_2S$, respectively.

Furthermore, it is worth noting that whatever the chemical composition of polystyrene derivatives, the adsorption level of Kedougou virus (Fig. 8b) is systematically higher than that for Babanki virus (Fig. 8a). For example, the maximal adsorption rate onto $PS\text{-}SO_2R_{20}$ ($R = 0.29$) equaled 5.4%/cm^2 for Babanki virus and 16.3%/cm^2 for Kedougou virus. These results suggested that the affinity of viral particles for functionalized polystyrene beads is higher for Kedougou than for Babanki virus. Both viruses are enveloped viruses and adsorb onto polymer surfaces via their envelopee glycoproteins. Even if belonging to two different families of the arbovirus group—Togaviridae and Flaviviridae—it is likely that their envelopee glycoproteins contain similar domains and should recognize the same chemical sites on polymer surface. However, the envelopee glycoprotein of Kedougou virus seems to develop a better complementarity to the bioactive sites present on the functionalized polystyrene beads than Babanki virus.

D. Discussion

In this study, we report interactions developed between two enveloped viruses—Babanki (Togaviridae) and Kedougou (Flaviviridae)—belonging to the arbovirus group and functionalized polystyrene beads. Results have shown that the virus adsorption level was dependent on the chemical composition of the polymer, characterized by the amino acid substitution percentage and the molar ratio $R = [COO^-/(COO^- + SO_3^-)]$. Furthermore, among the various polystyrene derivatives, $PS\text{-}SO_2R$ and $PS\text{-}SO_2D$ alone presented a maximal adsorption value for a precise chemical composition corresponding to amino acid substitution rate equaling 20% ($R = 0.29$) and 42% ($R = 0.75$), respectively. This result is in perfect agreement with those from previous studies demonstrating that random substitution of polystyrene by suitable chemical groups allows the creation of active sites along macromolecular chains that can specifically interact with biological components. These bioactive sites correspond to precise distributions of the different functionalized monomer units. The probability of occurrence of bioactive sites depends on the nature, number, and distribution of chemical groups, as well as the overall composition of the polymer. This probability displays a maximal value for a given chemical composition of the polymer. Thus, obtained results indicate that the probability of occurrence of active sites allowing biospecific interactions with both studied viruses presents a maximum for PS-SO_2R_{20} and for PS-SO_2D_{42}, suggesting that the envelopee glycoproteins of viruses probably recognize the same "active" site on the macromolecular chain with different affinities. It would be interesting to test these polymers with other arboviruses in order to determine the family or group specificity developed by our functionalized polystyrene derivatives.

Taking the adsorption levels exhibited by PS-SO_2R_{20} and PS-SO_2D_{42} into account makes it easy to calculate the required parameters—surface area and amount in milligrams of modified polystyrene beads—to develop a purification or concentration system of viral particles. These data are reported in Table 3: 61.3 mg of PS-SO_2R_{20} and 125 mg of PS-SO_2D_{42} are the respectively required amounts to use as stationary phase in order to specifically and completely remove Kedougou and Babanki viruses from transfusing plasma. The SEM image of Babanki virus adsorbed onto PS-SO_2D is presented in Fig. 9a and b. As seen, 40- to 70-nm isolated spherical particles can be attributed to viral particle, even though bigger structures in the range of 100 nm to 1 μm are also present, probably corresponding to clusters of viral particles.

V. CONCLUDING REMARKS

Capture of viruses is of great importance for biomedical applications and to develop preventive strategies. Indeed, this is the required first step before any others such as concentration, purification, identification, or detection.

Various methods were described including nonspecific and specific interactions between virus particle and surface. Research in this domain has hastened during the last decade simultaneously to the intensified request of rapid, simple,

TABLE 3 Surface Area and Quantity of Virus-Specific Polymer Needed to Adsorb 100% of Viral Particles

Polymer	Virus	Surface area (cm^2)	Quantity of polymer (mg)
PS-SO_2R_{20}	Babanki	18.52	185.2
	Kedougou	6.13	61.3
PS-SO_2D_{42}	Babanki	50	125
	Kedougou	29.67	74.2

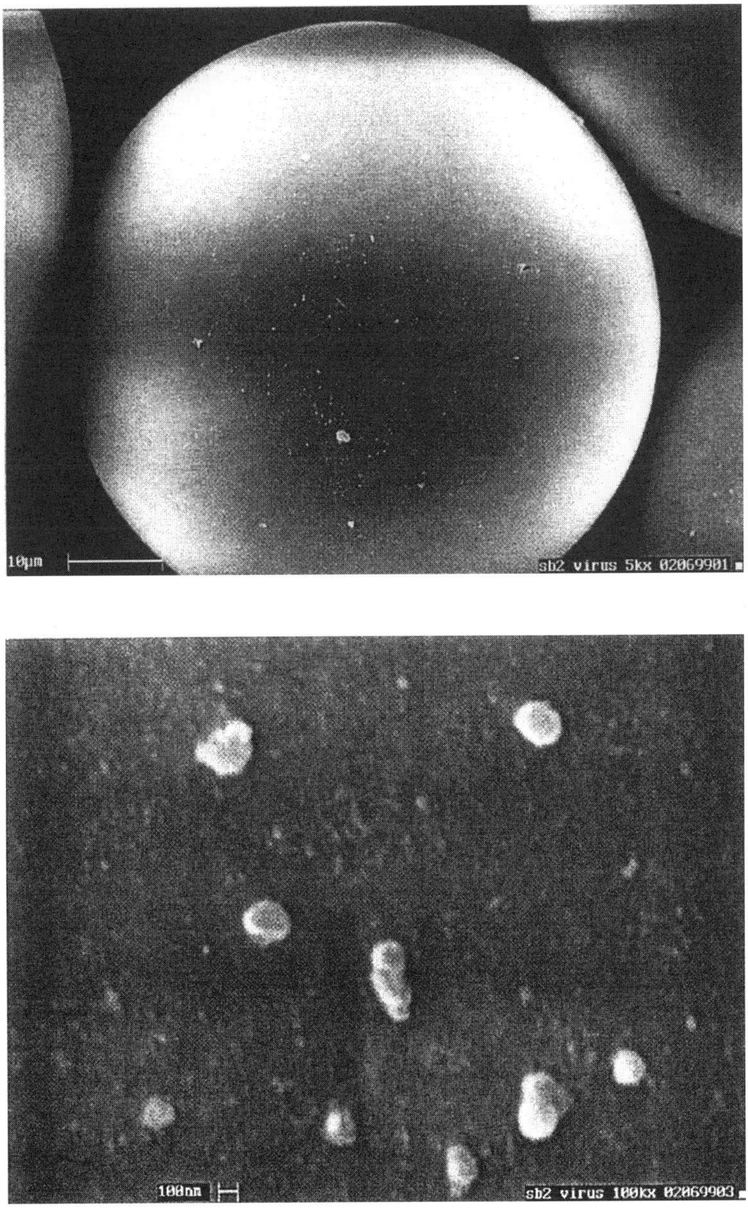

(a)

(b)

FIG. 9 Scanning electron microscope image of Babanki virus adsorbed onto polystyrene beads substituted with 48% aspartic acid. (a) Magnification × 5000; (b) magnification × 100,000.

sensitive, and reproducible techniques. Classical techniques carried out to detect and quantify viruses are mainly cell culture, electron microscopy, ELISA, and EIA. But they are time consuming, expensive, and of low sensitivity. In parallel there is a dramatic need to detect low amounts of "diluted" enteroviruses for virological control of drinking water as well as to get perfect safety of bioproducts. Therefore, removal or inactivation of viruses such as HIV or other retroviruses is absolutely required.

To overcome this major problem several methods involving nonspecific or specific interactions were set up. First methods were developed in order to replace chemical processes for virus concentration, removal, or inactivation and were characterized by the nonspecificity of interactions between viral particles and materials such as glass beads, hollow fibers, or filtration membrane. The encountered problems with these methods concerned the decrease in sensitivity and efficiency due to the nonspecific adsorption and competition of other molecules simultaneously to viruses. These main reasons led to the development of methods based on specific interactions between viral particles and materials. Most applications were in diagnosis and virological and immunological researches. In addition, it represents another tool to measure virus content in blood and to monitore the effectiveness of therapies for several diseases. Moreover, that should be useful in the prognosis of infected patients. Indeed, the measurement of plasma viral load in human immunodeficiency virus (HIV) infection is now unequivocal. Therefore, a rapid, accurate, and quantitative technique for the detection of adenoviral vectors and other viruses is required.

Polymeric materials in the shape of beads or particles are frequently used to immobilize biomolecules such as antibodies, enzymes, and specific ligands in order to realize detection, identification, purification, and/or concentration systems. Among them, cross-linked polystyrene beads are useful because of the easiness of their preparation. However, surface of beads is highly hydrophobic and lead to the nonspecific adsorption of other biomolecules together with virus particles. In order to develop specific interactions with viral particles, specific antibodies were coated onto polystyrene beads. However, the observed sensitivity of latex agglutination tests involving antibody-coated polystyrene beads was not as good as expected because of the nonspecificity of the antibody's link to latex particles. Covalent binding of antibodies specifically directed against virus seemed to be the best solution to increase the sensitivity of detection methods as far as the biological active site of the antibody is still well oriented. Another solution was the covalent binding of virus-ligand molecules such as lectins or sialyl oligosaccharide [33] to polystyrene beads. The inconvenience of the later method was the ligands properties of theses molecules toward a lot of glycoproteins, enzymes, hormones, toxins, bacteria, and so forth.

At least, another approach was the development of biomimetic polymers capable of interacting specifically with viruses. This was achieved by random substitution of polystyrene beads with suitable chemical groups such as sulfonate groups and amino acids. One biomedical application is the use of these modified polystyrene beads as stationary phase in plasmapheresis. However, these biospecific beads may also be useful as biosensors.

GLOSSARY

AGP	agar gel precipitin
BSA	bovine serum albumin
CH_2Cl_2	dichloromethane
CMV	cucumber mosaic virus
Con A	concanavalin A
CVP	canine parvovirus
DNA	desoxyribonucleic acid
EGF	epidermal growth factor
EIA	enzyme immunoassay
ELISA	enzyme-linked immunosorbent assay
Et_3N	triethylamine
FITC	fluorescein isothiocyanate
HAV	hepatitis A virus
HIV	human immunodeficiency virus
HSA	human serum albumin
HSO_3Cl	monochlorosulfonic acid
IBDV	infectious bursal disease virus
IgG	immunoglobulin G
LA	latex agglutination
MuLv	murine leukemia virus
PBMCs	peripheral blood mononuclear cells
PCR	polymerase chain reaction
PFU	particle-forming units
PP-PEI	polypropylene-polyethyleneimine
PS-SO_2D	aspartic acid–substituted polystyrene
PS-SO_2F	phenylalanine-substituted polystyrene
PS-SO_2N	asparagine-substituted polystyrene
PS-SO_2R	arginine-substituted polystyrene
PS-SO_2S	serine-substituted polystyrene
PS-SO_3Na	sodium sulfonate–substituted polystyrene
PVDF	poly(vinylidene fluoride)
PVY	potato virus Y
RIA	radioimmunoassay
RLV	Rauscher murine leukemia virus
RNA	ribonucleic acid

RT-PCR	reverse transcription–polymerase chain reaction
$TCID_{50}$/mL	50% tissue culture infective doses per mL
ToMV	tomato mosaic virus

REFERENCES

1. Sarrette, B.A.; Dznglot, C.D.; Vilagines, R. A new and simple method for the concentration and detection of viruses in water. Water Res. **1977**, *11*, 355–358.
2. Vilagines, P.; Sarrette, B.; Husson, G.; Vilagines, R. Optimal operating conditions for using sodocalcic glass wool to concentrate poliovirus type 1: application to drinking and waste waters analysis. J. Fr. Hydrol. **1992**, *23*, 101–117.
3. Bellara, S.R.; Cui, Z.; MacDonald, S.L.; Pepper, D.S. Virus removal from bioproducts using ultrafiltration membranes modified with latex particle pretreatment. Bioseparation **1998**, *7*, 79–88.
4. Hou, K.; Gerba, C.P.; Goyal, S.M.; Zerda, K.S. Capture of latex beads, bacteria, endotoxin, and viruses by charge-modified filters. Appl. Environ. Microbiol., **1980**, *40*, 892–896.
5. Oshima, K.H.; Evans-Strickfaden, T.T.; Highsmith, A.K. Comparison of filtration properties of hepatitis B virus, hepatitis C virus and simian virus 40 using a polyvinylidene fluoride membrane filter. Vox Sang. **1998**, *75*, 181–188.
6. Owada, T.; Motomura, T.; Miyashita-Ogawa, Y.; Kawada-Homma, M.; Onishi, M.; Matondo, P.; Terunuma, H.; Numazaki, Y.; Yamashita, S.; Yamamoto, N. Antibody masking renders HIV-1 resistant to cationic membrane filtration through alteration of its electrostatic characteristics. J. Virol. Meth. **2001**, *94*, 15–24.
7. Makino, M.; Ishikawa, G.; Yamaguchi, K.; Okada, Y.; Watanabe, K.; Sasaki-Iwaki, Y.; Manabe, S.; Honda, M.; Komuro, K. Concentration of live retrovirus with regenerated cellulose hollow fiber, BMM. Arch. Virol. **1994**, *139*, 87–96.
8. Sanekata, T.; Yoshida, Y.; Okada, H. Detection of rotavirus in faeces by latex agglutination. J. Immunol. Meth. **1981**, *41*, 377–385.
9. Hughes, J.H.; Tuomari, A.V.; Mann, D.R.; Hamparian, V.V. Latex immunoassay for rapid detection of rotavirus. J. Clin. Microbiol. **1984**, *20*, 441–447.
10. Brandt, C.D.; Arndt, C.W.; Evans, G.L.; Kim, H.W.; Stallings, E.P.; Rodriguez, W.J.; Parrott, R.H. Evaluation of a latex test for rotavirus detection. J. Clin. Microbiol. **1987**, *25*, 1800–1802.
11. Kohno, H.; Akihara, S.; Nishio, O.; Ushijima, H. Development of a simple and rapid latex text for ratovirus in stool samples. Pediatr. Int. **2000**, *42*, 395–400.
12. Andrade, J.D. Principles of protein adsorption. In *Surface and Interfacial Aspects of Biomedical Polymers*; Vol. 2, Protein Adsorption. Andrade, J.D., Ed.; Plenum Publishers: New York, 1985.
13. Bodeus, M.; Cambiaso, C.; Surleraux, M.; Burtonboy, G. A latex agglutination test for the detection of canine parvovirus and corresponding antibodies. J. Virol. Meth. **1988**, *19*, 1–12.
14. Esfandiari, J.; Klingeborn, B. A comparative study of a new rapid and one-step test for the detection of parvovirus in faeces from dogs, cats and mink. J. Vet. Med. B **2000**, *47*, 145–153.

15. Nakamura, T.; Kato, A.; Lin, Z.; Hiraga, M.; Nunoya, T.; Otaki, Y.; Ueda, S. A rapid quantitative method for detecting infectious bursal disease virus using polystyrene latex microspheres. J. Virol. Meth. **1993**, *43*, 123–130.
16. Iannelli, D.; D'Apice, L.; Cttone, C.; Viscardi, M.; Scala, F.; Zoina, A.; Del Sorbo, G.; Spigno, P.; Capparelli, R. Simultaneous detection of cucumber mosaic virus, tomato mosaic virus and potato virus Y by flow cytometry. J. Virol. Meth. **1997**, *69*, 137–145.
17. Samoylova, T.I.; Smith, B.F. Flow microsphere immunoassay-based method of virus quantitation. Biotechniques **1999**, *27*, 356–361.
18. Gonda, M.A.; Gilden, R.V.; Oroszlan, S.; Hager, H.; Hsu, K.C. Immunolatex spheres for cell and virion surface labeling in the electron microscope. Virology **1978**, *86*, 572–576.
19. Akashi, M.; Niikawa, T.; Serizawa, T.; Hayakawa, T.; Baba, M. Capture of HIV-1 gp120 and virions by lectin-immobilized polystyrene nanospheres. Bioconj. Chem. **1998**, *9*, 50–53.
20. Hayakawa, T.; Kawamura, M.; Okamoto, M.; Baba, M.; Niikawa, T.; Takehara, S.; Serizawa, T.; Akashi, M. Concanavalin A-immobilized polystyrene nonaspheres capture HIV-1 virions and gp120: potential approach towards prevention of viral transmission. J. Med. Virol. **1998**, *56*, 327–331.
21. Jozefowicz, M.; Jozefonvicz, J. Randomness and biospecificity: random copolymers are capable of biospecific molecular recognition in living systems. Biomaterials **1997**, *18*, 1633–1644.
22. Migonney, V.; Souirti, A.; Pavon-Djavid, G.; Ravion O.; Pfluger, F.; Jozefowicz, M. Biospecific polymers: recognition of phosphorylated polystyrene derivatives by anti-DNA antibodies. J. Biomater. Sci. Polym. Ed. **1997**, *8*, 533–544.
23. Imbert, E.; Souirti, A.; Menard, V.; Pfluger, F.; Jozefowicz, M.; Migonney, V. DNA-like and phosphorylated polystyrenes: characterisation, distribution of functional groups, and calcium complexation properties. J. Appl. Polym. Sci. **1994**, *52*, 91–97.
24. Siali, R.; Correia, J.; Pavon-Djavid, G.; Jozefowicz, M. Biofunctional polystyrene derivatives exhibit a high affinity for the epidermal growth factor receptor (EGF-R): Part I. J. Mater. Sci. Mater. Med. **1996**, 103–107.
25. Sebastien, K.; Pavon-Djavid, G.; Migonney, V.; Kazatchkine, M.; Jozefowicz, M. C3b-like random copolymers: biospecific interactions with complement receptor type one (CR1, CD35). Int. J. Biochromatogr. **1997**, *3*, 313–328.
26. Murphy, F.A. Emerging zoonoses. Emerg. Infect. Dis. **1998**, *4*, 429–435.
27. Childs, J.; Shope, R.E.; Fish, D.; Meslin, F.X.; Peters, C.J.; Johnson, K.; Debess, E.; Dennis, D.; Jenkins, S. Emerging zoonoses. Emerg. Infect. Dis. **1998**, *4*, 453–454.
28. Dalrympl, J.M.; Schlesinger, S.; Russel, P.K. Antigenic characterization of two Sindbis envelopee glycoproteins separated by isoelectric focusing. Virology **1976**, *69*, 93–103.
29. Rey, F.A.; Heinz, F.X.; Mandl, C.; Kunz, C.; Harrison, S.C. The envelope glycoprotein from tick-borne encephalitis virus at 2 Å resolution. Nature **1995**, *375*, 291–298.
30. Andrade, J.D.; Hlady, V.; Wei, A.P.; Gölender, C.G. A domain approach to the

adsorption of complex proteins: preliminary analysis and application to albumin. Croatica Chemica Acta **1990**, *63*, 527–538.
31. Smith, B.A.H.; Sefton, M.V. Thrombin and albumin adsorption to PVA and heparin-PVA hydrogels. I. Single protein isotherms. J. Biomed. Mater. Res. **1992**, *26*, 947–958.
32. Young, B.R.; Pitt, W.G.; Cooper, S.L. Protein adsorption on polymeric biomaterials. I. Adsorption isotherms. J. Colloid Interface Sci. **1988**, *124*, 28–43.
33. Tsuchida, A.; Kobayashi, K.; Matsubara, N.; Muramatsu, T.; Suzuki, T.; Suzuki, Y. Simple synthesis of sialyllactose-carrying polystyrene and its binding with influenza virus. Glycoconj. J. **1998**, *15*, 1047–1054.

6

Polymer Beads in Biomedical Chromatography

Separation of Biomolecules

ALI TUNCEL, ENDER ÜNSAL, S. TOLGA ÇAMLI, and SERAP ŞENEL Hacettepe University, Ankara, Turkey

I. INTRODUCTION

In most of the bioffinity chromatography applications involving the selective isolation of biomolecules, the target biomolecule is usually adsorbed onto the sorbent via a specific interaction taking place between the ligand of the sorbent and the target molecule in the aqueous medium. In this process, no nonspecific interaction between the base material of the sorbent and the target molecule is expected. Following this stage, the adsorbed target molecule is desorbed from the sorbent into a separate medium under controlled conditions. For this reason, in the development of a sorbent specific to a certain biomolecule, the nonspecific and specific interactions between the target molecule and the base material of the sorbent should be defined and then eliminated. Then the inertial microsphere surface is derivatized for introducing the ligand having a specific interaction ability with the target biomolecule. In this chapter, the nonspecific and specific adsorption properties and the chromatographic behaviors of the sorbents in the form of monodispersed latex particles are described. These sorbents were developed for the isolation and/or immobilization of proteins and nucleic acids.

II. NONSPECIFIC PROTEIN ADSORPTION BEHAVIOR OF LATEX PARTICLES

The nonspecific protein adsorption behaviors of monodispersed latex particles prepared with different surface chemistries were extensively investigated [1–4]. In the most of these studies, the polystyrene-based latex particles in submicrometer size range were prepared by the conventional emulsion polymerization of styrene or the emulsion copolymerization of styrene with the functional acrylic monomers e.g., 2-hydroxyethyl methacrylate (HEMA), acrylic acid, and acrylamide [1–3]. These latexes had relatively higher surface charge densities because strong acidic groups coming from the water-soluble initiator were present on the particle surface. For this reason, plain polystyrene latex particles exhibited relatively low nonspecific albumin adsorption capacities at the isoelectric point (IEP). On the other hand, the particles obtained by the copolymerization of styrene with functional acrylic monomers showed lower protein adsorption with respect to the plain polystyrene particles, particularly in the pH region (pH > 5) at which the albumin binding onto the particle surface was controlled by the hydrophobic interactions [2–4]. Okubo et al. comparatively investigated the nonspecific albumin adsorption behavior of polystyrene and polyalkyl methacrylate latexes produced by emulsifier-free emulsion polymerization, and found that more hydrophilic particles had lower nonspecific protein adsorption at the IEP [4].

In our study, the nonspecific protein adsorption behaviors of monodispersed polystyrene latex particles were investigated by selecting albumin as the model protein [5]. Polystyrene-based functional latexes 2.3 μm in size, i.e., poly(styrene), poly(styrene-acrylic acid), poly(styrene-2-hydroxyethyl methacrylate), and poly(styrene-*N*,*N*-dimethylaminoethyl methacrylate, were prepared by a two-stage seeded polymerization [5].

The surface charge density of the latex particles is one of the factors controlling the extent of nonspecific protein adsorption onto the particle surface [1–5]. In the case of latex particles prepared with the conventional emulsion or soapless emulsion processes, the latex surface should contain strongly acidic SO_4^- groups coming from the persulfate-type initiators commonly used in these processes [1–4]. These groups are effective on the nonspecific protein adsorption process in two ways. First, the sulfate groups occupy a certain total area on the surface of the each particle. This should be considered as a factor reducing the parking area of the protein molecules on the particle surface. Second, some unpredicted chemical interactions may occur between the strongly acidic sulfate groups and the amino acid units of the proteins. Contrary to the first case, the second one should be evaluated as a factor involving an increase in the extent of nonspecific protein adsorption. The protein adsorption occurs via the balance established between these two opposite effects [5]. However, in the preparation of the functional latexes used in our study, the polystyrene seed particles were prepared by the dispersion polymerization of styrene by using an oil-soluble initiator (i.e., 2,2′-azobisisobutyronitrile, AIBN). The same initiator was also used in the second-stage polymerization to obtain a surface layer dominantly comprised of functional acrylic monomer on the surface of the seed particles [5]. For this reason, relatively lower surface charge values were obtained in the ζ potential measurement [6]. Then, to observe the interaction of model protein with the polymeric surfaces having different chemistries, the effects of process variables (i.e., pH, ionic strength, and protein concentration) on nonspecific bovine serum albumin (BSA) adsorption were investigated [5]. This methodology allowed the investigation of protein adsorption onto the latex surface in the absence of interference originated from the strongly acidic sulfate groups [5].

Medium pH is one of the most important factors controlling the nonspecific protein adsorption onto the surface of biomaterials. In the interaction of latex particles with the proteins, usually a bell-shaped curve is obtained for the variation of equilibrium protein adsorption with medium pH [1–5]. The typical curves showing the variation of equilibrium BSA adsorption with pH are exemplified in Fig. 1 for the latexes with different surface chemistries [5]. As seen

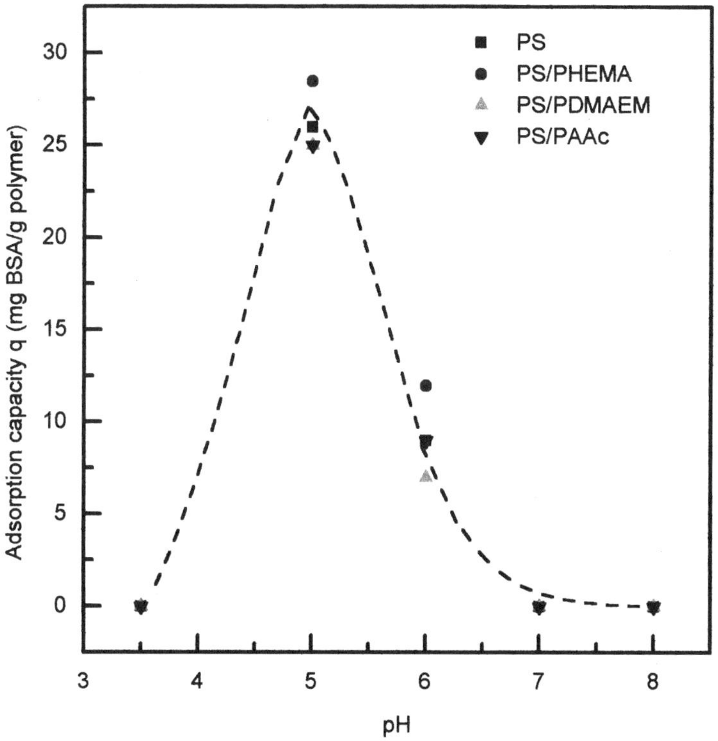

FIG. 1 Variation of equilibrium BSA adsorption capacity with the medium pH for the monodisperse polystyrene latex particles prepared with different surface chemistries. Ionic strength, 0.01; electrolyte, NaCl. (Data taken from Ref 5 with permission, Copyright © 1992, Elsevier Science.)

here, the maximal protein adsorption was observed at pH 5, since the IEP of BSA was very close to that point (e.g., 4.9). At the IEP, the solubility of BSA in the aqueous medium should be minimal since the net charge on the protein molecule was almost zero [5]. For this reason, the protein molecules tend to locate on the particle surface. For different latex types, the comparison of BSA adsorption capacities obtained at the IEP of BSA indicated that the adsorption was predominantly controlled by the solubility of BSA in the aqueous medium at this pH [5]. As a conclusion, the surface chemistry was not so effective for controlling the adsorption of BSA at the pH corresponding to the IEP of selected protein. The same conclusion was also valid for the latexes prepared by using persulfate-type water-soluble intiators [2–5]. At the IEP, the adsorption capacity defined based on the adsorbed amount of albumin per unit area was approximately five times higher for functional latexes prepared by the oil-soluble initiator with respect to the latexes prepared by persulfate-type initiator [1–5]. This comparison indicated that the sulfate groups coming from the water-soluble initiators are strongly effective for reducing protein adsorption onto the latex surface at IEP.

For BSA, the left side of the IEP locates in the acidic pH region at which the protein molecule is in the positively charged form due to the protonation of amine groups present in the amino acid units. Therefore, the nonspecific protein adsorption at the pH values lower than the IEP should be controlled by the electrostatic attraction forces between the positively charged protein molecules and negatively charged surface of the latex particles [5]. Based on this principle, considerable BSA adsorption was observed in the acidic pH region onto the latex particles prepared by the soapless emulsion polymerization [1–3]. The equilibrium adsorption capacities measured at pH 3 onto the polystyrene and poly(styrene-2-hydroxyethyl methacrylate) and poly(styrene-acrylamide) latexes prepared with persulfate-type initiators corresponded nearly half of the maximal equilibrium adsorption observed at the IEP [2,3]. However, nearly zero nonspecific albumin adsorption was obtained at pH 3.5 with the all functional latexes prepared by an oil-soluble initiator (Fig. 1) [5]. This comparison indicated that the electrostatic interaction via the attraction forces between the albumin molecules and the latex surface in the acidic pH region is reasonably weaker for the functional latex particles prepared with our protocol [5].

In the right side of IEP, BSA molecule should be in the negatively charged form. For this reason, an electrostatic repulsion exists between the BSA molecule and the negatively charged particle surface [1–4]. At pH values higher than the IEP of BSA, the protein adsorption onto the particle surface dominantly occurs by the hydrophobic interactions [1–4]. For this reason, the latex particles modified with the hydrophilic polar monomers usually exhibit lower protein adsorption with respect to their plain forms having more hydrophobic character [2,4,5]. The ζ potential measurements of the plain polystyrene and poly(styrene-2-hydroxyethyl methacrylate) latexes prepared by persulfate-type initiators indicated that the surface charge of the polystyrene latex was gradually reduced by the modification with a polar acrylic monomer like HEMA or acrylamide [1,2]. Therefore, both factors—the hydrophilization of particle surface and the reduction—occurred in the surface charge density of sulfate groups are probably responsible for the reduction of adsorption capacity of the polystyrene latex at pH values higher than the IEP of BSA. However, the polystyrene latex particles prepared by the dispersion polymerization exhibited nearly zero albumin adsorption both in the neutral and slightly alkaline medium [5]. The same behavior was also observed with the latex particles prepared with different surface chemistries by using an oil-soluble initiator [5].

The ionic strength is another important parameter controlling the nonspecific protein adsorption onto the surface of latex particles. An increase in the ionic strength usually involves a decrease in the adsorption capacity observed at the IEP of protein. This behavior is probably related to the better stabilization of protein molecule with the ions in the aqueous medium. This behavior was observed with the latexes produced both for water- and oil-soluble initiators [1–5]. In the case of latex particles carrying sulfate groups, an increase in the ionic strength involves higher protein adsorption at the pH values located both in the left and right side of IEP [1,3]. However, in the presence of latex particles prepared by the oil-soluble initiators, the increase in the ionic strength did not cause any significant change in the protein adsorption behavior observed both in the left and right size of IEP [5], and approximately zero albumin adsorption capacities were again observed at the higher ionic strengths in these pH regions [5].

Although the nonspecific protein adsorption onto the polystyrene-based latex particles prepared by an oil-soluble initator was higher at the IEP of the selected protein, these particles exhibited almost zero nonspecific protein adsorption over a wide pH range [5]. Following the definition of the nonspecific adsorption behavior of polystyrene-based latex particles, these materials were used for the synthesis of selective sorbents against proteins.

III. PROTEIN ISOLATION WITH THE NONPOROUS LATEX PARTICLES CARRYING SELECTIVE LIGANDS

The use of synthetic ligands in the synthesis of sorbents capable of selective protein isolation is an alternative and commonly preferred route against relatively expensive biological ligands. The reactive triazinyl dyes have been extensively investigated as ligands in the affinity separation of different proteins [7–15]. These dyes are usually known as inexpensive, readily available, biologically inert, and able to bind proteins in specific manner [13,14]. By utilizing their reactive groups (i.e., sulfonic acid, primary amine or hydroxyl), different triazinyl dyes were covalently attached to the surface of polymeric carriers like polyacrylamide, Sephadex, agarose, cellulose, silica, and glass via conventional chemical activation procedures. Among them, Cibacron blue F3G-A (CBF3G-A) is known as one of the most attractive ligands because it exhibits an affinity for albumin [13,14]. As a sorbent material, latex particles in the micrometer size range have a significant advantage due to their relatively large specific surface area which allow relatively higher equilibrium adsorption capacities particularly for large molecules such as proteins [13]. By considering this property, we designed some sorbents capable of selective protein isolation based on the triazinyl dye–attached latex particles.

In our study, CB-F3G-A carrying monodispersed latex particles was utilized as sorbent in the affinity purification of albumin [13]. For the synthesis of CBF3G-A-carrying sorbent, the 4-μm monodispersed polystyrene particles were obtained by the dispersion polymerization of styrene [13]. In order to prevent nonspecific protein adsorption onto the particle surface, the latex particles were coated with polyvinyl alcohol (PVA) by adsorption and then chemical cross-linking in the presence of terephthaladehyde [13]. In the last stage, the albumin-specific dye CBF3G-A was covalently attached to the PVA-coated polystyrene particles via the hydroxyl groups of PVA unconsumed during the cross-linking reaction [13].

The variation of equilibrium BSA adsorption capacity with the initial BSA concentration at the IEP of BSA and at 25°C is given in Fig. 2 [13]. As seen here, plain polystyrene particles exhibited some degree of nonspecific albumin adsorption at the IEP probably due to the hydrophobic interactions taking place between albumin and the particle surface [13]. In the presence of a hydrophilic PVA layer on the surface of polystyrene particles, the nonspecific adsorption was completely eliminated. In the case of CBF3G-A-carrying particles, high albumin adsorption capacities up to 60 mg BSA/g polystyrene could be achieved by the specific interaction of CBF3G-A with the albumin [13]. Albumin was desorbed from the CBF3G-A carrying polystyrene particles in an aqueous medium containing NaSCN at pH 8.0 with the yields ranging from 84% to 92% [13].

IV. MONODISPERSE-MACROPOROUS LATEX PARTICLES FOR AFFINITY CHROMATOGRAPHY

The affinity high-performance liquid chromatography (affinity HPLC) applications have been performed by using stationary phases based on macroporous particles functionalized with the ligands sensitive to the biomolecules. The macroporous particles carrying functional groups are particularly important for the synthesis of affinity HPLC packing materials since their functional groups can be utilized for the covalent attachment of various ligands with specific recognition abilities.

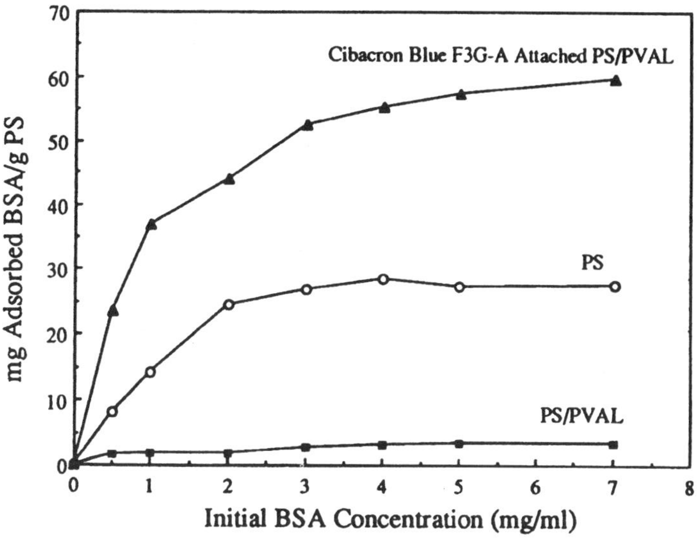

FIG. 2 Variation of equilibrium BSA adsorption capacity with the initial BSA concentration for the CBF3G-A attached latex particles PVA-coated latex particles, and plain polystyrene particles. (Reprinted with permission from Ref. 13, Copyright © 1992, Elsevier Science.)

The variation of BSA adsorption capacity of the macroporous particles having different surface chemistries with the initial BSA concentration is given in Fig. 3 [14]. As seen here, the BSA adsorption capacities of plain poly(S-DVB) and the particles modified with relatively hydrophobic acrylic monomers like epoxypropyl methacrylate (EPMA) and methyl methacrylate (MMA) provided high nonspecific albumin adsorption at the IEP of BSA [14]. The macroporous particles produced in the presence of relatively hydrophilic monomers like 2-hydroxyethylmethacrylate (HEMA) and methacrylic acid (MAAc) exhibited lower non-specific albumin adsorption capacities [14]. Therefore, the presence of hydrophilic units on the macroporous surface caused a decrease in the hydrophobic interaction between BSA and the particle surface. However, butyl methacrylate (BMA)–modified particles exhibited surprisingly low BSA adsorption capacity [14]. The macroporous particles carrying strongly entrapped or covalently bonded PVA chains on their surface had nearly zero nonspecific albumin adsorption [14]. This result is particularly important for the development of specific affinity HPLC sorbents exhibiting no nonspecific protein adsorption behavior.

The monodisperse-macroporous particles produced by including MMA, BMA, HEMA, or MAAc in the repolymerization stage and the plain poly(S-DVB) parti-

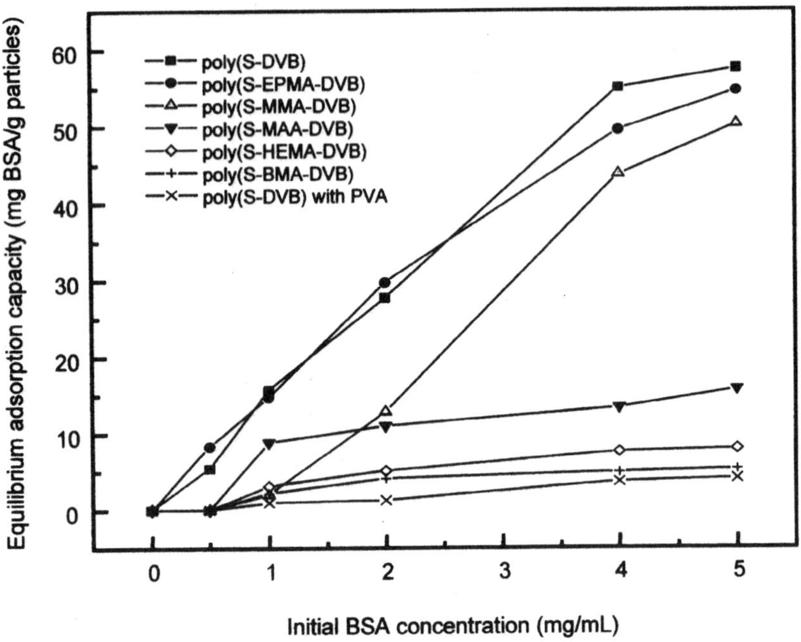

FIG. 3 Variation of BSA equilibrium adsorption capacity of uniform macroporous particles prepared with different surface chemistries with the initial BSA concentration. (Reprinted with permission from Ref. 14, Copyright © 2002, John Wiley and Sons, Inc.)

cles provided desorption ratios higher than 80% (w/w) [14]. These results indicated that albumin was physically adsorbed onto the surface of these particles. However, the low desorption ratio with the EPMA-functionalized particles (about 5%) was evaluated as an evidence of the irreversible chemical interaction between epoxypropyl groups of the particles and the primary amine groups of BSA [14].

The use of macroporous particles as affinity supports in the chromatographic separation of proteins has attracted significant attention. The interactions of PVA-coated macroporous poly(S-DVB) particles were investigated by Leonard et al. [10]. Procion yellow HE3G-carrying macroporous particles were tried as the stationary phase for lysozyme and human serum albumin (HSA) separation [11]. Triazinyl dye–carrying perfluoropolymer supports in the particulate form were used as sorbents for the specific isolation of different proteins [12]. However, all particle types tried for protein separation were produced in the polydispersed form. By considering the potential advantages of the macroporous particles in the monodispersed form as a column-packing material in the chromatographic applications, we attempted to develop an affinity HPLC support material by starting from the monodispersed poly(S-DVB) particles.

A triazinyl dye acting an affinity ligand for albumin, CBF3G-A was immobilized on the surface of the PVA-carrying monodisperse-macroporous particles [15]. PVA-functionalized monodisperse-macroporous poly(S-DVB) particles of 6.25 μm were obtained by a multistage dispersion polymerization by including PVA as the steric stabilizer [15]. The representative electron micrographs of the particles are given in Fig. 4 [15]. The median pore diameter and the specific surface area of the particles were 104 nm and 24.8 m^2/g, respectively [15]. The presence of PVA chains on the surface was shown by the comparison of the FTIR-DRS (Fourier transform infrared-diffuse reflectance spectroscopy) spectrum of PVA-functionalized particles with that of the plain poly(S-DVB) parti-

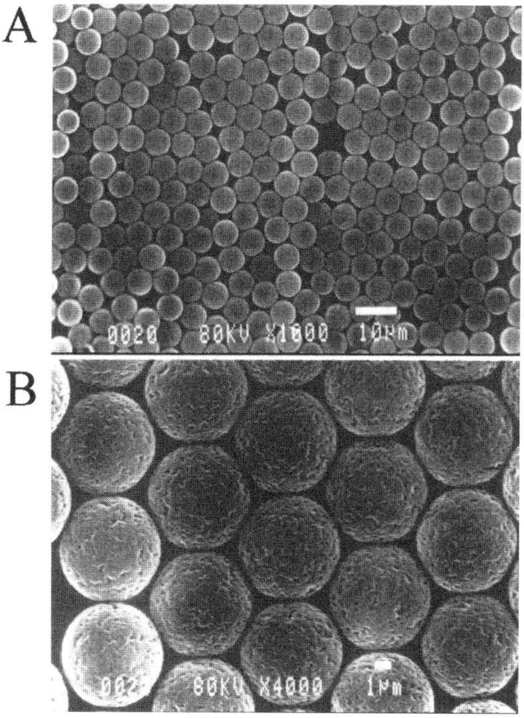

FIG. 4 Representative electron micrographs showing the size distribution and the detailed surface morphology of PVA functionalized, monodisperse-macroporous particles. Magnification (A) x1000, (B) x4000. (Reprinted with permission from Ref. 15, Copyright © 1999, VSP International Science Publishers.)

cles [15]. For the covalent attachment of CBF3G-A onto the PVA chains, a route similar to that used for nonporous particle was followed [13,15].

For the monodisperse-macroporous particles with different surface chemistries, the variation of equilibrium BSA adsorption capacity with the initial BSA concentration is presented in Fig. 5 [15]. As seen in the figures, the plain poly(S-DVB) particles exhibited some nonspecific BSA adsorption probably due to the relatively hydrophobic character of the particle surface. However, no significant nonspecific BSA adsorption with the PVA-functionalized poly-(S-DVB) particles indicated that the presence of a PVA layer on the surface strongly inhibited the hydrophobic interaction [15]. Albumin adsorption capacities up to 100 mg BSA/g particles were achieved with the CBF3G-A-attached particles due to the pseudospecific interaction between BSA and immobilized dye molecules [15].

It should be noted that the equilibrium BSA adsorption capacities of 60 mg/g could be obtained with the CBF3G-A-attached uniform polystyrene particles in the nonporous form [13]. Larger specific surface area of the macroporous material was probably responsible for the higher BSA adsorption [15]. In the

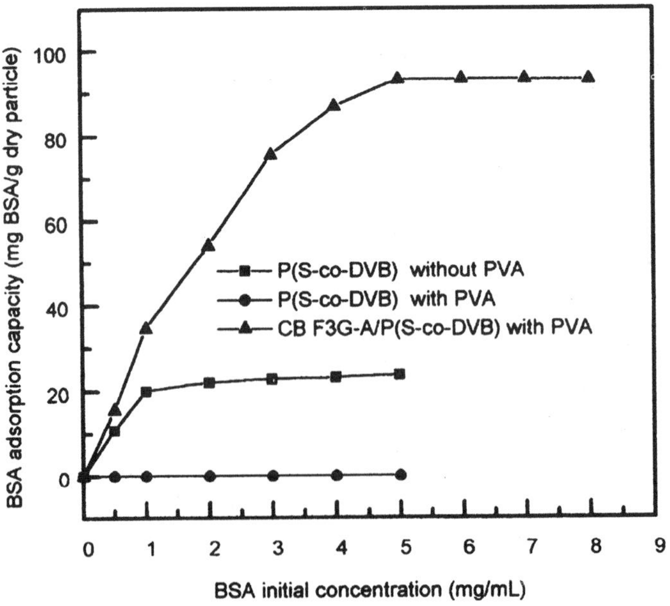

FIG. 5 Variation of equilibrium BSA adsorption capacity with the initial BSA concentration for the dye attached, PVA functionalized and plain poly(S-DVB) particles. (Reprinted with permission from Ref. 15, Copyright © 1999, VSP International Science Publishers.)

desorption experiments performed at pH 8 in the presence of 1 M NaSCN, desorption yields of approximately 90% (w/w) of adsorbed BSA could be obtained [15].

V. POLYDISPERSE SUPPORT MATERIALS FOR PROTEIN ISOLATION

As described in the literature, polyethylene glycol (PEG) is known as one of the most widely used passivation agents preventing the nonspecific protein adsorption onto the surface of biomaterials [16–18]. It has been also known that most of the protein molecules tend to be adsorbed onto the solid surfaces with hydrophobic character [17,18]. The hydrophobic interactions between the protein molecule and biomaterial surface play an important role in the adsorption of proteins onto the solid surface. By considering these properties, a PEG-based sorbent in the form of polydispersed hydrogel particles was designed for controlling the extent of nonspecific protein adsorption [18].

Microporous poly(PEG-co-divinylbenzene)–poly(PEGMA-co-DVB) hydrogel beads were obtained by suspension polymerization [18]. The hydrophobicity of the sorbent was adjusted by controlling the feed ratio of PEGMA to DVB [18]. Hence, it was aimed at controlling the extent of nonspecific protein adsorption. Suspension polymerizations with different PEGMA/DVB feed ratios provided gel beads with an average size in the range of 110–145 μm with the polydispersity indexes varying between 1.12 and 1.21 [18]. The albumin adsorption isotherms derived at pH 5.0 and 25°C are given in Fig. 6 [18]. As expected, the increase in the feed concentration of PEGMA resulted in a significant decrease in nonspecific albumin adsorption [18].

In the next stage of the study, an albumin-specific dye, CBF3G-A, was covalently attached to the surface of the gel beads via the chemical reaction between Cl groups of CBF3G-A and terminal hydroxyl groups of poly(PEGMA-co-DVB) beads [18]. The representative optical micrographs of the plain and CBF3G-A-attached gel beads prepared with the PEGMA/DVB feed ratio of 5.0: 0.5 mL/mL are exemplified in Fig. 7 [18]. Albumin adsorption capacities up to 30 mg/g could be achieved with the CBF3G-A-attached form of the poly-(PEGMA-co-DVB) gel beads prepared with a PEGMA/DVB feed ratio of 1.5: 4.0 mg/mL [18].

VI. POLYMERIC SUPPORTS FOR DNA IMMOBILIZATION

DNA-immobilized sorbents have been widely used in the management of some autoimmune diseases involving the removal of anti-DNA antibodies from plasma [19–25]. Complexation of phosphate segments of DNA with the amine

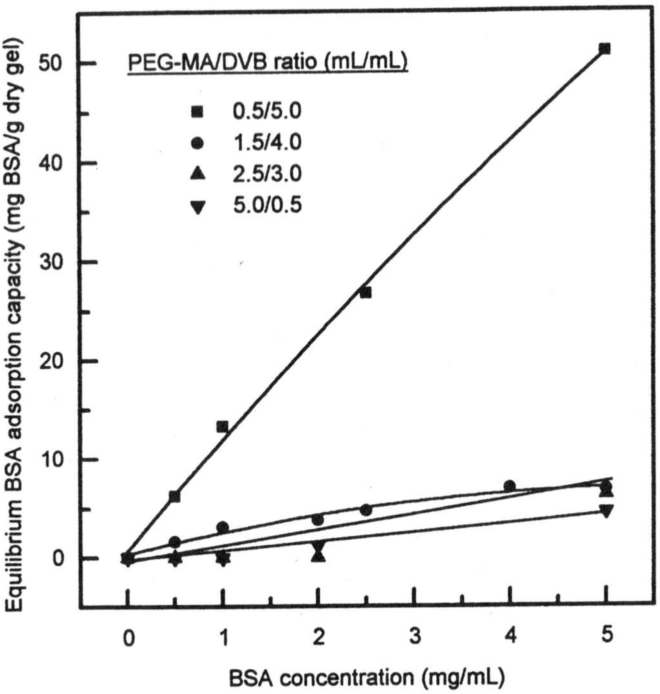

FIG. 6 Albumin adsorption isotherms derived at pH 5.0 and 25°C, with the poly-(PEGMA-co-DVB) gel beads prepared with different PEGMA/DVB feed ratios. (Reprinted with permission from Ref. 18, Copyright © 2000, Springer-Verlag GmbH & Co. KG.)

groups of the support materials is a commonly preferred route for the immobilization of DNA on the solid supports. For this reason, the support materials carrying primary and secondary amine functionalities have been used for DNA immobilization. A magnetic sorbent based on amine-functionalized agarose was successfully tried for DNA immobilisation. Kato and Ikada proposed plasma-treated poly(ethylene terephthalate) (PET) fibers as carriers for the immobilization of DNA [20]. Elaissari et al. developed a sorbent material for RNA adsorption based on the thermosensitive poly(*N*-isopropylacrylamide) particles carrying cationic groups [21]. In this study, the conditions providing the maximal RNA adsorption was investigated and an isolation protocol based on the successive adsorption/desorption of RNA was developed [21]. Core shell magnetic latex particles bearing poly(*N*-isopropylacrylamide) in the shell were also used for the nucleic acid extraction purification and concentration [22].

A B

FIG. 7 Representative optical micrographs of the plain and CBF3G-A attached gel beads prepared with the PEGMA/DVB feed ratio of 5.0/0.5 mL/mL. (Reprinted with permission from Ref. 18, Copyright © 2000, Springer-Verlag GmbH & Co. KG.)

We also attempted to synthesize DNA-carrying support materials based on the poly(*p*-chloromethylstyrene) (PCMS) particles. As described in the literature, a chloromethyl functionality is a reactive group against primary and secondary amines [23,24]. For this reason, polyamine- or polyimine-type ligands are suitable candidates for the immobilization on the polymeric support materials carrying chloromethyl groups. In such a case, the primary and secondary amine groups not utilized for the binding of polymeric ligand onto the support material should exhibit an ionic interaction affinity against the phosphate segments of DNA molecules. In our studies, we selected polyethyleneimine (PEI) as the ligand having an interaction ability with DNA [23,24]. Hence, PEI was covalently immobilized onto the PCMS particles via the reactions between chloromethyl groups and primary or secondary amine groups in PEI [23,24]. Owing to the higher reactivity of chloromethyl group against secondary amine groups, the formation possibility of the first reaction should be higher [23].

In our study, the DNA-sensitive ligand PEI was covalently attached to the PCMS-based particles (average size 186 μm) produced by suspension polymerization [23]. Comparison of surface morphologies of the plain and PEI-attached PCMS particles is given in Fig. 8. As seen here, the roughness of the bead surface almost disappeared as a result of by PEI binding [23]. Hence, a tailor-made particle structure composed of an inner PCMS-based core surrounded by

A

B

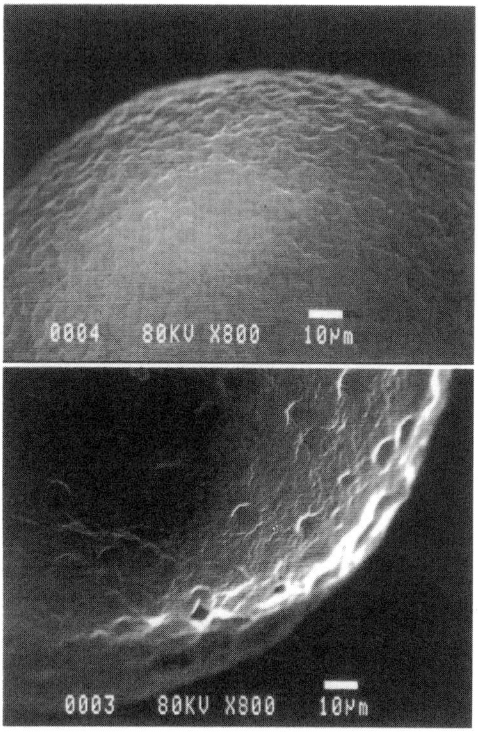

FIG. 8 Comparison of surface morphologies of the plain and PEI-attached PCMS particles. (Reprinted with permission from Ref. 23, Copyright © 2000, Elsevier Science.)

a PEI layer probably forming flexible polycation chains in the aqueous medium was obtained [23].

The variation of equilibrium DNA adsorption capacity for plain and PEI-attached PCMS beads at pH 7 and at 25°C are given in Fig. 9. The tailor-made structure of the support material provided extremely high DNA adsorption [23]. The possible interaction mechanism between PEI and DNA is given in Fig. 10. As seen here, DNA was adsorbed onto the PCMS beads probably by the H-bond formation between phosphate groups of DNA and primary amine groups of immobilized PEI [23]. The desorption experiments showed that only a fraction lower than 5% (w/w) of adsorbed DNA could be released from the PEI-carrying PCMS beads (23). This result indicated that the ionic interaction between the amine/imine groups of PEI and phosphate groups of DNA was sufficiently strong to obtain a stable DNA coating on the surface of the PCMS particles [23].

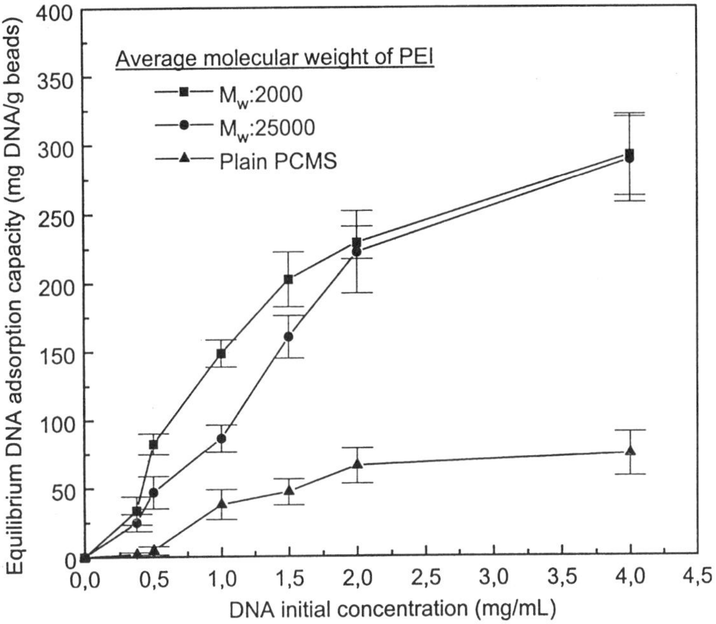

FIG. 9 Variation of equilibrium DNA adsorption capacity for plain and PEI-attached PCMS beads. (Reprinted with permission from Ref. 23, Copyright © 2000, Elsevier Science.)

On the other hand, a similar route was also followed for the synthesis of DNA-responsive uniform latex particles. In this study, uniform (1.75-μm) PCMS beads were obtained by the dispersion polymerization of CMS [24]. PEI (M_r 25.000) was covalently attached to the PCMS particles via the direct reaction between chloromethyl and primary amine or imine groups [24]. The DNA adsorption experiments performed at pH 7.4 and 25°C showed that irreversible DNA binding up to 40 mg DNA/g particles was possible for the PEI-carrying PCMS particles [24]. This interaction resulted in a change in the absorbance of the aqueous latex dispersion containing DNA at a certain concentration in the visible region. Hence, the responsive behavior against DNA concentration was quantified by a spectrophotometric procedure [24]. The variation of absorbance of the aqueous latex dispersion–DNA mixture with the DNA concentration is exemplified in Fig. 11. The aggregation behavior of PEI-carrying PCMS latex is probably due to the binding of some PCMS particles onto the same DNA molecule as shown in Figure 12 [24].

For the immobilization of DNA, another carrier based on monodispersed polycationic gel beads of 3 mm was obtained by the suspension polymerization

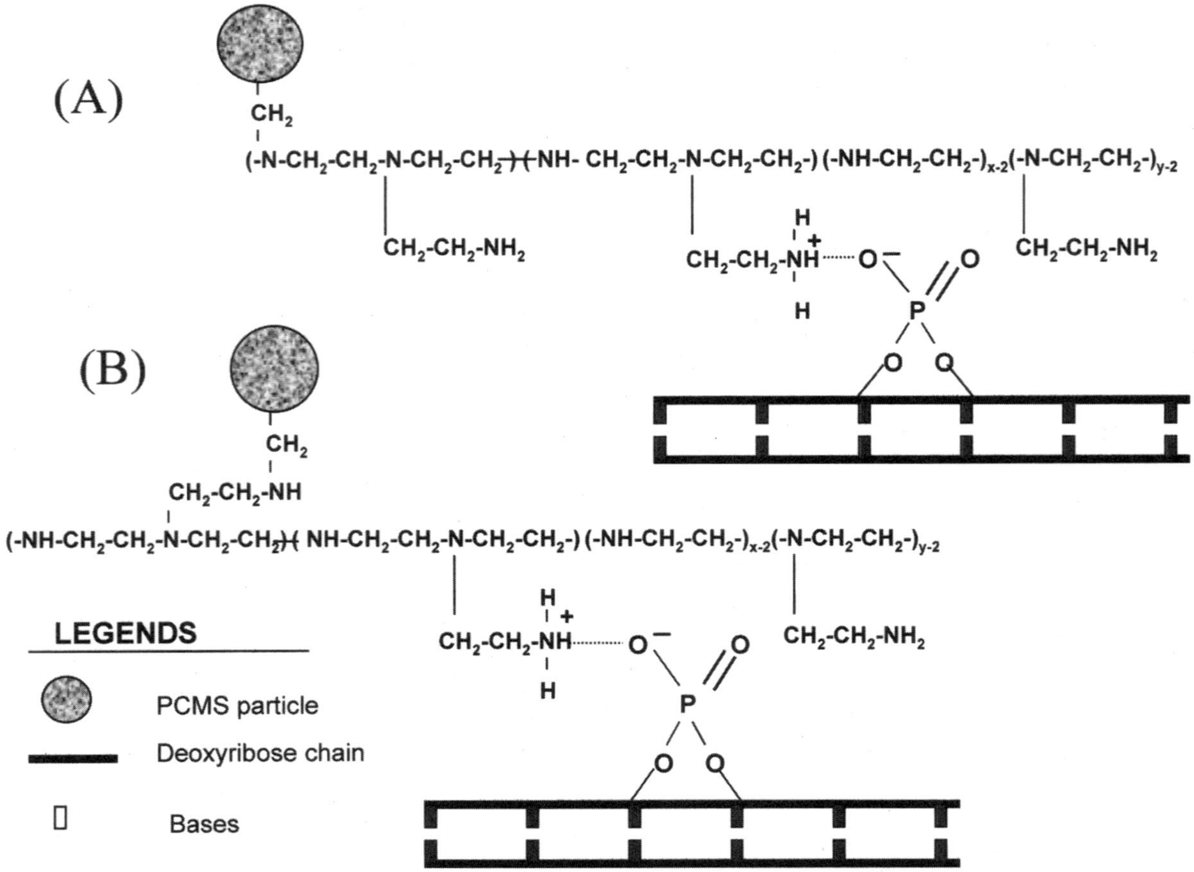

FIG. 10 Possible interaction mechanisms between immobilized PEI and DNA. (Reprinted with permission from Ref. 23, Copyright © 2000, Elsevier Science.)

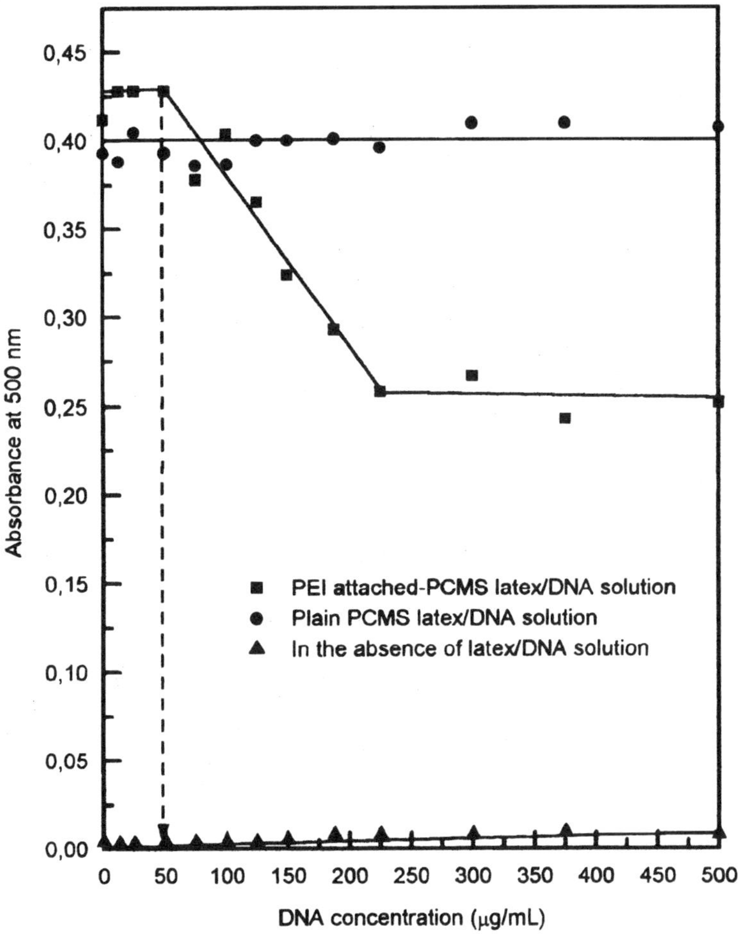

FIG. 11 Variation of absorbance for the PEI-attached PCMS latex-DNA mixtures and plain PCMS latex-DNA mixtures with the DNA concentration. (Reprinted with permission from Ref. 24, Copyright © 2001, VSP International Science Publishers.)

of *N*-3-dimethylaminopropylmethacrylamide (DMAPM) [25]. The variation of equilibrium DNA adsorption capacity with the initial DNA concentration at pH 7.4 and 25°C is shown in Fig. 13. As seen in the figure, the equilibrium DNA adsorption capacities up to 55 mg DNA/g dry beads could be achieved via the ionic interaction between the dimethylaminopropyl groups of the support material and phosphate bonds of DNA [25]. In the desorption experiments, yields lower than 5% (w/w) were obtained in Tris-HCl buffer solution including 1 M

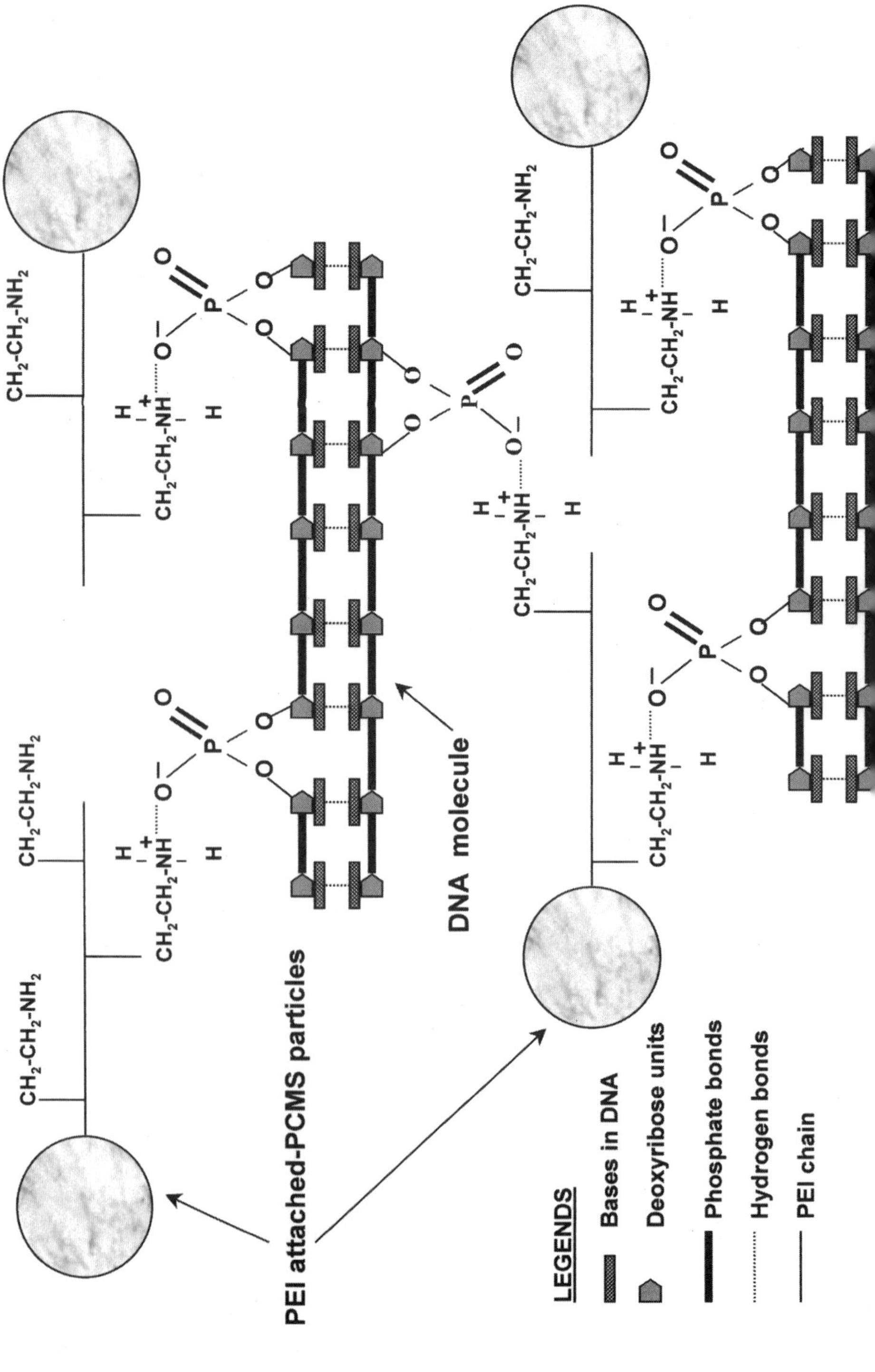
PEI attached-PCMS particles
DNA molecule
CH2-CH2-NH2
CH2-CH2-NH
LEGENDS
Bases in DNA
Deoxyribose units
Phosphate bonds
Hydrogen bonds
PEI chain

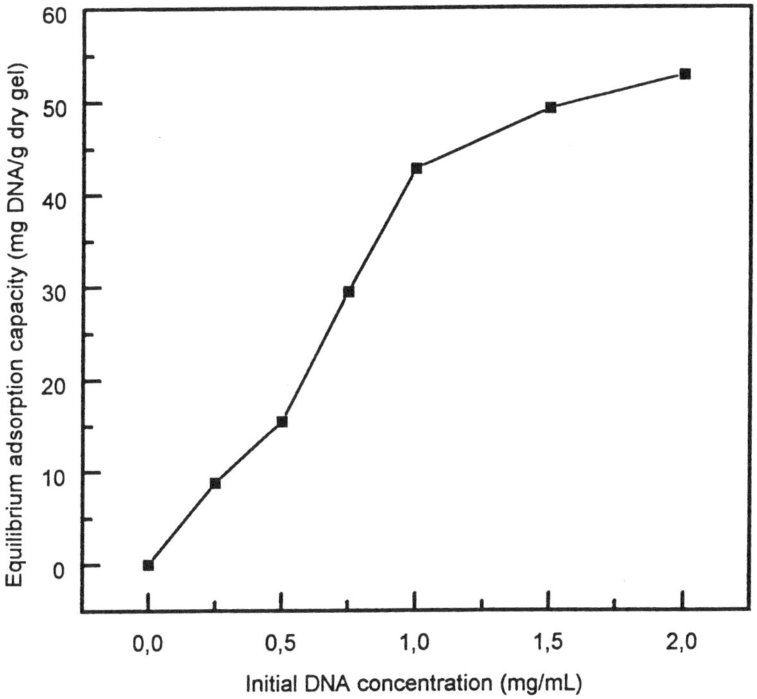

FIG. 13 Variation of equilibrium DNA adsorption capacity of uniform poly(DMAPM) gel beads with the initial DNA concentration. (Reprinted with permission from Ref. 25, Copyright © 2000, John Wiley and Sons, Inc.)

NaCl [25]. These results indicated that the ionic interaction was sufficiently strong to obtain a stable DNA binding on the surface of the gel beads [25].

VII. CHROMATOGRAPHIC APPLICATIONS OF MONODISPERSE-MACROPOROUS PARTICLES

As mentioned before, the monodisperse-macroporous particles are suitable as chromatographic packing materials, particularly in the HPLC applications. Ugel-

---

FACING PAGE

FIG. 12 Binding of several PCMS particles onto one DNA molecule via the interaction between primary amine groups of PEI and phosphate groups of DNA. (Reprinted with permission from Ref. 24, Copyright © 2001, VSP International Science Publishers.)

stad et al. were the first to propose the use of monodisperse-macroporous styrene-divinylbenzene-based particles as a chromatographic support in the separation of proteins by reversed-phase liquid chromatography (RPLC) [26]. Styrene-divinylbenzene-based particles produced by a shape-template polymerization were also successfully tried as chromatographic packing material in size exclusion chromatography (SEC) [27,28]. The SEC calibration curves were almost identical for all of the beads produced with different concentrations of the cross-linking agent [27]. However, the more cross-linked particles exhibited better mechanical and hydrodynamic properties in chromatographic studies [27]. The same particles were also used as a stationary phase in protein separation RPLC [28]. These macroporous particles were obtained by using a porogen mixture composed of a low molecular weight organic agent and the linear polystyrene derived from the seed latex [28]. Frechet's group developed a shape-template polymerization method for the production of monodisperse-macroporous glycidyl methacrylate beads [29–32]. The effects of production conditions (i.e., the type and concentration of the swelling agent, inert porogenic solvent, and the cross-linking agent) on both the size and porosity properties and the chromatographic behavior of the columns including these particles were extensively investigated [29–32]. The pore size distribution of these particles could be controlled by using a chain transfer agent in the repolymerization stage [29]. Poly(dihydroxypropyl methacrylate-co-ethylene glycol dimethacrylate) particles obtained by the acidic hydrolysis of poly(glycidyl methacrylate-co-ethylene glycol dimethacrylate)–poly(GMA-co-EGDMA) beads were tried as packing material in normal-phase HPLC [32]. The columns obtained with the hydrolyzed particles exhibited an excellent chromatographic performance in the separation of positional isomers of benzene derivatives and hydrophilic poly(ethylene oxide)s differing only in their chain length [32]. The monodisperse-macroporous particles with reasonably polar character were obtained in the form of poly-(methacrylic acid-co-ethylene glycol dimethacrylate) and poly(2-hydroxyethyl methacrylate-co-ethylene glycol dimethacrylate) [33]. The chiral supports obtained from these beads were successfully used as stationary phase for the enantiomeric separation of amino acids by normal-phase HPLC [33]. Recently, thermoresponsive chromatographic supports in form of poly(*N*-isopropylacrylamide)-grafted macroporous poly(ethylene glycol dimethacrylate) particles were obtained by Frechet et al. [34]. In their study, temperature-dependent SEC calibration curves were obtained with the produced supports by using dextran standards [34]. A pore size–specific functionalization process was applied to the monodisperse and macroporous poly(GMA-co-EGDMA) particles for the synthesis of a separation media for the complete separation of complex samples that require a combination of size exclusion or ion exchange with reversed-phase chromatographic modes [35]. In the synthesis of the separation medium, the large pores of the poly(GMA-co-EGDMA) particles were selectively hydrolyzed to diol groups whereas the small pores were modified by highly hydrophobic octadecylamine [35]. The pore size–specific functionalized particles were successfully used in the complete separation of samples including both hydrophilic proteins and hydrophobic drugs [35]. A similar study on the pore size–specific functionalization of poly(GMA-co-EGDMA) particles by a different protocol was also performed by the same group [36].

In our studies on the production of monodisperse-macroporous particles suitable for HPLC applications, poly(styrene-co-divinylbenzene)–poly(S-co-DVB)–based materials were predominantly investigated [37–40]. Following the extraction with tetrahydrofuran (THF), the poly(styrene-co-divinylbenzene) particles with different size and porosity properties were slurry charged into the stainless steel HPLC columns (30 × 7.8 mm) under a pressure of 150 atm. The chromatographic performance of these columns was tested both in SEC and RPLC. In the chromatographic studies performed with the SEC mode, the polystyrene standards in the MW range of 2100–1,460,000 were injected into the columns containing monodispersed polystyrene-divinylbenzene beads. The liquid chromatograms obtained with the poly(S-co-DVB) beads 7 μm in size with an average pore size of approximately 200 nm were exemplified in Fig. 14. Here THF

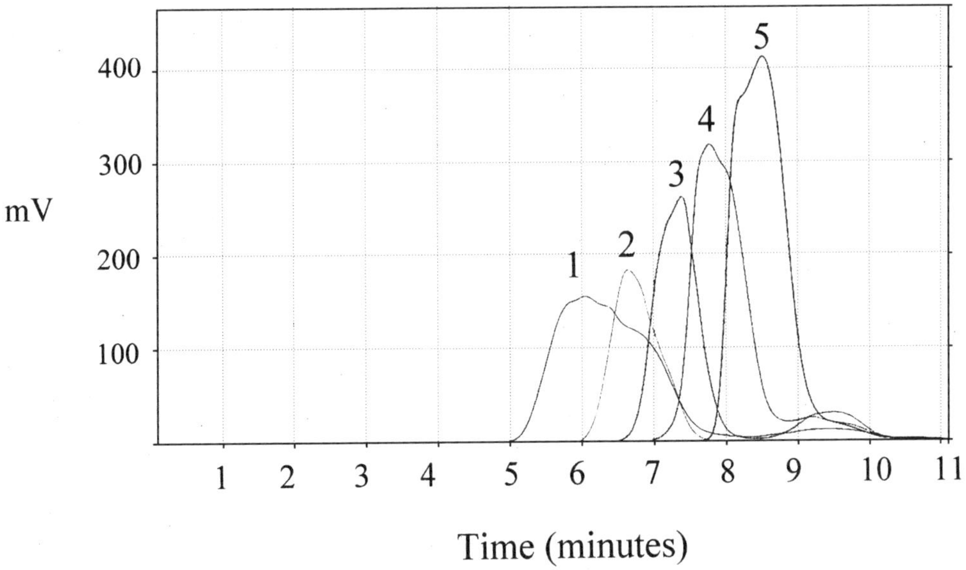

FIG. 14 Liquid chromatograms of polystyrene standards in the average MW range of 2100–1,460,000. Order of elution: 1, MW 1,460,000; 2, MW 288,000; 3, MW 92,000; 4, MW 19,000; 5, MW 2100.

was used as the mobile phase with a volumetric flow rate of 1 mL/min. The polystyrene samples were monitored with a UV-Vis detector at a wavelength of 254 nm.

The calibration curves obtained with the evaluation of these chromatograms are given in Fig. 15. Here the curves obtained with the columns including monodisperse-macroporous poly(S-co-DVB) beads produced with two different cross-linking agent (i.e., DVB) feed concentrations were exemplified. The curves were almost linear for the entire range of MW tried. Thus, one concludes that the tried SEC columns can be utilized for the determination of average MW in the range of 2100–1,460,000. The variation of column back pressure with the mobile phase flow rate should be considered as another important parameter strongly affecting the flow regime in the chromatographic column. For this reason, the variation of column back pressure with the mobile-phase flow rate was investigated by selecting a commercial column containing monodisperse-macroporous particles in the polydispersed form as a reference material. The variation of column back pressure with the flow rate of mobile phase is comparatively shown for monodispersed and polydispersed poly(S-co-SVB) particles

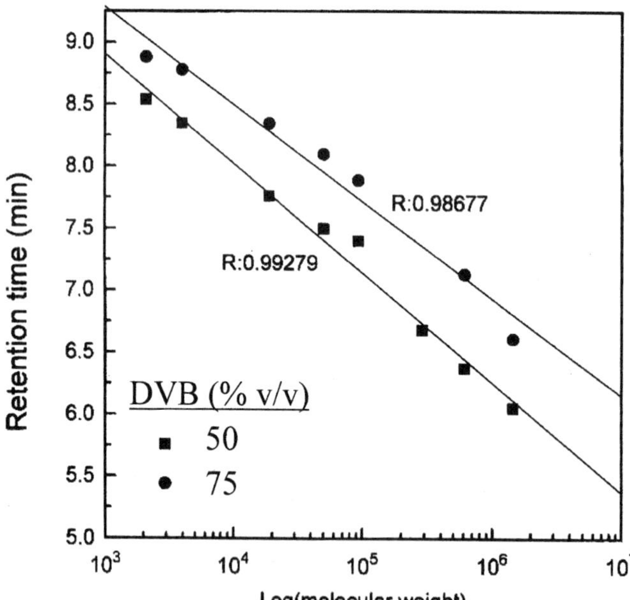

FIG. 15 Calibration curve indicating the variation of average MW of polystyrene standards with the retention time within the SEC column containing monodisperse-macroporous 7-µm particles.

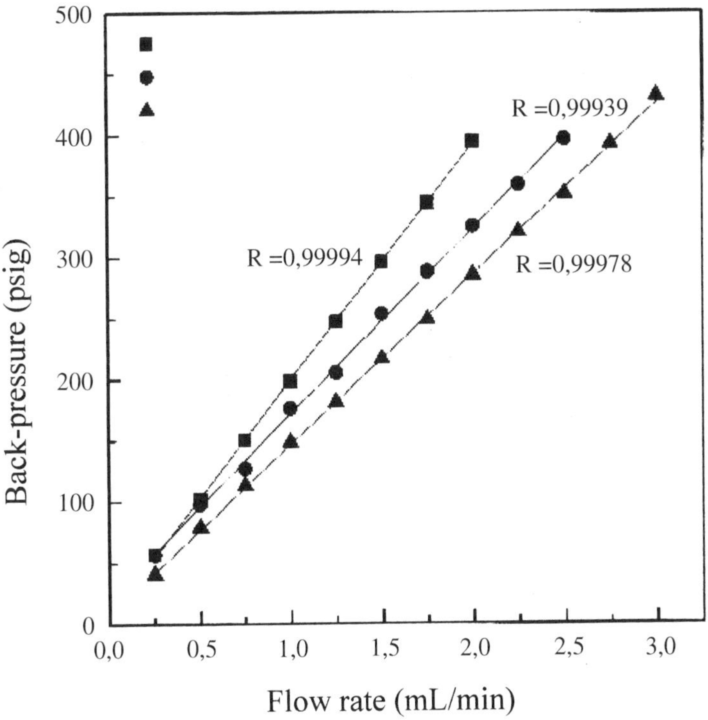

FIG. 16 Variation of column back pressure with the mobile phase flow rate. Mobile phase:THF.

in Fig. 16. In the studied range of flow rate, highly linear relations were obtained for both columns. The columns containing monodispersed particles provided lower back pressures at constant flow rate.

The chromatographic performance of the poly(S-co-DVB) particles was also investigated in the RPLC mode for the separation of proteins. Here a protein mixture that was very similar to those used by different researchers, including four proteins with different hydrophilicities (i.e., cytochrome *c*, lysozyme, albumin, and insulin), was used. Figure 17 exemplifies the liquid chromatograms of the protein mixture taken with acetonitrile/water gradient from 30% to 70% (v/v) acetonitrile in 40 min. Here the 7-μm poly(S-co-DVB) particles were used as the column-packing material. The chromatograms were taken with a UV-Vis detector at 280 nm. As seen here, the chromatograms with high resolutions could be achieved by using monodisperse and macroporous poly(S-co-DVB)

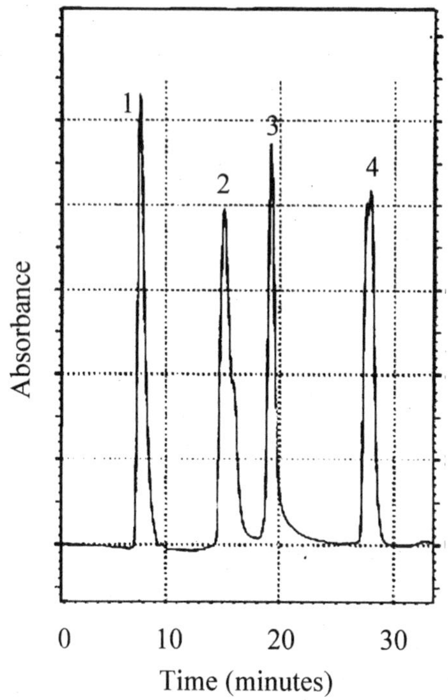

FIG. 17 Sample liquid chromatogram of a protein mixture including four different proteins. UV-Vis detector at 280 nm, Acetonitrile/water gradient from 30% to 70% acetonitrile in 40 min. Flow rate: 1 mL/min. Order of elution: 1, Cytochrome *c*; 2, Lysozyme; 3, albumin; 4, insulin.

beads. It should also be noted that these high resolutions are valid for a broad range of mobile phase flow rate (i.e., 1–4 mL/min).

REFERENCES

1. Shirahama, H.; Takeda, K.; Suzawa, T. Adsorption of bovine serum albumin onto polystyrene latex: effect of coexistent electrolyte anions. J. Colloid Interface Sci. **1987**, *109*, 552–556.
2. Shirahama, H.; Suzawa, T. BSA adsorption onto HEMA copolymerized latex. J. Colloid Interface Sci. **1985**, *104*, 416–421.
3. Shirahama, H.; Shikawa, T.; Suzawa, T. Participation of electrolyte cations in albumin adsorption onto negatively charged polymer latices. Colloid Polym. Sci. **1989**, *267*, 587–594.
4. Okubo, M.; Azume, I.; Yamamoto, Y. Preferential adsortion of bovine serum albu-

min dimer onto polymer microspheres having a heterogeneous surface consisting of hydrophobic and hydrophilic parts. Colloid Polym. Sci. **1990**, *268*, 598–603.
5. Tuncel, A.; Denizli, A.; Abdelaziz, M.; Ayhan, H.; Piskin, E. Albumin adsorption onto large size monodisperse polystyrene latices having functional groups on their surfaces. Clin. Mater. **1992**, *11*, 139–144.
6. Tuncel, A.; Kahraman, R.; Piskin, E. Monosize polystyrene lattices carrying functional groups on their surfaces. J. Appl. Polym. Sci. **1994**, *51*, 1485–1498.
7. Lowe, C.R.; Glad, M.; Larrson, P.O.; Ohlson, S.; Small, D.A.P.; Atkinson, T.; Mossbach, K. J. Chromatogr. 299:157, 1975.
8. Clonis, Y.D.; Atkinson, A.A.; Bruton, C.J.; Lowe, C.R. Eds. *Reactive Dyes in Protein and Enzyme Technology*; Macmillan: Basingstoke, UK, 1987.
9. Compagnini, A.; Fisichella, S.; Foti, S.; Maccarone, G.; Saletti, R.; J. Chromatogr. A **1996**, *736*, 115.
10. Leonard, M.; Fournier, C.; Delacherie, E. J. Chromatogr. B Biomed. Appl. **1995**, *664*, 39.
11. Nash, D.C.; Chase, H.A. J. Chromatogr. A **1997**, *776*, 65.
12. McCreath, G.E.; Oven, R.O.; Nash, D.C.; Chase, H.A. J. Chromatogr. A **1997**, *773*, 73.
13. Tuncel, A.; Denizli, A.; Purvis, D.; Lowe, C.R.; Piskin, E. Cibacron blue F3G-A attached monosize polyvinyl alcohol coated polystyrene microspheres for specific albumin adsorption. J. Chromatogr. **1993**, *634*, 161–168.
14. Camli, T.; Tuncel, M.; Senel, S.; Tuncel, A. Functional, uniform and macroporous latex particles: preparation, electron microscopic characterization and nonspecific protein adsorption properties. J. Appl. Polym. Sci. **2002**, *84*, 414–429.
15. Camli, S.T.; Senel, S.; Tuncel, A. Cibacron blue F3G-A attached uniform and macroporous poly(styrene-co-divinylbenzene) particles for specific albumin adsorption. J. Biomater. Sci. Polym. Ed. **1999**, *10*, 875–889.
16. Lee, J.H.; Kopeckova, P.; Kopecek, J.; Andrade, J.D. Biomaterials **1990**, *11*, 455–464.
17. Harper, G.R.; Davies, M.J.; Davis, S.S.; Tadros, T.F.; Taylor, D.C.; Irving, M.A.; Waters, J.A. Biomaterials **1991**, *12*, 695–700.
18. Tuncel, A. Suspension polymerization of polyethyleneglycol methacrylate: a route for spherical swellable gel beads with controlled hydrophilicity and functionality. Colloid Polym. Sci. **2000**, *278*, 1126–1138.
19. Bruce, I.J.; Davies, M.J.; Howard, K.; Smethurst, D.E.; Todd, M. Magnetizable solid phase supports for purification of nucleic acids. J. Pharm. Pharmacol. **1996**, *48*, 147–149.
20. Kato, K.; Ikada, Y. Biotechnol. Bioeng. **1996**, *51*, 581–592.
21. Elaissari, A.; Holt, L.; Meunier, F.; Voisset, C.; Pichot, C.; Madrand, B.; Mabilat, C. J. Biomater. Sci. Polym. Ed. **1999**, *10*, 403–420.
22. Elaissari, A.; Rodrigue, M.; Meunier, F.; Herve, C. Hydrophilic magnetic latex for nucleic acid extraction purification and concentration. J. Magnet. Magnet. Mater. **2001**, *225*, 127–133.
23. Unsal, E.; Bahar, T.; Tuncel, M.; Tuncel, A. DNA adsorption onto polyethylenimine-attached poly(p-chloromethylstyrene) beads. J. Chromatogr. A **2000**, *898*, 167–177.

24. Elmas, B.; Camli, S.T.; Tuncel, M.; Senel, S.; Tuncel, A. DNA-responsive uniform latex particles based on p-chloromethylstyrene. J. Biomater. Sci. Polym. Ed. **2001**, *12*, 283–296.
25. Tuncel, A. Unsal, E. Çiçek, H. pH-sensitive uniform gel beads for DNA adsorption. J. Appl. Polym. Sci. **2000**, *77*, 3154–3161.
26. Ellingsen, T.; Aune, O.; Ugelstad, J.; Hagen, S. Monosized stationary phases for chromatography. J. Chromatogr. **1990**, *535*, 147–161.
27. Galia, M.; Svec, F.; Frechet, J.M.J. Monodisperse polymer beads as packing material for high performance liquid chromatography: effect of divinylbenzene content on the porous and chromatographic properties of poly(styrene-co-divinylbenzene)-beads prepared in the presence of linear polystyrene as a porogen. J. Polym. Sci. A Polym. Chem. Ed. **1994**, *32*. 2169–2175.
28. Wang, Q.C.; Svec, F.; Frechet, J.M.J. Fine control of the porous structure and chromatographic properties of monodisperse macroporous poly(styrene-co-divinylbenzene) beads prepared by using polymer porogens. J. Polym. Sci. A Polym. Chem. Ed. **1994**, *32*, 2577–2588.
29. Smigol, V.; Svec, F. Preparation and properties of uniform beads based on macroporous glycidylmethacrylate-ethylene dimethacrylate copolymer: use of chain transfer agent for control of the pore size distribution. J. Appl. Polym. Sci. **1993**, *48*, 2033–2039.
30. Horak, D.; Smigol, V.; Labsky, J.; Svec, F.; Pilar, J. An epr study of the effect of suspension polymerization conditions on the properties of glycidylmethacrylate-ethylene dimethacrylate beads. Polymer **1992**, *33*, 2051–2056.
31. Smigol, V.; Svec, F. Synthesis and properties of uniform beads based on macroporous copolymer glycidyl methacrylate-ethylene dimethacrylate: a way to improve separation media for HPLC. J. Appl. Polym. Sci. **1992**, *46*, 1439–1448.
32. Petro, M.; Svec, F.; Frechet, J.M.J. Monodisperse hydrolysed poly(glycidyl methacrylate-co-ethylene dimethacrylate) beads as a stationary phase for normal phase HPLC. Anal. Chem. **1997**, *69*, 3131–3139.
33. Lewandowski, K.; Svec, F.; Frechet, J.M.J. Polar monodisperse, reactive beads from functionalized methacrylate monomers by staged templated suspension polymerization. Chem. Mater. **1998**, *10*, 385–391.
34. Hosoya, K.; Sawada, E.; Kimata, K.; Araki, T.; Tanaka, N.; Frechet, J.M.J. In situ surface selective modification of uniform size macroporous polymer particles with temperature responsive N-isopropylacrylamide. Macromolecules **1994**, *27*, 3973–3976.
35. Smigol, V.; Svec, F.; Frechet, J.M.J. High performance liquid chromatography of complex mixtures using monodisperse-dual chemistry polymer beads prepared by a pore-size specific functionalization process. Anal. Chem. **1994**, *66*, 2129–2138.
36. Smigol, V.; Svec, F.; Frechet, J.M.J. Two-dimensional high performance liquid chromatography using monodisperse polymer beads containing segregated chemistries prepared by pore size specific functionalization. Single column combinations of size exclusion or ion exchange with reversed phase chromatography. Anal. Chem. **1994**, *66*, 4308–4315.
37. Tuncel, A.; Tuncel, M.; Salih, B. Electron microscopic observation of uniform

macroporous particles. I. Effect of seed latex type and diluent. J. Appl. Polym. Sci. **1999**, *71*, 2271–2290.

38. Tuncel, A.; Tuncel, M.; Ergun, B.; Alagöz, C.; Bahar, T. Carboxyl carrying large uniform latex particles. Colloids Surf. A Physicochem. Eng. Asp. **2002**, *197*, 79–94.
39. Tuncel, A.; Tuncel, M.; Cicek, H.; Fidanboy, O. 2-Hydroxyethylmethacrylate carrying uniform, porous particles: preparation and electron microscopy. Polym. Int. **2001**, *51*, 75–84.
40. Senel, S.; Çamh, S.T.; Tuncel, M.; Tuncel, A. Nucleotide adsorption-desorption behaviour of boronic acid functionalized uniform-porous particles. J. Chromatogr. B **2002**, *769*, 283–295.

7

Interaction of Proteins with Thermally Sensitive Particles

HARUMA KAWAGUCHI Keio University, Yokohama, Japan

DAVID DURACHER and ABDELHAMID ELAISSARI
CNRS-bioMérieux, Lyon, France

I. INTRODUCTION

A. Significance of Study on Protein Adsorption

Proteins are poly(amino acid)s that are essential to the maintenance of organisms and the exhibition of biological functions. For example, when a person is bleeding, fibrinogen, a kind of globulin, changes to fibrin, which attaches to platelets and contributes to hemostasis. Cell-adhesive proteins stick to a specific receptor, so-called integlin, that exists on the cell membrane and activates the cell. Transcription factors are crucial proteins that regulate transcription by attaching to a specific site of double-stranded DNA. The above-mentioned examples indicate that the adsorption of proteins to certain targets is the key matter for biofunctions.

Protein adsorption is sometimes utilized to mask a surface of materials so as to change the surface property. Some hydrophobic surfaces are converted to hydrophilic ones as a result of being coated with albumin, and thus obtain nonthrombogenicity.

B. Features of Particles as Adsorbents and Advantages of Polymer Particles

One gram of submicrometer particles has a surface area of several square meters. Due to this huge surface area, reliable and reproducible results regarding protein adsorption onto them were obtainable. The polymeric particles have another advantage as an adsorbent. They have tremendous variety in terms of the chemical and physical structure of the surface as well as the particle size. Namely, desirable surface chemistry and morphology of particles are available from sophisticated particle-forming polymerization and surface modification methodologies.

When particles are dispersed in an aqueous medium, they distribute evenly in the medium. Suppose that particles having a diameter of d nm (are) dispersed with the volume fraction of φ. The interparticle distance h is expressed by the following equation [Eq. (1)]. Therefore, the finely dispersed particles have many chances to meet proteins in the aqueous medium.

$$h(\text{nm}) = [(0.76/\varphi)^{1/3} - 1]\, d \tag{1}$$

These are the reasons polymeric particles have been widely used in protein adsorption studies.

C. Mechanism and Kinetics of Protein Adsorption onto Particles

There were some inconsistencies in the results on protein adsorption reported by different researchers. This is attributed to the different conditions for adsorption, i.e., different surface properties, different concentration of particles, different dispersion media, etc. However, some general facts have been observed as presented below.

1. Mechanism of Protein Adsorption

The factors controlling protein adsorption were comprehensively studied by Haynes, Norde, et al. [1]. They concluded that protein adsorption is controlled by the following factors: (1) electrostatic force, (2) hydrophobic interaction force, (3) hydrogen bonding force, (4) conformational change of a protein molecule, and (5) change of dissociation upon adsorption. These are illustrated in Fig. 1 and explained in the following sections.

(*a*) *Electrostatic Force.* The electrostatic force plays the most significant role in adsorption. Proteins are amphoteric compounds. Therefore, the value and sign of their electrical charge and so the amount of adsorption depends on the

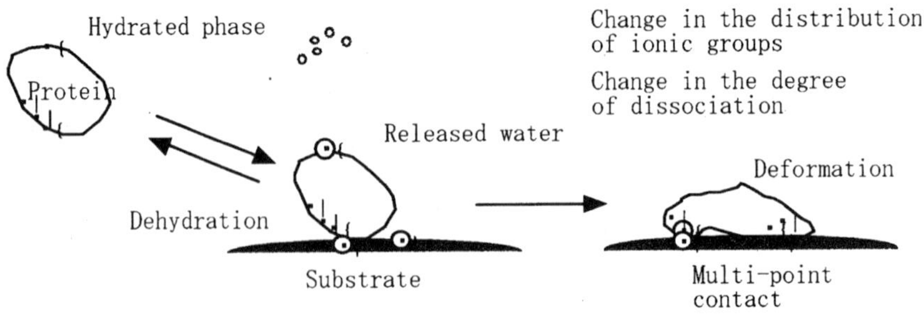

FIG. 1 Events following protein adsorption onto a substrate.

pH of the medium. There are two conditions under which the maximal adsorption is realized. In one case, the maximal adsorption was observed at the pH where the particle surface and the proteins have opposite charges so conferring strong electroattractive force. In the other case, proteins adsorb the maximum at the isoelectric point (IEP) of the protein. This happens because the cross-section area per protein molecule becomes least when the protein molecules adsorbed adjacently have no charge, so that they have the least repulsive force between them. Immunoglobulin molecules were supposed to close their antibody-binding fragments (Fab) tightly and occupy the minimal surface area of carrier at the IEP [2].

Which of the former or the latter of the above-mentioned two cases takes place depends on the ionic strength of particles. It must be mentioned that in such a system the adsorption of proteins is accompanied by redistribution of charged groups.

In the adsorption system in which electrostatic interaction is dominant, the increase in ionic strength or compression of the electric double layer leads to a decrease in adsorption. Therefore, the proteins once adsorbed in the medium of low ionic strength can be desorbed from the particles by the addition of salts.

(*b*) *Hydrophobicity.* Generally speaking, a protein is better adsorbed on a hydrophobic surface than a hydrophilic one. This is due to hydrophobic interaction of the protein and the surface, i.e., both prefer hydrophobic atmosphere in an aqueous phase and touch each other.

It is believed that excellent bioinertness of hydrophilic materials results from less adsorption of proteins on the hydrophilic surface. Coating of particle surface with polyethylene glycol has been one of the most common methods to make the surface bioinert [3]. However, an extremely hydrophilic surface is rather susceptible to protein adsorption, although the reason for this is unknown. Tanaka et al. argued that a moderately hydrophilic polymer surface adsorbs the lowest proteins [4]. This was attributed to a water phase with specific structure located on the polymer surface. Wavering of hydrated chains of the surface is also believed to be effective for the restriction of nonspecific protein adsorption [5].

(*c*) *Hydrogen Bond.* Protein adsorption is sometimes affected by hydrogen bonding. Proteins in water are hydrogen-bonded to the water molecules. Hydrogen bonding also works on some substrates, i.e., between the substrate and water or between the substrate molecules. When protein approaches the substrate surface, rearrangement of hydrogen bonds can take place among the protein, water, and the substrate.

(*d*) *Conformation.* Every protein shelters its hydrophobic part in the core and possesses its stable conformation in the solution. As the protein meets with a hydrophobic substrate, it may not need to keep the native conformation any longer and change it if the protein is soft. In other words, a protein having low native state stability is apt to suffer the breakdown of the native conformation when adsorbed. From the perspective of thermodynamics, such a conformational change can occur when the increase in entropy is compensated by the increase in enthalpy.

As easily expected, a larger molecule can have more contact points with the substrate and therefore receives more benefit from the conformational change upon adsorption.

(*e*) *Degree of Dissociation.* The change of dissociation of ionizable groups in protein molecules upon adsorption was determined by Haynes et al. [1]. As described in Ref. 1, electrostatic interaction has the most significant role in protein adsorption, but the interaction can be adjusted by self-regulation of protein dissociation. This means that the degree of dissociation is different between dissolved and adsorbed proteins, depending on the nature of substrate as well as the ionic strength.

2. Kinetics of Protein Adsorption

Protein adsorption behavior is generally described by the Langmuir isotherm if the protein concentration is not very high. Despite this, the adsorption is seldom reversible. It is attributed to the nature of proteins, i.e., the great size and flexibility of protein molecules. A protein molecule first contacts at one point with the substrate, and the number of contacting points increases with time. When the protein molecules have numerous contacts with the substrate, they become difficult to leave from the substrate and thus the adsorption becomes irreversible.

As the protein concentration increases, adsorption becomes saturated and multilayered adsorption of protein takes place. Therefore, the adsorption isotherm of such systems has a plateau range before the second ascent.

The adsorption kinetics of multiprotein systems is worth discussing. In competitive adsorption of two or more proteins, replacement of proteins proceeds obeying a certain role, i.e., the Vroman effect, in which the protein of low binding constant but high concentration is preferentially adsorbed but is gradually replaced by the protein of low concentration but high binding constant. This is the main reason for a large molecule to occupy most adsorption sites finally [6].

3. Balance of Contribution of Different Factors to Protein Adsorption

Yoon et al. prepared two series of functionalized particles. One series consisted of carboxylated particles having different carboxyl group content, prepared by controlling carboxylation reaction [7]. The other consisted of sulfonated particles. In both systems, adsorption minimum, corresponding to the smallest plateau of Langmuir-Freundlich isotherm, was observed at a certain density of functional groups. The transition point was regarded as the point at which hydrophobic interaction and hydrogen bonding (in the carboxylated system) or electrostatic interaction (in the sulfonated system) was balanced. Comparison of the plateau values of adsorption revealed that the adsorption was most affected by electrostatic force and least by hydrophobic interaction among three. In contrast, a kinetic factor in the Langmuir-Freundlich equation, the K value in the equation, was maximal at the same transition point. The results showed that adsorption by electrostatic force was slowest.

4. Application of Protein Adsorption

The above-mentioned protein adsorption behavior can be utilized for bioengineering and biotechnology applications such as protein separation. For example, cationic particles can selectively collect acidic proteins (proteins having low isoelectric points) from the mixture of proteins at a low pH because only the acidic protein has no electrorepulsive force with the particle under this condition [8]. If the adsorption of proteins onto cationic particles is carried out at a high pH, various kinds of proteins are adsorbed including acidic protein. But, with adding salts, other proteins are gradually desorbed and the acidic protein remains on the particle surface to the last. Such protein separation has so far been done with gel column chromatography equipped with a diethylaminoethyl Sepharose gel column. Cationic particle dispersion was superior to gel column chromatography in terms of the selectivity and yield of the desired protein.

Continuous separation of proteins having different isoelectric points was carried out with the use of a stirred cell charged with carboxylated and sulfonated particles [9].

D. Thermally Sensitive Polymer Particles as a Protein Adsorbent

As mentioned above, several kinds of particles were used as the adsorbents for proteins to study the effect of the surface properties of particles on the adsorption. However, if the particles are environmentally sensitive, one kind of particle is enough for the study because the surface property can be changeable with the environmental conditions.

The pH-sensitive particles, such as carboxylated and/or aminated particles, are representative environmentally sensitive particles that have actually been used as protein adsorbents. However, they have a crucial disadvantage as protein adsorbents at different pHs because proteins are also pH sensitive. Thus, if adsorption measurements are carried out using pH-sensitive particles and pH-sensitive proteins, the results are seldom analyzed although they are sometimes meaningful. In contrast with this, thermally sensitive particles are superior in terms of simplicity if used in a relatively narrow temperature range for protein adsorption.

Single thermally sensitive particles have different surface properties, e.g., hydrophilicity, softness, electric potential, and so on, and thus adsorb proteins in different modes as a function of temperature. The details will be presented in the following sections.

II. PREPARATION OF THERMALLY SENSITIVE PARTICLES

The adsorption of protein onto colloidal particles has been widely investigated, discussed, and reported as evidenced by the numerous published papers and extensive reviews. Based on these reports the adsorption of proteins is mainly related to the hydrophobic character and the charge density of the considered polymer support. In fact, the amount of proteins absorbed onto polystyrene latexes generally used in biomedical diagnostics was found to be high. Then, in order to control the proteic materials' immobilization onto colloidal polymer support, the adsorption should be well controlled. The adsorption of proteins can be dramatically reduced or totally suppressed when the adsorption was investigated on hydrophilic surfaces. However, the total reduction of protein adsorption may also reduce the covalent immobilization of such biomolecules when necessary. Then the challenge is to elaborate materials on which the adsorption can be controlled as a function of a given external stimulus. To satisfy such an objective, various polymerization conditions and methodologies have been investigated. The basic considerations needed to target appropriate the morphology are summarized in Fig. 2. In this domain, poly(*N*-isopropylacrylamide) [poly(NIPAM)]–based microgel particles have been elaborated [10] and evaluated in investigations of protein adsorption as a function of various parameters [11].

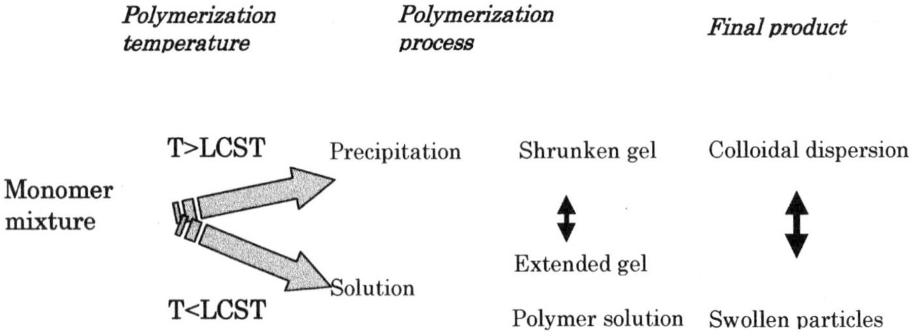

FIG. 2 Polymerization of *N*-alkyl(meth)acrylamide monomer derivatives in aqueous phase.

The preparation of poly(NIPAM)-based particles bearing the volume phase transition temperature (T_{VPT}) were prepared under three morphologies; microgel, core-shell and composite (Fig. 3). In this chapter, only the pertinent results on the poly(NIPAM)-based particles used in protein adsorption study is presented below.

A. Thermally Sensitive Microgel Particles

The microgel particles were prepared using classical batch precipitation polymerization of NIPAM, methylene-bis-acrylamide (MBA) as a cross-linker, a radical initiator (i.e., potassium persulfate, 2,2′-azobisamidinopropane dihydrochloride) [10,12], and functional monomer (acrylic acid, aminoethyl methacrylate hydrochloride, etc.) [13]. Various systematic studies have focused on the polymerization mechanism by studying the effect of each reactant involved in the polymerization process. The most interesting result in the elaboration of such microgel particles was (1) the high polymerization rate, since only 30 min is needed to reach total polymerization conversion, and (2) the amount of water-soluble polymer influenced by several parameters, such as initiator and functional comonomer concentration and polymerization temperature. The synthesis of thermally sensitive microgel particles bearing reactive or functional groups has been reported using versatile comonomers such as acrylic acid, aminoethyl methacrylate hydrochloride, aminopropyl methacrylamide hydrochloride, vinylbenzylisothiouronium chloride, acrylonitrile. As a general remark, all the reported results revealed the narrow size distribution of elaborated particles, which was attributed to the rapid nucleation period leading to a constant particle number throughout the polymerization process.

B. Thermally Sensitive Core-Shell Particles

Core-shell particles (bearing thermally sensitive polymer in the shell) were first reported by combining emulsion and precipitation polymerization processes [14,15]. The amount of NIPAM over styrene was reported to be the key parameter controlling the colloidal particle size and the layer thickness of the hydro-

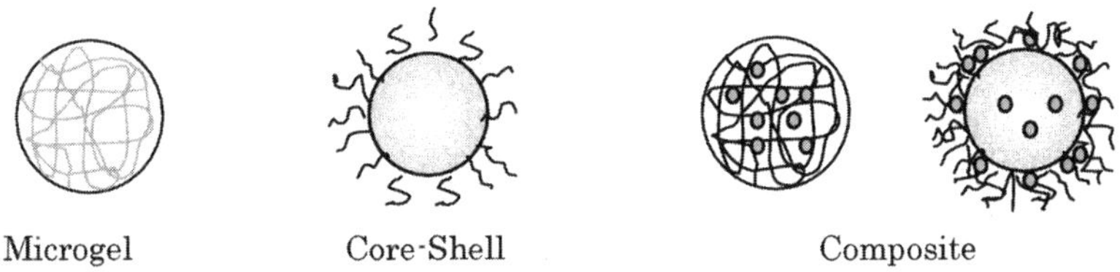

FIG. 3 Morphology illustration of thermally sensitive colloidal particles.

philic shell. The styrene copolymerization with NIPAM monomer enhances both the conversion and the polymerization rate. This has been attributed to two things: the solubility of water of the NIPAM and the high propagation speed constant (k_P) of NIPAM in comparison with that of styrene. The hydrodynamic particle size decreases as the NIPAM monomer increases in the polymerization recipe. The polymerization mechanism of such a system remains questionable even if various attempts have been made. The functionalization of such core-shell latexes was performed by investigating various processes such as batch polymerization and shot-grow process as recently reported by Duracher et al. [16]. In this domain, cationic amino-containing NIPAM–styrene copolymer latex particles were reported using aminoethyl methacrylate hydrochloride and the resulting particles were clearly characterized [17–19]. Two major results were interestingly pointed out from this functionalized core-shell latex: the increase of functional charged monomer in the polymerization recipe lead to (1) a thinner shell layer and (2) a high amount of water-soluble polymer. The behavioral result was mainly attributed to the radical transfer effect of such amino-containing monomer [16]. It is interesting to note that when the functionalization was performed using a shot-grow process, the morphology of the final particles was found to be dependent on the conversion of the batch process (batch polymerization of styrene and NIPAM) at which the shot was performed. In fact, the surface roughness (raspberry-like structure) was observed when the shot was performed at low batch conversions (30–60%), whereas smooth morphologies were obtained when the shot was investigated at high batch conversions (above 70%). The morphologic behavior has been examined using SEM and atomic force microscopy. The thermal sensitivity character of the resulting colloidal particles was directly related to the amount of NIPAM monomer incorporated in the shell layer and the amount of functional monomer used in the elaboration [16].

C. Thermally Sensitive Composite Particles

Composite thermally sensitive particles were elaborated by chemical grafting of reactive poly(NIPAM) chains on colloidal silica particles or by encapsulation of iron oxide nanoparticles. Based on the high interest in magnetic colloidal particles in the biomedical field, various approaches have been developed as evidenced from the reported papers. The first work in this domain was reported by Kawaguchi et al. [20] who produced iron oxide nanocrystals in the preformed cross-linked poly(NIPAM) microgel particles. The amount of iron oxide incorporated was reported to be low so as to induce fast magnetic separation of such composite materials. Then thermoflocculation was induced to enhance magnetic separation. To increase the iron oxide amount in the colloidal composite particles, a new method has recently been reported by Sauzedde et al. [21,22] derived from the early method developed by Furusawa et al. [23]. This approach is divided into two steps: (1) adsorption of iron oxide nanoparticles negatively charged onto polystyrene core cationic poly(NIPAM) shell, and (2) encapsulation and functionalization of adsorbed iron oxide particles via precipitation polymerization using NIPAM monomer, MBA as cross-linker, potassium persulfate, and itaconic acid as functional monomer. The elaborated composite particles contain at least 20% iron oxide material, which favors their magnetic separation.

The colloidal properties of such thermally sensitive particles were found to be dramatically affected by the environmental temperature. The hydrodynamic particle size decreased with increasing the incubation temperature, and the volume phase transition temperature was found to be in the poly(NIPAM) lower critical solution temperature (LCST) region. Consequently, the surface charge density (i.e., electrophoretic mobility) increased dramatically as the temperature increased above the T_{VPT}.

III. PROTEIN ADSORPTION ON THERMALLY SENSITIVE PARTICLES

A. Feature of Thermosensitive Particles

As mentioned in the previous section, single thermosensitive particles show a variety of surface properties. Their features are shown in Fig 4. Thermosensitive particles change in their hydrophilicity, softness, electric potential by virtue of the transition temperature of thermosensitive components in the particle. In the case of poly(NIPAM) particles, they are hydrophilic below the transition tem-

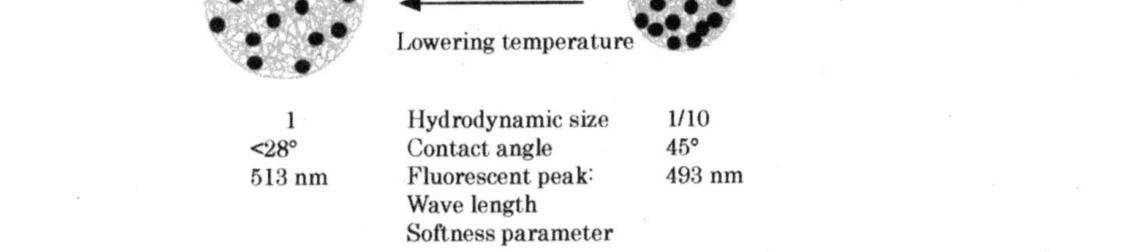

FIG. 4 Schematic illustration of the effect of temperature on the colloidal properties of thermally sensitive particles.

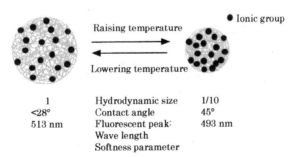

perature (32°C) as shown with the absorption peak of 1-anilinonaphthalene-8-sulfonic acid (ANS) at 515 nm or with the contact angle less than 28° [24], but become hydrophobic (or less hydrophilic) at a temperature higher than the transition temperature of poly(NIPAM) as shown with the absorption of ANS at 493 nm or with the contact angle of 45°. The absorption peak at 493 nm at high temperature indicates that the hydrophilicity of poly(NIPAM) atmosphere at a temperature close to that of acetone. Based on electrophoresis, the apparent surface charge of particles is low or negligible below the transition temperature but becomes high above 32°C. This was attributed to the change of density of ionic groups, e.g., sulfate group originated from the initiator fragment binding to the polymer chain end. Ohshima et al. argued that the electrophoretic mobility of hydrogel particles is a function of the volume density of the ionic group but not the surface density [25]. Ohshima's equation included a term for softness of the thermosensitive gel phase. The experimental results indicated that softness significantly decreased with increasing temperature. The effect of temperature on hydrodynamic particle size and electrokinetic properties is illustrated in Fig. 5.

As hydrophilicity and surface charge are two most important factors influencing the proteic material immobilization, the adsorption of proteins onto poly-(NIPAM) particles is significantly affected by the incubation temperature.

One point to consider is the permeation of proteins into the particles leading

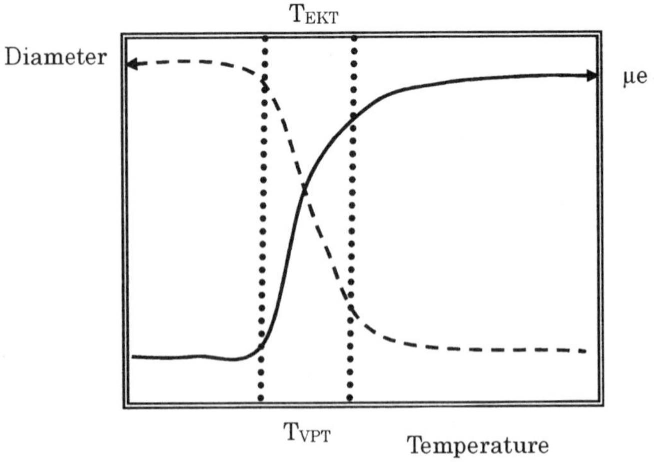

FIG. 5 Schematic illustration of hydrodynamic particle size and electrophoretic mobility as a function of temperature. T_{VPT} and T_{EKT} are the volume phase transition temperature and the electrokinetic transition temperature, respectively. μ_e is the absolute value of the electrophoretic mobility.

to protein absorption. If the particle has a hairy or loosely cross-linked structure, not only adsorption but also absorption can take place. In the case of thermally sensitive particles whose degree of swelling changes sharply at a certain temperature, swollen particles may have the same situation and invite proteins inside if the cross-link density is not so high; however, shrunken particles do not accept protein absorption even if they are cross-linked very loosely. Anionic group-containing poly(NIPAM) gel particles selectively absorbed cationic proteins at temperatures lower than 32°C, and this technology was applied to bioseparation [26].

B. Protein Adsorption on PNIPAM Microgel Spheres

Monodispersed, submicrometer poly(NIPAM) microgel particles were first obtained by Pelton et al. [10]. Onto such latex particles was adsorbed albumin (IEP 4.9) and globulin (IEP 6.8) at different temperatures (20°C and 45°C) [11].

The adsorbability of proteins onto poly(NIPAM) microspheres was found to depend on temperature. Below the LCST of poly(NIPAM) or the volume phase transition temperature of poly(NIPAM) particles in an aqueous medium, that is, around 32°C, the particles adsorbed less protein. This was attributed to the fact that the particles hold a large amount of water in the shell layer and their surface is hydrophilic enough to suppress the adsorption of proteins. On the contrary, above 32°C the particles deswell and their surface becomes hydrophobic and, consequently, susceptible to adsorption of a large amount of protein. Electrostatic force plays an important role for the protein adsorption on poly(NIPAM) particles. Particles were prepared by use of two ionic initiators, persulfate and amidinopropane hydrochloride, having anionic and cationic charges on the surfaces. The charges were distributed dilutedly in the inside of particles below 32°C but concentrated above 32°C. Concentration of ionic groups above 32°C resulted in stronger electrostatic interaction with particle surfaces. The HIV-1 capsid P24 protein adsorption onto thermally sensitive amino-containing polystyrene core poly(NIPAM) shell is presented in Fig. 6 to illustrate the influence of temperature.

The adsorption of lysozyme (IEP 11.0), which is a hydrophilic protein, is unique. Lysozyme was adsorbed in a small amount onto poly(NIPAM) particles at 37°C even though the electrostatic force between the negatively charged particle and positively charged protein was favorable for the adsorption. This was attributed to the extremely high hydrophilicity of lysozyme [27].

C. Effect of Hydrophobicity and Electric Force on Protein Adsorption as a Function of Temperature

As mentioned above, the main driving forces for protein adsorption are electrostatic force and hydrophobic interaction. These are strongly dependent on tem-

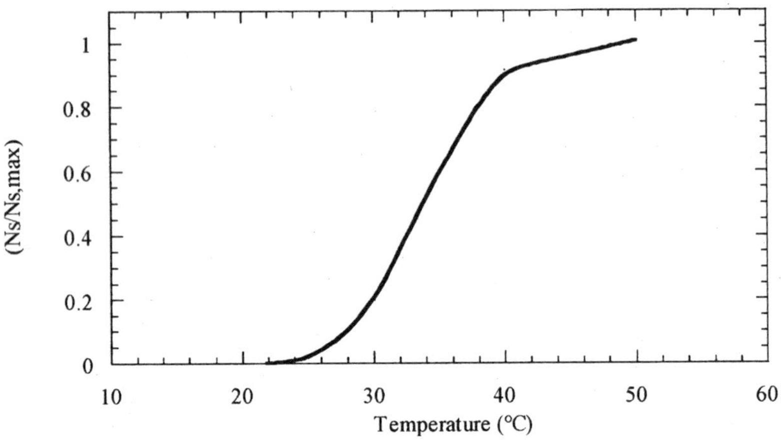

FIG. 6 Reduced amount ($N_S/N_{S,MAX}$) of HIV-1 capsid P24 protein absorbed onto thermally sensitive polystyrene core cationic cross-linked poly(NIPAM) shell as a function of temperature (at pH 6.1, 10 mM phosphate buffer).

perature for thermally sensitive particles. Therefore, many studies were done on the temperature dependence of protein adsorption on thermally sensitive particles referring to the conditions affecting the microenvironment and surface properties of the particles.

For example, the high adsorption of human serum albumin (HSA) on poly-(NIPAM) particles at 40°C and low adsorption at 20°C were explained as the effect of temperature-dependent hydrophobicity of the particles, as already mentioned [11,28]. On the other hand, the effect of pH adsorption was attributed to the electrostatic effect. That is, at pH 4.7, which was close to the IEP of HSA (4.9), the adsorption was maximal and decreased with increasing pH, becoming negligible at pH 8.6. This suggested that the electrostatic force between protein molecules and particles was the determining factor for the adsorption.

The electrostatic force depends on the ionic strength. Therefore, it is expected that the adsorption would be controlled with ionic strength. The effect of ionic strength on adsorption was determined in the solution at pH 4.7 and 8.6. The effect was opposite between two pHs, i.e., adsorption decreased with increasing ionic strength at pH 4.7 and increased at pH 8.6. These results reflected a screening effect by free ions. The above results were obtained at 40°C. However, this was not the case at 20°C where the effect of ionic strength was small. This was because charges were already buried in swollen particles at 20°C [29].

Sugiyama et al. examined protein adsorption on thermally sensitive particles that are not composed of poly(NIPAM) [30]. Several copolymers of hydroxypropyl methacrylamide (HPMA) and alkyl methacrylate [RMA, typically methyl methacrylate (MMA)] exhibit a clear response to temperature change if the ratio of two comonomers is in a suitable range. Some kinds of polymers were prepared using different initiators in the presence or absence of a third comonomer, (methacryloyloxy)ethylphosphorylcholine (MPC). The surface potential was controlled by the kind of initiators and addition of MPC. The equilibrium degree of hydration of one of the copolymers prepared was 35% at 28°C but 7% at 43°C. The adsorption experiments were carried out using two proteins, bovine serum albumin (Alb, at pH 5.6) and human serum gamma globulin (Glo, at pH 6.4) at 24°C and 43°C.

Their results are shown in Fig. 7, which indicates that the particles at 42°C adsorb more proteins than those at 25°C, except for the case in which globulin was adsorbed on positively charged particles. The major results support the above-mentioned general concept that particles become hydrophobic at a temperature higher than the transition temperature and susceptible to adsorption due to hydrophobic interaction with proteins. The exception was supposed to result

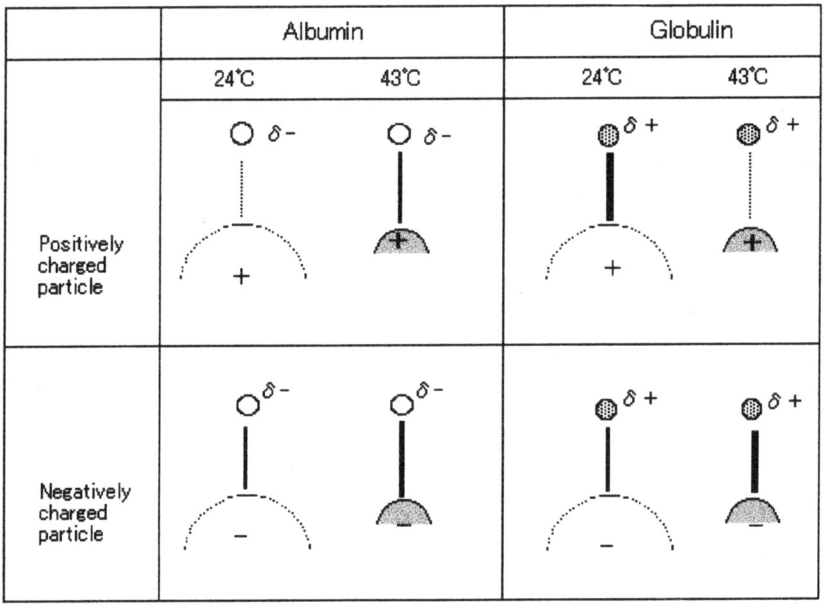

FIG. 7 Schematic representation of albumin and globulin on poly(HPMA-MMA) part. The thickness of bars between a protein molecule and a particle indicates the amount of protein absorbed. Boldface + and – indicate a more strongly effective charge than the plain letters. Gray particles indicate higher hydrophobicity and hardness than white ones.

from the electrorepulsive force between slightly positive protein and positive particles.

The second point that Fig. 7 shows us is that the higher the surface charge, the more proteins were adsorbed regardless of the charge being positive or negative without exceptions. This fact indicates that electric force contributes to the adsorption more or less positively. This might mean that the distribution of ionic groups or dissociation of proteins changes when the protein meets the charged particles to generate electrostatic force between them.

D. Sharpening the Temperature Dependence of Protein Adsorption

When poly(NIPAM) microgel and cross-linked shell-carrying particles were used as the adsorbent of proteins, the temperature dependence of adsorption was not sharp. The amount of adsorption started to increase at a temperature lower than the transition temperature and continued to increase above the transition temperature. This continuous increase of adsorption was related to gradual shrinkage of particles, and the latter was attributed to (1) the gradual formation of NIPAM units cluster starting from about 20°C and (2) restriction of simultaneous collapse at the transition temperature by scattered cross-linking points. The latter was canceled by removing cross-linking points in the particles, i.e., un-cross-linked poly(NIPAM) chains in the particles are expected to collapse simultaneously at the transition temperature. To confirm this, graft polymerization of NIPAM onto a core particle was carried out to prepare hairy particles having no chemical cross-linking points. However, these particles, were not sufficient to prevent physical cross-linking, and exhibited gradual increase in protein adsorption with increasing temperature. Therefore, graft polymerization of NIPAM with a small amount of ionic monomer, e.g., vinyl imidazole, was performed in order to give repulsive force to graft chains. The resulting hairy particles were confirmed to exhibit discontinuous protein adsorption at the transition temperature of poly(NIPAM) [31].

IV. DESORPTION STUDY

The colloidal properties of such stimulus-responsive colloidal particles were principally temperature dependent and reversible. Consequently, the desorption of preadsorbed protein should also be temperature dependent as illustrated below (Fig. 8). In the following sections, the effects of various parameters on protein preadsorbed onto thermally sensitive latexes are presented and discussed. The first study in this domain investigated the effect of temperature on the preadsorbed amount above the volume phase transition temperature (such as 40°C).

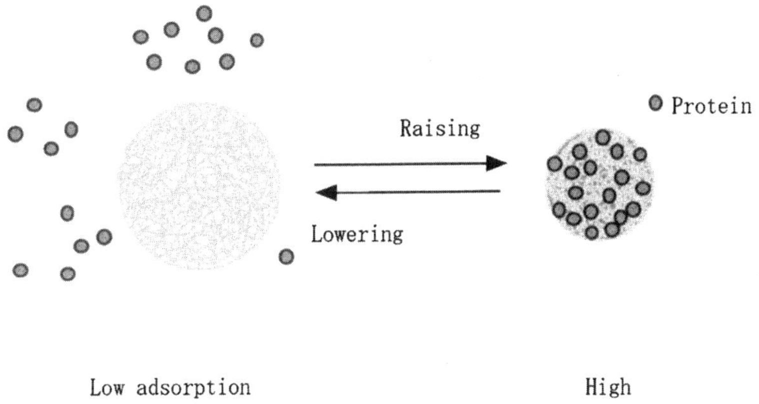

FIG. 8 Schematic illustration of the effect of temperature on protein absorption–desorption.

A. Influence of Adsorption Incubation Time on the Desorption of Proteins

The first experiments on the desorption of proteins revealed the complexity of the adsorption–desorption process. In fact, adsorption is principally amplified when performed above the T_{VPT} [11]. Desorption was found to be effective by lowering of the incubation temperature. In addition, the desorbed protein was directly related to the incubation time during the adsorption process above the T_{VPT} as reported below in the case of cationic thermally sensitive latexes (Fig. 9).

As evidenced in Fig. 6, total desorption (below the T_{VPT}) was observed when adsorption was performed over a short incubation time using, for example, amino-containing polystyrene core poly(NIPAM) shell, whereas only 80% of the adsorbed amount was released when the adsorption was conducted over 2 h. A similar tendency has also been confirmed by Kawaguchi et al. [11] using human gamma globulin (HGG) protein adsorption onto anionic poly(NIPAM) microgel particles. The observed behavior has been discussed on the basis of two phenomena: (1) the possible diffusion and mechanical entrapment of proteins in the loosely cross-linked parts and (2) the denaturation of adsorbed proteins during the adsorption above the T_{VPT} and in the contact of low hydrated (or dehydrated) colloidal particles. Until now, this problem wasn't totally clarified and more work is needed.

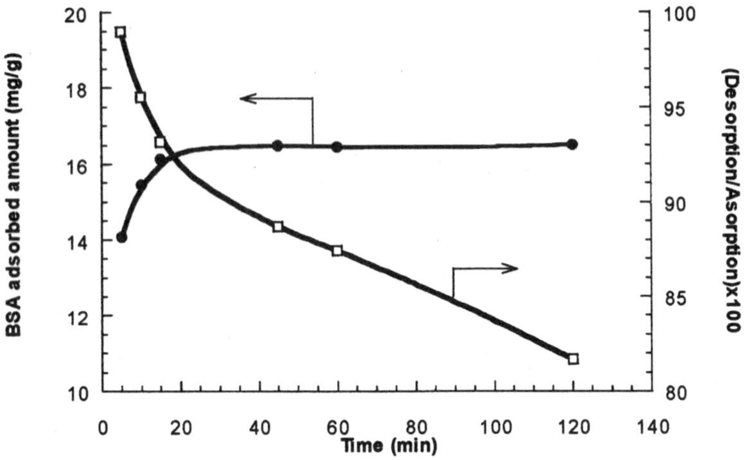

FIG. 9 Protein desorbed amount onto amino-containing polystyrene core poly(NIPAM) shell vs. adsorption incubation time. Conditions: adsorption at pH 6.1 in the phosphate buffer and at 40°C, and desorption at 20°C in the same buffer [32].

In order to control and to understand the adsorption–desorption mechanism, the desorption processes were investigated in systematic studies as a function of various physical parameters.

B. Effect of pH on the Protein Desorption

The effect of pH on desorption (Fig. 10)] was investigated by performing the adsorption under favorable conditions and above the T_{VPT}. The reported results clearly revealed the contribution of electrostatic interactions. In fact, enhancement of attractive electrostatic interactions effected principally by changing the pH leads to low or negligible desorption, as expected in the case of oppositely charged species. In contrast, the desorbed amount was found to be high when both protein and colloidal particles exhibited the same charged nature induced by the pH of the medium.

C. Effect of Ionic Strength on the Protein Desorption

The effect of ionic strength on the desorption of preadsorbed proteins above the transition temperature (i.e., 40°C) was also discussed and found to be of great interest as reported by Duracher [32]. In fact, the desorption investigated (in the case of an oppositely charged system at a given pH) as a function of salt concentration was found to dramatically increase from 50% to 95% of the desorbed amount. This was attributed to the screening effect of electrolytes which reduces

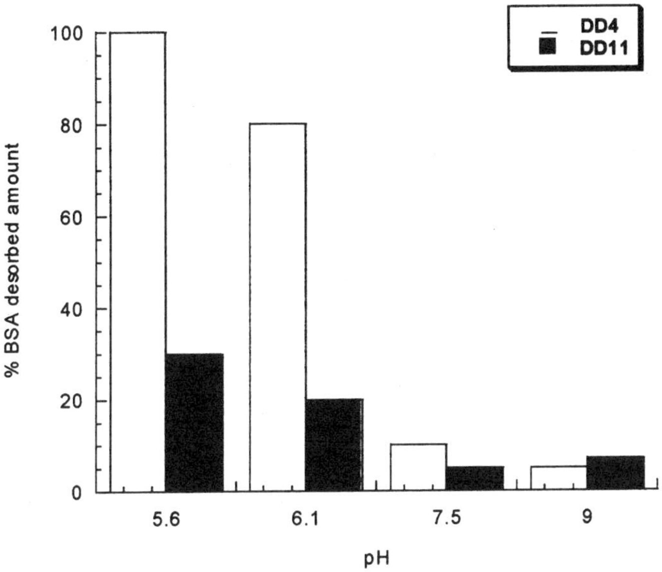

FIG. 10 Protein desorbed amount (wt %) onto amino-containing polystyrene core cross-linked poly(NIPAM) shell. Adsorption–desorption conditions: adsorption at a given pH in the phosphate buffer (10 mM) and at 40°C (for 2 h), and desorption by cooling of sample at 20°C. (The hydrophilic thickness layer of DD4 latex is higher than that of DD11) [32].

the attractive electrostatic interactions as shwon in (Fig. 11). As for classical polyelectrolytes, the adsorption and desorption processes were dramatically affected when pH and salinity were modified as a consequence of alteration in the attractive electrostatic interactions. The only neglected point in the case of thermally sensitive poly(NIPAM)-based colloidal particles was the effect of ionic strength on the poly(NIPAM) chain conformation or basically the effect of ionic strength on the gel swelling ability [17].

V. PROTEIN COMPLEXATION ONTO THERMALLY SENSITIVE MICROGELS

Based on the effect of temperature on the protein adsorption, the oriented immobilization of proteins has been also investigated using *N*-(vinylbenzylimino)diacetic acid (IDA) containing poly(*N*-isopropyl methacrylamide) microgel particles [32,33] as illustrated in Fig. 12. The aim of this work was to control the orientation and the direct immobilization of any proteic material bearing at least

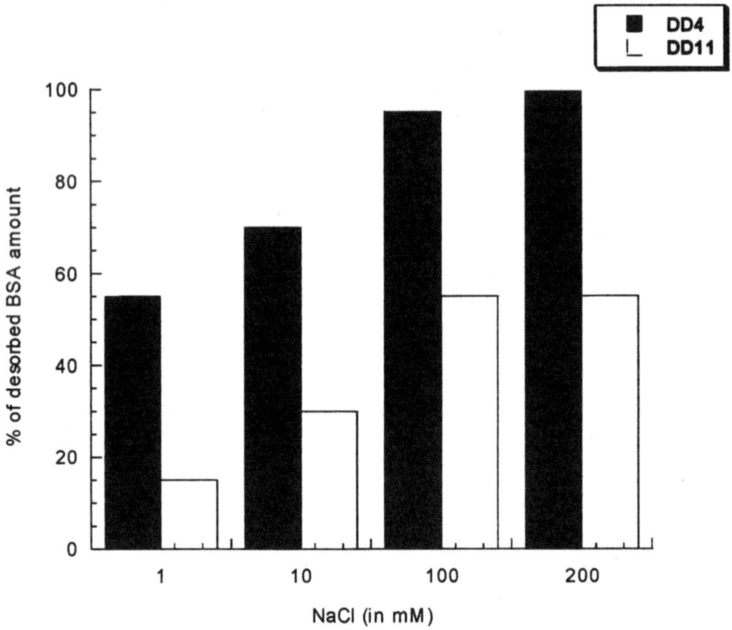

FIG. 11 Protein desorbed amount (wt %) onto amino-containing polystyrene core cross-linked poly(NIPAM) shell versus adsorption incubation time. Conditions: adsorption at pH 6.1 in the phosphate buffer (10 mM) and at 40°C (for 2 h), centrifugation and redispersion of the coated particles at 20°C in phosphate buffer (pH 6.1, 10 mM) at a given NaCl concentration (the incubation was performed over 17 h). (The hydrophilic thickness layer of DD4 latex is higher than that of DD11) [32].

six histidine tags. Then, by controlling the pH and salinity of the medium above the transition temperature, the adsorption efficiency enhances the affinity between the protein and the shrunken particles and, consequently, the complexation yield via additional divalent metals such as Zn^{2+}, Cu^{2+}, and Ni^{2+}. To favor the release of the specifically immobilized proteins onto chelating compound, the pH and the temperature are the key parameters of the immobilization process [32]. Such a process can be used in future immunoassays using modified antibodies (antibodies bearing histidine tag). For illustration, immobilization of the modified HIV-1 recombinant protein (RH24K) onto di-carboxylic-containing poly(NIPAM) microgel particles is given as a function of pH in Fig. 13.

VI. CHEMICAL GRAFTING

The covalent binding of biomolecules is generally preferred to the classical adsorption process. In fact, chemical grafting is needed to avoid desorption phe-

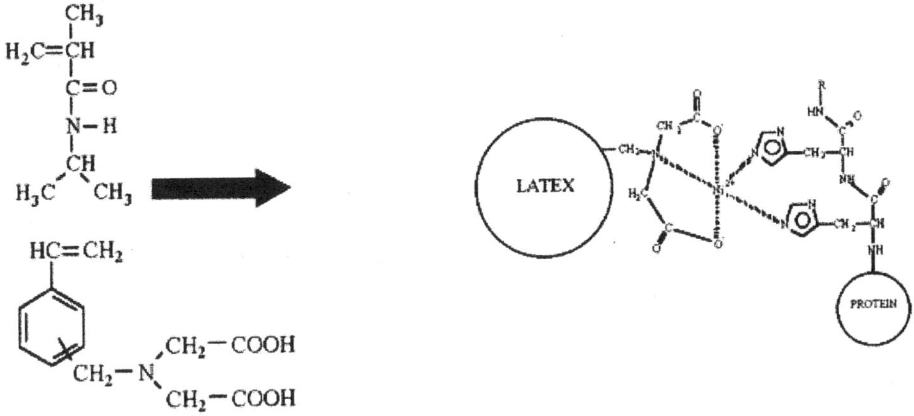

FIG. 12 Schematic illustration of modified protein complexation onto chelating compound–containing particles.

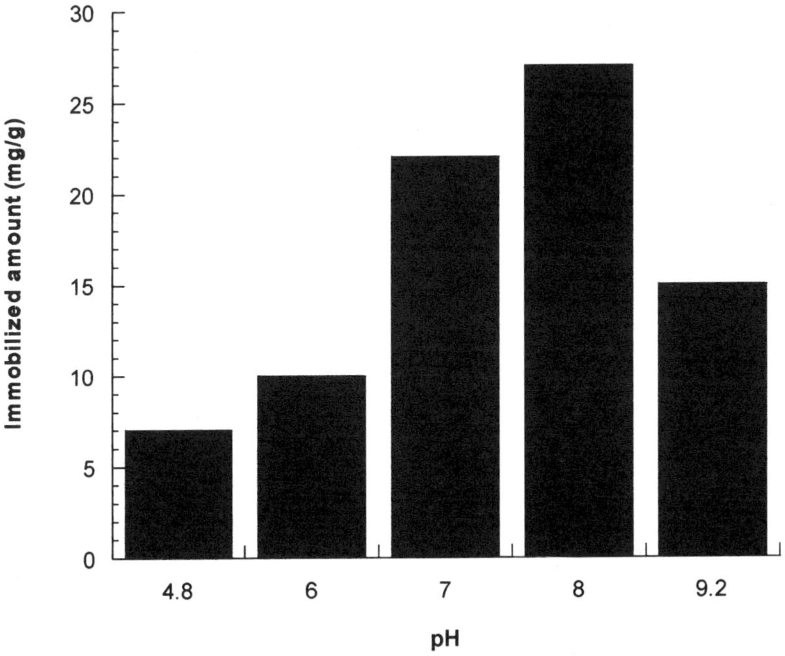

FIG. 13 Amount of immobilized RH24K onto chelating poly(NIPMAM) particles as a function of pH.

nomena when adsorption is used. Since the adsorption of proteic materials onto hydrophilic, thermally sensitive particles is mainly temperature dependent, covalent coupling has been performed using reactive copolymers [33] and as a function of incubation temperature. As a general result, the chemical grafting yield was enhanced when the adsorption of proteins are first performed. The only work reported in this special field has been investigated using amino-containing thermally sensitive particles, HIV-1 capsid p24 protein, and reactive copolymer poly(maleic anhydride-alt-methyl vinyl ether) (PMAMVE) [32], as is illustrated in Fig. 14. This method was found to be easy and efficient for protein immobilization and orientation onto such stimuli-responsive polymer particles. In addition, the nonspecific immobilization of proteins can be totally monitored.

VII. CONCLUSION

The adsorption and the desorption of proteic materials onto thermally sensitive colloidal particles (*N*-isopropylacrylamide-based monomer) have been investigated and discussed as a function of numerous parameters, such as temperature, salinity, pH, surface charge density, and charge nature (anionic and cationic). In addition to the adsorption–desorption studies, the immobilization of proteins via covalent coupling method has been also explored. The pertinent results in those domains are summarized in the following concluding remarks.

The adsorption of protein biomolecules was found and reported to be principally temperature dependent. In fact, the amount of protein adsorbed increased

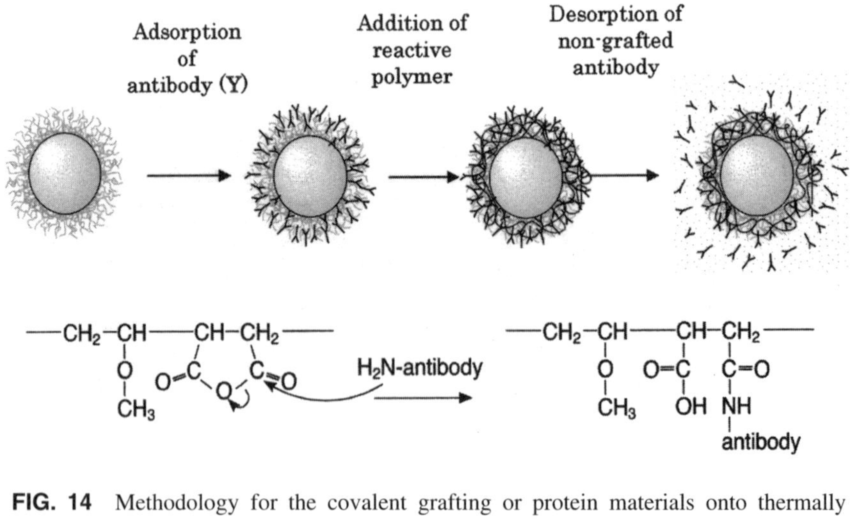

FIG. 14 Methodology for the covalent grafting or protein materials onto thermally sensitive latex particles.

as the incubation temperature rose from below to above the T_{VPT}. The observed behavior has been discussed on the basis of two factors: (1) the possible adsorption via hydrophobic interaction attributed to the dehydration process of poly-(NIPAM) chains and (2) the electrostatic interaction between charges involved in the adsorption process as evidenced from the effect of pH and salinity. Concerning the effect of hydration and dehydration in protein adsorption, it has been shown in numerous studies that the adsorption of protein onto highly hydrophilic surfaces was low and in some cases negligible whatever the charge, whereas the adsorption onto charged hydrophobic surfaces has been reported to be high, with salinity, pH, and charge nature dependent. The reported results concerning protein adsorption onto poly(NIPAM)-based particles as a function of pH and salinity above the T_{VPT} (i.e., above the LCST of corresponding linear polymer) can only be discussed in terms of electrostatic interactions. Anyway, in the case of reported thermally sensitive colloidal particles, to point out the driven forces involved in the protein adsorption, the interaction between protein and non-charged poly(NIPAM) gel or chains should be investigated as a function of temperature.

The desorption study of preadsorbed proteins onto thermally sensitive polymer particles has been not extensively investigated as few investigators have focused on this aspect. The reported results revealed (1) the effect of adsorption time (above the T_{VPT}) on the desorption efficiency below the volume phase transition temperature, and (2) the effect of salinity and pH on the protein desorbed amount below the T_{VPT}. As a result of those investigations, the desorption was found to be total when the adsorption was performed in a short time without pH and salinity modification. In contrast, for long-term adsorption, the desorption was enhanced by reducing the attractive electrostatic interactions.

The chemical immobilization or controlled complexation of proteic material can be obtained by virtue of the effect of temperature, which governs the affinity between both species.

REFERENCES

1. Haynes, C.A.; Norde, W. Colloids Surf. B Biointerfaces **1994**, *2*, 517.
2. Bagchi, P.; Birnbaum, S.M. J. Colloid Interface Sci. **1981**, *83*, 460.
3. Basinska, T. J. Biomater. Sci. Polym. Ed. **2001**, *12*, 1359.
4. Lacasse, F.X.; Filion, M.C.; Phillips, N.C.; Escher, E.; McMullen, J.N.; Hilgen, P. Pharm. Res. **1998**, *15*, 312.
5. Tanaka, M. Colloids Surf. A **2001**, *193*, 145.
6. Vroman, L.; Adams, A. L.; Fischer, G. C.; Muhoz, P. C. Blood, **1980**, *55*, 156.
7. Yoon, J.-Y.; Kim, J.-H.; Kim, W.-S. Colloids Surf. A Phys. Eng. Asp. **1999**, *153*, 413.
8. Sumi, Y.; Shiroya, T.; Fujimoto, K.; Wada, T.; Handa, H.; Kawaguchi, H. Colloids Surf. B **1994**, *2*, 419.

9. Lee, J.H.; Yoon, J.-Y.; Kim, W.-S. Biomed. Chromatogr. **1998**, *12*, 330.
10. Pelton, R.H.; Chibante, P. Colloids Surf. **1986**, *20*, 247.
11. Kawaguchi, H.; Fujimoto, K.; Mizuhara, Y. Colloid Polym. Sci. **1992**, *270*, 53.
12. Kawaguchi, H.; Kawahara, M.; Yaguchi, N.; Hoshino, F.; Ohtsuka, Y. Polym. J. **1988**, *20*, 903.
13. Meunier, F.; Elaissari, A.; Pichot, C. Polym. Adv. Tech. **1995**, *6*, 489.
14. Hoshino, F.; Fujimoto, T.; Kawaguchi, H.; Ohtsuka, Y. Polym. J. **1987**, *19* (2), 241.
15. Hoshino, F.; Kawaguchi, H.; Ohtsuka, Y. Polym. J. **1987**, *19* (10), 1157.
16. Duracher, D.; Sauzedde, F.; Elaïssari, A.; Perrin, A.; Pichot, C. Colloid Polym. Sci. **1998**, *276*, 219.
17. Duracher, D.; Sauzedde, F.; Elaïssari, A.; Pichot, C.; Nabzar, L. Colloid Polym. Sci. **1998**, *276*, 920.
18. Nabzar, L.; Duracher, D.; Elaissari, A.; Chauveteau, G.; Pichot, C. Langmuir **1998**, *14*, 5062.
19. Castanheira, E.M.S.; Martinho, J.M.G.; Duracher, D.; Charreyre, M.T.; Elaïssari, A.; Pichot, C. Langmuir **1999**, *15*, 6712.
20. Kawaguchi, H.; Fujimoto, K.; Nakazawa, Y.; Sakagawa, M.; Ariyoshi, Y.; Shidara, M.; Okazaki, H.; Ebisawa, Y. Colloids Surf. A **1996**, *109*, 147.
21. Sauzedde, F.; Ganachaud, F.; Elaïssari, A.; Pichot, C. J. Appl. Polym. Sci. **1997**, *65*, 2331.
22. Ganachaud, F.; Sauzedde, F.; Elaïssari, A.; Pichot, C. J. Appl. Polym. Sci. **1997**, *65*, 2315.
23. Furusawa, K.; Nagashima, K.; Anzai, C. Colloid Polym. Sci. **1994**, *272*, 1104.
24. Fujimoto, K.; Nakajima, Y.; Kashiwabara, M.; Kawaguchi, H. Polym. Int. **1993**, *30*, 237.
25. Ohshima, H.; Makino, K.; Kato, T.; Fujimoto, K.; Kondo, T.; Kawaguchi, H. J. Colloid Interface Sci. **1993**, *159*, 512.
26. Sassi, A.P.; Shaw, A.J.; Han, S.M.; Blanch, H.W.; Prausnitz, J.M. Polymer **1996**, *37*, 2151.
27. Fujimoto, K.; Mizuhara, Y.; Tamura, N.; Kawaguchi, H. Int. Mater. Syst. Str. **1993**, *4*, 184.
28. Elaissari, A.; Bourrel, V. J. Magnet. Magnet. Mater. **2001**, *225*, 151.
29. Sauzedde, F.; Elaissari, A.; Pichot, C. Macromol. Symp. **2000**, *151*, 617.
30. Sugiyama, K.; Mitsuno, S.; Shiraishi, K. J. Polym. Sci. Polym. Chem. Ed. **1997**, *35*, 3340–3357.
31. Kawaguchi, H.; Isono, Y.; Sasabe, R. ACS Symp. Ser. **2002**, *801*, 307.
32. Duracher, D. Thesis, Université Claude Bernard-Lyon 11, 1999.
33. Duracher, D.; Elaissari, A.; Pichot, C. Macromol. Symp. **2000**, *150*, 297).

8

DNA-Like Polyelectrolyte Adsorption onto Polymer Colloids

Modelization Study

SERGE STOLL Université de Genève, Geneva, Switzerland

I. INTRODUCTION

Due to their fascinating, complex, and important influence on solution properties, mixtures of polyelectrolyte chains (e.g., synthetic polymers, polysaccharides, DNA) and oppositely charged colloids (e.g., organic and inorganic particles, proteins, micelles, vesicules) have stimulated a great deal of interest in soft condensed matter [1,2], biology [3], environmental chemistry [4], and industrial applications [5]. Applications in the field of water treatment as flocculating/water-insoluble mixtures, adhesion, food technology, and powder processing are numerous [6–8], and extension to gene therapy and bioengineering is now under consideration [9,10]. In environmental chemistry, interactions between inorganic colloids and biopolymers and complexation processes are expected to control the coagulation of colloidal material in suspension and the fate and transport of trace polluants associated with them [4,11].

However, the long-range attractive and/or repulsive character of electrostatic interactions between polyelectrolytes and colloids, solution chemistry, chemical composition of the different species, geometry and concentration of both polyelectrolytes and colloids, competitive adsorption, flocculation processes, and so forth, give these solutions very specific properties that are partially understood. So far, little is known regarding the rational use of polyelectrolytes with oppositely charged colloidal particles. By considering the polyelectrolyte charge fraction and salt concentration, Haronska et al. [12] proposed a theoretical model (based on a self-consistent approach) for the complex formalism of flexible polyelectrolytes and oppositely charged spheres. A particular emphasis on the influence of the finite size of both polyelectrolyte and sphere was considered. The theoretical results were compared with the adsorption of polymethacrylate

(PMA) on a cationic micronetwork, and it was demonstrated that the criterion for critical adsorption shows a different behavior for small and large curvature of the sphere. Dobrynin et al. [13] developed a scaling theory of polyelectrolyte adsorption at an oppositely charged surfaces. At low surface charge densities, they predicted two-dimensional adsorbed layers with thickness determined by the balance between electrostatic attraction to the charged surface and chain entropy, whereas at high surface charge densities, they expected a three-dimensional layer with a density profile determined by the balance between electrostatic attraction and short-range monomer–monomer repulsion. The adsorption of weakly charged polyelectrolytes at planar and oppositely charged surfaces was modeled by Linse and Shubin [14,15] using a mean-field lattice theory for flexible polyelectrolytes in solution. They demonstrated that, in most cases, as salt concentration is increased the adsorbed amount is reduced but the thickness of the adsorbed layer is increased. Using a combination of variational procedures and ground-state dominance approximation as well as off lattice MC simulations, Von Goeler et al. [16,17] derived explicit formula for the dependence of adsorption characteristics of polyelectrolytes onto curved surfaces (spheres and cylinders) on temperature, Debye screening length, polyelectrolyte charge density, molecular weight, and curvature. Adsorption was found to be promoted by lowering temperature, chain length, and salt concentration and by increasing the radius and the surface charge density of the sphere. Thermodynamic as well as kinetic factors controlling the stability of colloid–polymer mixtures have been addressed experimentally by Dubin and coworkers [18] by considering interactions between polyelectrolytes and micelles. Critical conditions for adsorption vs. pH and ionic concentration were reported, showing that the required critical surface charge density of the micelles necessary for adsorption was proportional to the inverse Debye screening length κ.

Recently, the overcharging issue or charge inversion of the colloidal particles has attracted significant attention and debate. Indeed, in some conditions, the adsorption of a charged polymer (or a macroion) on an oppositely charged colloid can induce charge inversion of the colloid. Mateescu et al. [19] have shown that overcharging increases with the diameter of the colloid (continuously or through multiple transitions), until a total collapse of the polyelectrolyte takes place. Nguyen et al. [20,21] have investigated complexation for both salt-free and salty solutions and demonstrated that polyelectrolyte winds around the macroion, its turn repealing each other and forming an almost equidistant solenoid. They also demonstrated that adding monovalent salt makes charge inversion stronger, exceeding 100% in some cases. By studying an idealized model for the adsorption of weakly charged polyelectrolyte, Gurovitch and Sens [22] predicted that the connectivity between the charges of the polymer leads to an overcharge of the colloidal particle.

The behavior of polyelectrolyte–particle complexes is usually described by considering the ionic concentration of the solution, polyelectrolyte charge linear density, particle charge, and radius. An additional and key parameter is the chain flexibility (in particular for DNA or polysaccharides like polyelectrolytes), which includes both chain stiffening due to electrostatic monomer–monomer repulsions (electrostatic persistence length) and stiffness of the underlying chain backbone (intrinsic flexibility). Sintes et al. [23] investigated the problem of adsorption of a single semiflexible polymer chain onto a planar, homogeneous surface using MC simulations. Adsorption characteristics were studied at different temperatures for chains with various degrees of stiffness. They demonstrated that the stiffer chains adsorb more onto the surface and the adsorption takes place at a higher temperature when compared to that of a flexible chain. The effect of copolymer sequence distribution and stiffness on the adsorption-desorption transition temperature was also examined using Monte Carlo (MC) by Chidambaram and Dadmun. [24] Netz and Joanny [25,26] recently provided a full complexation phase diagram for a stiff polyelectrolyte in the presence of an oppositely charged sphere. Both the effects of added salt and chain stiffness were taken into account. They demonstrated that for intermediate salt concentration and high sphere charge, a strongly bound complex, where the polymer completely wraps around the sphere, is obtained and that the low salt regime is dominated by monomer-monomer repulsions leading to a characteristic hump shape where the polymer partially wraps around the sphere with two polymer arms extended parallel and in opposite directions from the sphere. In the high salt regime, they suggested that the polymer partially wraps the sphere. They also provided an adsorption criterion, then derived the density distribution of the adsorbed polyelectrolyte to calculate the particle radius below which adsorption does not occur because of curvature effects. Adsorption/desorption limits as well as charge reversal mechanism were investigated revealing that an increase in the chain stiffness promote chain desorption. Park et al. [27] showed that spontaneous overcharging in polyelectrolyte-colloid complexes was accounted for in the Poisson-Boltzmann approximation and increasing the rigidity of the polyelectrolyte was leading to undercharging of the complex. Mateescu et al. [19] demonstrated the existence of a "wrapping transition" between a slightly bent conformation of the polyelectrolyte close to the sphere and a conformation where the polyelectrolyte wraps or collapses around the sphere. This "wrapping transition" is expected to be dependent on the salt concentration and sphere diameter.

Investigations using MC simulations have corroborated some experimental and theoretical conclusions. Chain flexibility, linear charge density, and micelle radius were considered by Wallin and Linse [28–30] to investigate the behavior of a polyelectrolyte–micelle complex. They concentrated on free-energy calculations to determine the key parameters influencing critical aggregation concentration. Critical aggregation concentration was found to increase with decreasing chain stiffness, linear charge density, and micelle radius. The complexation between a linear flexible polyelectrolyte and several oppositely charged macroions was recently examined by Jonsson et al. [31] with focus on electrostatic interactions. The composition and the structure of the complex as well as conformational data of the polyelectrolyte were obtained. The oligomerization of lysozyme in aqueous solution was also investigated by MC simulations [32] as a function of protein concentration, pH, and electrolyte screening. It was observed that increasing the protein concentration, or decreasing the electrostatic repulsion between protein molecules by either reducing the protein charge or increasing the ionic strength, promoted cluster formation. Structure factors and equilibrium constants obtained were compared to those obtained experimentally and were found to capture the experimentally obtained effects of pH and ionic strength.

Owing to the important potential of computer simulations to provide qualitative and quantitative means of understanding the factors that could influence DNA-like polyelectrolyte and colloidal polymer–particle interactions, we describe here an MC approach to get insight into the behavior of a flexible, semiflexible, and rigid polyelectrolyte with the presence of an oppositively charged colloid [33,34]. As the ionic concentration is expected, via screening effects, to play a key role in controlling both chain conformation (via the electrostatic persistence length) and polyelectrolyte–particle interaction energy, we also focused on it. A simple model with a uniformly charged hard sphere to mimic a colloidal particle and a pearl necklace chain consisting of connected point charges is used. Since a Debye-Hückel approach is considered, which is an established level of description, ions and counterions are implicitly considered. The adsorption–desorption limit, which is a key parameter for technical applications of polyelectrolyte–particle mixtures, is investigated. The polyelectrolyte conformations are analyzed prior to and after adsorption, the polymer interfacial structure is investigated along with the particle surface coverage and polyelectrolyte adsorbed amount. Snapshots of equilibrated conformations are also provided to achieve qualitative views of the polymer–particle complexes. The adsorption of a polyelectrolyte on a colloid is presented with special attention on the effect of (1) the polyelectrolyte length, (2) colloid size, (3) polymer intrinsic flexibility and salt concentration on the adsorption–desorption limit, interfacial structure, and charge inversion phenomena.

II. MODEL DESCRIPTION

A pearl necklace model is used to generate off-lattice three-dimensional polyelectrolyte chains. They are represented as a succession of N freely jointed hard spheres, and each sphere is considered to be a physical monomer of radius $\sigma_m = 3.57$ Å with a negative charge equal to −1 on its center. The fraction of ionized monomers f is adjustable (by placing negative charges equal to −1 at the center of the monomers), and the bond length is constant and equal to the Bjerrum [35] length $l_B = 7.14$ Å.

The particle or polymer/colloid is represented as an impenetrable, uniformly charged sphere with a variable radius σ_p so as to get full insight into particle curvature effects. The particle surface charge is assumed to be concentrated into a point charge located at its center and adjusted so as to keep a constant surface charge density. The solvent is treated as a dielectric medium with a relative dielectric permittivity constant ε_r taken as that of water at 298 K, i.e., 78.5. The total energy E_{tot} (k_BT units) for a given conformation is the sum of repulsive electrostatic interactions between monomers, attractive electrostatic interactions between the chain and the particle E_{el}, and the chain stiffness or bending energy E_{tor}. Hard-core interactions E_{ev} are also considered to include both monomer and particle excluded volumes.

$$E_{tot} = E_{ev} + E_{el} + E_{tor} \tag{1}$$

Hard-core repulsions between i and j physical units lead to excluded volume interactions using

$$E_{ev} = \sum_{i<j} u_{ev}(r_{ij}) \tag{2}$$

with

$$u_{ev}(r_{ij}) = 0 \qquad \text{when } r_{ij} > \sigma_i + \sigma_j \tag{3}$$

$$u_{ev}(r_{ij}) = \infty \qquad \text{when } r_{ij} \leq \sigma_i + \sigma_j \tag{4}$$

where σ_i represents the radius of unit i (which can be the monomer or the particle) and r_{ij} the distance between the centers of the two units.

All pairs of charged monomers within the polyelectrolyte interact with each other via a screened Debye-Hückel long range potential:

$$u_{el}(r_{ij}) = \frac{z_i z_j e^2}{4\pi\varepsilon_r\varepsilon_0 r_{ij}} \exp(-\kappa r_{ij}) \tag{5}$$

where z_i represents the amount of charge on unit i.

Monomers interact with the particle according to a Verwey-Overbeek potential:

$$u'_{el}(r_{ij}) = \frac{z_i z_j e^2}{4\pi\varepsilon_r\varepsilon_0 r_{ij}} \cdot \frac{\exp[-\kappa(r_{ij} - \sigma_p)]}{1 + \kappa\sigma_p} \tag{6}$$

Free ions are not included explicitly in the simulations but their overall effects on monomer–monomer and monomer–particle interactions are described via the dependence of the inverse Debye screening length κ^2 [m^{-2}] on the electrolyte concentration according to

$$\kappa^2 = 1000e^2 N_A \sum_i \frac{z_i^2 C_i}{\varepsilon_0 \varepsilon_r k_B T} \tag{7}$$

It is worth noting here that all pairwise interactions are calculated without taking into account cutoff distances, and the entropy of counterion release is only captured on a linear level. The intrinsic chain stiffness is adjusted by a square potential with variable amplitude to vary its strength. This gives the bending energy:

$$E_{tor} = \sum_{i=2}^{N} k_{ang}(\alpha_i - \alpha_0)^2 \tag{8}$$

where $\alpha_0 = 180°$ and α_i represents the angle achieved by three consecutive monomers $i - 1$, i, and $i + 1$. k_{ang} [$k_B T/\text{deg}^2$] defines the strength of the angular potential or chain stiffness.

III. MONTE CARLO PROCEDURE

Monte Carlo simulations are performed according to the Metropolis algorithm in the canonical ensemble. In this method, successive "trial" chain configurations are generated to obtain a reasonable sampling of low-energy conformations [36]. After applying elementary movements that are randomly selected, the Metropolis selection criterion is employed to either select or reject the move. If the change in energy ΔE resulting from the move is negative, the move is selected. If ΔE is positive, the Boltzmann factor p,

$$p = \exp\left[\frac{-\Delta E}{kT}\right] \tag{9}$$

is computed and a random number z with $0 \leq z \leq 1$ is generated. If $z \leq p$, the move is selected. When $z > p$, the trial configuration is rejected and the previous configuration is retained and considered as a "new" state in calculating ensemble averages. That conformation is the one that is perturbed in the next step. The perturbation process is continued a specified number of times (a typical run requires several million perturbations) until the conformation is energy minimized and equilibrated.

To generate new conformations, the monomer positions are randomly modified by specific movements. These movements include three "internal" or elementary movements (end-bond, kink-jump, and crankshaft, respectively), the pivot, and the reptation, respectively (Fig. 1). The use of all these movements is very important in ensuring the ergodicity of the system as well as the convergence toward minimized conformations. One important challenge and problem consist of allowing the energy of the complex structure to be minimized gradually without trapping the structure in a local energy minimum. This problem is of particular importance when compact conformations have to be achieved or when large polyelectrolyte chains are considered owing to the fact that a few monomer–monomer contacts can lead to the formation of "irreversible" bonds that freeze the complex structure. Some MC refinements are thus necessary to overcome the formation of structures in local minima and increase the chances of success when sampling new conformations. Anneal MC can be used to gradually minimize the complex structures by altering the temperature from an initial temperature to a final one and vice versa, so as to increase the chances of success when sampling conformations. However, as this method requires a large amount of CPU time, it should be used for the formation of dense structures only. Another approach that we are using consists, at each step of the MC procedure, of using the most effective movements (internal, pivot, or reptation) to achieve a more rapid and total exploration of the configurational space of the chains [37]. To check the efficiency and validity of our model and algorithms, the formation of dense globules was first experienced by considering that the computational efficiency of MC simulations is known to decrease dramatically

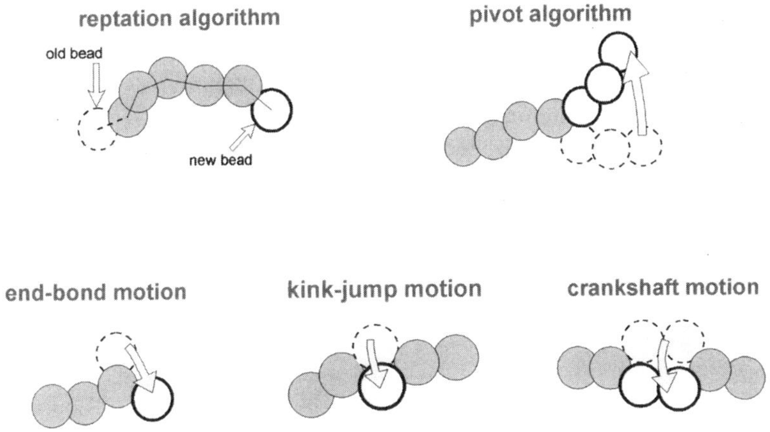

FIG. 1 To generate new conformations, the monomer positions of the polyelectrolyte chains are randomly modified by specific movements. These movements include three "internal" or elementary movements (end-bond, kink-jump, and crankshaft, respectively), the pivot, and the reptation.

with system compactness. To generate a collapsed chain reptation is the most efficient algorithm to rapidly increase, in a first step, the degree of chain compactness. Then to achieve dense and collapsed spherical conformations, as reptation acceptance rate is poor, the combination of reptation and internal movements gives good results. On the other hand, to obtain extended structures, pivot movements are the more efficient and can be used alone. To achieve a necklace structure, i.e., a global stretched conformation composed of compact beads, the chain must first pass through a stretched structure. This is rapidly achieved, for example, with the pivot algorithm. A combination of reptation and internal movements is then necessary to form beads along the chain backbone. It should be noted that to sample the self-avoiding walk chain's conformational space, the three algorithms have a good acceptance rate. However, the pivot appears to be the most efficient procedure to sample new conformations, whereas if internal movements only are employed, a large number of conformations are required to eventually achieve a significant root mean square difference with the starting reference conformation.

To investigate the formation of polyelectrolyte–particle complexes, the central monomer of the chain is initially placed at the center of a large three-dimensional spherical box having a radius equal to $2N\sigma_m$ and the particle is randomly placed in the cell. The polyelectrolyte and the oppositely charged particle are then allowed to move (a random motion is used to move the particle). After each calculation step, the coordinates of both the particle and monomers are translated in order to replace the central monomer of the polyelectrolyte in the middle of the box. It should be noted that the chain has the possibility of diffusing further away and leaving the particule surface during a simulation run (so the polyelectrolyte desorption process can be investigated). After relaxing the initial conformation through 10^6 cycles (equilibration period), chain properties are calculated and recorded every 1000 cycles. Owing to the large number of possible situations to investigate with regard to the modifications of k_{ang}, N, and C_i, the application of this model with regard to the actual processor speeds has currently limited the chain length to 100 monomer units (200 for the isolated chain).

To quantitatively characterize polyelectrolyte configurations in solution and at the surface such as the mean square radius of gyration $<R_g^2>$, the mean square end-to-end distances $<R_{ee}^2>$ are calculated after the equilibration period [38]. Then to determine the position of the chain monomers along the coordinate normal to the surface, spherical layers around the surface are defined. The thickness of each layer, excepting the first one, is set to one monomer radius. Because of strong excluded volume effects between the monomers and the surface, the thickness of the first layer is arbitrarily increased to two monomer radii. To characterize the conformation of the adsorbed chain, the monomer fraction in tails, loops, and trains is considered. Two parameters, the particle surface coverage θ and the adsorbed amount of polyelectrolyte Γ, are also calculated to characterize the particle surface. Surface coverage is defined as the fraction of the particle surface covered with the monomers that are present in the first layer:

$$\theta = \frac{a_0 N^*}{a_{surf}} \tag{10}$$

where a_0 is the projected area of one monomer, N^* the number of adsorbed monomers lying in the first layer, a_{surf} the surface of the particle, and the adsorbed amount Γ defined as

$$\Gamma = \theta/\theta_{\max} \tag{11}$$

where $\theta_{max} = Na_0/a_{surf}$ is the maximal fraction of particle surface area that can be covered by a fully adsorbed polyelectrolyte chain.

IV. CONFIGURATIONAL PROPERTIES OF ISOLATED POLYELECTROLYTE CHAINS

A. Flexible Chains

Chain structural changes prior to adsorption are usually discussed by first considering isolated polyelectrolyte chains. Electrolyte concentration ranges here from $C_i = 0$ M ($\kappa^{-1} = \infty$) to $C_i = 1$ M ($\kappa^{-1} = 3$ Å), chain lengths adjusted to $N =$ 25, 50, 100, 140, 160, and 200 monomers (negatively charged, $f = 1$), and the particle surface charge density set to + 100 mC.m^{-2} (central point charge $Q =$ +100) which is representative of typical values observed for natural and synthetic charged colloids. In order to achieve a qualitative picture of chain conformations, equilibrated structures have been extracted from MC calculation runs and presented as three-dimensional projections in Table 1 vs. C_i and N. Polyelectrolyte conformations are linked to the intensity of the electrostatic repulsions between the monomers. By decreasing the ionic concentration, rodlike structures are favored in order to minimize the energy of the chain. It should be noted that 'rodlike' does not mean a straight pole but rather an object that is highly oriented. When Coulombic screening increases, polyelectrolytes become less stretched; if the Debye screening length κ^{-1} is smaller than the distance between two charges, the polymer recovers the self-avoiding walk limit. Scaling concepts can be applied to polyelectrolytes and the scaling exponent theoretically varies from $\nu = 3/5$ (SAW limit) to $\nu = 1$ (rodlike limit) when the following relationship is considered:

$$<R_g^2> \sim N^{2\nu} \tag{12}$$

TABLE 1 Monte Carlo Equilibrated Conformations of Isolated Polyelectrolyte Chains as a function of the Chain Monomer Number N and Ionic Concentration C_i[a]

C_i [M] / N	0	0.001	0.01	0.1	1
25					
100		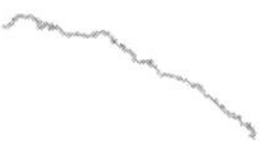			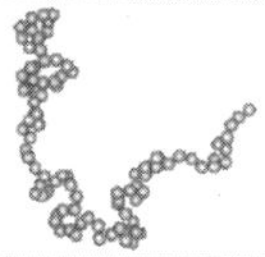
200				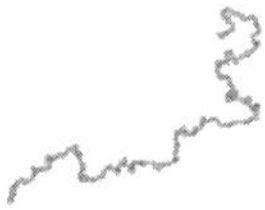	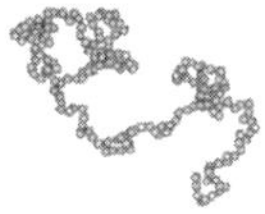

[a]By decreasing ionic strength, long-range electrostatic interactions promote the formation of elongated conformations whereas swollen conformations are achieved by increasing C_i.

This relationship is usually used to check model hypothesis as well as the validity of algorithms. The ratio $r = \langle R_{ee}^2 \rangle / \langle R_g^2 \rangle$, which provides a more quantitative description of chain extension, has also been investigated. A random walk is identified by a value of $r = 6$ and a fully stretched polymer is characterized by $r = 12$. When a SAW chain is considered r is close to 6.3. The effects of κ^{-1} and N on r are presented in Fig. 2; r increases with κ^{-1} and reaches a plateau value when $\kappa^{-1} > R_{ee}$. Whatever the ionic concentration, polyelectrolytes never reach the limit $r = 12$. This is due to the fact that thermal energy always causes movements and, in turn, local chain deformations. When κ^{-1} is less than the distance of separation between charges along the chain backbone, r converges to 6.3. Short chains have the largest deviation from the ideal value of r, which is expected to be more representative of chains with infinite sizes. Subsequently, short chains are less stretched in the salt-free case and more swollen than the long chains in the SAW region. Thus, due to finite size effects, the curves cross each other for different values of N. High degrees of polymerization cause an increase of the repulsions between the monomers, and as a result more extended conformations are obtained.

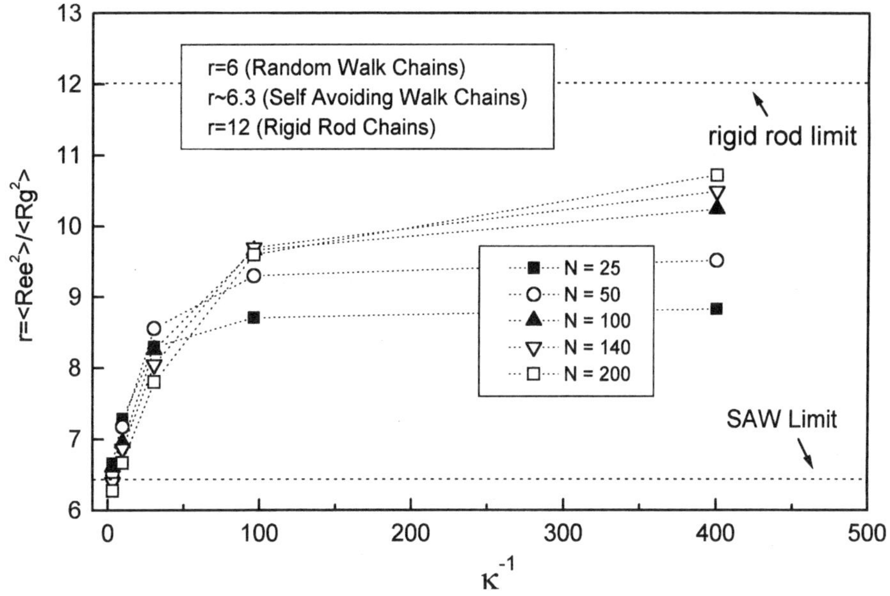

FIG. 2 $r = \langle R_{ee}^2 \rangle / \langle R_g^2 \rangle$ as a function of κ^{-1} at various chain lengths. Fully stretched conformations are never reached. Data on the *y* axis at 400 are for the salt-free case. On the one hand, in the high salt concentration regime, the SAW limit ($r = 6.3$) is achieved. On the other hand, the totally stretched chain (rigid rod) $r = 12$ is not obtained due to the thermal energy of the chain, which causes movements and local chain deformation. It is worth noting that increasing the degree of polymerization leads to greater chain stretching because of the increase of the mean electrostatic energy per monomer.

B. Semiflexible Chains

Equilibrated conformations of isolated semiflexible polyelectrolyte chains are presented vs. C_i and k_{ang} for comparison with flexible conformations. Electrolyte concentration ranges from $C_i = 0$ ($\kappa^{-1} = \infty$) to $C_i = 1$ M ($\kappa^{-1} = 3$ Å), and k_{ang} values are adjusted from 0 to 0.02 k_BT/deg^2, the chain length being equal to 200 monomers ($f = 1$). Table 2 demonstrates that the semiflexible polyelectrolytes exist as extended and SAW configurations controlled by both the intensity of the electrostatic repulsions between the monomers and intrinsic stiffness k_{ang}. When $k_{ang} = 0$ k_BT/deg^2, i.e., when fully flexible chains are considered, a decrease in the ionic concentration causes a gradual spreading of the polyelectrolyte dimensions, so that rodlike structures are achieved to minimize the electrostatic energy of the chain. When $C_i = 1$ M, chain stiffness has full effect.

TABLE 2 Monte Carlo Equilibrated Conformations of Isolated Polyelectrolytes ($N = 200$) as a function of the Intrinsic Polyelectrolyte Stiffness k_{ang} and Ionic Concentration C_i[a]

k_{ang} [k_BT/deg^2] / C_i [M]	0	0.0005	0.001	0.02
0	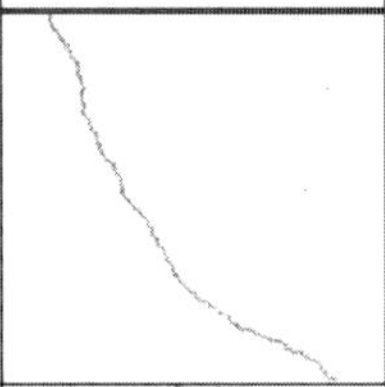	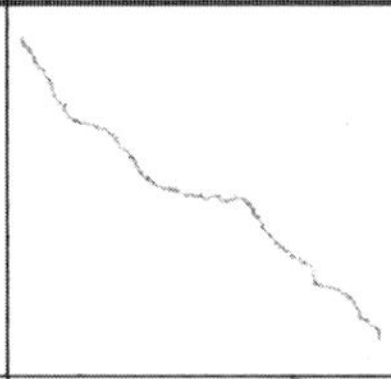		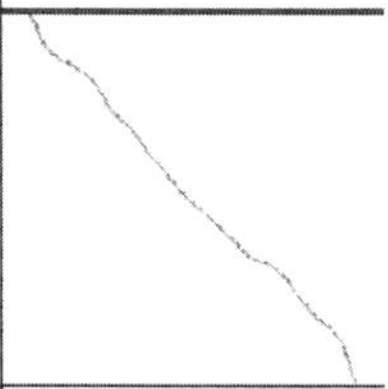
10^{-2}	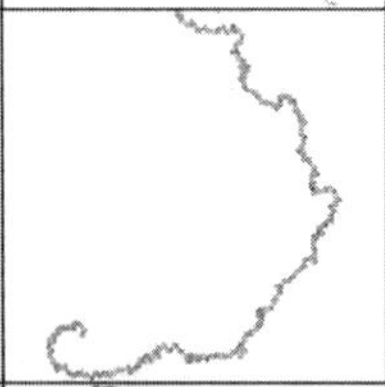	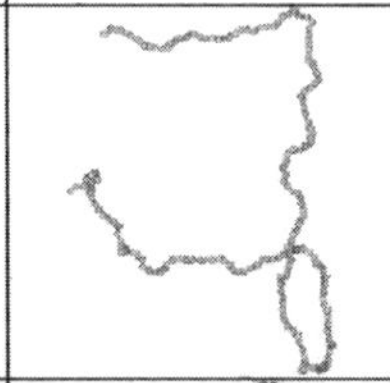	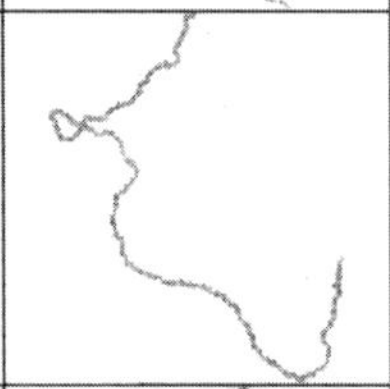	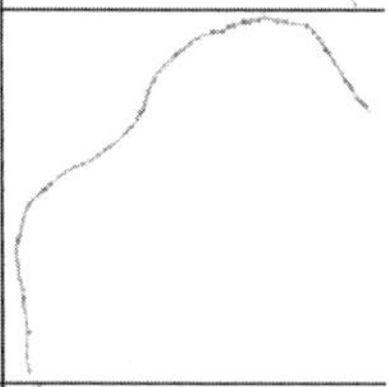
1	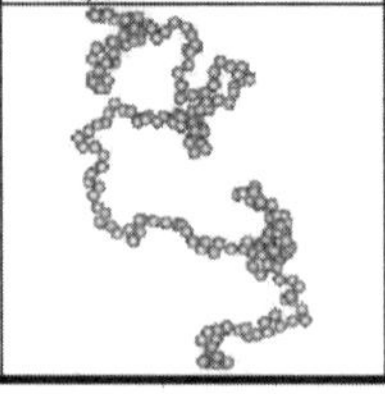	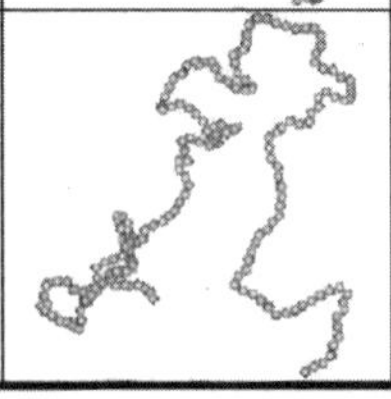	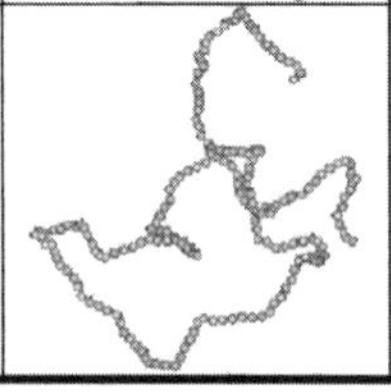	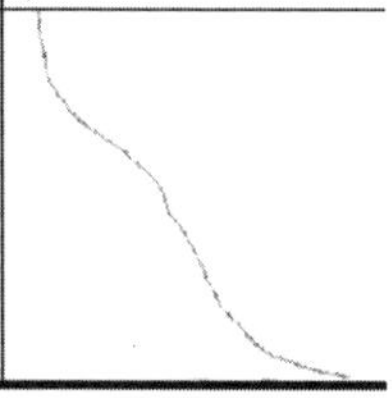

[a]When electrostatic interactions between monomers are screened ($C_i = 1$ M), the increase in chain stiffness induces the formation of rigid domains that are connected to each other by more flexible regions.

However, a different picture of chain stretching has to be considered. The increase of intrinsic stiffness locally destroys a large amount of chain entropy and results in the formation of rigid domains connected to each other by flexible bonds (see, for example, Table 2, $C_i = 1$ M and $k_{ang} = 0.001$ k_BT/deg^2). As shown in Fig. 3 where $\langle R_{ee}^2 \rangle$ variations are presented as a function of k_{ang} for different C_i values, electrostatic interactions and intrinsic stiffness influence each other but at different scales. Although long-range electrostatic repulsions have full effects when $C_i = 0$ M, it is still possible to increase the chain dimension through local geometrical constraints. It is worth noting that (1) both electrostatic and intrinsic rigidity are required to achieve straight poles and (2) chain dimensions

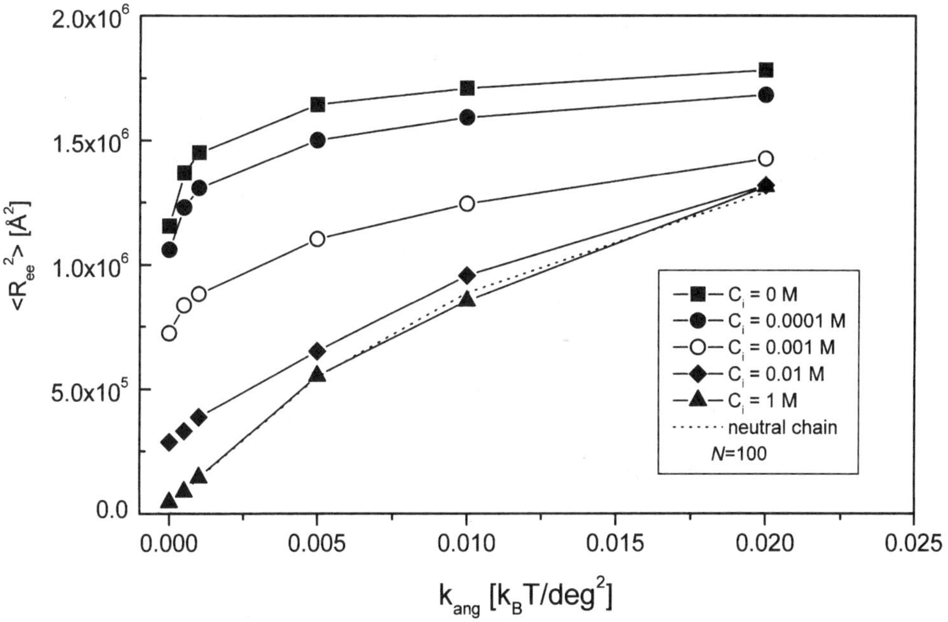

FIG. 3 Mean square end-to-end distance $<R_{ee}^2>$ as a function of the chain intrinsic rigidity k_{ang} at different ionic concentrations C_i. The $<R_{ee}^2>$ variation for a neutral chain is represented by the dotted line. Rigid rods are achieved by both increasing the electrostatic repulsions between the monomers (large-scale effects) and intrinsic stiffness (local effects).

of screened rigid polyelectrolytes ($C_i = 1$ M and $k_{ang} = 0.02\ k_BT/\text{deg}^2$) and unscreened flexible polyelectrolyte ($C_i = 0$ M and $k_{ang} = 0\ k_BT/\text{deg}^2$) are similar.

V. POLYELECTROLYTE–PARTICLE COMPLEXES

A. Influence of the Polymer Length

1. Equilibrated Conformations

To investigate adsorption processes in the polyelectrolyte–particle system, a spherical charged particle ($Q = +100$, $\sigma_p = 35.7$ Å) is added to the equilibrated polyelectrolyte chain ($N = 100$, $\sigma_m = 3.57$ Å, $f = 1$, $l_B = 7.14$ Å) so that they are close with each other. Then the polyelectrolyte–particle complex is enclosed in a spherical cell and allowed to relax. Equilibrated conformations of the complex as a function of C_i and N are presented in Table 3. It should be noted that the

TABLE 3 Equilibrated Conformations of the Polyelectrolyte–Particle Complex as a function of N and C_i[a]

C_i [M] / N	0	0.01	0.3	1
25	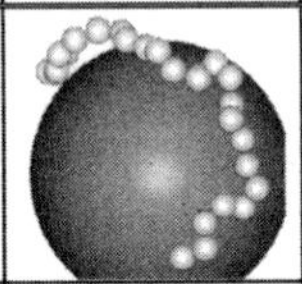	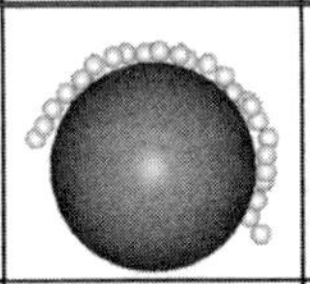	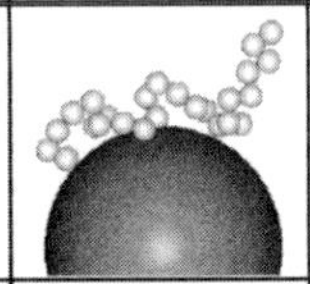	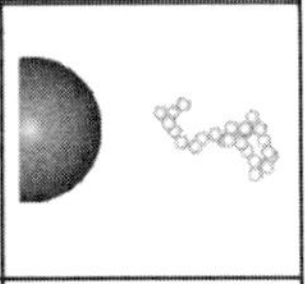
50	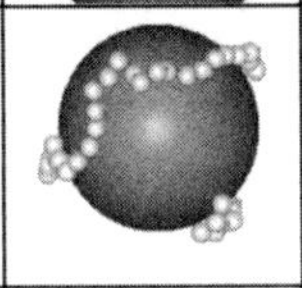	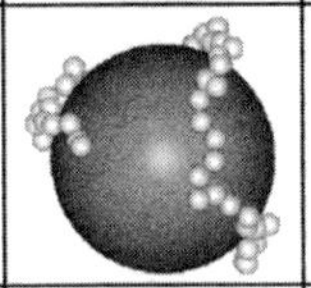	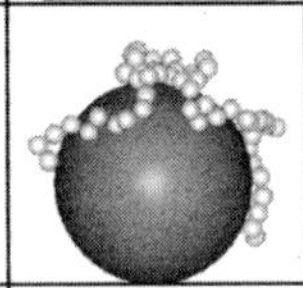	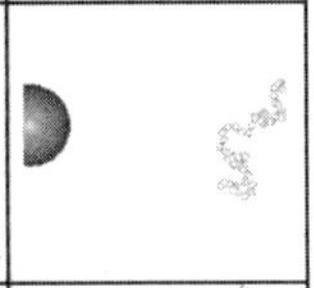
140	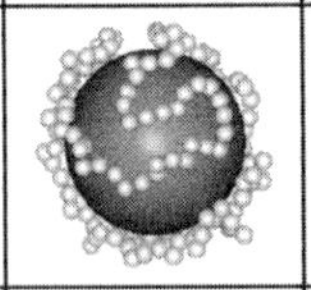	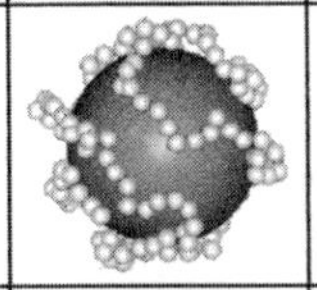	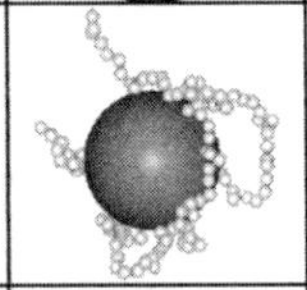	
160		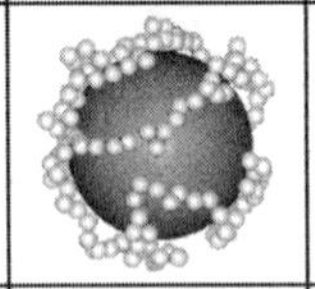	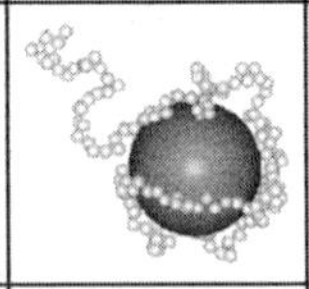	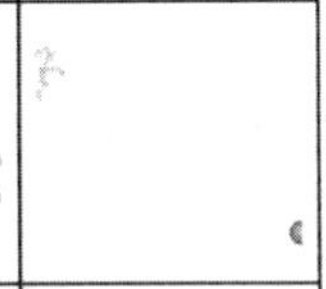
200	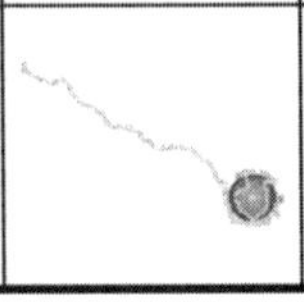	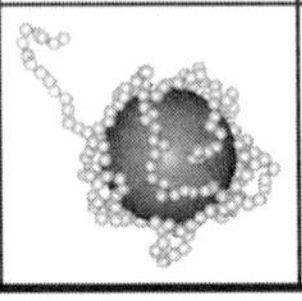	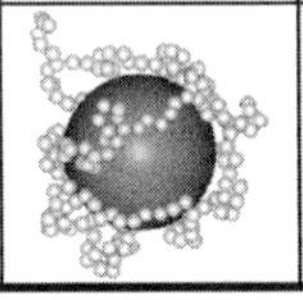	

[a]Because of the lack of available space different scales have been used to represent the polyelectrolyte–particle complexes.

extended polymer chain conformations appear smaller than their actual size as they have been reduced for reproduction (N constant = 100).

2. Adsorption-Desorption Limit

It can be clearly seen from Table 3 that no adsorption is observed when $C_i \geq 1$ M. Attractive surface–polymer interactions in this domain are not strong enough to overcome the entropy loss of the polymer due to its confinement near the particle. In order to determine the adsorption–desorption limit (a chain is considered as adsorbed when it is in contact with the particle during for more than 50% of the simulation time), the ionic concentration is adjusted between 0.3 M and 1 M for each chain length. The plot of the critical ionic concentration C_i^c (ionic strength at the adsorption–desorption limit) as a function of N is presented in Fig. 4. To overcome the entropy loss per monomer due to adsorption, it is demonstrated that stronger electrostatic attractions, with decreasing ionic concentration, are needed to adsorb short polyelectrolyte chains. C_i^c increases with N from 0.34 M when $N = 25$ to 0.4 M when $N = 200$. When chains longer than 100 monomers are considered, it is important to note that C_i^c increases slowly to reach a plateau value close to 0.4 M. The critical electrostatic energy E_s^c associated with this limit for a single monomer in contact with the particle surface ranges from $-1.18\ k_BT$ when $N = 25$ to $-1.03\ k_BT$ when $N = 200$. Hence, adsorption is achieved when the attractive energy is greater than the thermal energy, i.e., $1\ k_BT$. These results are in accordance with the picture of polymer adsorption on flat surfaces. Indeed, in dilute solutions of polydispersed polymers, long chains are found preferentially on the surface because less translational entropy (per unit of mass) is lost compared to the short ones, while they gain approximately the same (total) adsorption energy even if the adsorption

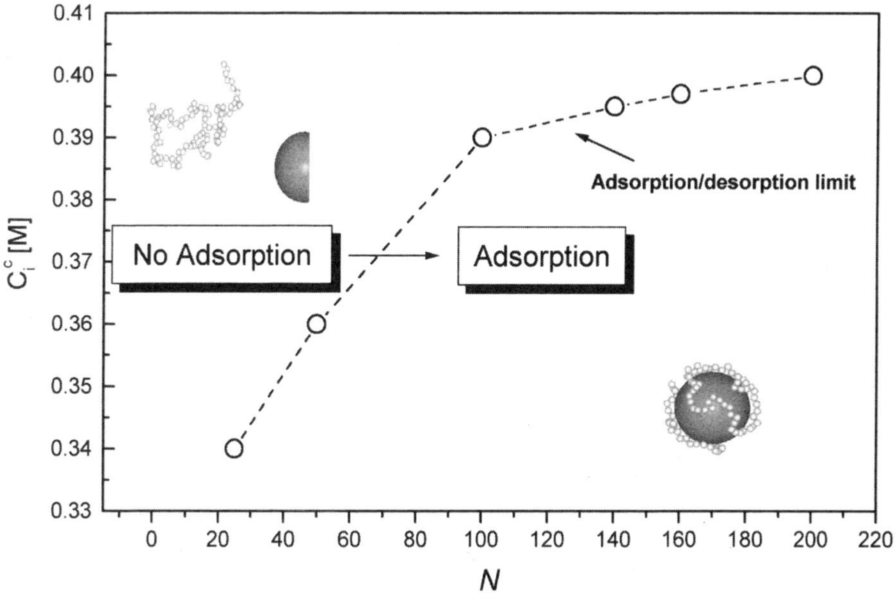

FIG. 4 The adsorption–desorption limit is expressed by the variations of the critical ionic concentration C_i^c as a function of N.

energy per segment is the same. When $N = 25$ adsorption approaches that on a nearly planar surface. In this case, the chain can fully spread on the surface with dimensions close to its dimensions in a free solution. When chain length is increased up to $N = 140$, polyelectrolytes wrap around the particle to optimize the number of contacts. The conformation of the polymer is then dictated by the particle size and is subject to the highest level of deformation (the maximum is observed when $N = 140$). By increasing further chain length, excluded electrostatic volume prevents any additional monomer adsorption on the surface via the formation of an extented tail in solution. Due to the formation of that protruding tail in solution, the ratio $<R_g^2>_{ads}/<R_g^2>_{free}$ increases to 1. Complexes with two tails were not observed during our simulations according to the size ratio between the particle diameter and polyelectrolyte length.

3. Trains, Loops, and Tails

The number of monomers in trains, loops, and tails as a function of N and C_i is presented in Fig. 5a and 5b, respectively, to give insight into the structure of the interfacial region. When the chain length is increased (Fig. 5a) with C_i greater than 0.01 M, the total number of monomers in trains and loops increases monotically. When C_i is less than 0.01 M and N is greater than 140 monomers, the number of monomers in trains and loops does not change because any additional monomer is expelled in tails. Beyond that critical chain length N^c, intrachain repulsion outweigh the attraction between the monomers and the particle to form tails. It is worth noting that the length of the tails increases linearly with N above N^c. When short chains are considered, monomers are mainly in trains, whereas a few are present in loops. Short chains thus have a tendency to flatten more because this lowers the energy most.

By increasing C_i up to 0.03 M (Fig. 5b), the electrostatic excluded volume is decreased, allowing the particle to attract more monomers. Then, by further increasing C_i to the critical adsorption–desorption limit C_i^c, the number of monomers in trains decreases while the number of monomers in loops and tails increases to reach a maximum. When $C_i > C_i^c$ polymer desorption is observed so that the number of monomers in trains, loops, and tails decreases rapidly.

4. Amount of Adsorbed Monomers: Surface Coverage

Further insight into the adsorption properties of polyelectrolytes is gained by examining the amount of adsorbed polymer Γ and the particle surface coverage θ. The influence of C_i and N on Γ (Fig. 6) clearly demonstrates that the polyelectrolyte's capacity to be adsorbed on the particle surface decreases with the increase of C_i and N. When $N \geq 140$ Γ reaches a maximum value in the range 0.1 M $> C_i >$ 0.01 M, the exact value being dependent on the polyelectrolyte contour length. By further increasing C_i, desorption takes effect and Γ decreases. Variations of surface coverage θ (i.e., the total number of adsorbed monomers in the

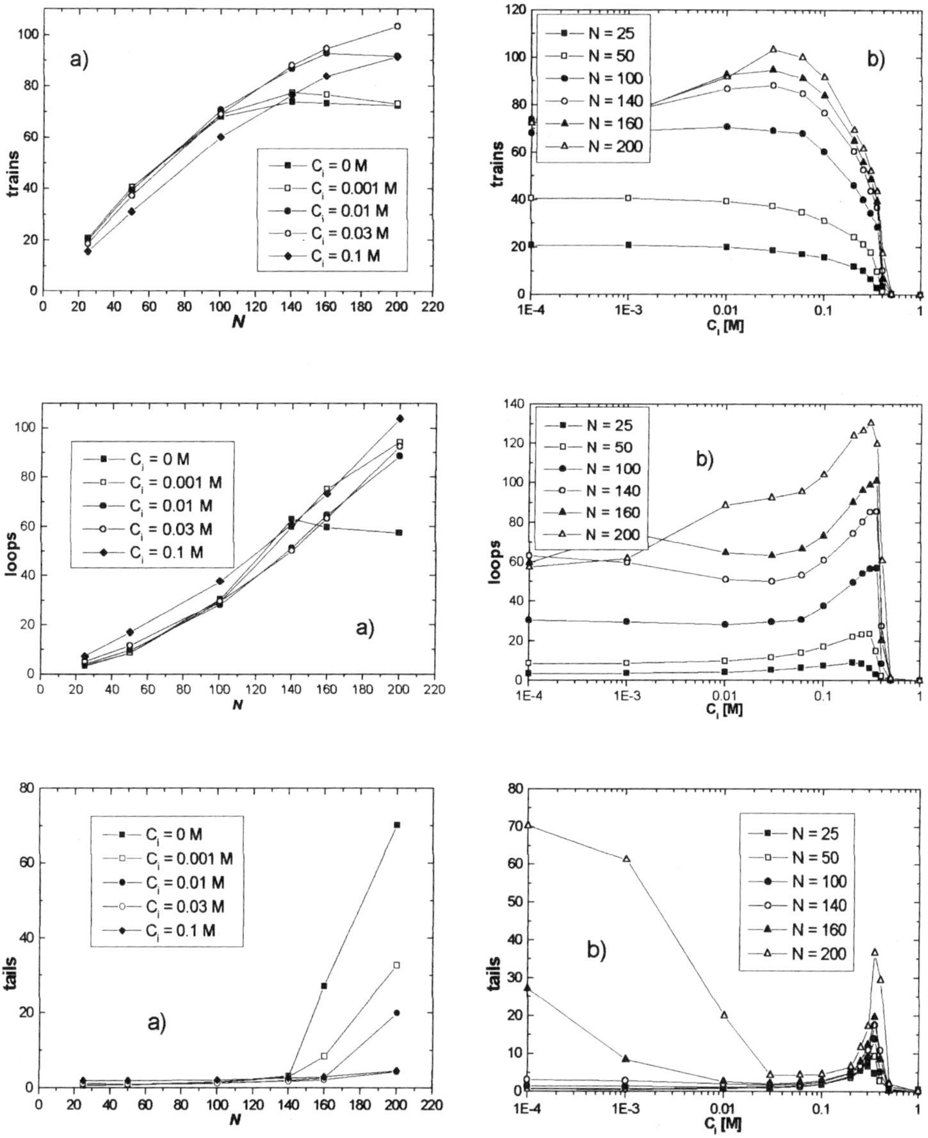

FIG. 5 Quantitative description of the interfacial region. Number of monomers in trains, loops, and tails as a function of N and C_i.

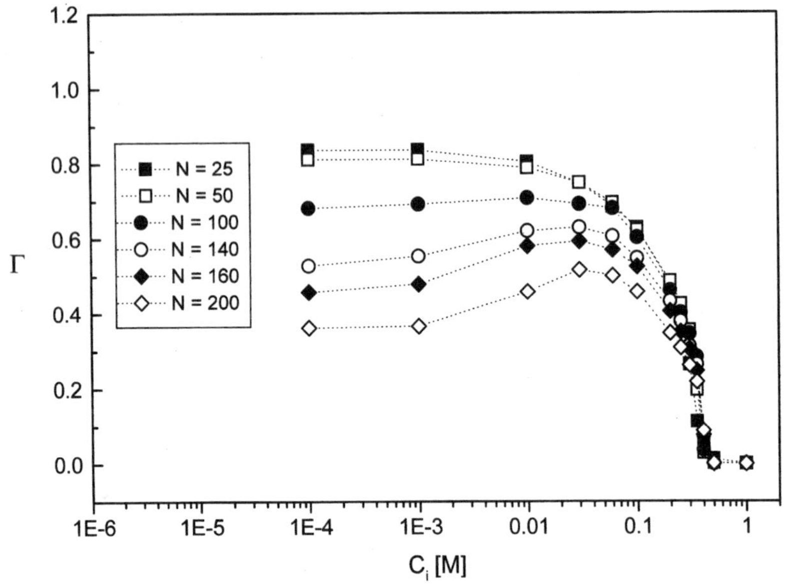

FIG. 6 Adsorbed amount Γ of polymer as a function of N and C_i. Data with the coordinates $\log(C_i) = -4$ correspond to the salt-free case.

first layer) as a function of ionic strength and for different degrees of chain polymerization are presented in Fig. 7. When $N < N^c$, θ is monotically decreasing with the ionic concentration while a maximal value is achieved when C_i is close to 0.03 M and $N \geq N^c$. These results clearly demonstrate that particle surface coverage and the amount of adsorbed polymer are not simple monotonic functions of N and C_i when the polyelectrolyte is large enough to form a tail in solution.

5. Overcharging

Theoretically, collapsed monomers are those that participate in the overcharging process and are either in contact with the surface of the particle or belong to one of the tightly packed multiple layers. With simulations, this limit is subjective and often fixed arbitrarily owing to the difficulties in deciding where the adsorbed layer ends. In this section, when overcharging is discussed, the monomers are either adsorbed (N^{ads}) or in tails (N^{tail}), so that the number of adsorbed monomers is defined as $N^{ads} = N - N^{tail}$.

In Fig. 8, we plot as a function of N the variation of the number of collapsed monomers N^{ads} in the salt-free case. Our MC simulations and the theoretical prediction of the Nguyen-Shklovskii (NS) model [19,20] are presented, and a

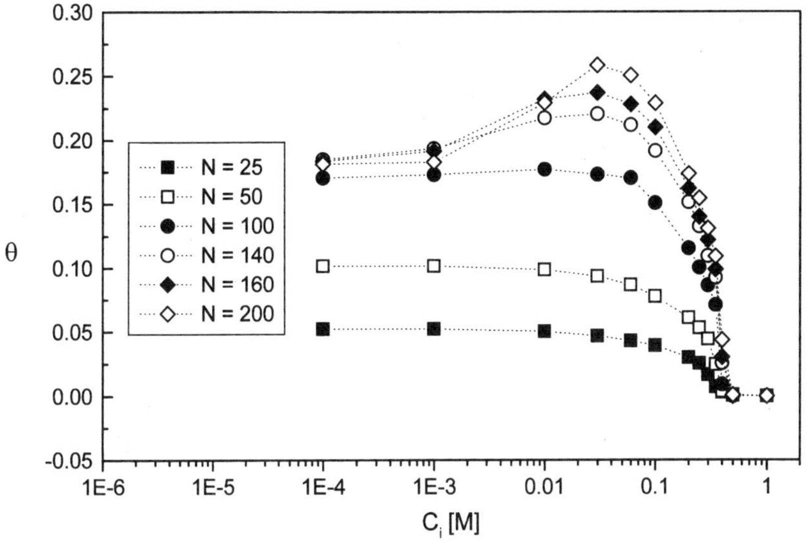

FIG. 7 Particle surface coverage θ as a function of N and C_i. Data with the coordinates $\log(C_i) = -4$ correspond to the salt-free case.

qualitative and quantitative good agreement is found between the two models. A numerical solution giving a first-order transition at $N^c = 151$ with the formation of a protuding tail composed of 19 monomers is obtained in perfect agreement with our observations.

With increasing the size of the polyelectrolyte several key results are demonstrated: (1) the chain is fully collapsed on the charged particle as long as $N \leq Q$ and $N \approx N^{ads}$ and the complex is undercharged; (2) when $N > Q$ more monomer adsorbs on the particle surface than is necessary to neutralize it, and the complex is overcharged. Accumulation of monomers close to the surface continues up to $N^c = 151$; (3) beyond that critical number of monomers, a protuding tail in solution appears. The NS model predicts a first-order transition at this point followed by a small decrease in N^{ads}, with N rapidly reaching a plateau value. Our MC simulations follow this behavior perfectly.

We now consider the overcharging issue in the presence of added salt. All calculations were performed with $C_i \leq 0.03$ M, i.e., in the adsorption domain, before the desorption process takes place. The number of collapsed monomers as a function of N and C_i is reported in Fig. 9. By increasing the screening of the electrostatic interactions, it is clearly demonstrated that the electrostatic volume of the monomers decreases, thus allowing the adsorption of a greater number of monomers on the particle surface. This result supports the NS model,

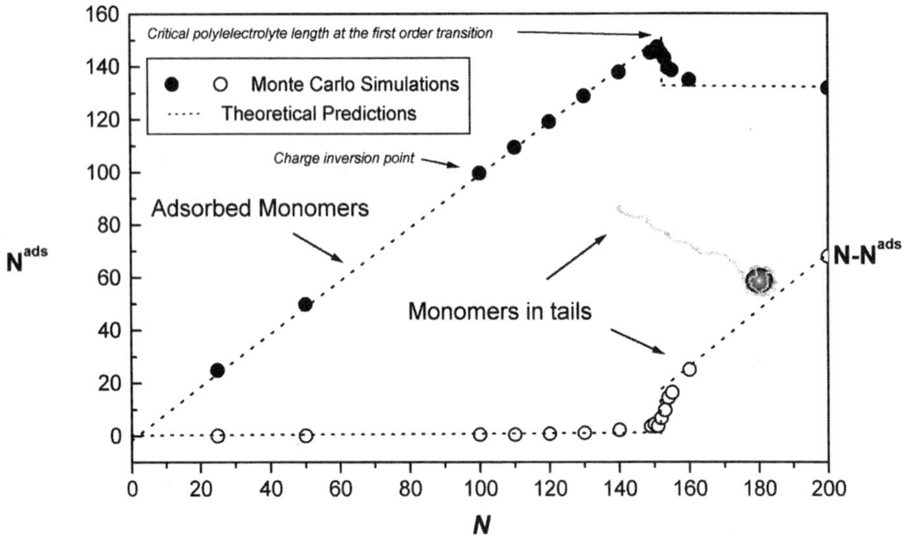

FIG. 8 Number of collapsed monomers N^{ads} and monomers in tails ($N - N^{ads}$) as a function of the total number of monomer in the chain N. Monte Carlo data (circles) are in good agreement with theoretical predictions (dotted lines) of Nguyen and Shklovskii.

which predicts an increase of charge inversion as κ^{-1} decreases. In addition to the increase of charge inversion with C_i, Fig. 9 demonstrates that the position of the first-order transition is increased with increasing polyelectrolyte chain length. In particular, when $C_i = 0.01$ and 0.03 M, it is worth noting that a chain composed of 200 monomers is not large enough to induce the formation of a tail.

B. Influence of the Size of the Polymer Colloid

1. Equilibrated Conformations, Adsorption–Desorption Limit

Equilibrated conformations of particle–chain complexes (with $N = 100$, $\sigma_m =$ 3.57 Å, $f = 1$, $l_B = 7.14$ Å) as a function of ionic concentration and size ratio are presented in Table 4. It can be seen that the adsorption–desorption limit is dependent on both particle size and ionic concentration. Adsorption is promoted by increasing particle size because of the concomitant decrease in chain entropy loss during adsorption on flat surface and increase of the particle charge to keep the surface charge density constant (+100 mC.m^{-2}). As the the particle size is

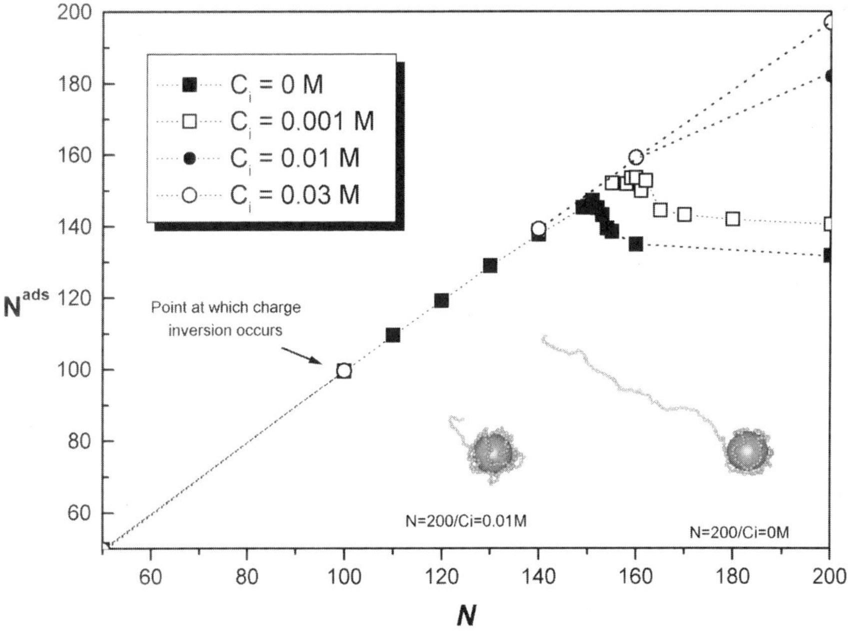

FIG. 9 Number of collapsed monomers N^{ads} as a function of the total number of monomers in the chain N, for different values of the ionic concentration. Due to an increase of the particle capacitance N^{ads} increases with C_i.

increased, the charge of Q is also changed as an effect of the constant surface charge density imposed. No adsorption is observed when $C_i \geq 1$ M (the attractive surface–polymer interactions in this domain are not strong enough to overcome polyelectrolyte confinement near the particle), whereas adsorption is always observed when $C_i \leq 0.01$ M. To calculate the limit between the adsorption and desorption domain and critical particle size σ_p^c for each particle size, the ionic concentration C_i was progressively and carefully decreased and monitored to satisfy the adsorption criteria (in contact at least with one monomer for more than 50% of the simulation time). A plot of the critical particle radius σ_p^c (the critical value below which no adsorption occurs) as a function of κ is presented in Fig. 10. No adsorption or desorption is observed in the region situated at the right of the curve.

TABLE 4 MC Equilibrated Conformations of Polyelectrolyte–Particle Complexes as a function of the Ionic Concentration C_i and Particle/Monomer Size Ratio σ_p/σ_m[a]

C_i [M] / σ_p/σ_m (Q)	0	0.01	0.1	0.3	1
2 (4)	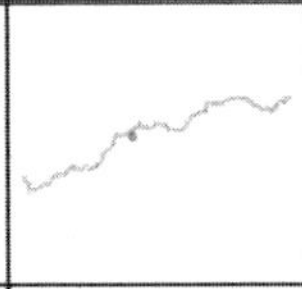	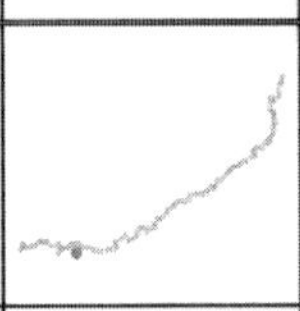	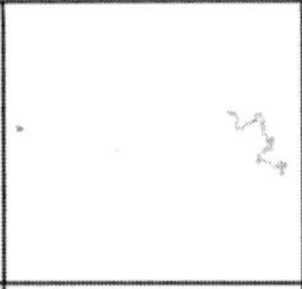		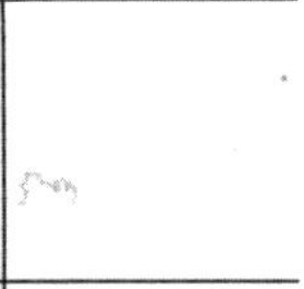
5 (25)	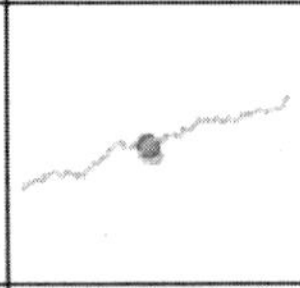	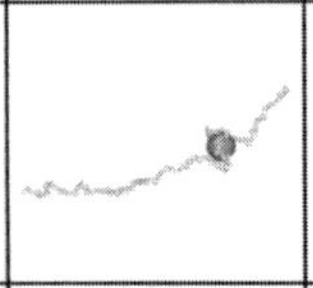	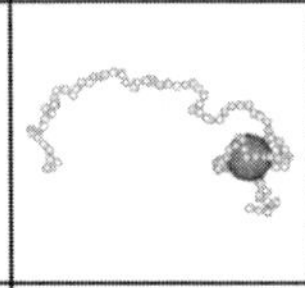		
10 (100)	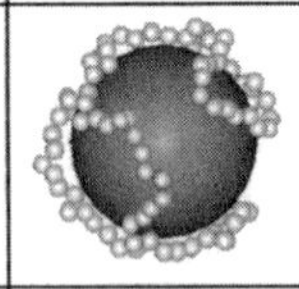	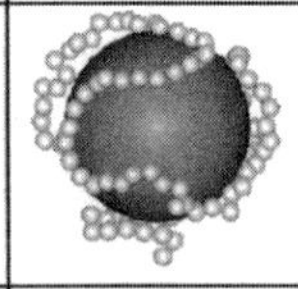	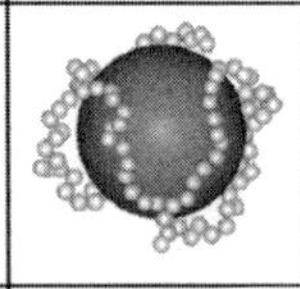	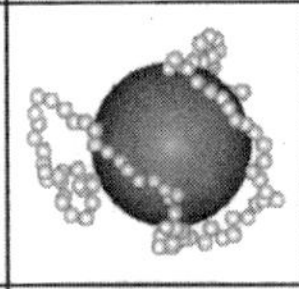	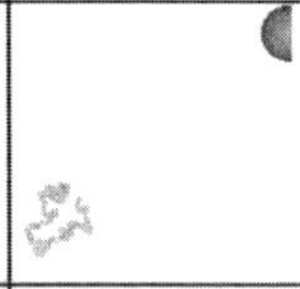
15 (225)	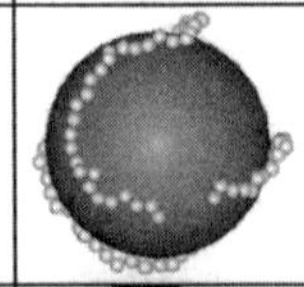	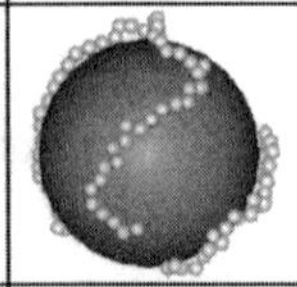	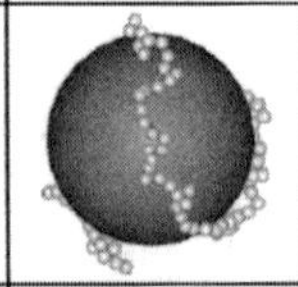	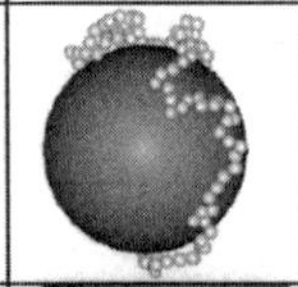	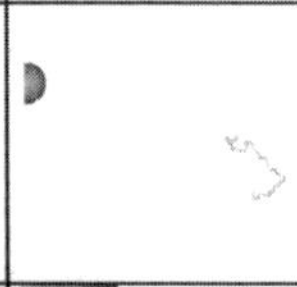
50 (2500)	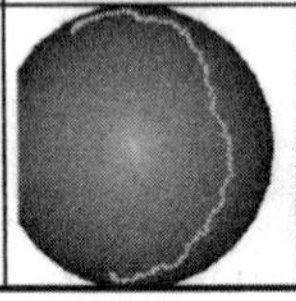	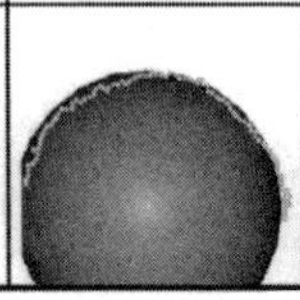	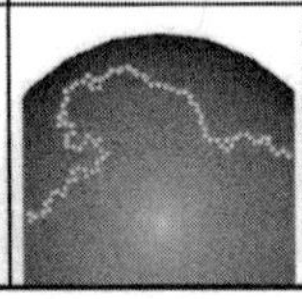	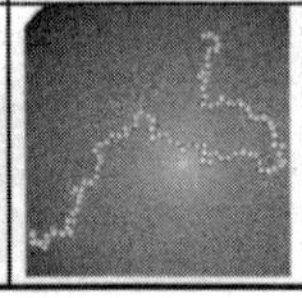	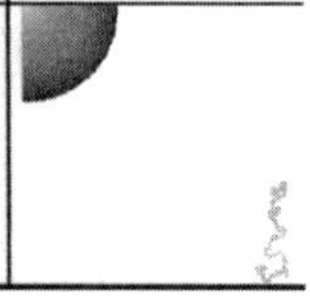

[a]The polymer size is constant and equal to $N = 100$. The central point charge Q of the particle is adjusted so as to keep a constant surface charge density. Depending on the size of the particle, three regions may be defined. Small particles do not permit adsorption of all polyelectrolyte monomers owing to the confinement energy of the chain. As a result, extended tails are formed in solution. The adsorbed polyelectrolyte conformation is quite similar to the isolated polyelectrolyte conformation. When the particle radius is large enough the polyelectrolyte collapses on it to form "tennis ball" conformations. By further increasing particle size, polyelectrolytes can spread to the same extent as on a flat surface. Finally, by increasing the ionic concentration, the charged polymer becomes more or less bonded to the particle surface. When screening is important, no adsorption is observed.

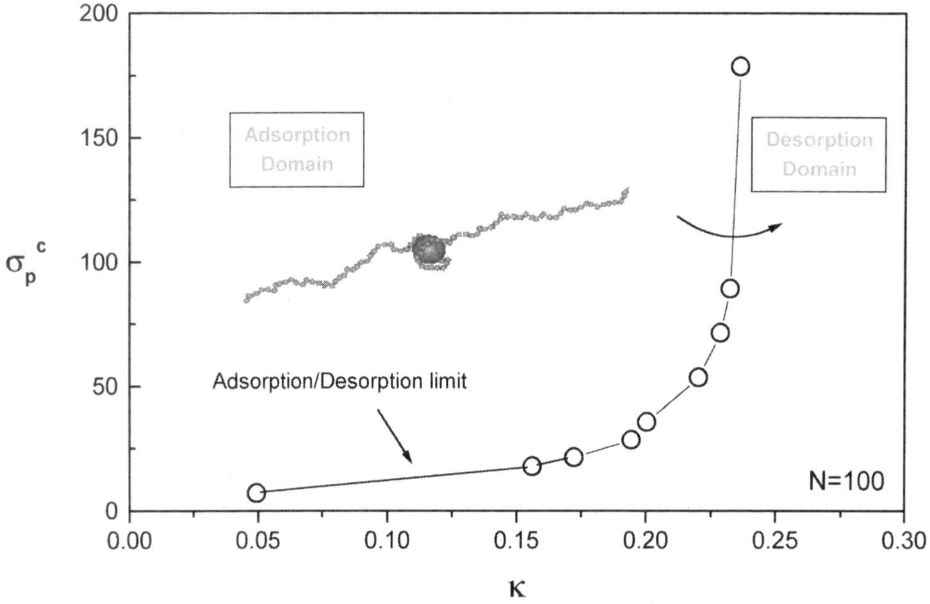

FIG. 10 Plot of the critical particle radius σ_p^c versus κ, the inverse of the Debye length ($N = 100$). It is shown that by increasing κ, adsorption is achieved by increasing the size of the particle.

2. Polyelectrolyte Conformation at the Particle Surface

When the polyelectrolyte is adsorbed at the particle surface, its conformation is expected to be different from its conformation in solution without the influence of the particle. To get an insight into the extent of change, the mean square radius of gyration $\langle R_g^2 \rangle$ of the polyelectrolyte is plotted in Fig. 11 as a function of C_i for different σ_p/σ_m values. $\langle R_g^2 \rangle$ for the isolated chain is also represented for comparison. When $\sigma_p/\sigma_m \leq 2$, the polyelectrolyte conformation is poorly affected by the presence of a small particle. By increasing the σ_p/σ_m ratio up to 5 and 6, respectively, the particle surface becomes large enough to adsorb a few monomer segments and start producing loops and trains. Maximal deformation is achieved when $C_i = 0.\ 1$ M. Below this concentration the electrostatic repulsion between the monomers is strong enough to limit the adsorption process through the formation of extended tails in solution. When $\sigma_p/\sigma_m = 8$, maximal deformation is achieved whatever the ionic concentration. The surface area is now large enough that the polyelectrolyte can wrap around the sphere to optimize the number of contacts. When $\sigma_p/\sigma_m > 8$, the conformation and size of the polyelectrolyte is now dictated by the size of the particle. By increasing the

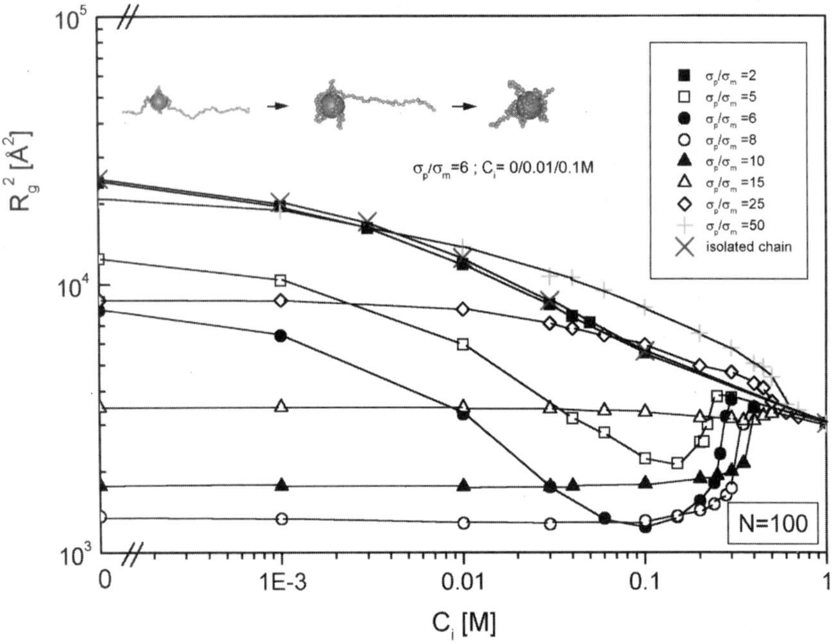

FIG. 11 Mean square radius of gyration $<R_g^2>$ of the polyelectrolyte as a function of C_i for different σ_p/σ_m values. $<R_g^2>$ for the isolated chain is also represented for comparison.

particle size further, a nearly planar surface limit is reached. In this case, the chain can spread fully on the surface with a dimension close to its dimensions in a free solution.

3. Number of Monomers in Trains, Loops, and Tails

The number of monomers in trains, loops, and tails as a function of σ_p/σ_m and different C_i values is presented in Fig. 12 to provide insight into the structure of adsorbed layer. The number of monomers in trains increases rapidly when particle size increases from $\sigma_p/\sigma_m = 2$ to 10. Then, by further increasing the σ_p/σ_m ratio to 50, it reaches a plateau value that is close to the total number of monomers of the polyelectrolyte (in particular when $C_i = 0$ M). Thus, in the presence of large particles the chains have a tendency to flatten and to be fully extended on the particle surface since this lowers the free energy most. Under such conditions, the electrostatic excluded volume of the monomers does not play a key role in the limitation of the amount of adsorbed monomer on the particle surface.

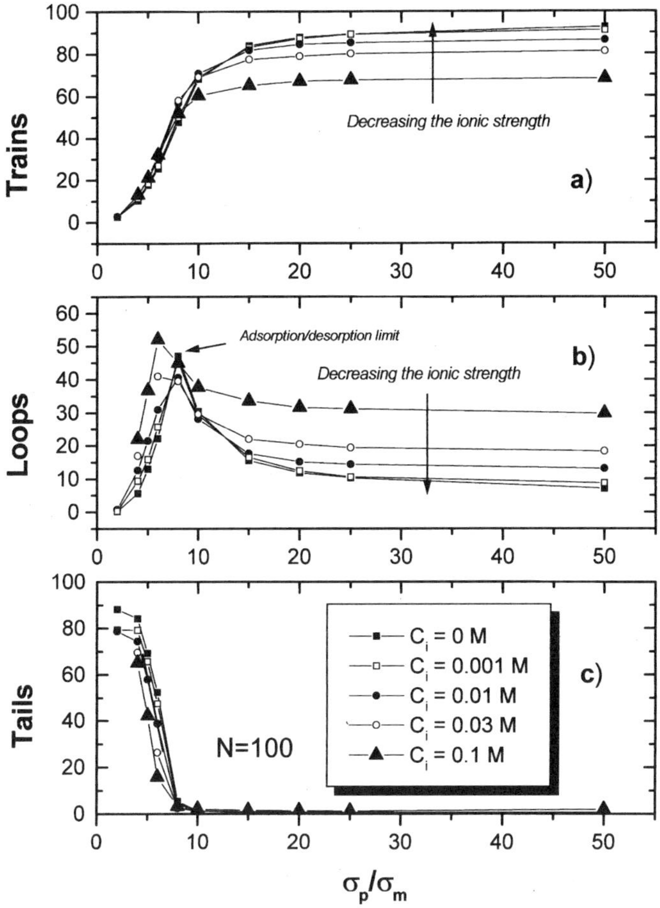

FIG. 12 Quantitative description of the adsorbed monomer layer; number of monomers in trains, loops, and tails vs. σ_p/σ_m and C_i. The monomer fraction in trains increases with the σ_p/σ_m ratio, then reaches a plateau value, whereas the monomer fraction in tails decreases with σ_p/σ_m The monomer fraction in loops exhibits a maximal value when the surface area of the particle is large enough to attract only a few monomers.

When increasing C_i, and when the particle surface is large enough ($\sigma_p/\sigma_m >$ 10), the number of monomers in trains is less than in the salt-free case due to the formation of loops and tails. When curvature effects become significant ($8 < \sigma_p/\sigma_m < 15$), the fraction of monomers in trains decreases whereas that in the loops increases, the monomers being transferred to loops but not to tails. Figure 12b illustrates that the number of monomers in loops reaches a maximum when σ_p/σ_m is close to 8, the ratio at which maximal chain deformation occurs. By decreasing the σ_p/σ_m ratio ($\sigma_p/\sigma_m < 8$), monomer desorption becomes critical and the monomers are transferred in tails.

Further insight into the adsorption properties of the polyelectrolyte on the surface is gained by examining particle surface coverage or the amount of adsorbed polymer. The effect of C_i and σ_p/σ_m presented in Fig. 13 clearly demonstrates that the amount of adsorption Γ generally increases as the σ_p/σ_m ratio increases and C_i diminishes. Nonetheless, it is important to note that Γ does not take on monotonic behavior because of the competition between the attractive monomer–particle interactions and an increase in the electrostatic excluded volume of the polyelectrolyte at the particle surface. The monomer excluded electrostatic volume is expected to limit the number of monomers confined close to the particle surface in the low-salt regime and when $4 \leq \sigma_p/\sigma_m \leq 15$.

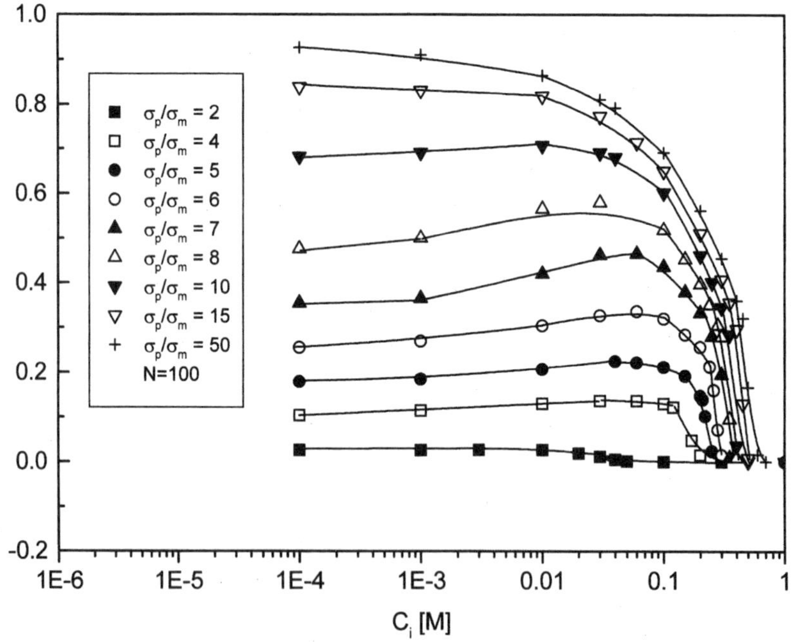

FIG. 13 Adsorbed amount Γ of the polyelectrolyte vs. C_i and σ_p/σ_m. Γ increases by increasing the size of the particle and reaches a maximal value in the low-salt regime. However, although monomer–particle interactions are promoted in the low-salt regime, the excluded electrostatic volume between monomers at the particle surface strongly limits the number of adsorbed monomers in particular when $\sigma_p/\sigma_m = 10, 8, 7, 6, 5$. Data with the coordinates $\log(C_i) = -4$ correspond to the salt-free case.

4. Overcharging

Here we mainly consider situations where charge inversion occurs, i.e., when the polyelectrolyte is sufficiently adsorbed so that the net charge of the complex changes sign. We shall begin with the salt-free case. In Fig. 14 the net charge of the complex $|Q^*|$, which is equal to $|Q - N^{ads}q|$, is represented as a function of particle charge Q. Our results are in reasonably good agreement with the NS model, where the net inverted charge is predicted to be proportional to $Q^{1/2}$ (and subsequently to the σ_p/σ_m ratio, which is also presented in Fig. 14). According to our MC simulations, when $\sigma_p/\sigma_m \leq 6$, two-tailed configurations are achieved with two polymer arms extended in opposite directions from the particle, overcharging is not observed. Data points situated above the theoretical curve overestimate the charge inversion. The excess of monomers at the macroion surface is due to latency in the first-order phase transition and tail formation. By further increasing the size of the chain above a critical value, a certain amount of the

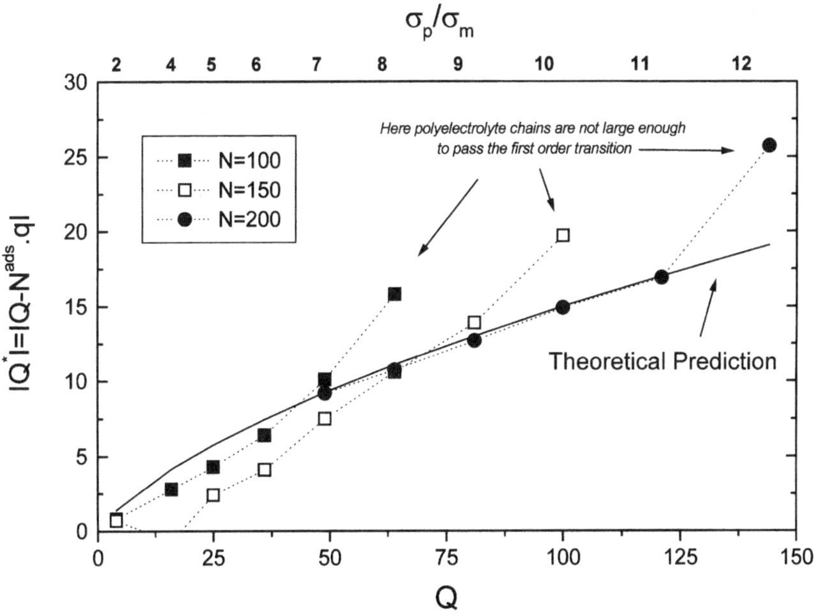

FIG. 14 Net charge $|Q^* = |Q - N^{ads}\, q|$ variation and overcharging intensity as a function of particle charge and σ_p/σ_m ratio when $C_i = 0$M. When the polyelectrolyte size is large enough, Monte Carlo data are found asymptotically to be in good agreement with the Nguyen-Shklovskii model. Points situated below the theoretical curve underestimate overcharging. Polyelectrolyte chains that are not large enough to produce a tail in solution overestimate overcharging.

adsorbed monomers is spontaneously ejected (removed) from the particle surface to form a tail radially outward from the center of the particle. A decrease of $|Q^*|$ is then observed. The effect of the presence of a monovalent salt is now discussed by considering Fig. 15, which reports the $|Q^*|$ values as a function of Q for C_i values equal to 0.001, 0.01, and 0.03 M, respectively. Charge inversion increases with charge screening, clearly demonstrating that monomers condense more on the particle in a salty solution. Once again, our MC results are in good agreement with the analytical theory, which predicts a linear dependance of $|Q^*|$ with Q. It is worth noting that when $\sigma_p/\sigma_m > 6$ and $N = 100$, the data do not fit the theoretical model because of the limited size of the polyelectrolyte, which is not large enough to form a tail in solution.

C. Chain Rigidity Effects

1. Adsorption–Desorption Limit, Complex Configurations

Equilibrated configurations as a function of C_i and k_{ang} are presented in Table 5. No adsorption is observed when $C_i \geq 1$ M whereas adsorption is always ob-

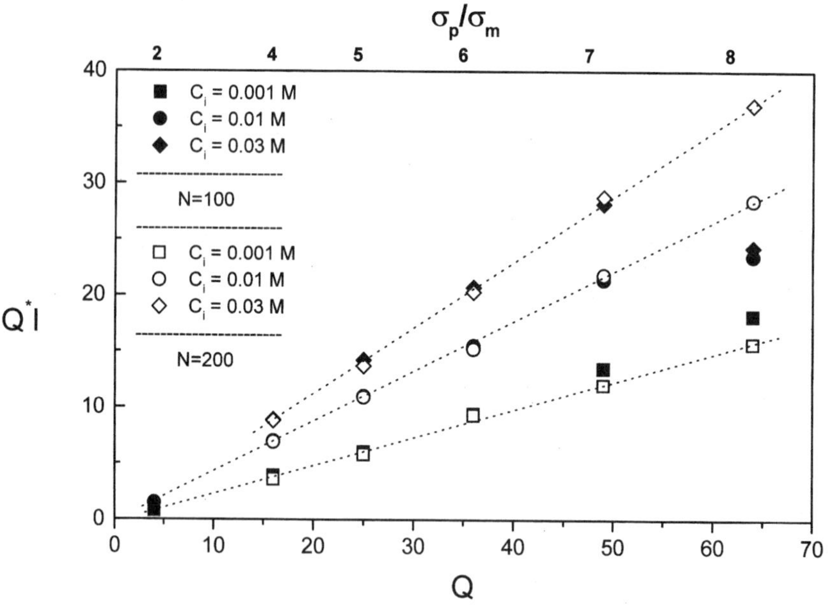

FIG. 15 Net charge $|Q^*| = |Q - N^{ads} q|$ variation as a function of particle charge, σ_p/σ_m ratio, and ionic concentration C_i. Charge inversion increases with charge screening. When $N = 100$ and $\sigma_p/\sigma_m = 8$, the data do not fit the linear variations because the chain is too small to attain the asymptotic conditions of the Nguyen-Shklovskii analytical theory.

TABLE 5 Monte Carlo Equilibrated Conformations of Semiflexible Polyelectrolyte–Particle Complexes as a function of C_i and k_{ang} [a]

C_i [M] / k_{ang} [k_BT/deg^2]	0	0.01	0.1	0.3	1
0	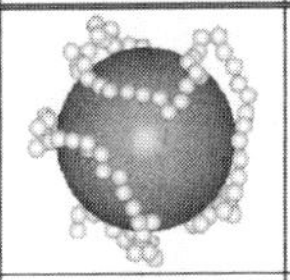	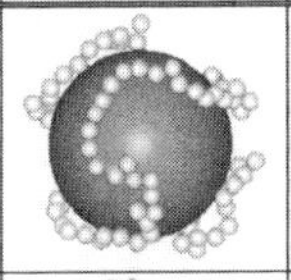	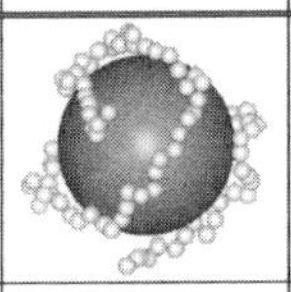	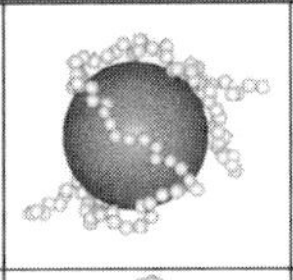	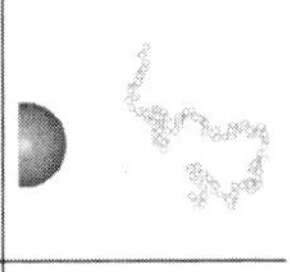
0.001	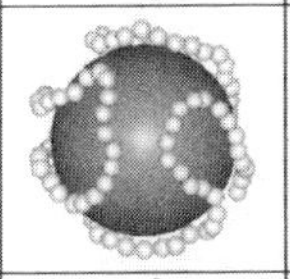	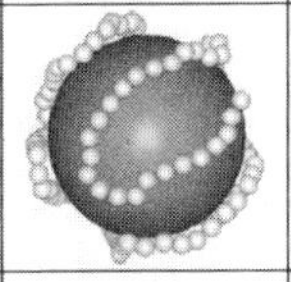	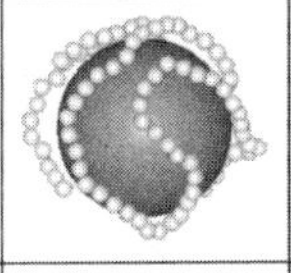	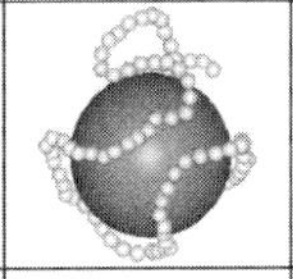	
0.005	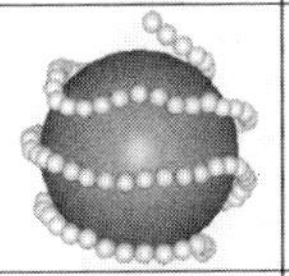	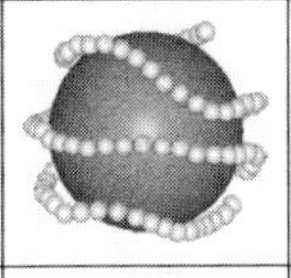	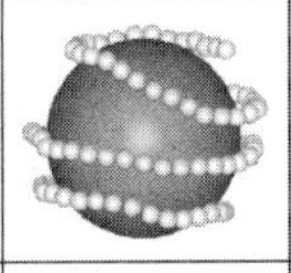	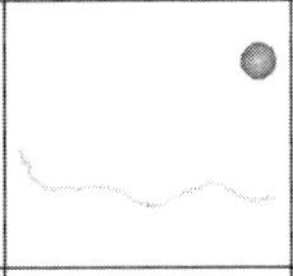	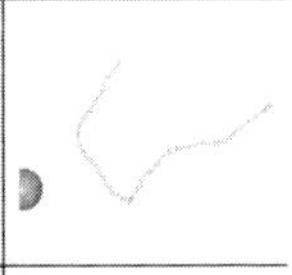
0.01	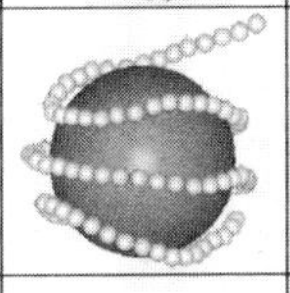	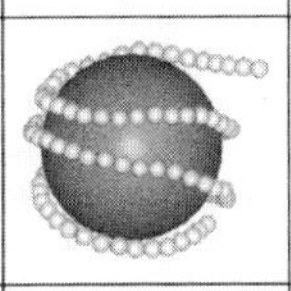	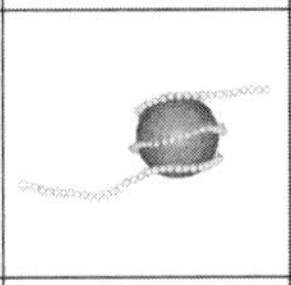	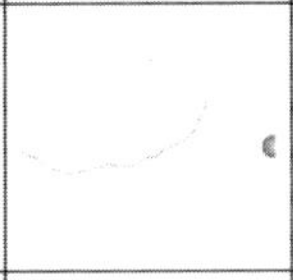	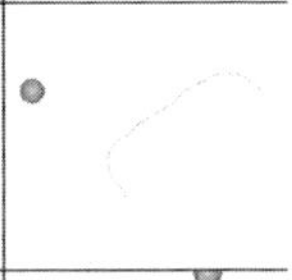
0.02	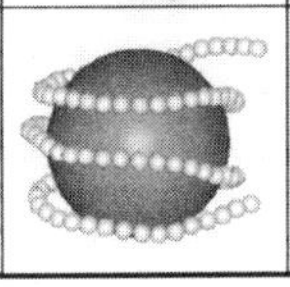	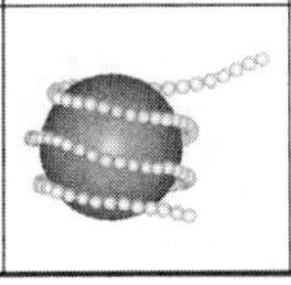	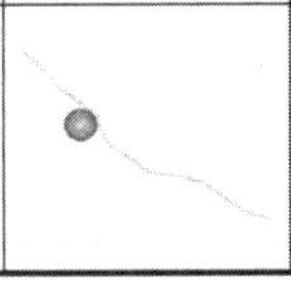		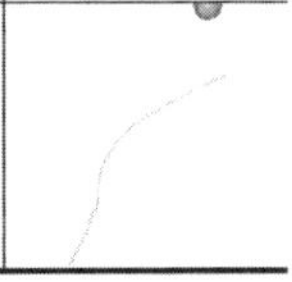

[a] By increasing the chain stiffness, solenoid conformations are progressively achieved at the particle surface.

served when $C_i \leq 0.1$ M. At $C_i = 0.3$ M, Table 5 clearly demonstrates that adsorption is strongly controlled by the value of k_{ang}. Hence, adsorption is promoted by (1) decreasing the chain stiffness (and subsequently the required energy to confine the semiflexible polyelectrolyte at the particle surface) and (2) decreasing the ionic concentration (thus increasing the electrostatic attractive interactions between the monomer and the particle surface that we consider as the driving force for the adsorption). Because of charge screening in the high-salt regime, monomer–particle interactions are not large enough to overcome the polyelectrolyte confinement near the particle.

A quantitative picture of adsorption–desorption is achieved by calculating the critical ionic concentration C_i^c vs. k_{ang} at which the adsorption–desorption process is observed (Fig. 16). The adsorption–desorption limit is determined by monitoring the ionic concentration required to satisfy the adsorption–desorption criteria (for a constant k_{ang} value). We found that C_i^c is rapidly decreasing with k_{ang} from 0.38 M to a plateau value at 0.22 M when $k_{ang} > 0.01\ k_B T/\mathrm{deg}^2$. In that plateau region, as the polyelectrolyte conformation is similar to that of a rigid rod, the polyelectrolyte is in contact with the particle surface with a few consecutive monomers only, with two polymer arms extended in opposite directions from the particle. We also calculated vs. C_i the critical chain stiffness k_{ang}^c at which adsorption–desorption is observed (inset of Fig. 16). We found that $k_{ang}^c \sim \kappa^{-\delta}$ (with $\delta = 14.5$), revealing that the increase of chain stiffness rapidly promotes chain desorption on curved surfaces.

The monomer distribution at the particle surface is largely controlled by the value of k_{ang}. When the chain flexibility is important ($k_{ang} \leq 0.001\ k_B T/\mathrm{deg}^2$), "tennis ball" conformations are achieved, whereas when rigid chains are consid-

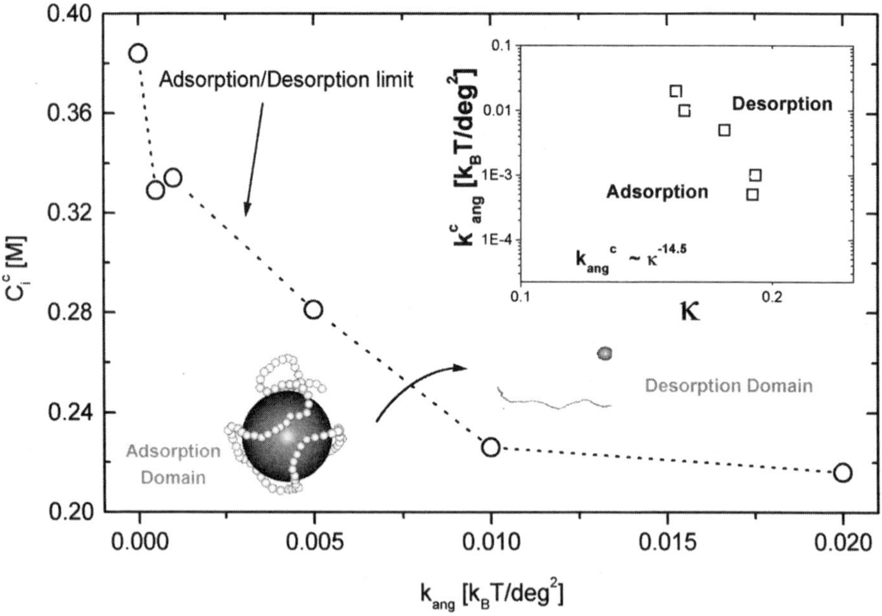

FIG. 16 Plot of the critical ionic concentration C_i^c vs. k_{ang} at the adsorption–desorption limit. The critical stain stiffness value k_{ang}^c, at which desorption is observed, is shown in the inset vs. κ.

ered ($k_{ang} > 0.001\ k_BT/\text{deg}^2$), the intrinsic flexibility forces the polyelectrolyte to adopt solenoid conformations such as those predicted by the analytical model of Nguyen and Shklovskii. Both the strong electrostatic repulsions between the neighboring turns and intrinsic chain rigidity keep the turns parallel to each other and a constant distance between them. It is worth noting that the polymer ends are not adsorbed at the particle surface in all cases.

When $C_i > 0.01$ M, large changes in the chain dimensions are now observed with increasing the chain stiffness. As long as $k_{ang} \leq 0.001 k_BT/\text{deg}^2$, loops and tail are promoted (Table 5; $C_i = 0.3$ M and $k_{ang} = 0.0005 k_BT/\text{deg}^2$), resulting in an increase of the thickness of the adsorption layer. When $k_{ang} \geq 0.005 k_BT/\text{deg}^2$ and with increasing C_i, the polyelectrolyte starts to leave the surface by winding off. Extended tails in solution are formed concomitantly with a decrease in the number of turns of the solenoid and monomers in trains (Table 5; $C_i = 0.1$ M and $k_{ang} = 0.01 k_BT/\text{deg}^2$). By further increasing the ionic concentration or chain intrinsic flexibility, the polyelectrolyte becomes tangent to the particle surface with dimensions close to its free unperturbed dimensions.

2. Number of Monomers in Trains, Loops, and Tails

The number of monomers in trains, loops, and tails as a function of k_{ang} and C_i is presented in Fig. 17a–c. Monomer fraction in trains has a maximal value when $k_{ang} = 0.005 k_BT/\text{deg}^2$, i.e., when semiflexible chains are considered. Below that value, i.e., when flexible chains are considered, the solenoid conformation is lost and monomers move from the first layer to the others to form loops. Above that value, when rigid chains are considered, most of the monomer entropy is lost and ordered solenoid conformations are obtained. By decreasing the attractive monomer–particle interactions, rigid chains desorb to form tails.

Competition between the attractive electrostatic energy and the increase in the torsional energy to control the final complex structure is well illustrated when $C_i = 0.1$ M. By increasing k_{ang}, we obtain a transition from a disordered and strongly bound complex to a situation where the polymer touches the particle over a finite length while passing by the formation of a solenoid structure. Complexes exhibit a significant monomer loop fraction only when $k_{ang} < 0.001 k_BT/\text{deg}^2$. In the other cases, the fraction of monomers in train decreases while that of the tails increases, monomers being transferred to tails but not to loops. The amount of chain adsorption Γ, which is commonly used experimentally to derive adsorption isotherms, is presented in Fig. 18. Γ is decreasing with increasing ionic concentration because of monomer desorption. On the other hand, the amount of monomer adsorption has a maximal value when $k_{ang} = 0.005 k_BT/\text{deg}^2$ (i.e., for semiflexible chains) according to the formation of a solenoid without parts extending in the solution. Because of the displacement of the adsorption–desorption limit with k_{ang}, curves cross each other.

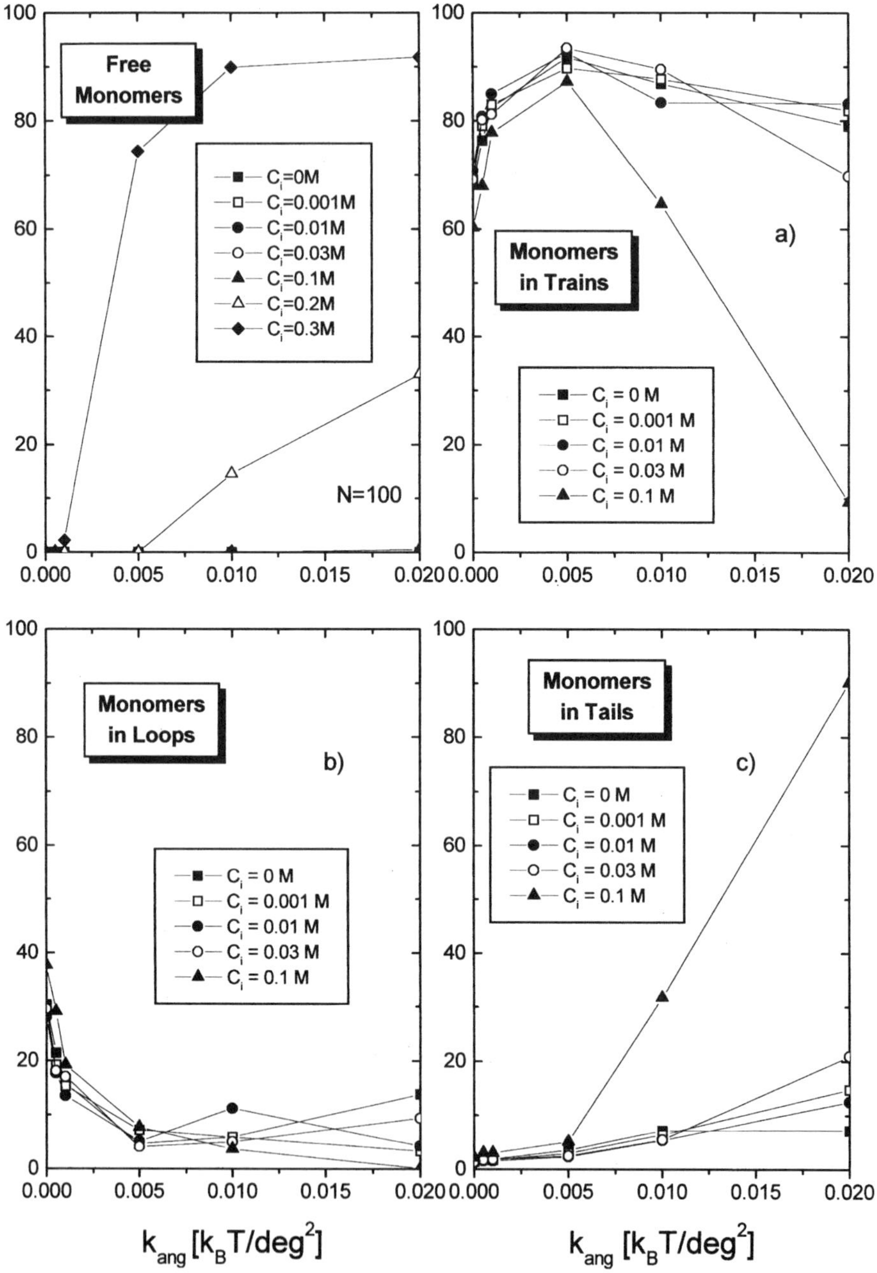

FIG. 17 Number of monomers in trains, loops, and tails vs. k_{ang} at various ionic concentration values ($N = 100$). Monomers in loops are promoted by decreasing the chain rigidity whereas monomers in trains exhibits a maximal value for semiflexible chains (i.e., when $k_{ang} = 0.005\ k_B T/\text{deg}^2$).

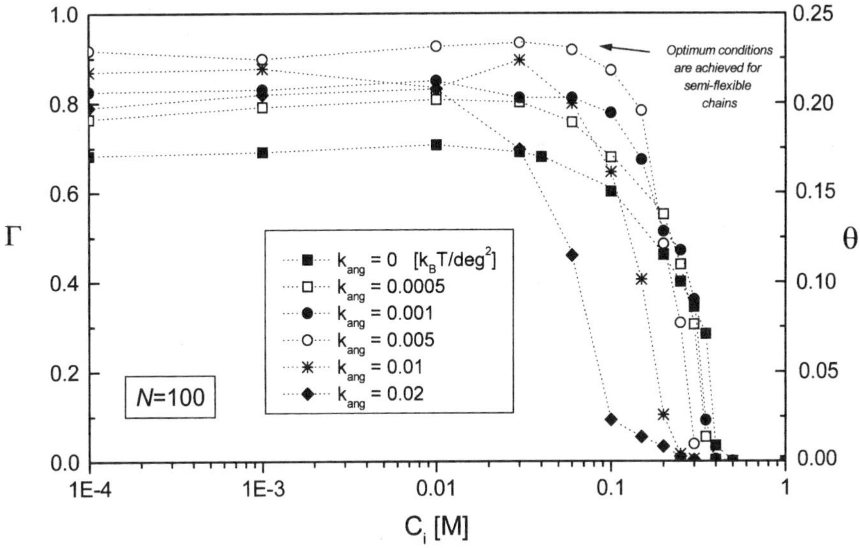

FIG. 18 Adsorbed amount Γ and surface coverage θ of the polyelectrolyte and particle, respectively, vs. C_i for different k_{ang} values. Γ reaches a maximal value for the semiflexible chain $k_{ang} = 0.005$ k_BT/deg^2 (and with decreasing C_i). Above that value both chain rigidity (by forcing the chain to be extended in solution) and flexibility (monomers are transfered to loops) limit the amount of adsorbed monomers in the first adsorption layer.

D. Linear Charge Density Effects

The LCD, which is related to the charge number and charge fraction on the polyelectrolyte backbone, is an important factor controlled by the solution chemistry and polyelectrolyte concentration.

The influence of the LCD on polyelectrolyte adsorption has been investigated. In Fig. 19 the mean square radius of gyration and mean square end-to-end distance are plotted against the LCD. The analysis shows two different regimes corresponding to nonadsorbed (A) and adsorbed (B) polyelectrolytes, respectively. In regime A, one observes high $\langle R_g^2 \rangle$ values ($>3 \times 10^3$ [Å^2]), which are identical to those obtained for isolated chains under similar conditions. In regime B, these values are decreasing rapidly ($\sim 2 \times 10^3$ [Å^2]) immediately after polyelectrolyte adsorption. It is worth noting that for small ionic strengths (0–0.01 M) the polymer is always adsorbed on the surface of the colloid. When C_i is in the range 0.1–0.3 M, the curves show an increase in the $\langle R_g^2 \rangle$ value which is the same as for the free chains (regime A); then a sharp transition is observed to the lower values of $\langle R_g^2 \rangle$ (regime B) corresponding to adsorbed conforma-

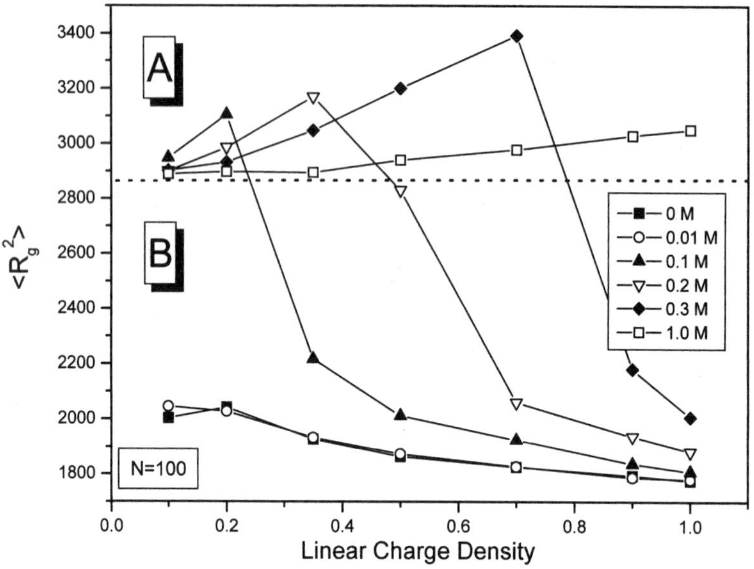

FIG. 19 Mean square radius of gyration $<R_g^2>$ (a) and expansion factor $r = <L_{ee}^2>/<R_g^2>$ (b) vs. the linear charge density (LCD) at different ionic concentrations for the polyelectrolyte in the presence of the oppositely charged particle. Nonadsorbed (A) and adsorbed (B) regimes are delimited by dashed lines.

tions. Finally, when $C_i = 1.0$ M the curve exhibits the same increase in the $<R_g^2>$ value as for the free chains. Since there is no adsorption at this ionic strength, no transition in the $<R_g^2>$ value is observed. Similar trends are also observed for the $<L_{ee}^2>$ values.

A qualitative picture of the polyelectrolyte–particle complexes has also been achieved by representing equilibrated structures as a function of C_i and LCD (Table 6). The absorption–desorption limit is moved from the higher to the lower ionic strengths with decreasing LCD. Hence, when the linear charge density is small, adsorption is promoted by increasing the attractive interactions between monomers and particle via the ionic strength. For example, when $C_i = 1.0$ M the polymer is never adsorbed on the colloid whereas when $C_i = 0.1$ M adsorption occurs only when LCD > 0.3.

1. Conformation of the Adsorbed Polyelectrolytes

The conformation of adsorbed polymers are usually described in terms of trains, loops, and tails. In Fig. 20a–c we present the average number of monomers in trains, loops, and tails as a function of C_i and LCD for the colloid–polyelectrolyte complex.

TABLE 6 Monte Carlo Equilibrated Conformations of Isolated Charged Polymers as a Function of Ionic Concentration C_i and Polyelectrolyte Linear Charge Density (LCD)[a]

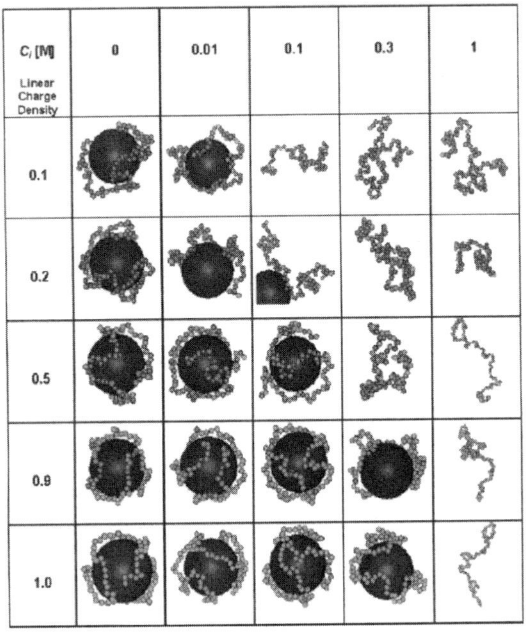

[a]Bright monomers represent charged monomers. The LCD is clearly controlling the adsorption/desorption limit and polymer conformation at the particle surface. When screening is important, no adsorption is observed.

(*a*) *Trains.* In the absence of trains, according to our definition of polymer adsorption, the polyelectrolyte is never adsorbed. This is observed when C_i = 1.0 M. In the other situations, after a critical value of the LCD has been reached, one can observe a monotonic growth of monomers in trains with the increasing LCD. At a given LCD, with decreasing ionic strength the number of trains is growing, which means that polyelectrolyte chains adopt flat conformations at the interface.

(*b*) *Loops.* According to Fig. 20b, globally the number of monomers in loops decreases with increasing LCD and with decreasing C_i. Here one can also observe a sharp transition between the nonadsorbed and adsorbed regimes. A maximal value is reached just after adsorption; then a smooth decrease with increasing LCD is observed. The monomers in loops are then transferred to trains.

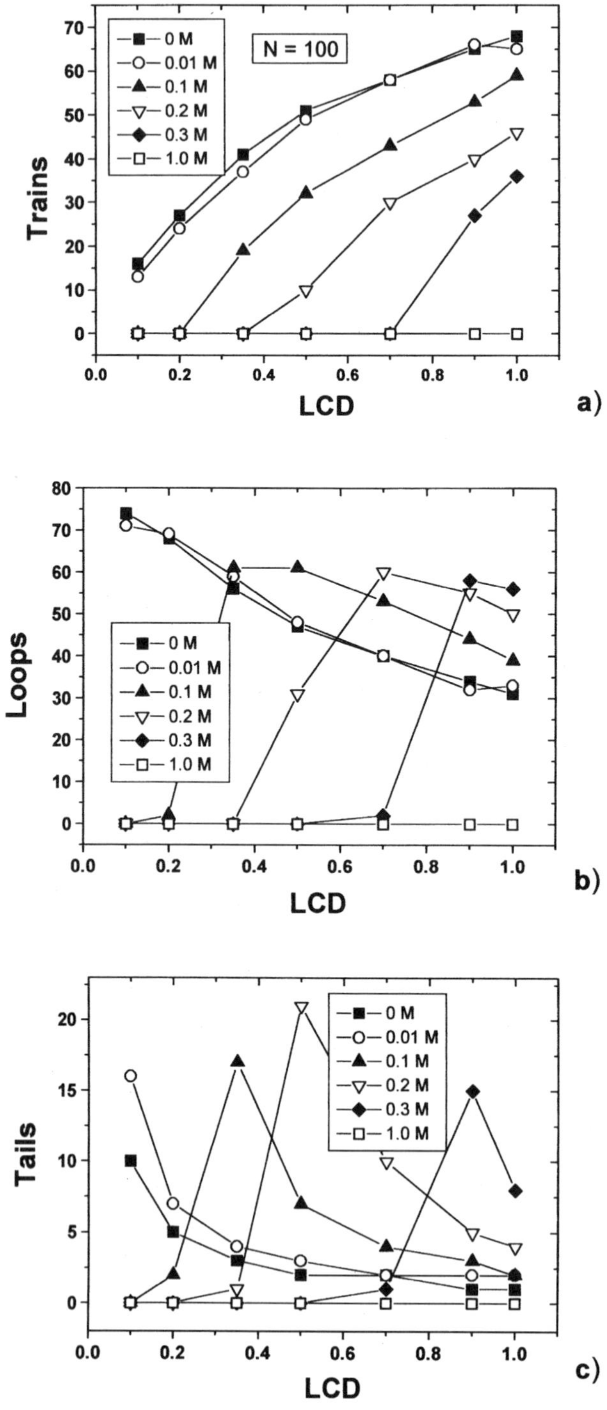
0 M
0.01 M
0.1 M
0.2 M
0.3 M
1.0 M
N = 100
Trains
LCD
a)
0 M
0.01 M
0.1 M
0.2 M
0.3 M
1.0 M
Loops
LCD
b)
0 M
0.01 M
0.1 M
0.2 M
0.3 M
1.0 M
Tails
LCD
c)

(*c*) *Tails.* When $C_i = 0$ and 0.01 M, the number of monomer in tails monotonically decreases with the increase in LCD. In the other cases, a maximal value is obtained at the adsorption–desorption limit.

As the polymer is only weakly adsorbed at the small values of LCD and low ionic strength, an important balance is observed in favor of loops and tails instead of configurations where the monomers are in trains. As the three parameters are correlated, the proportional changes can be easily observed; with increasing LCD the amount of monomers in trains increases whereas the amount of monomers in tails and loops decreases. At higher values of LCD the number of monomers in trains reaches a maximum and finally decreases slowly, further increasing the values of LCD.

2. Adsorption–Desorption Limit

We estimated the critical values of adsorption at different salt concentrations by adjusting the linear charge density. Examination of the contact probability parameter P_c values vs. LCD (Fig. 21) shows a sharp transition that seems to be more gradual with increasing ionic strength. As small changes in the LCD values generate very important changes in the absorption ratio, we calculated the critical linear charge density (LCD^{crit}) at which adsorption–desorption is observed. We plotted the critical LCD adsorption value vs. C_i in Fig. 22 and observed a linear dependence with a slope equal to 2.5. The LCD^{crit} dependence vs. κ is shown in the inset of Fig. 22. The LCD^{crit} is related to κ according to $LCD^{crit} \approx \kappa^y$ where $y = 2$ was estimated by plotting log(LCD) vs. log(κ).

VI. OUTLOOK AND CONCLUSION

The effects of ionic concentration, colloid size, polyelectrolyte length, charge, and intrinsic flexibility on the conformation of a polyelectrolyte like DNA in the presence of an oppositely charged polymer have been investigated using an MC approach. Monte Carlo simulations constitute a rewarding and invaluable approach. In fact, it has been shown that computer simulations and theory can isolate in good agreement the molecular factors that control polyelectrolyte conformation in solution and at the interface, the adsorption limit, and overcharging issue, and thus can be used to address the optimization of colloid–polymer mixtures and guide new experiments.

FACING PAGE

FIG. 20 Number of monomers in trains (a), loops (b), and tails (c) as a function of the linear charge density (LCD) at different ionic concentration for the polyelectrolyte–particle complex.

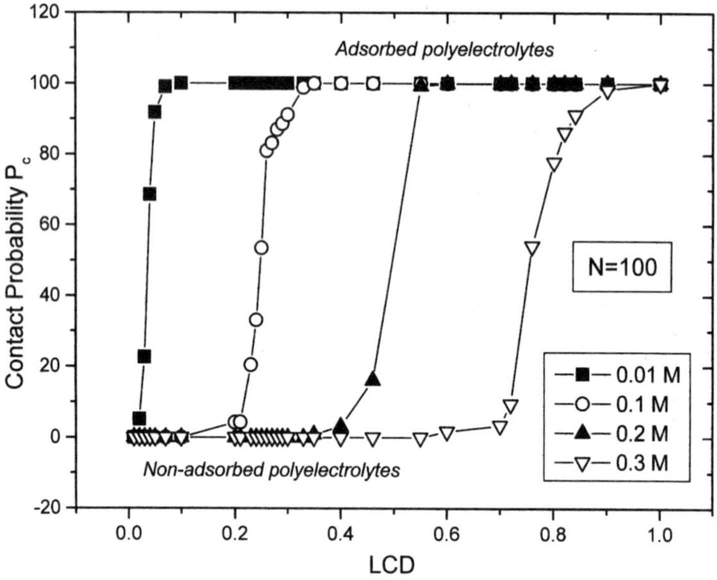

FIG. 21 Contact probability parameter P_c vs. the linear charge density (LCD) between the polyelectrolyte and the particle.

Adsorption occurs if the loss of entropy of the chain is at least compensated by the gain of energy. Thus, most favorable conditions to achieve adsorption are obtained by considering long chains. Our simulations point out the importance of two competing effects when the ionic concentration increases; on the one hand, the particle capacitance increases and so much monomer can be adsorbed; on the other hand, the electrostatic attraction between the particle and the monomer becomes less important, giving the monomers and the polyelectrolyte the opportunity to leave the particle surface. Polyelectrolyte conformational changes resulting from adsorption are found to be dependent on the polyelectrolyte size. Maximal chain deformation is achieved in the low screened limit and when the ratio of the mean end to end distance of the free polyelectrolyte to the particle radius is close to 20. By further increasing the chain length, excluded electrostatic volume prevents any additional monomer adsorption on the surface. As a result, an extended tail is formed in solution. Our MC results demonstrate that the complexation between a polyelectrolyte and a charged sphere can lead to overcharging when $N > N^c$. We find a perfect agreement with the NS model both in the salt-free case and in the case of added salt. Variations in the number of adsorbed monomers as a function of the total number of monomers support

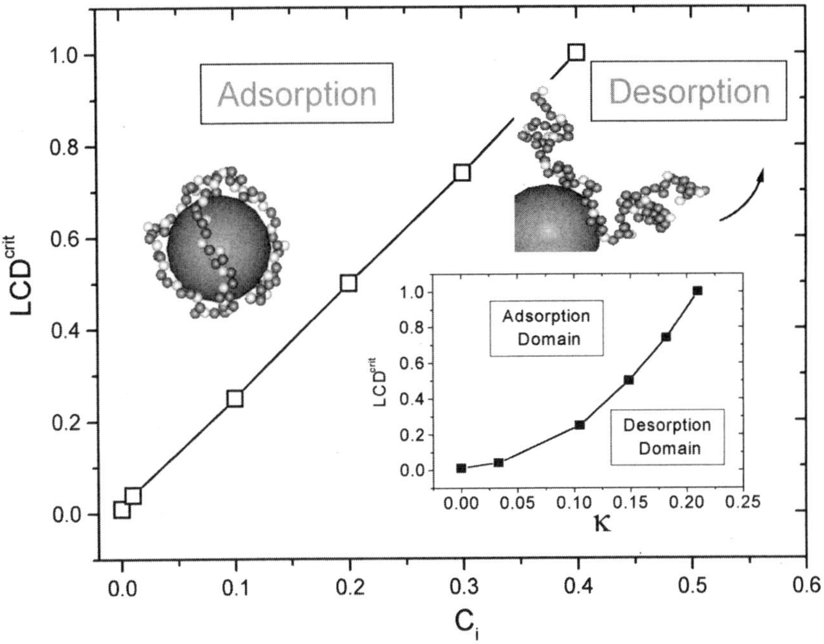

FIG. 22 Critical value of the linear charge density LCD^{crit} as a function of C_i. In the inset LCD^{crit} is plotted against κ. Adsorption and desorption domains are delimited by the corresponding curves.

a first-order transition with the spontaneous formation of a protuding tail in solution.

Adsorption of charged polymers is not only controlled by ionic concentration but also by particle diameter. Surface curvature effects clearly limit the amount of adsorbed monomers; large particles allow the polyelectrolyte to spread on the surface; small particles limit the number of adsorbed monomers that may be attached to it. When small particles are considered, the low-salt regime is dominated by the monomer–monomer repulsions forcing the polyelectrolyte to form extended tails in opposite directions. When the particle size is equal to or greater than the radius of gyration of the charged polymers, polyelectrolytes can wrap fully around the particle. Under such conditions, trains are favored whereas loops become more frequent when increasing ionic strength before the adsorption–desorption limit. MC results also demonstrate that the complexation between a polyelectrolyte and a charged sphere can lead to overcharging when the polymer size is large enough. In the case of added salt, our simulations point out the importance of charge inversion with the increase of ionic concentration because of the increase of particle capacitance as well as the key role of the charged polymer size.

The influence of polyelectrolyte intrinsic stiffness and ionic concentration on the adsorption–desorption limit, polyelectrolyte conformation, and monomer distribution on the particle surface was presented. Chain stiffness modifies the conformation of isolated and adsorbed chains by locally destroying a large amount of monomer entropy. As a result, chain stiffness influences the amount of adsorbed monomers, monomer distribution at the particle surface, and adsorption–desorption limit. The amount of adsorbed monomers has a maximal value for the semiflexible chains in the low-salt-concentration regime. Under such conditions, the polyelectrolyte is strongly adsorbed at the particle surface as a solenoid and the confinement energy does not contribute to the formation of tails in solution. When the chain intrinsic stiffness is small, tennis ball–like conformations are achieved. At the opposite, when rigid chains are considered the polyelectrolyte becomes tangent to the particle. Hence, adsorption is promoted by decreasing the chain stiffness or decreasing the salt concentration for a given chain length.

Polyelectrolyte adsorption is promoted by increasing its LCD or/and decreasing the ionic concentration to promote electrostatic attractive effects. Calculating the adsorption–desorption limit, we found the LCD^{crit} scales as $LCD^{crit} \sim \kappa^2$. By focusing on the variations of the LCD and probability of contact, we found a sharp transition at the adsorption–desorption limit that appears to be more gradual by increasing the ionic strength. Adsorption occurred when the mean attractive energy per monomer is more negative than $-0.4kT$. By analyzing the monomer fraction in loops, tails, and trains, we demonstrated that at small values of the LCD and low ionic strength, a large amount of monomer is present in loops and tails. By increasing the LCD, the amount of monomers in trains reaches a maximal value and polyelectrolytes adopt flat conformations at the particle surface.

The simulations reported here are a preliminary step toward a more precise modeling of the problem to get insight into the behavior of more concentrated polymer solutions (systems with several chains) and thus flocculation/stabilization processes of polymer–particle mixtures. A simple model involving one chain interacting with one particle has been described, but it can be extended to more concentrated systems involving several chains (and/or colloidal particles). Nonetheless, the computational description of adsorption processes is still part of a great challenge. A Debye-Hückel approach involving one charged polymer interacting with one spherical particle has been investigated, but it can be extended to more precise modeling of the phenomena (e.g., by including explicit counterions, hydrophobic interactions), polydisperse systems (sizes and shapes), and more concentrated solutions. We hope that the observations made in this study are particularly useful for choosing an appropriate polymer or biopolymer for applications such as steric stabilization of dispersed particles.

ACKNOWLEDGMENT

Many of the ideas presented here derived directly from discussions with J. Buffle, E. Pefferkorn, B. Shklovskii, and others. Such ideas are based on results obtained by graduate and postgraduate students over the years, in particular P. Chodanowski and M. Brynda. We acknowledge the financial support of the Swiss National Foundation.

GLOSSARY

Γ	amount of adsorbed polyelectrolyte
θ	particle surface coverage
$<R_{ee}^2>$	mean square end to end length
$<R_g^2>$	mean square radius of gyration
N	number of monomers of the polyelectrolyte chain
κ^{-1}	Debye screening length
C_i	ionic concentration
z_i	amount of charge on unit i
k_{ang}	chain stiffness
r	polyelectrolyte expansion factor
Q	charge of the particle
Q^*	net charge of the polyelectrolyte–particle complex
σ_m	monomer radius
σ_p	radius of the particle
σ	charge surface density of the particle
f	fraction of ionized monomers

REFERENCES

1. Barrat, J.L.; Joanny, J.F. In *Advances in Chemical Physics;* Prigogine, I., Rice, S.A., Eds.; Theory of Polyelectrolyte Solutions, Volume 94; Wiley & Sons: New York, 1996.
2. Flory P.J. *Principles of Polymer Chemistry*; Cornell University Press: Ithaca, NY, 1992.
3. Xia, J.; Dubin, P.L. In *Macromolecular Complexes in Chemistry and Biology;* Dubin, P., Bock, D., Eds.; Protein–Polyelectrolyte Complexes; Springer-Verlag: Berlin, 1994.
4. Buffle, J.; Wilkinson, K.J.; Stoll, S.; Filella, M.; Zhang, J. Environ. Sci. Technol. **1998**, *32*, 2887.
5. Hara, M. *Polyelectrolytes: Sciences and Technology*; Marcel Dekker: New York, 1993.
6. Napper, D.H. *Polymeric Stabilization of Colloidal Dispersions;* Academic Press: New York, 1983.
7. Finch, C.A. *Industrial Water Soluble Polymers;* The Royal Society of Chemistry: Cambridge, 1996.

8. Walker, H.W.; Grant, S.B. Colloids Surf. A **1998**, *135*, 123.
9. Rädler, J.O.; Koltover, I.; Salditt, T.; Safinya, C.R. Science **1997**, *275*, 810.
10. Darnell, J.E.; Lodish, H.; Baltimore, D. *Molecular Cell Biology*, 3rd Ed.; Scientific American Books: New York, 1995.
11. Vermeer, A.W.P.; Leermakers, F.; Koopal, L. Langmuir **1997**, *13*, 4413.
12. Haronska, P.; Vilgis, T.A.; Grottenmüller, R.; Schmidt, M. Macromol. Theory Sim. **1998**, *7*, 241.
13. Dobrynin, A.V.; Deshkovski, A.; Rubinstein, M. Macromolecules **2001**, *34*, 3421.
14. Linse, P. Macromolecules **1996**, *29*, 326.
15. Shubin, V.; Linse, P. Macromolecules **1997**, *30*, 5944.
16. von Goeler, F.; Muthukumar, M.J. Chem. Phys. **1994**, *100*, 7796.
17. Kong, C.Y.; Muthukumar, M.J. Chem. Phys. **1998**, *109*, 1522.
18. Feng, X.H.; Dubin, P.L.; Zhang, H.W.; Kirton, G.F.; Bahadur, P.; Parotte, J. Macromolecules **2001**, *34*, 6373.
19. Mateescu, E.M.; Jeppesen, C.; Pincus, P. Europhys. Lett. **1999**, *46*, 493.
20. Nguyen, T.T.; Grosberg, Y.; Shklovskii, B.I. J. Chem. Phys. **2000**, *113*, 1110.
21. Nguyen, T.T.; Shklovskii, B.I. Physica A **2001**, *293*, 324.
22. Gurovitch, E.; Sens, P. Phys. Rev. Lett. **1999**, *82*, 339.
23. Sintes, T.; Sumithra, K.; Straube, E. Macromolecules **2001**, *34*, 1352.
24. Chidambaram, S.; Dadmun, M.D. Comput. Theor. Polym. Sci. **1999**, *9*, 47.
25. Netz, R.R.; Joanny, J.-F. Macromolecules **1998**, *31*, 5123.
26. Netz, R.R.; Joanny, J.-F. Macromolecules **1999**, *32*, 9026.
27. Park, S.Y.; Bruinsma, R.F.; Gelbart, W.M. Europhys. Lett. **1999**, *46*, 454.
28. Wallin, T.; Linse, P. Langmuir **1996**, *12*, 305.
29. Wallin, T.; Linse, P. J. Phys. Chem. **1996**, *100*, 17873.
30. Wallin, T.; Linse, P. J. Phys. Chem. B **1997**, *101*, 5506.
31. Jonsson, M.; Linse, P. J. Chem. Phys. **2001**, *115*, 3416.
32. Carlsson, F.; Malmsten, M.; Linse, P. J. Phys. Chem. B **2001**, *105*, 12189.
33. Chodanowski, P.; Stoll, S. J. Chem. Phys. **2001**, *115*, 4951.
34. Chodanowski, P.; Stoll, S. Macromolecules **2001**, *34*, 2320.
35. Manning, G.S. J. Chem. Phys. **1969**, *51*, 924.
36. Kremer, K.; Binder, K. Comput. Phys. Rep. **1988**, *7*, 259.
37. Chodanowski, P.; Stoll, S. J. Chem. Phys. **1999**, *111*, 6069.
38. Flory, P.J. *Statistical Mechanics of Chain Molecules*; Hanser: Berlin, 1989.

9

Amino-Containing Latexes as a Solid Support of Single-Stranded DNA Fragments and Their Use in Biomedical Diagnosis

FRANÇOIS GANACHAUD Université Pierre et Marie Curie–CNRS, Paris, France

CHRISTIAN PICHOT and ABDELHAMID ELAISSARI
CNRS-bioMérieux, Lyon, France

I. INTRODUCTION

The principle of medical diagnostics is based on the detection and quantification of chemical (cholesterol, urea) or biological molecules (antigens, viruses, bacteria, etc.) in samples of blood, urine, cerebrospinal fluid, and cellular extracts. These samples are complex media comprising high concentrations of salts and different biological molecules among which is sought the analyte. The diagnostic is carried out in two steps: (1) the *capture* step consists of immobilizing a biological molecule on a solid carrier capable of recognizing selectively the analyte; (2) after eliminating the residual serum, the analyte is quantified by enzymatic reaction, chemiluminescence, or radioactivity; this is called the *detection* step. Research and development in vitro diagnostics henceforth aim at satisfying the following four requirements: specificity, sensitivity, rapidity, and cost.

The in vitro tests currently carried out in analysis laboratories are based on the specific recognition of antigens and antibodies. These immunoassays consist of dosing either the antibodies produced by the human body in response to an antigen or the antigens themselves. The diagnosis can thus be carried out only once if a sufficient concentration of antibody has been generated by the organism. In case of infectious diseases such as AIDS (acquired immunodeficiency syndrome) or hepatitis where early detection and regular monitoring are required, such tests are not satisfactory (it takes the human body 3 months to generate anti-HIV antibodies associated with AIDS).

On the other hand, the detection of a pathogenic organism via the genetic material contained in its DNA (deoxyribonucleic acid) allows for a rapid and specific diagnosis as soon as illness appears. The DNA that makes up the human genome consists of two complementary strands, whose nucleic bases interact in pairs via "stacking"-type forces and hydrogen bonds to form a double-helix structure. Strand pairing or hybridization is facilitated at neutral pH and high ionic strength. Conversely, these chains can be separated and denatured by increasing temperature and pH, which break the hydrogen bonds.

Genetic diagnoses make use of DNA's property of molecular recognition. A fragment of the DNA of a given virus can be captured by hybridization with a complementary sequence immobilized onto a specific support. These specific assays became a reality principally because of milestone development of molecular biology techniques such as the polymerase chain reaction (PCR), a powerful method for amplifying small quantities of DNA extracted from the genetic material of the virus. Another key improvement was the possible synthesis of short DNA fragments (oligodesoxyribonucleotides, ODNs) using an automatic synthesizer.

A key parameter in diagnosis assays is the support used to extract the targeted biomolecule (capture step). Colloids rather than flat solid supports are generally preferred, the former exhibiting very large specific surface and being easily separated from the biological medium by centrifugation, filtration, or magnetic separation. Among these, colloid polymers attracted major interest because they can be tailored to fit the requirements of the targeted application (concentration of biomolecules, solid support in biomedical diagnostics, drug targeting, etc.) by controlling the particle size and their surface characteristics (polarity, charge, presence of reactive functions such as amine, carboxylic acid, and aldehyde groups).

The immobilization of DNA fragments or ODN on solid supports may be achieved either by physical adsorption or via chemical grafting. In the last few years, several works have focused on the adsorption of ODNs on latex particles. A priori, this method of immobilization is simpler because it only depends on the intermolecular forces existing between the solid surface and the ODN molecules. Furthermore, it is possible to control this adsorption process by varying the conditions of the medium, e.g., by adding divalent ions such as Mg^{2+} and Ca^{2+} to induce ion bridges in DNA/anionic particles systems. However, adsorption process shows two major drawbacks: (1) the immobilizing process is reversible (desorption of molecules with time, competitive adsorption with surfactant, instability of the latex/ODN conjugate); (2) ODN conformation may affect the capture efficiency of the targets via the hybridization process.

ODN grafting on the functionalized latex surface contrarily presents the benefit of creating stable latex–ODN conjugates. In addition, medium conditions can be chosen so that the ODN protrudes in aqueous phase (desorption buffer) to permit fast and quantitative hybridization/capture of DNA. Few studies of ODN grafting onto latex carriers have been reported in the literature, partly for industrial property reasons. One of the most frequently applied techniques consists of immobilizing biotinylated ODN onto latex particles bearing adsorbed streptavidin molecules (biotin–streptavidin complex is extraordinarily strong). Chemical grafting is another way to immobilize ODN bearing a functional spacer arm (amino-link spacer) onto reactive colloidal particles. Particular mention is made of the work of Kawaguchi et al. [1], who describe chemical grafting of ODN onto poly(glycidyl methacrylate)-containing polystyrene latex particles and their utilization for specific capture of DNA fragments.

The global strategy designed for preparing colloid–nucleic acid conjugates for the purpose of diagnostic application is divided in three steps as illustrated in Fig. 1 for amino-containing latex particles and ODN-bearing amino-link spac-

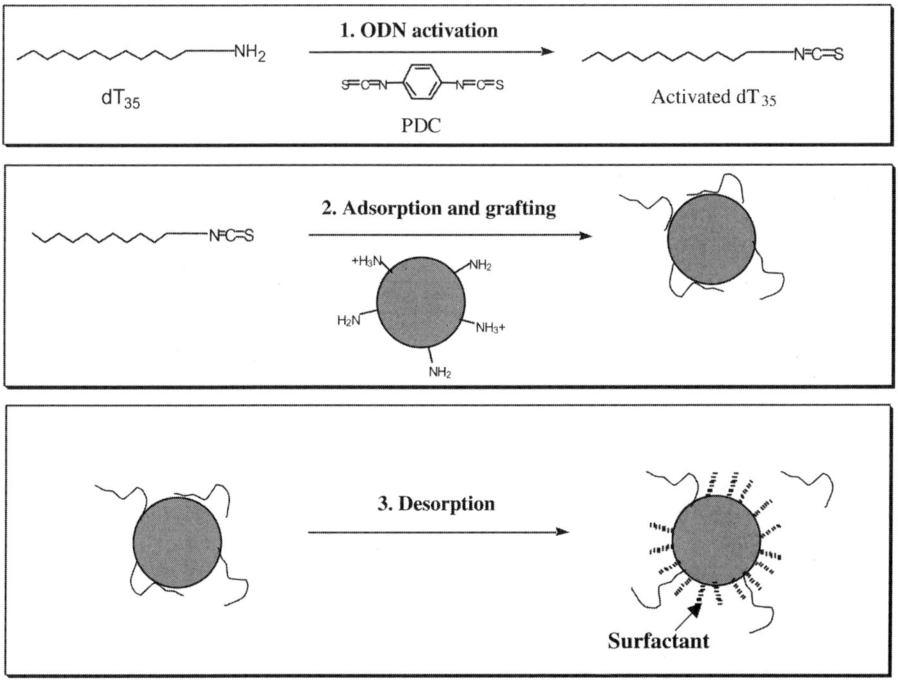

FIG. 1 Methodology used for the covalent grafting of dT_{35} onto amine-containing polystyrene latex particles. Step 1: activation reaction of the dT_{35} by PDC. Step 2: adsorption and grafting of the activated dT_{35} onto aminated latex particles. Step 3: desorption of noncovalently grafted dT_{35} molecules.

ers: (1) activation step of the functionalized ODN, (2) adsorption and covalent of the activated ODN onto considered colloidal dispersion, and (3) desorption study to control the release of nonchemically grafted ODN.

The present chapter is divided into six separate parts where general discussion and appropriate examples combine:

1. Key properties of ideal colloidal polymer particles as a solid support of ODN for nucleic acid diagnostic purposes are described. The different synthesis processes and characterizations performed specifically on amino-containing polystyrene latexes are reviewed.
2. With a view to immobilizing ODN probes on latex particles via chemical grafting process, prior understanding of the interactions between these two entities is vital. The influence of various parameters on ODN adsorption onto latex particles are investigated as a systematic study.
3. In a third part, special conditions leading to complete ODN desorption are discussed. This part is the key step needed to discriminate between the ODN chemically grafted from those simply adsorbed.
4. This step reports on the covalent grafting of chemically activated oligonucleotides onto latex particles. The influence of various covalent binding conditions are presented and discussed.
5. The fifth part deals with the examination of ODN conformation immobilized onto reactive latex particles using small-angle neutron scattering (SANS).
6. This last part describes some fine applications of latex–ODN conjugates in biomedical diagnostic.

II. DESCRIPTION OF AN IDEAL REACTIVE CARRIER

This part first shows the elaboration of appropriate latex particles for the desired biomedical application. Various investigations in biomedical diagnostic studies applying latex particles as a solid phase support did not take into account the basic properties of colloidal dispersion such as particle size, size distribution, hydrophilic–hydrophobic properties, surface charge density, accessibility of reactive compounds, and, finally, colloidal stability of the particles.

The systematic research of the appropriate carrier is illustrated in this chapter by using amine-containing latex particles in a view of oligonucleotide immobilization [2–4]. Cationic polystyrene latexes were prepared via emulsifier-free emulsion polymerization of styrene radical polymerization. The functionality was brought at the interface both by using a functional initiator (2-2′ azobisamidinopropane dihydrochloride, V-50) and a functional comonomer, e.g., vinylbenzene amine chloride (VBAH), or aminoethyl methacrylate hydrochloride (AEMH) (Fig. 2). These functions (amidine and amine) are pH responsive with pK_a around 8 and 10, respectively. Amine rather than amidine group incorpora-

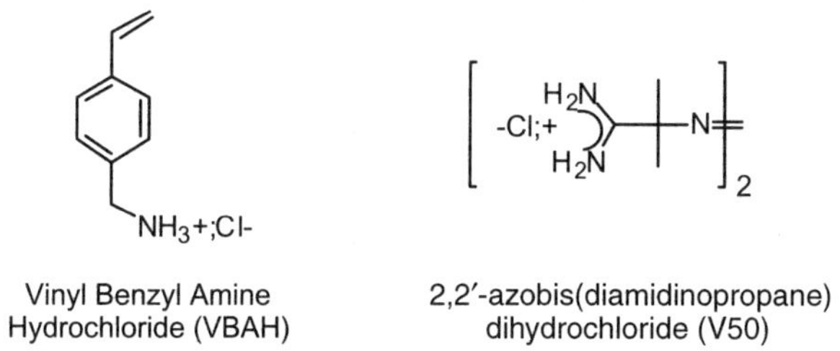

FIG. 2 Comonomer (VBAH) and initiator (V50) used in soap-free emulsion polymerization of styrene.

tion on the particle surface is preferred, the former being stable against hydrolysis in basic medium and serving as a reactive compound for ODN chemical immobilization.

For the same quantity of colloid material, the smaller the particle size, the greater the latex-specific area and the larger the ODN content immobilized. However, particles should not be too small (i.e., smaller than 100 nm) in order to avoid centrifugation separation problems. Using such an emulsion polymerization process, the particle size ranged between 100 and 500 nm depending on the amino-containing comonomer content (Fig. 3). Particle size may vary as

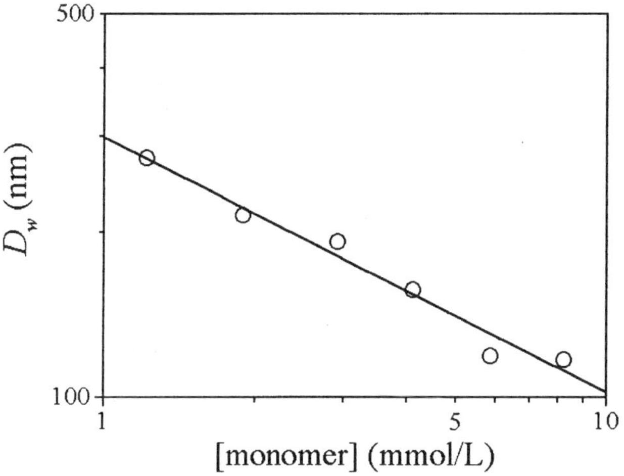

FIG. 3 Log-log plot of particle size vs. functional comonomer (VBAH) concentration in batch emulsion polymerization of stryrene. (From Ref. 3.)

well with the polymerization temperature and the amount of water-soluble initiator, so that systematic studies are necessary to prepare appropriate latex particles.

The addition of tiny quantities of a charged comonomer, either anionic or cationic, in a batch emulsion polymerization of styrene is a way to control the particle size. Incorporation of the functional comonomers in the polystyrene chains mainly takes place during the nucleation step in the aqueous phase (Fig. 4). Its efficiency depends on two intrinsic criteria: the partition coefficients and reactivity of both monomers in water. In addition, reactivity ratios showed that styrene incorporation is favored, ensuring the formation of copolymers rather than water-soluble homopolymer formation. The amphiphilic character of the copolymers formed endows surfactant properties and stabilizes the particle precursors (such as for classical styrene emulsion polymerization scheme) and then more particles are created. This facilitates not only a reduction in the final size

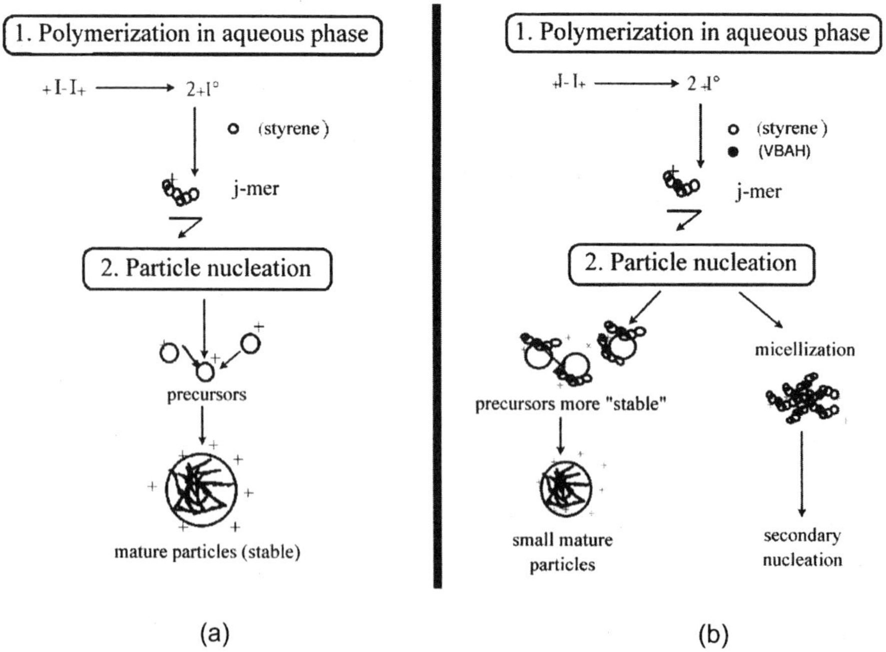

FIG. 4 Mechanism of nucleation in soap-free emulsion polymerization of styrene without (a) and with (b) VBAH monomer.

of the particles but also an increase in both the polymerization rate and the overall conversion.

In the presence of ionic comonomer, the particles are generally of uniform size since the nucleation process arises in a very short time, as is usually the case in traditional emulsion polymerization. However, too high a concentration of functional monomer enlarged the particle distribution in addition to increasing the average particle size. Indeed, excess amphiphilic species compared to the surface developed by particle precursors aggregate into micelles and form new particles (Fig. 4). Such a continuous nucleation process results in a polydispersed final colloid because new particles do not grow at the same speed as original ones. Again, even if the physicochemical properties and reactivity of a functional monomer are known, experimental study is the only means of defining the nucleation threshold in the presence of a given functional monomer.

For any biomolecule immobilization, the surface polarity is of great importance. In fact, the hydrophilic and hydrophobic balance of the interface and the charge density govern the affinity between both species (biomolecules and surface latex particles). It was established that the latex surface polarity is indeed principally related to the nature of the functional groups tethered on the particle surface [5]. Particle surface polarities of amino, carboxyl, and sulfate latexes were measured by contact angle measurements. Increasing pK_a of the tethered surface groups dramatically diminishes the surface polarity. Recent theoretical studies showed that counterion condensation at the interface depends on the nature of the functional groups [6]. Strong acids, such as sulfate groups, entail a high surface polarity, whereas amino groups do not significantly affect the polystyrene surface hydrophobicity. This tendency is indirectly confirmed by investigating the comonomer amount necessary for reaching a given size. Small amounts of strong acid monomers such as sodium styrene sulfonate deeply decrease particle size, whereas high amounts of amine-containing monomers are necessary to obtain similar results. However, on final examination the surface charge densities are equivalent between colloids bearing various charged groups [7].

It is obvious that the surface charge density affects the biomolecule adsorption (i.e., adsorbed amount and adsorption affinity) and consequently the covalent binding efficiency. As described above, high surface charge density may also induce high surface hydration, which lowers the immobilization of hydrophobic biomolecule such as proteins and antibodies.

High charge content (up to 40 μeq/g) can be reached in the shot-growth process. Furthermore, discrepancies of the type of surface functions are attributable to the varieties of titration techniques: colorimetric dosing with SPDP gives the overall density of the charges whereas other approaches allow quantification of the total fraction of amine functions incorporated (nuclear magnetic resonance, fluorescent dosing of the supernatant) or localized on the surface (fluorescent titration on particles). A material balance can thus be done.

In batch process, all particles contain some amidine and amine groups, whose balance and overall content varies with the functional comonomer concentration in the recipe (Fig. 5). No charges were buried inside the latex according to differential titration between the serum and the particles [2,4]. Charge content and particle size can not be independently controlled in a batch process. It was of concern to fix the particle size while varying the comonomer interfacial content by postfunctionalizing the latex surface, either by seeded or shot-addition processes.

In the former, the seed is first synthesized without functional monomer but with a ionic surfactant to control the particle size. The latex is then washed and concentrated by several centrifugation–redispersion cycles (first step is carried out at low solid content, and surfactant must be removed prior seeded process). In a second stage, a given quantity of styrene is introduced beforehand to swell the external layer of the seed followed by addition of a mixture of styrene, functional monomer, and initiator either once (batch process) or at several regular intervals (semicontinuous process). The latter method is particularly recommended to avoid undesirable secondary nucleation.

The shot-addition procedure first consists of carrying out classical batch recipe with a given amount of functional monomer that controls the size of the "seed." At high conversion (≈90%), a mixture of styrene, comonomer, and

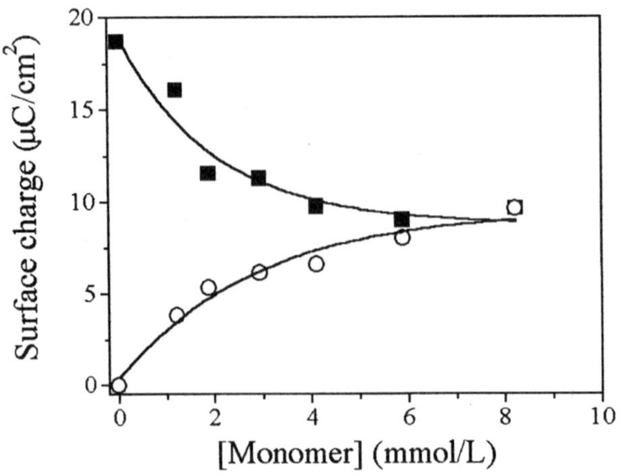

FIG. 5 Overall surface charge (■) and amino surface charges (○) for batch polymerization of styrene with various VBAH monomer content. (From Ref. 4.)

possibly initiator is introduced, either in one step or continuously, with a stoichiometric ratio of comonomer to styrene. Such a process is particularly suited to ensure highly functionalized latex particles; because the seed preparation is easy to perform, secondary nucleation can be avoided (as long as the shot addition is performed above 70% conversion). Finally, it is not time consuming.

The addition of monomers in one or more shots seems to have low effect on the efficiency of incorporating the functional comonomer (VBAH) on the surface of the seed particles. The shot-growth process provides much better yields of incorporation of functional monomer (about 50% as opposed to 10–20% for seeded latexes). Consequently, polyelectrolytes created during batch process and shot-growth polymerization are lower than for a seeded process. Some limitations can, however, be drawn even for the shot-growth process. Late styrene/VBAH shot must be done at high conversion to ensure that all styrene will effectively swell the particles. Furthermore, the quantities of functional monomer to be introduced in the second step must be controlled since, as for batch reactions, adding too much comonomer leads to the formation of new particles most likely by micellar nucleation.

The colloidal stability of latex particles before, during, and after the ODN chemical grafting step is required. The colloidal instability of bare particles may be caused by various parameters, such as high salinity of the medium or pH variations, which decrease the electrostatic repulsions by charge screening or neutralization, respectively. During the grafting step, the presence of free biomolecules in the dispersed medium can also induce bridging flocculation of the colloidal particles. Finally, the colloidal stability of the particles during their application may be totally critical for the reproducibility and the specificity of the investigated biomedical application.

III. ADSORPTION AND DESORPTION OF ODN ONTO LATEX PARTICLES

A. Generalities About Adsorption of Polyelectrolytes onto Solid Supports

DNA or RNA derivatives are polyelectrolytes in nature. This section thus recalls some fundamental aspects of polyelectrolyte adsorption and conformation onto charged colloidal dispersions.

A given polyelectrolyte is considered as being adsorbed when, after a mixing period with solid surface, one of its units is in direct contact with the surface. The adsorption is considered as a thermodynamic balance between free polyelectrolytes and adsorbed ones. However, desorption is generally a slow process governed by the reduction in the attractive forces involved in the adsorption process.

When the interactions between the polyelectrolytes and the charged support are weak (such as in a system bearing the same charges), the maximal amount of adsorbed polyelectrolytes is very low and increases with ionic strength, since the presence of electrolytes decreases the electrostatic repulsion forces between the polyelectrolytes and solid support. If the interactions between the polyelectrolytes and the charged surfaces are strong (oppositely charged system polymer/surface), two possibilities may exist: either a low increase or a substantial decrease of adsorption with ionic strength is observed. In the first case, the increase of the ionic strength reduces the repulsive electrostatic repulsions between the adsorbed polyelectrolytes and the free polyelectrolytes in solution; thus the amount adsorbed increases slightly. In the second case, a competition between the adsorption of polyelectrolytes and the ions present in solution occurs which participates in macromolecule desorption. Experimental facts remain the best referee.

The buffer plays a considerable role in adsorption phenomena by modulating the solubility of the polymer in the adsorption medium [8]. The less the polymer is soluble in the considered medium, the higher the adsorption. The pH, salinity, and temperature are then the drastic parameters to control while performing polyelectrolyte adsorption.

B. Methodology

Most of the studies were performed on a poly(thymidylic acid) (dT_{35}) oligonucleotide whose structure is given below (Fig. 6), though concepts derived here mainly apply to other structured ODNs. The oligonucleotide bears an amine group ($_2HN\text{-}(CH_2)_6\text{-}$) at the 5′ position, which permits grafting reactions. In some cases, the oligodeoxyribonucleotide was labeled with ^{32}P or with chemically grafted fluorescein isothiocynate (FITC).

The quantification of immobilized (adsorbed or grafted) nucleic acids was principally determined from the residual nucleic acids in the supernatant (depletion immobilization methodology). The amounts of immobilized nucleic acids were determined using the following equation which allows the calculation of the amount of immobilized oligonucleotide, A_S (mg/g):

$$A_s = V\left(\frac{C_i - C_f}{m}\right) \qquad (1)$$

where V (mL) is the final volume of the solution, C_i (mg/mL) and C_f (mg/mL) are the initial and final concentrations of ODN in the solution, respectively, and m (g) is the mass of adsorbent. The total amount of immobilized ODN is generally expressed in mg/m^2, to overcome the differences in particle sizes among various latexes.

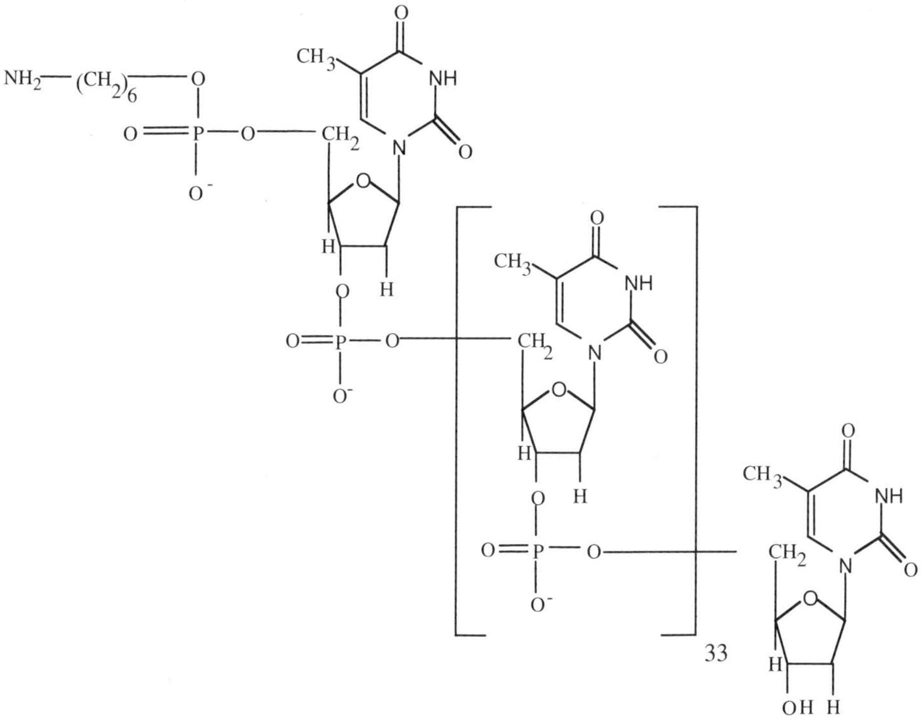

FIG. 6 Structure of polythymidylic acid (dT_{35}) bearing aliphatic primary amine.

C. Adsorption Kinetics

The adsorption kinetics of oligonucleotides onto latexes is essential to evaluate the incubation time needed to reach the adsorption equilibrium. Both positively and negatively charged latexes (latex bearing sulfate groups and latex bearing amine and amidine groups, respectively) were tested [9].

The adsorption of ODN onto positively charged polymer particles was found to be rapid as expected due to their opposite charges, whereas the adsorption rate on negatively charged latex particles was low. The observed kinetics results have mainly been discussed in terms of attractive or repulsive electrostatic forces, though in some cases weaker hydrophobic interactions play a nonnegligible role in the adsorption process, especially on negatively charged latexes [10]. In addition, complementary experiments were also carried out on coated cationic polystyrene particles, e.g., where a noncharged surfactant was first adsorbed on

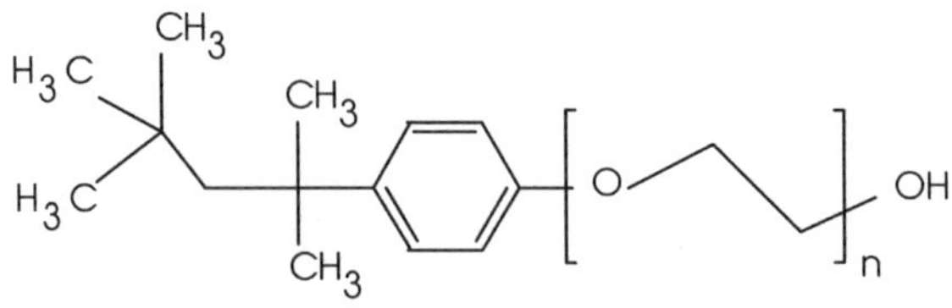

FIG. 7 Structure of Triton X-405 surfactant [*p*-(1,1,3,3-tetramethylbutyl)phenoxypoly-(oxyethylene glycol)] with an average *n* value of 40.

the particles to modify the surface polarity and cationic charge accessibility (Triton X-405; see Fig. 7) [11]. In this case, the adsorption rate was low compared to bare cationic polystyrene latex particles because of the reduction in attractive forces involved in the adsorption process (Fig. 8). In addition, the noncharged layer on the particle surface acts as an adsorption barrier responsible for the decrease of both the adsorption rate and the amount of adsorbed oligonucleotides.

D. Adsorption Isotherms

Adsorption isotherms show the variation of the quantity of macromolecules adsorbed per surface unit (or by mass) of the adsorbent as a function of the concentration in solution at the adsorption equilibrium state. Langmuir's model

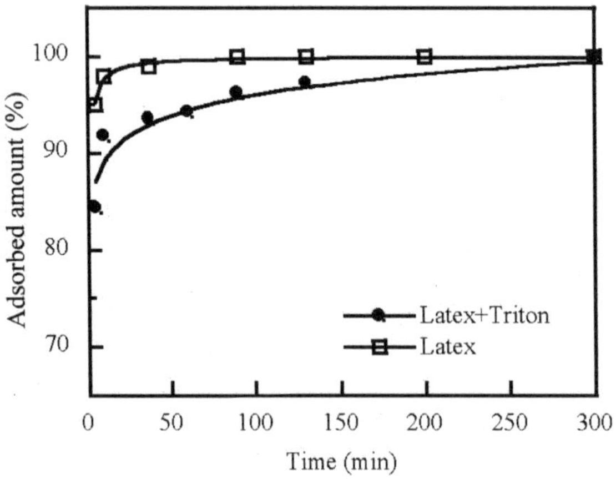

FIG. 8 Normalized adsorbed amount of adsorbed oligonucleotide (dT_{35}) as a function of time (at 20°C, 0.1 mg mL^{-1} of dT_{35}, 0.1 mg mL^{-1} of latex particles, 10^{-2} M ionic strength, and pH 5). (From Ref. 11.)

[12], first defined to describe the adsorption of gas on solid supports, is often adapted to macromolecule adsorption onto colloidal particles. It is based on the following hypotheses: (1) the surface is homogeneous; (2) a single molecule is adsorbed per adsorption site (i.e., monolayer of interfacial molecules); (3) there are no lateral interactions between the adsorbed macromolecules; (4) there is no competitive adsorption; (5) the adsorption is reversible (equilibrium between adsorption and desorption).

By considering the adsorption equilibrium and the Langmuir adsorption hypothesis, the adsorbed amount (A_s) and the macromolecule concentration equilibrium (C_{eq}) are related to the adsorption affinity constant (K) via the following equation:

$$\frac{A_{s,\max}}{A_s}.C_{eq} = \frac{1}{K} + C_{eq} \tag{2}$$

where $A_{s,max}$ is the maximal quantity of the macromolecule adsorbed onto solid surface. The determination of K constant provides information on the affinity of the macromolecule for the adsorbent surface, whereas the plateau of the isotherm corresponds to the saturation of the surface under the adsorption conditions (expressed as $A_{s,max}$). Although this model fits such isotherms quite well, interpretations are restricted as neither the equilibrium adsorption nor the conformation of the macromolecules at the interface is taken into account.

After kinetics, adsorption isotherms of nucleic acids onto latexes as a function of various parameters, such as pH, salinity, and temperature, are the starting point of any adsorption study. Both cationic and anionic polystyrene latexes were examined by Elaissari et al. [10]. The adsorption isotherms using anionic latexes particles were found to exhibit low affinity compared to the cationic latexes [9,10], using oligonucleotides of 27 and 30 nucleotides, respectively (Fig. 9a). The adsorption isotherms exhibit two marked domains: (1) a rapid increase in the adsorbed amount for low bulk concentration of oligonucleotide and then (2) a slight increase of the amount adsorbed before reaching a plateau value.

The isotherms of oligonucleotides adsorption onto cationic polystyrene particles bearing noncharged surfactant (Triton X-405, 0.1 mg/m^2) have been also investigated (Fig. 9b). The presence of the interfacial surfactant on the colloidal particles affects the adsorption affinity and the maximal amount of adsorbed ODNs. In fact, low affinities and adsorbed amounts were observed principally at basic pH rather than at acidic pH, whereas high affinities are observed in the case of bare cationic latexes.

E. Effect of pH

The investigation of the effect of pH on the adsorption of charged macromolecules onto charged solid support is of paramount importance. In fact, the pH

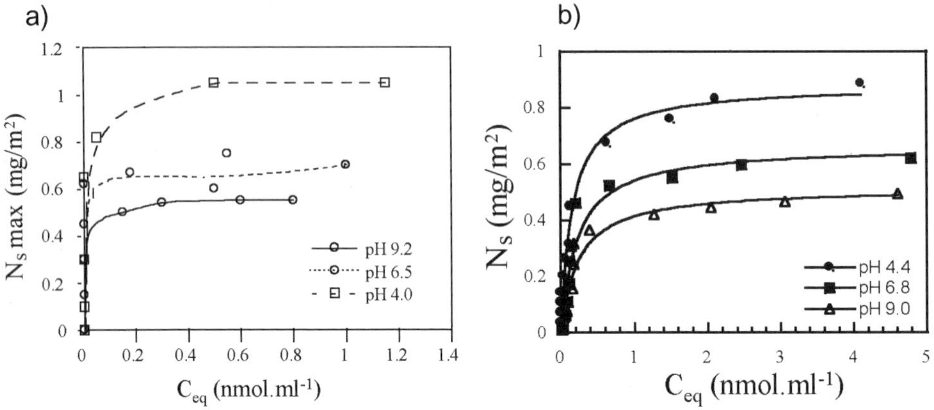

FIG. 9 (a) Adsorption isotherms of oligonucleotide onto bare latex particles. Samples were mixed and incubated for 2 h at 20°C, 10^{-2} M ionic strength at a given pH. (b) Adsorption isotherms of oligonucleotide onto precoated latex particles (by adsorbing small amount of Triton-X405). Samples were mixed and incubated for 2 h at 20°C, 10^{-2} M ionic strength at a given pH. (From Refs. 11, 15.)

affects the charge density and charge distribution of the considered support without altering the nucleic acid charges.

The adsorption of oligonucleotides onto negatively charged sulfate-polystyrene latexes was found to be low and marginally pH dependent. The behavior is mainly related to the negligible effect of pH on the strong acids borne by the latex. On the opposite, the adsorption of ODNs onto cationic (amidine or amine) polystyrene latexes is dramatically affected by the pH of the medium irrespective of oligonucleotide sequence (Fig. 10). The observed behavior is attributed to a decrease of cationic surface charge density on the latex particles while increasing the pH.

The effect of pH on oligonucleotide adsorption onto cationic polymer particles can be dramatically perturbed or at least modified while coating the latex surface with nonionic surfactant. The low adsorbed amount of Triton X-405 (~0.1 mg/m^2) has a negligible effect on the oligonucleotide adsorption below pH 7, i.e., in the high attractive electrostatic interactions domain. On the other hand, above pH 7, the surfactant influence is more marked since the adsorption of oligonucleotides is dramatically reduced and under some conditions the ODN adsorption is totally suppressed (Fig. 11). The surfactant layer clearly acts as an adsorption barrier leading to reduced ODN amounts on the surface. Apart from charge screening, implementing a certain degree of hydrophilicity of the particle surface may also reduce the van der Waals forces, such as stacking interaction between aromatic bases of ODN and the polystyrene support.

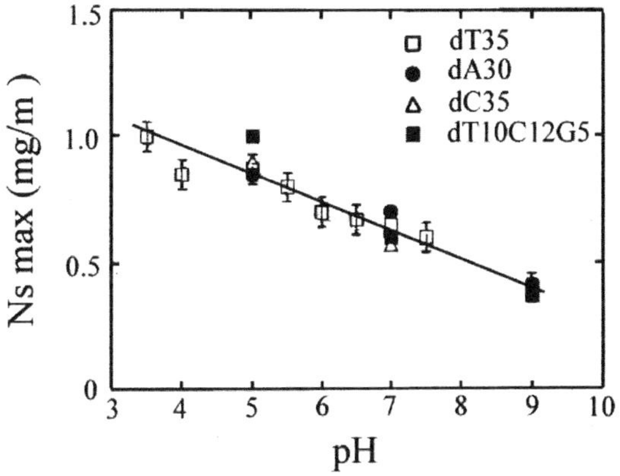

FIG. 10 Maximal adsorbed dT_{35}, dC_{35}, dA_{30}, and $dT_{10}dC_{12}dG_5$ on cationic latex particles as a function of pH at 20°C and 10^{-3} M ionic strength. (From Ref. 15.)

F. Effect of Ionic Strength

According to the polyelectrolyte character of oligonucleotides, the effect of ionic strength should not be neglected in the understanding of the driven forces involved in the adsorption process. It is imperative when investigating the effect of salt to preserve the colloidal stability of the latex particles. The first study on the influence of ionic strength has been dedicated to the ODN adsorption onto

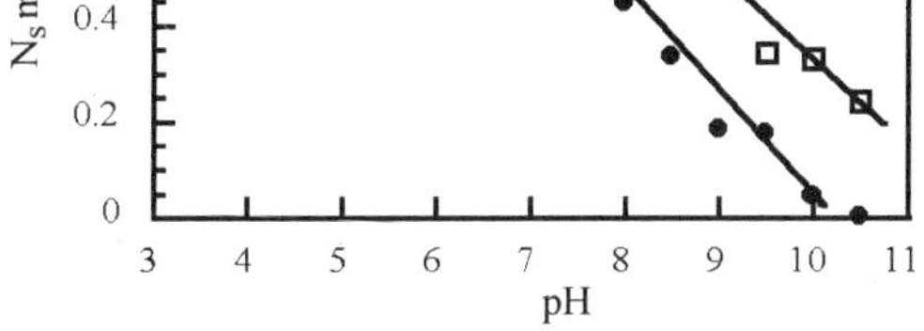

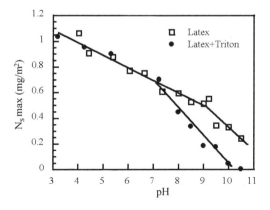

FIG. 11 Effect of pH on the maximal adsorbed dT_{35} onto bare and precoated latex particles (by adsorbing a small amount of Triton-X405) as a function of pH at 20°C and 10^{-3} M ionic strength. (From Ref. 11.)

sulfate polystyrene latexes. Surprisingly, the salinity has a marginal effect on the ODN adsorbed amount [10].

On the contrary, the increase in salt concentration has a nonmarked effect while using cationic latexes. A small increase in ODN adsorption was observed for bare cationic particles, whereas surface ODN contents decreased in the presence of the interfacial nonionic surfactant layer (Fig. 12). The behavior observed has been mainly attributed to the reduction of attractive electrostatic interactions. In addition, increasing the salt concentration may enhance the contribution of hydrophobic forces [13].

G. Exchange Kinetics

The exchange process in the adsorption of macromolecules onto colloidal particles has been investigated by Pefferkorn et al. [14] and was related to the establishment of adsorption equilibrium, which was found to be dependent on incubation conditions (temperature, pH, salinity, solvent nature, etc.). Similarly, the exchange processes and kinetics of oligonucleotides adsorption using dT_{35} and radiolabeled dT_{35}-^{32}P were performed on bare and coated cationic polystyrene latexes. The radiolabeled ODN was first adsorbed and the supernatant (or fraction of the supernatant) containing free ODN molecules was replaced by nonradiolabeled ODN. After incubation time, the radiolabeled ODNs were detected either in the supernatant or directly on the particles.

The exchange kinetics investigated at basic pH in which the attractive electrostatic interactions are low were found to increase slightly as a function of

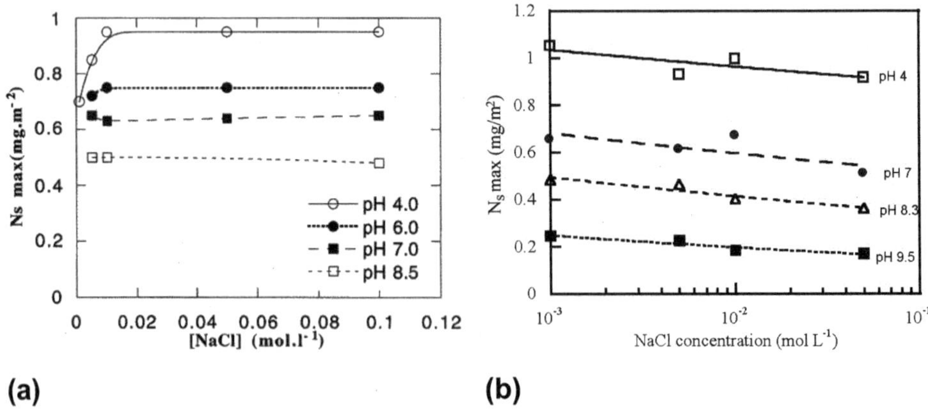

FIG. 12 Effect of ionic strength on the maximal adsorbed dT_{35} onto (a) bare and (b) precoated latex particles (by adsorbing a small amount of Triton-X405) as a function of pH. (From Refs. 11, 15.)

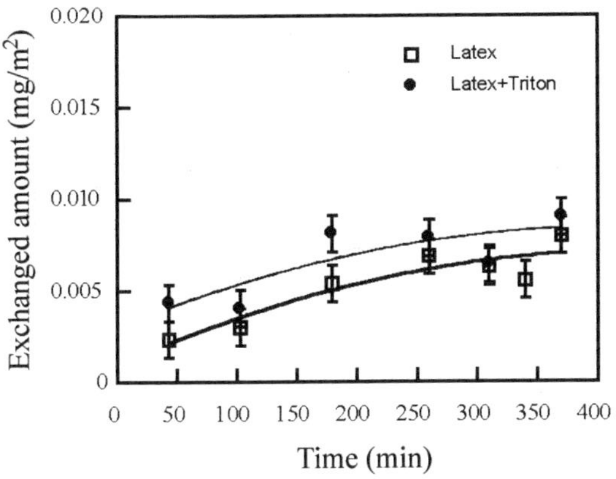

FIG. 13 Amount of dT_{35} exchanged as a function of time. The adsorbed amount was 0.5 mg m^{-2} and 0.3 mg m^{-2}onto bare and precoated latex respectively (at 20°C, pH 9.2, and 10^{-2} M ionic strength). The exchange was performed using surfactant-free buffer. (From Ref. 11.)

time before reaching a plateau (Fig. 13). Exchange between adsorbed and free oligonucleotides doubtless exists but remains small. The initial exchange rate determined from the slope of the exchanged ODN amount vs. time was found to be 1.3 μg m^{-2} h^{-1}.

H. Thermodynamic Aspects

The adsorption of single-stranded DNA or RNA fragments (i.e., oligonucleotides) onto colloidal particles bearing negative or positive surface charges turned out to depend mainly on electrostatic forces. The attractive electrostatic interactions between the cationic support and the ODNs favor their immobilization, whereas for anionic latexes for which hydrophobic interactions slightly compete with electrostatic repulsions, very few chains adsorb on the particle surface.

Similar to polymer adsorption onto solid support, the ODN adsorbed amount can be related to the adsorption energy (ΔG) of each monomer constructing the ODN chain and the ODN bulk concentration (C) as expressed by the following relationship [15]:

$$N_s \approx kC.\exp(-n\ \Delta G) \tag{3}$$

where n is the number of bases and k is a constant depending on the nature of the system and the experimental conditions. The adsorption energy of each molecule can be considered as the sum of all contributions ($\Phi_{\text{electrostatic}}$ and $\Phi_{\text{hydrophobic}}$ energies):

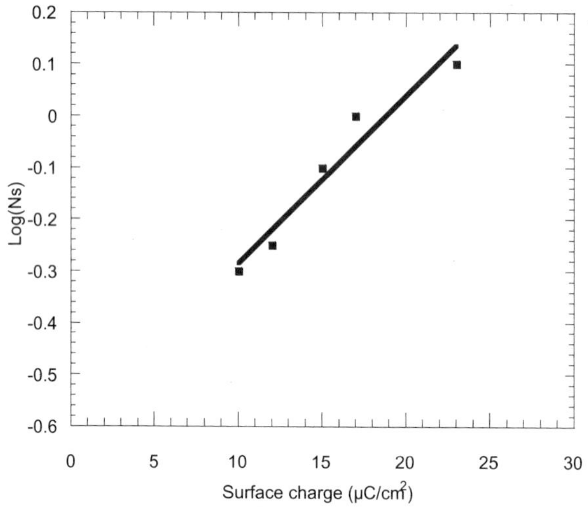

FIG. 14 Maximal adsorbed dT_{35} (Ns in mg.m^{-2}) on latex particles as a function of surface charge density (pH 5.0, 25°C, and ionic strength 10^{-2}). (From Ref. 16.)

$$n\ \Delta G \approx \Phi_{hydrophobic} + \Phi_{electrostatic} \tag{4}$$

The hydrophobic adsorption energy is neglected so that only the electrostatic term is considered. This electrostatic adsorption energy can be expressed as the product of oligonucleotide charge (σ_{oligo} assumed constant) and the surface charge density (σ_{latex}) of the colloidal support as expressed below:

$$\Phi_{electrostatic} \approx \sigma_{oligo}.\sigma_{latex} \tag{5}$$

To point out the relationship between ODN adsorption and pH, the plot of $\log(N_s)$ vs. the surface charge density (i.e., ζ potential at a given pH) is reported in Fig. 14. Extrapolation of linear variation behavior at zero charge density (i.e., zero ζ potential) gives the maximal ODN adsorbed amount reached via exclusive hydrophobic interactions ($\sim$0.3 mg/m^2) [10].

IV. DESORPTION STUDY

A. Methodology

Two methods are generally used to investigate the desorption of preadsorbed macromolecules: (1) the desorbed amount of ODNs is studied as a function of washing steps using an appropriate desorption medium (in this case, the continuous phase is totally exchanged); (2) the desorption is performed by slightly changing the adsorption medium by diluting the dispersion (or changing the salinity or the pH) in order to affect the adsorption equilibrium. The first method is particularly suited to distinguish between chemically grafted ODNs and physically adsorbed ones.

The desorption in the case of sulfate latexes was found to be total after a few washes irrespective of the desorption condition (i.e., pH, salinity, etc.) [16]. In this case, serum replacement was found to be efficient. We report here exclusively on desorption of ODNs preadsorbed on cationic latex particles.

B. Effect of Nonionic Surfactant on the Desorption of Oligonucleotide

Since the adsorption of oligonucleotides onto coated cationic latex particles is reduced irrespective of pH and ionic strength, ODN desorption was first investigated by washing the latex particles bearing adsorbed ODN (at pH 5, 10 mM ionic strength) with a borate buffer (10 mM, pH 9.2) containing 1 wt % Triton X-405 surfactant. In this case, the washing step was performed by removing and replacing 25% v/v of the supernatant. The desorption yield was clearly affected compared to surfactant-free buffer as illustrated in Fig. 15, in which the residual adsorbed amount is reported as a function of washing steps. Desorption yield at acidic pH still remains low because the supernatant was not totally

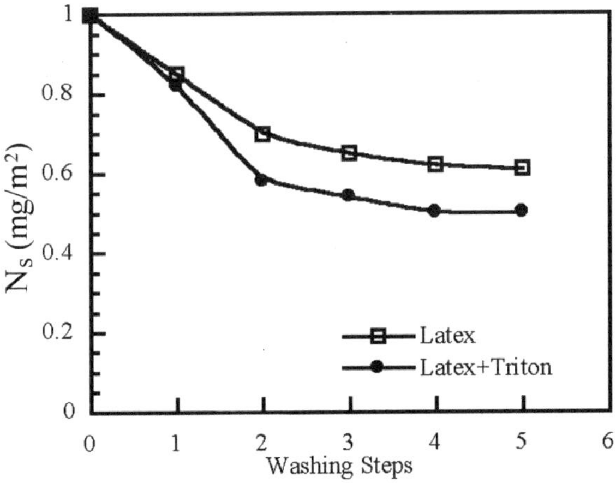

FIG. 15 Residual adsorbed dT_{35} onto bare and precoated latex particles as a function of washing steps. Adsorption was performed onto bare latex particles at pH 5 and 10^{-2} M ionic strength. (Precoated particles: colloidal particles bearing a small amount of adsorbed Triton-X405.) (From Ref. 11.)

replaced. Consequently, a plateau with a nonnegligible residual amount of adsorbed ODN is reached, even after five washes. Anyway, the utilization of buffer containing noncharged surfactant enhances the desorption efficiency, which may be attributed to surface modification property (due to the surfactant adsorption) and to slight exchange between ODNs and free surfactant molecules.

C. Effect of pH on the Desorption of Oligonucleotide

In order to reduce the attractive electrostatic forces, the desorption was then investigated as a function of pH using basic buffer containing 1 wt % Triton X-405. Remaining adsorbed ODN amounts after two washes using the above basic buffer at a given pH are reported in Fig. 16. As clearly illustrated in the figure, the amount of residual oligonucleotide on cationic latex particles decreased dramatically with increasing the pH of the buffer after only two washes. The residual average adsorbed amount (~0.4 mg/m^2) is nonnegligible even at high basic pH (such as pH 11). In fact, the amino-containing latex particles exhibit slightly cationic character even at pH 10, as evidenced from the electrokinetic study and attributed to the high pK_a of primary amine groups on the particles.

D. Effect of Ionic Strength on the Desorption of Oligonucleotide

To reduce the attractive electrostatic interactions between both species (ODN and cationic latex particles), the influence of salt concentration was investigated. The effect of ionic strength on the desorption process is reported in Fig. 17, in

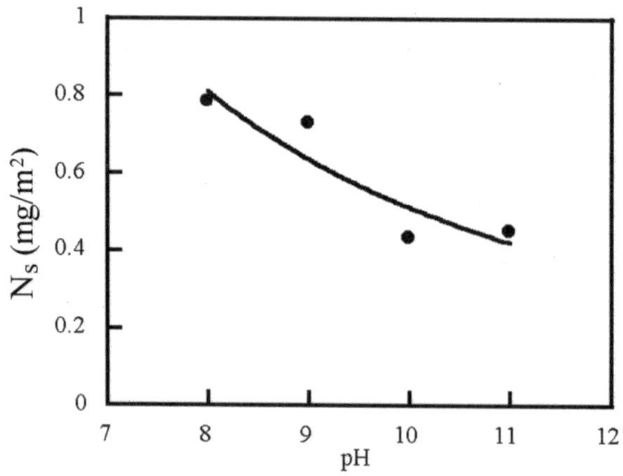

FIG. 16 Residual adsorbed dT_{35} onto bare latex particles as a function of pH. Adsorption was performed on bare latex particles. The desorption was carried out after two washes using buffer (at a given pH, 10^{-2} M ionic strength, and 1% Triton X-405). (From Ref. 11.)

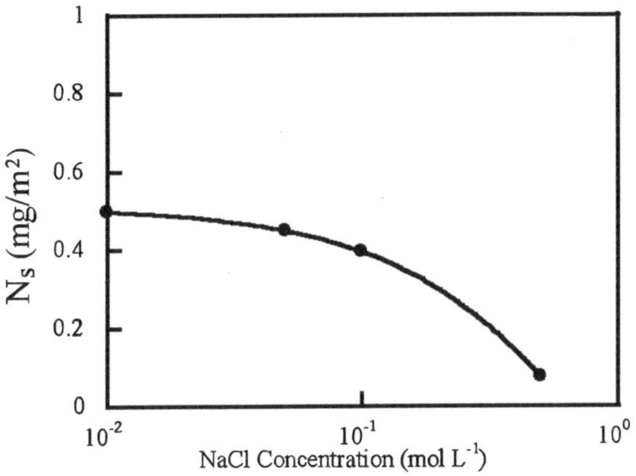

FIG. 17 Residual adsorbed dT_{35} onto bare latex particles as a function of ionic strength. Adsorption was performed on bare latex particles with Ns = 1.0 mg m^{-2} at pH 5 and 10^{-2} M ionic strength. The desorption was carried out after two washes using buffer (at various ionic strengths, pH 10, and 1% Triton X-405). (From Ref. 11.)

which the residual adsorbed amount of preadsorbed ODN (at pH 5) is reported as a function of added NaCl concentration (buffer pH 10, 1 wt % Triton X-405). The desorption was found to be dramatically enhanced at high salt content (after two washes only). The screening effect of the surface charges (due to the salt) and the hydrophilic (because of the surfactant layer) character of latex surface enhance the repelling process of the ODNs. Then the combination of pH, ionic strength, and competitive agent (i.e., noncharged surfactant) leads to good control of the desorption phenomena. In fact, the pH and ionic strength reduce the attractive forces, whereas the surfactant acts as a competitor and as an energy barrier affecting the adsorption of free ODN.

V. CHEMICAL IMMOBILIZATION OF OLIGONUCLEOTIDES ONTO AMINO-CONTAINING LATEX PARTICLES

A. Methodology

The covalent binding of reactive ODN onto functionalized latex particles can be reached via two possible methodologies (direct and indirect process). Due to the synthesis facility, the commonly used ODNs exhibit amine spacer arm at the 5′ position. The possible chemical reactions are summarized in Table 1. Direct covalent immobilization is generally favored to avoid the activation step which induces undesired phenomena such as steps multiplication, low colloidal stability, uncontrolled chemical activation process, possible hydrolysis, and in

TABLE 1 Possible Chemical Reaction Between Reactive ODN Bearing Amine Spacer Arm and Functionalized Support

Reactive support	Remarks
Amine	Activation step
Carboxylic	Activation of support
Aldehyde	Direct coupling
Ester	Direct coupling
Epoxy	Direct coupling

some cases secondary reactions. In this section, the covalent coupling via two steps is presented in order to show the complexity involved in such mythology in which systematic studies are needed.

B. Activation of ODNs

The amino-functionalized ODN is first activated to react with amino groups on latex amine surface. The PDC is first chemically grafted to the amino-linker borne on ODN 5′ extremity; the remaining isothiocyanate group should react with amino groups at the latex surface (Fig. 18) [17]. Capillary gel electrophoresis technique is mainly used to separate ODNs differing by only one nucleotide in their length. Electrophoregrams of activated dT_{35} ODNs showed a large shift on its retention time compared to the nonactivated one, which could not solely be explained by one PDC addition.

MALDI-TOF mass spectrometry confirmed this substantial increase in molecular weight. An average of four PDC/ODN chains were found, indicating that amines located at 3′ position (N-3) on the thymine cyclic moiety react with PDC. The chain distribution for activated ODNs was also broader than that of nonactivated ones, reflecting that the number of PDC reactions on nucleic bases varies significantly on the same ODN sample.

Enzymatic hydrolysis of the ODNs followed by reversed-phase HPLC analysis provided direct proof of base activation, as showed in Fig. 19. Apart from dT and dC (used as a marker) peaks, two new peaks showed up at high retention times belonging to more hydrophobic molecules. One was attributed to the 5′ amino-link arm substituted with PDC (confirmed by coinjection), whereas the second peak was assigned to the thymine nucleoside bearing one PDC molecule. Peak integration and comparison with their expected values (according to MALDI-TOF molar mass) confirmed these peak allocations. The reaction of the amine groups of the aromatic bases with the activating agent (PDC) was then clearly evidenced.

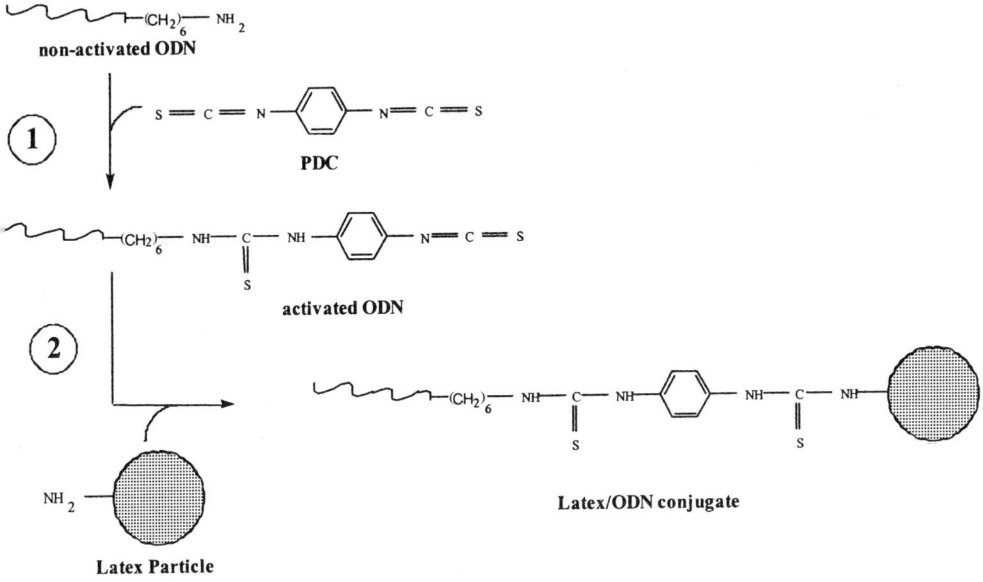

FIG. 18 Methodology of covalent immobilization of ODNs onto aminated latex particles. Step 1: activation reaction of the ODNs by PDC. Step 2: grafting of the activated ODNs onto aminated latex particles.

C. Influence of the Activated ODN Concentration on Covalent Grafting Yield

The covalent grafting yield was found to be appreciable when the base activation is high. This result attests that PDC side reactions on the amine groups of the nucleic bases contribute to the ODN grafting efficiency onto amino-containing latex particles. Figure 20 shows the maximal grafted ODN as a function of the number of bases activated, independent of the ODN structure. A maximal amount of 0.45 mg/m^2 was found, which is quite high compared to the amount of adsorbed ODN in similar medium conditions.

Poly(thymidylic acid) exhibiting various chain lengths (i.e., 13, 18, 20, and 35 nucleotides) was activated and grafted onto polystyrene latexes. The percentage of activated bases was roughly the same for all ODNs, except for the short sequence. The amount of covalently immobilized ODN increased upon increasing the chain length to reach a similar plateau value as discussed above. For long ODN chains, the potential sites available for the ODN grafting reaction with the latex surface amine groups are higher.

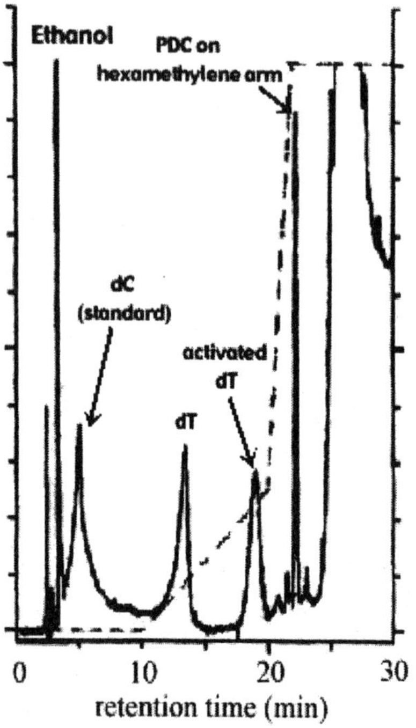

FIG. 19 Reversed-phase HPLC of digested dT_{35} ODN activated by PDC. The dash line represents the gradient applied for HPLC (methanol from 0 to 95% in phosphate buffer). (From Ref. 17.)

D. Influence of Adsorption on Chemical Grafting Efficiency

The adsorption of ODN during chemical grafting experiences appears to be vital in order to bring the ODN into contact with the surface of the latex particles. A priori, the conditions facilitating adsorption and covalent grafting are incompatible. In fact, the adsorption is more efficient under high cationic charge densities (protonated amines at acidic pH) whereas grafting is better facilitated when the amine functions are no longer protected. Thus, a compromise has to be found between these different operating conditions.

As previously reported, adsorption of dT_{35} is a very rapid process. Similarly, covalent grafting was shown to be very fast presumably because of multisite grafting. Meanwhile, the amount of overall immobilized activated dT_{35} increased slightly with time as shown in Fig. 21. The covalent grafted amounts were determined after desorption investigations as discussed above. The amount of

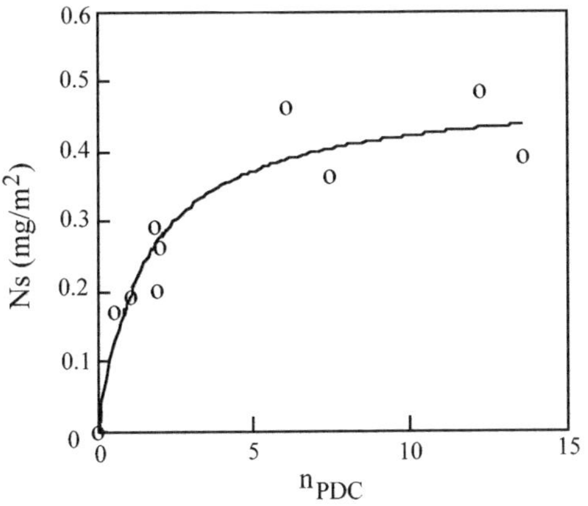

FIG. 20 Influence of the number of PDCs per ODN chain on the amount of ODN covalently linked to the polystyrene latex. (From Ref. 17.)

immobilized oligonucleotide decreased upon increasing pH (from 8 to 11), due to the vanishing of the cationic surface charges on the particles (Fig. 22). However, this decrease is less marked than in the case of the adsorption process. The behavior revealed principally the contribution of electrostatic interaction to the chemical grafting process.

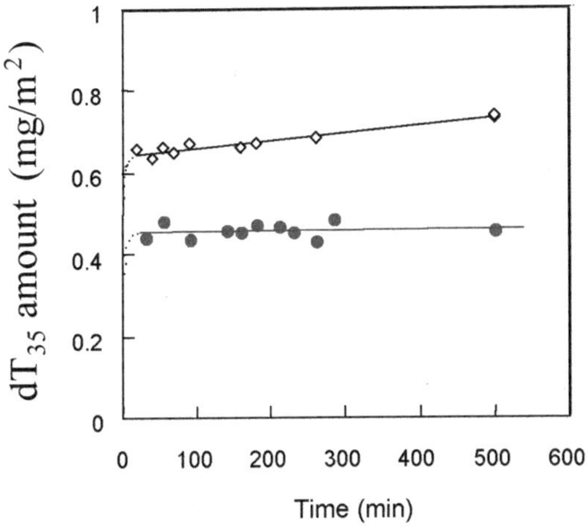

FIG. 21 Kinetics of adsorption and grafting of dT_{35} on latex particles in 10 mM sodium borate buffer, pH 9.3. (◇) immobilized dT_{35}; (●) covalently grafted dT_{35}. (From Ref. 26.)

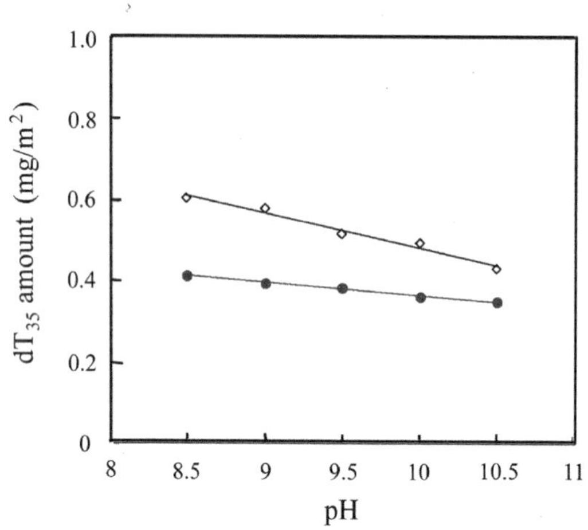

FIG. 22 Influence of pH on the adsorption and grafting of dT_{35} onto latex particles. (◇) immobilized dT_{35}; (●) covalently grafted dT_{35}. (From Ref. 26.)

E. Influence of the Surface Charge Density

The effect of the latex surface density on dT_{35} grafting was studied using various latexes. As expected, the greater the cationic charge on the latex surface, the higher the dT_{35} adsorption through electrostatic interactions, as illustrated in Figure 23. The extrapolation to zero charge provides 0.35 mg/m^2 immobilized

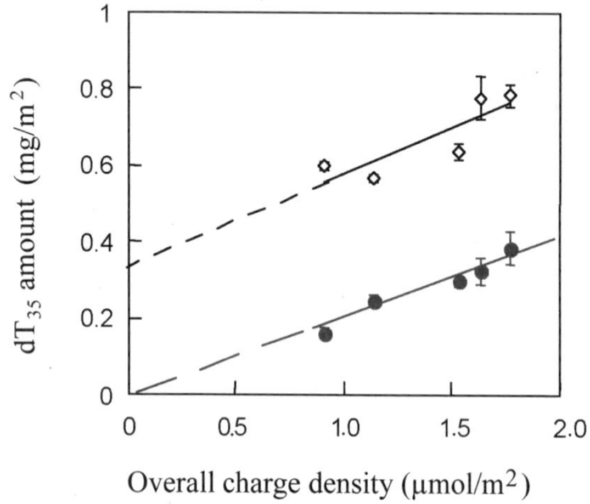

FIG. 23 Influence of the overall charge density of the latexes on the adsorbed and grafted dT_{35} amounts. (◇) immobilized dT_{35}; (●) covalently grafted dT_{35}. (From Ref. 26.)

dT_{35}, slightly higher than the value for nonactivated dT_{35}. In addition, both amine and amidine functions were able to react with the activated dT_{35} molecules, as denoted by the linear relationship between the amount of grafted dT_{35} and surface charge density (Fig. 23). The amount of desorbed dT_{35} was almost constant, revealing that both functions (amine and amidine) exhibited the same reactivity toward the activated ODN. In any case, to target a high grafted amount of activated ODN, high surface charge density is needed.

The influence of a dT_{35} sequence added as a spacer arm to any ODN sequence results in enhancement of the covalent binding efficiency [17] and, consequently, the hybridization yield.

VI. CONFORMATION OF IMMOBILIZED OLIGONUCLEOTIDES ONTO LATEX SURFACE

A. Generalities on the Conformation of Polyelectrolytes

The conformation of polyelectrolytes in *adsorbed* or grafted state is thermodynamically unstable since it can change in time and space, though without desorption of the macromolecule. Globally, the conformation is the result of two factors: (1) electrostatic interaction from an attractive field exerted by the charges on the solid surface and (2) the intra- and interchain electrostatic repulsions between segments of the same charge. This effect depends essentially on the charge density of the macromolecule and on the amount adsorbed.

The conformation of a flexible polyelectrolyte in a good solvent is generally described by a "train-loop-tail" model [8]. Adsorption trains are composed of segments entirely adsorbed on the surface; loops have two points of contact with the surface while tails are chain-end protrusions. These conformations depend on the intrinsic nature of both the macromolecule (particularly its molar mass) and the colloid (nature and charge density, surface polarity, etc.), as well as the adsorption medium (pH, ionic strength, buffer composition, temperature). Note that predicting the conformation of the chains at the interface as a function of molar masses remains a complex phenomenon to describe [18].

Two extreme conformations are generally considered to describe macromolecule adsorption on solid surfaces. If strong interactions occur between the macromolecule and the surface, segments of the polyelectrolyte are all in contact with or close to the surface, resulting in a so-called flattened conformation. If, however, the surface and/or the polyelectrolytes are weakly charged, the conformation of the macromolecular chain depends on its affinity with the adsorption sites. This leads to the formation of sporadic contacts between the polyelectrolyte and the surface, producing a loop-and-tail conformation. The chains thus compose a diffused interfacial phase whose concentration profile depends on the affinity of the solvent with the polymer.

B. Small-Angle Neutron Scattering Analysis

The conformation of the immobilized ODN molecules (adsorbed or grafted) on the deuterated polystyrene latexes was analyzed using SANS on three types of sample: bare latex, latex bearing adsorbed ODN, and latex carrying chemically grafted ODN [19].

Since the neutron scattering of the latex particles was not completely nil and the quantities of ODN adsorbed or grafted were very low, each measurement required considerable counting time. However, a systematic approach during data processing combined with knowledge of given variables (length of scattering of certain species, quantities of ODN immobilized) made it possible to draw out a trend on ODN surface conformation.

The latexes carrying the adsorbed or grafted ODN were then analyzed at acidic pH (i.e., pH 5), where the scattered intensity $I(q)$ is shown as a function of scattering vector (q) in Fig. 24. The curves in the figure are obtained from a model in which the thicknesses of the ODN layer are 20, 50, 80, and 120 Å, respectively. The results show that under these conditions the conformation of adsorbed ODN is neither "brush-like" (which would be the case for $h = 150$ Å) nor completely flat (for $h = 10$ Å). A mean thickness from 50 Å to 70 Å correlated with the experimental results depending on the ODN-immobilized amount. Varying the medium's ionic strength and pH had no effect on adsorbed or grafted ODN layer thickness.

The conformation of ODN adsorbed onto anionic and cationic polystyrene latexes particles was first reported by Walker et al. [20,21]. Preliminary determination of the ODN layer thickness on latex surface by photon correlation spec-

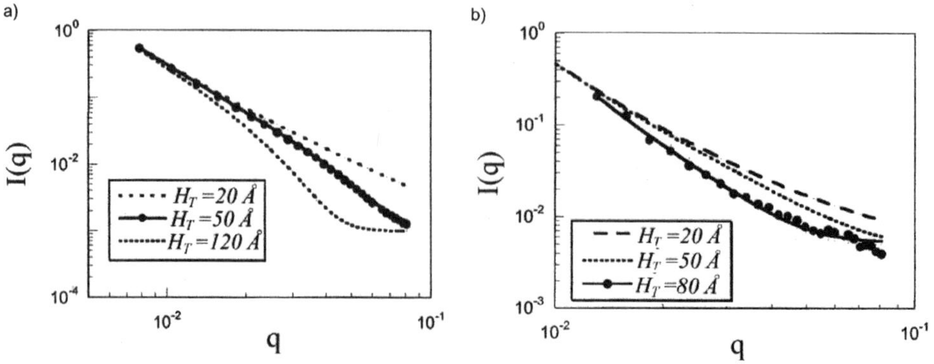

FIG. 24 Scattering intensity $I(q)$ vs. q in log-log scale of deuterated polystyrene particles where dT_{35} was (a) adsorbed at pH 5 and 0.01 M NaCl, and (b) covalently grafted at pH 9.2 and 10^{-2} M NaCl. (From Ref. 19.)

troscopy failed, due to the poor sensitivity of the method and using such small polyelectrolytes. A second method issued from biology, i.e., hydroxyl radical footprinting, was adequately used. It consists of cleaving ODNs by a radical source and analyzes the length and distribution of the obtained fragments by gel-blot migration. In the present case, cleaving did not occur for ODN segments wholly adsorbed on the surface, which is a way to differentiate trains from loops and tails. The conformation of ODN adsorbed on the surface of cationic or anionic polystyrene latexes shown to be flat in both cases. ODN can be adsorbed either by its charges (electrostatic forces) or by its nucleic bases (stacking effect between the aromatic compounds of the bases and styrene), so that ODN and the colloidal support are in close contact.

Recently, Charreyre et al. [22] investigated the conformation of ODN chemically grafted on an amino-containing polystyrene latex using fluorescence energy transfer. A couple of fluorescent molecules, fluorescein (donor) and tetramethylrhodamine (acceptor), are grafted to the 3′ end of the ODN and on amino groups on the latex surface, respectively. The energy transfer between acceptor and donor varies with the mean distance between the two fluorophores, i.e., between the ODN and the surface, so that a semiquantitative study is envisaged. Prior to grafting reactions, latexes were stabilized by nonionic surfactant (Triton X-405) which, even after extensive centrifugation/redispersion steps, was not fully eliminated (see discussion on desorption). In these conditions, a "brush"-type structure was observed at a basic pH (pH 10), whereas at neutral or weak acid pH (pH 6) conformation is mostly flat.

VII. HYBRIDIZATION REACTIONS AND ODN/LATEX CONJUGATE APPLICATIONS

A. Enzyme-Linked Oligosorbent Assay (ELOSA)

ELOSA, which is a typical technique equivalent to enzyme-linked immunosorbent assay (ELISA), permits specific capture and detection of a target DNA or RNA sequence. The principle of this technique is shown in Fig. 25a for flat solid support and Fig. 25b for colloidal particles. A capture ODN (ODN-1) with an appropriate sequence is immobilized on a solid support. This ODN specifically captures the complementary single-stranded nucleic acid (DNA or RNA) via hybridization process. To point out the specific capture of the target, a second ODN-bearing HRP enzyme (ODN-2) is then hybridized on a given part of the target. The quantification is then performed via enzyme oxidation leading to a colored or fluorescent product. The fluorescence intensity is proportional to the captured target concentration (Fig. 25b).

Major research in biomedical diagnostic has consisted of increasing the capture and the detection of nucleic acid targets without any polymerase chain

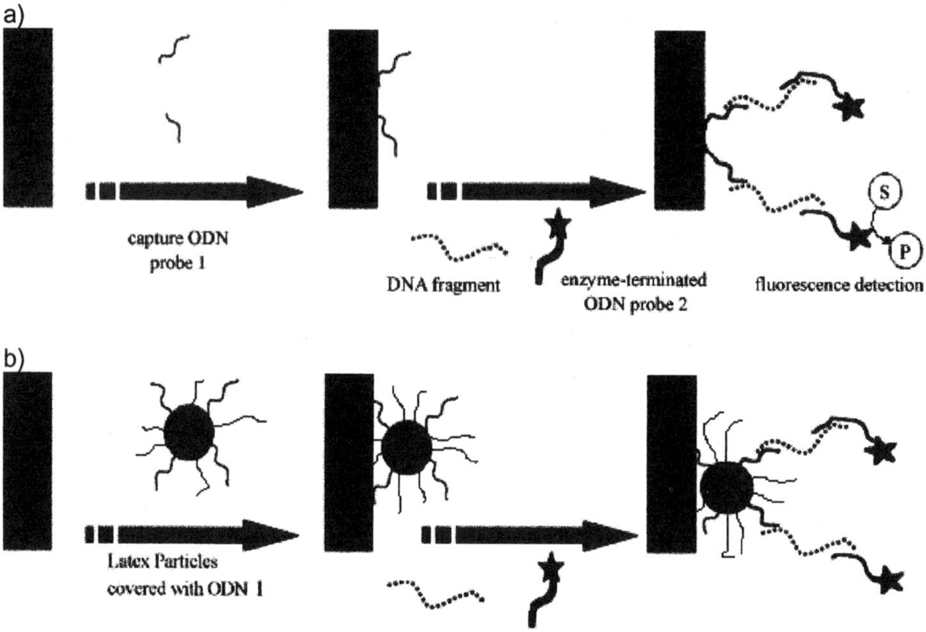

FIG. 25 Enzyme-linked oligosorbent assay (ELOSA). (a) Basic system (ODN probe 1 adsorbed on the plate). (b) Modified ELOSA procedure (latex/ODN probe 1 conjugate adsorbed to the plate).

reaction (PCR) or reverse transcriptase polymerase chain reaction (RT-PCR) amplification step. The efficiency of latex–ODN conjugate in ELOSA has been recently validated using an automated diagnostic apparatus (Vidas from bioMérieux S.A.) [23]. Such methodology has been derived from the interesting results obtained using linear polymers on which single-stranded DNA fragments were chemically grafted to amplify the sensitivity of ELOSA [24].

Figure 26 shows some results obtained using latex–ODN conjugate in ELOSA test in which the amplified signal from enzymatic reaction of the ODN/latex particle ratio. The latex–ODN conjugates are adsorbed onto the solid support of the automated system; then the capture and detection of a model nucleic acid target are investigated as illustrated in Fig. 25b. The sensitivity result observed using latex–ODN conjugate was appreciable and was enhanced by combining reactive polymers/latex particles/ODN in both capture and detection steps [23,25].

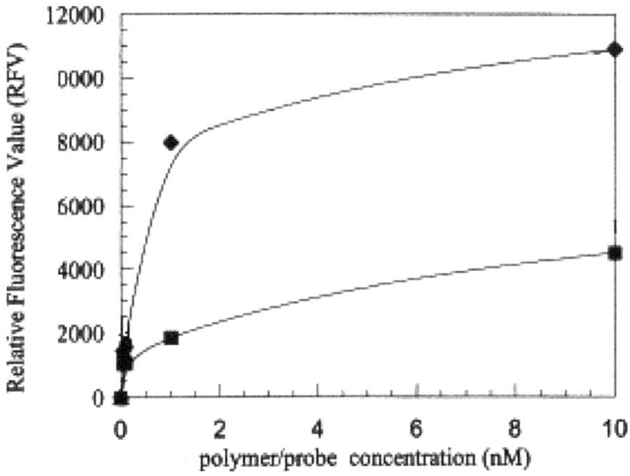

FIG. 26 Evaluation of two-dimensional latex assemblies obtained with covalent conjugates (◆) and conjugates obtained by ODN adsorption (●). (From Ref. 23.)

VIII. CONCLUSION

The purpose of this chapter is to identify the conditions favoring and controlling the elaboration of a latex–ODN conjugate for biomedical diagnostic applications. In brief, the capture of a target DNA or RNA by a latex–ODN system is possible when the chemical immobilization and the interfacial conformation of the considered ODN were well controlled after appreciable systematic studies. In fact, the sensitivity of such system is mainly related to the ODN grafted amount, to the extended conformation and accessibility of presumably ended immobilized ODN, and, finally, to the medium conditions used in the hybridization process. This requires consideration of at least three criteria:

1. The physicochemical properties of the considered biomolecules and, in this case, the single-stranded DNA fragments need to be addressed by investigating the effect of pH, salt concentration, ionic nature, and temperature on the solution characteristics of such polyelectrolytes, for instance.
2. The design of the appropriate carrier for a given biomedical applications and the needed specificities should be well defined. The detailed list can contains particle size, size distribution, nature and density of functional group, surface polarity, colloidal and chemical stability, and, finally, morphology. The established particle identities govern not only the polymerization process but also the appropriate recipe.

3. Before any chemical immobilization of the considered ODN, the adsorption should first be investigated in a systematic study. In fact, the adsorption–desorption results are of paramount interest in order to point out the driven forces involved in the interaction process.
4. Finally, the chemical coupling of ODN can be performed by taking into account the adsorption–desorption results. When the activation step is needed, detailed chemical analysis of the activated ODN should be particularly considered.

When the latex particles–ODN conjugates are elaborated, the optimal conditions leading to both high specific capture and good detection of the target DNA or RNA open a new research area of great interest.

REFERENCES

1. Kawaguchi, H.; Asai, A.; Ohtsuka, Y.; Watanabe, H.; Wada, T.; Handa, H. Nucleic Acids Res. **1989**, *17* (5), 6229.
2. Ganachaud, F.; Mouterde, G.; Delair, T.; Elaissari, A.; Pichot, C. Polym. Adv. Tech. **1995**, *6*, 480.
3. Ganachaud, F.; Sauzedde, F.; Elaïssari, A.; Pichot, C. J. Appl. Polym. Sci. **1997**, *65*, 2315.
4. Sauzedde, F.; Ganachaud, F.; Elaïssari, A.; Pichot, C. J. Appl. Polym. Sci. **1997**, *65*, 2331.
5. Ganachaud, F.; Bouali, B.; Veron, L.; Lantéri, P.; Elaissari, A.; Pichot, C. Colloid Surf. A Physicochem. Eng. Asp. **1998**, *137*, 141.
6. Fernandez-Nieves, A.; Fernandez-Barbero, A.; de las Nieves, F.J. Langmuir **2000**, *16*, 4090.
7. van Streun, K.H.; Belt, W.J.; Piet, P.; German, A.L. Eur. Polym. J. **1991**, *27*, 931.
8. de Gennes, P.G. *Scaling Concept in Polymer Physics*. Cornell University Press: Ithaca, NY, 1979.
9. Elaissari, A.; Pichot, C.; Delair, T.; Cros, P.; Kurfurst, R. Langmuir **1995**, *11*, 1261.
10. Elaissari, A.; Cros, P.; Pichot, C.; Laurent, V.; Mandrand, B. Colloids Surf. **1994**, *83*, 25.
11. Ganachaud, F.; Elaissari, A.; Pichot, C. Langmuir **1997**, *13*, 7021.
12. Andrade, J.D. *Surface and Interface Aspect of Biomaterial Polymers*. Plenum Publishers: New York, 1985; Volume 2.
13. Walker, H.W.; Grant, S.B. Colloids Surf. **1996**, 3837T.
14. Pefferkorn, E.; Carroy, A.; Varoqui, R. J. Polym. Sci. **1985**, *23*, 1997.
15. Ganachaud, F.; Elaissari, A.; Pichot, C.; Laayoun, A.; Cros, P. Langmuir **1997**, *13*, 701.
16. Elaissari, A.; Chauvet, J.P.; Halle, M.A.; Decavallas, O.; Pichot, C.; Cros, P. J. Colloid Interface Sci. **1998**, *202*, 2252.
17. Ganachaud, F.; Laayoun, A.; Chaix, C.; Delair, T.; Pichot, C.; Elaissari, A. J. Dispers. Sci. Technol. **2001**, *22*(5), 473.
18. Varoqui, R.; Johner, A.; Elaissari, A. J. Chem. Phys. **1991**, *94*, 6873.

19. Elaïssari, A.; Chevalier, Y.; Ganachaud, F.; Delair, T.; Pichot, C. Langmuir **2000**, *16*, 1261.
20. Walker, H.W.; Grant, S.B. Langmuir **1996**, *12*, 3151.
21. Walker, H.W.; Grant, S.B. J. Colloid Interface Sci. **1996**, *179*, 552.
22. Charreyre, M.T.; Tcherkasskaya, O.; Winnik, M.A.; Hiver, A.; Delair, T.; Pichot, C.; Mandrand, B. Langmuir **1997**, *13*, 3103.
23. Delair, T.; Meunier, F.; Elaissari, A.; Charles, M.-H.; Pichot, C. Colloids Surf. A Physicochem. Eng. Asp. **1999**, *153*, 341.
24. Erout, M.N.; Troesch, A.; Pichot, C.; Cros, P. Bioconj. Chem. **1996**, *7*, 568.
25. Charles, M.-H.; Charreyre, M.T.; Delair, T.; Elaïssari, A.; Pichot, C. S.T.P. Pharma Sci. **2001**, *11*(4), 241.
26. Ganachaud, F.; Elaissari, A.; Pichot, C. J. Biomater. Sci. Polym. Ed. **2000**, *11*, 931.

10

Covalent Immobilization of Peptides onto Reactive Latexes

JULIO BATTISTONI and SILVINA ROSSI Facultades de Química y Ciencias, Instituto de Higiene, Montevideo, Uruguay

I. INTRODUCTION

A. Background

Agglutination of particles was one of the earliest techniques used to detect antigen–antibody reactions. After mixing sample and reagent (as a colloidal suspension), the signal that reveals the reaction is the production of aggregates visible to the naked eye. Agglutination occurs either because when the molecule in the liquid phase binds its counterpart on the solid phase, it produces the linkage of particles (so the molecule must be at least bivalent), or because it generates a destabilization of the colloidal suspension by modifying the particle electric double layer (i.e., once bonded to the surface, the molecule decreases the surface potential).

The earliest particles used included charcoal, red blood cells, and bentonite. The classic immunoassay for the diagnosis of syphilis by agglutination of cholesterol particles has been used for many years in spite of serious limitations concerning its specificity. The continued preference for agglutination tests is due to certain advantages: they are easy and quick to perform, and do not require skilled personnel or sophisticated instruments. The first assays to be developed were "card tests" or "tube tests"; the reaction was performed on a card or in a test tube and the agglutination rated with the naked eye. With the introduction of radio- and enzyme immunoassays, which provide greater sensitivity, other techniques to evaluate the agglutination were developed. Nephelometric or turbidimetric measurement of agglutination not only improved sensitivity but also allowed the assays to be automated.

Since van den Hul and Vanderhoff's pioneer studies [1,2] on surface and size characterization of synthetic polymer colloidal particles, and their subsequent application to the detection of rheumatoid factors by Singer and Plotz [3], the use of such particles was quickly extended to a great many immunoassays [4,5]. Replacing "natural" particles with polymeric ones provides great advantages, especially the possibility of tailoring particles according to the exigencies of the assay under development. Moreover, synthetic particles are immunologically inert, whereas natural particles frequently give rise to nonspecific interactions. Latex particles were widely used in basic colloidal physics research as model colloids [6–10], and the development of this discipline has led to a better understanding of the physical chemical behavior of the latex–protein conjugate, especially where colloidal stability and the process of agglutination reactions are concerned [11]. A great many different polymers have been studied (polyvinylbenzene, polymethacrylates, polystyrene, etc.), but polystyrene has been most extensively used for physical immobilization.

B. Immobilization of Immunoreagents

Classic physical adsorption of macromolecules as an immobilization procedure has some important advantages: it is easy to carry out, very reproducible, and suited to the production of large quantities of reagent. Immobilization occurs by the hydrophobic effect upon simple mixing of the latex with a solution of the macromolecule, under the right conditions of pH and ionic strength. This procedure has the additional advantage that a single type of solid phase particle can be used to immobilize a wide range of macromolecules, especially proteins. The most popular polymeric latexes, such as polystyrene, have highly hydrophobic surfaces and charged groups ($-OSO_3^-$) which stabilize the colloidal suspension. These groups are derived from the reagents used during synthesis (emulsion polymerization, where persulfate ions are used as a source of free radicals to initiate polymerization). Carboxyl groups may also appear by degradation of the sulfate ions (Kolthoff's reaction) [12]. In summary, a typical polystyrene latex possesses a highly hydrophobic surface with a low charge density, of the order of 3–5 $\mu C/cm^2$. This charge has little or no influence on the establishment of final immobilization interactions; occasionally, when a mixture of proteins is being adsorbed, the surface charges can favor diffusion to the surface of a particular molecule [13], so that the final composition on the latex surface may be different from the initial composition of the protein mixture, especially if the amount of protein is well above what is required to saturate the latex surface.

However, physical adsorption does have some disadvantages—which rule out the development of some immunoassays—such as the following: (1) Some molecules are difficult to immobilize on hydrophobic surfaces because they are themselves highly hydrophilic, as in the case of polysaccharides and highly glycosylated proteins. Here one must resort to latex with highly charged and hydrophilic surfaces, which allow a large number of electrostatic interactions to occur. These interactions are fundamentally dependent on the pH and ionic strength of the medium, and if these parameters vary desorption of adsorbed molecules may occur, particularly if the conditions for adequate stability of the binding of the adsorbed molecule do not coincide with optimal pH and ionic strength for the antigen–antibody reaction. (2) The orientation acquired by molecules upon immobilization may prevent their recognition. Such orientation is linked to the distribution of hydrophobic patches (which provide the molecule's anchorage to the surface), with respect to its dominant epitopes (Fig. 1). It is difficult to manipulate the orientation of attachment solely by varying the physicochemical properties of the medium during immobilization. (3) Finally, physical adsorption is almost certain to cause denaturing of the macromolecules, since the adsorption process can alter the equilibrium of the forces that maintain secondary and tertiary structure [14–17]. The extent of denaturing may be quite variable, and it depends on the structural rigidity of the molecule [18] (Fig. 2) and the solid phase area available per molecule to be adsorbed (Fig. 3). Rigidity of the molecule depends essentially on the number of intramolecular disulfide bridges; non-covalent interactions also support the native structure, but they are more easily disrupted by the immobilization process. The larger the solid phase area available per molecule to be adsorbed, the greater will be the unfolding of the macromolecule, making more binding contacts with the surface, and the greater the extent of denaturation.

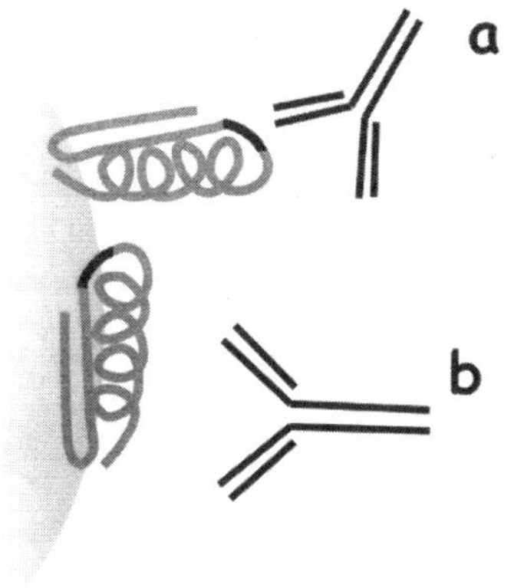

FIG. 1 Physical adsorption involves hydrophobic interactions between the surface and hydrophobic patches in the protein. It is a random process wherein the protein orientation is not the same for all of the immobilized molecules. As schematized in the figure, the way the protein is immobilized affects antibody recognition. Thus, in (a) antigen–antibody reaction is allowed, whereas in (b) the epitope is not exposed to the solution and the antigen–antibody reaction is avoided.

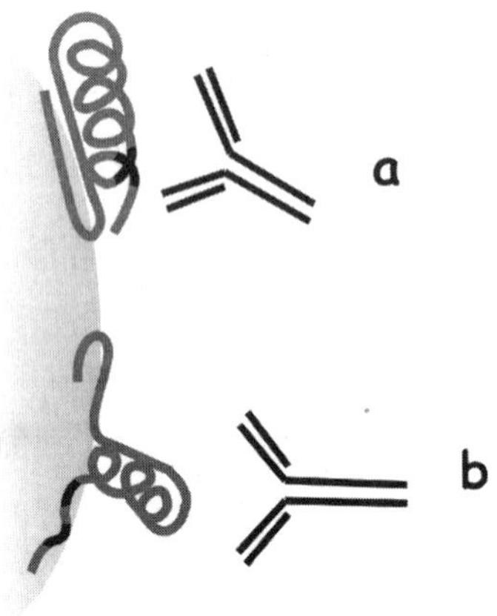

FIG. 2 Physical adsorption may involve several "interaction points." The flexibility of the molecule favors interaction through multiple points. However, damage of the tertiary and/or quaternary structure of the protein increases with the number of such interactions. Conformational epitopes are very sensitive to denaturation or loss of the three-dimensional structure. After immobilization, the conformational epitope can be (a) preserved, so that antigen–antibody reaction is allowed, or (b) lost, so that antigen–antibody reaction is avoided.

In addition to the problems resulting from a decrease or annulment of recognition because of denaturation on the surface, there may be other problems independent of immunoreactivity. Denaturing particularly affects conformational epitopes, those in which the immunoglobulin recognizes fragments of the macromolecule that are separated in its linear sequence, but close together in the folded molecule. Depending on whether epitopes are mostly linear or conformational, denaturing will have a greater or lesser effect on final reactivity. Even if recognition is not affected, the immobilized and denatured protein molecule is a source of nonspecific signals, since nonspecific protein–protein interactions may occur, particularly by means of hydrophobic regions that have been exposed by denaturation, quite independently of the antigen–antibody reaction. Such interactions occurring with proteins present in the sample will contribute to the nonspecificity of the assay. Again, particle agglutination techniques must achieve optimal colloid stability, and self-agglutination caused by denaturation is one of the most serious problems to appear frequently in their development. The denatured protein on the particle surface interacts with that on other particles in a manner similar to that described above.

In the case of antibodies, it has been shown that less than 5% of the immobi-

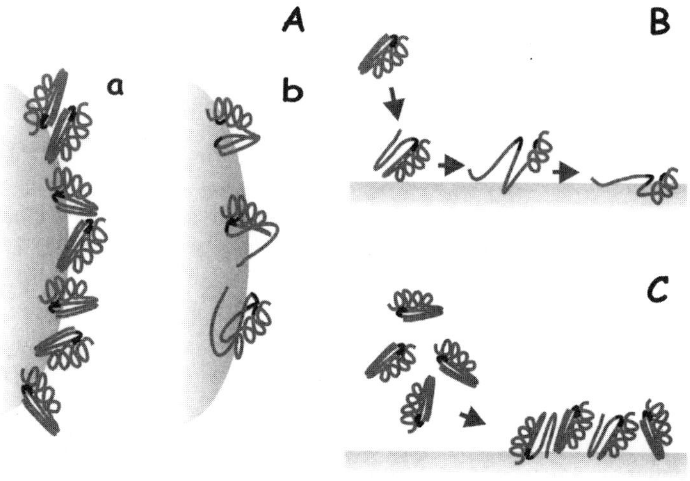

FIG. 3 Denaturation depends on the available area per protein molecule. Panel A illustrates that tight packaging of the molecules on the surface prevents denaturation, especially for flexible molecules (a), while loose packaging favors it (b). Panels B and C show a schematic representation of the effect of protein concentration—low and high, respectively—during immobilization.

lized protein is active [19] in the sense of being capable of recognizing antigen. This low activity may be ascribed as much to immobilization in a way that sterically hinders access to the molecule's paratopes as to modification or denaturation of the recognition site by the immobilization process.

In any immunoassay—especially those that are dependent on the stability of a colloid—it is important to measure precisely the quantity of bound immunoreagent per unit area. With this figure, important parameters such as the limit of detection and the specificity of the reagent can be modulated. Therefore, all denaturation processes undermine the rational development of new reagents and also often result in the waste of expensive materials.

II. IMPROVEMENT OF ASSAYS BY THE DEVELOPMENT OF NEW IMMUNOREAGENTS AND SURFACES

In order to overcome the limitations described in the introduction, new immunoassay techniques have been developed, with new methods for improving antigen reactivity and new surfaces to obviate the problems of physical adsorption.

A. Improvement of Specificity by Improving Immunoreagents

Classical biochemical methods for purifying antigens (such as a pathogen) from biological material give rise to samples that frequently contain other macromolecules, which are of no value in the assay and which interfere with it considerably because they are not inert. They have a negative effect on the detection limit and the specificity of the test. A large part of the improvement in immunoassay performance has been due to improvements in the reagents as a result of attempts to immobilize only those molecules that are involved in the reaction of interest, thus avoiding the problems mentioned above. This has been achieved by using monoclonal antibodies, and/or molecularly defined antigens, or fragments of antigens. With the advent of genetic engineering [20], it became possible to synthesize perfectly defined proteins, which gave a highly specific signal in immunoassays. This versatile technology allows expression in bacteria of protein fragments of diagnostic value, which may contain linear or conformational epitopes; sequence modifications can also be introduced to facilitate immobilization, whether by adsorption or by covalent linkage.

Recently, synthetic peptides have become another alternative in improving immunoassays [21]; these are linear amino acid polymers, made by chemical synthesis [22]. The study of these compounds did not arise from the need to improve immunoassays, but from basic research into the structure of epitopes [23]. Peptides of diagnostic value are B epitopes, i.e., they react with antibodies generated against proteins (immunogens) recognized as foreign to the organism (antigens) (Fig. 4).

From the point of view of assay production, incorporating synthetic peptides into immunoassays has definite advantages. The cost of reagents may be lower because tedious purification processes starting from complex biological materials are avoided and/or because the availability of very versatile automatic peptide synthesizers has allowed extraordinary decreases in the price of peptides in recent years. Production processes have also been simplified (economically as well as technically) and have benefited from the use of more stable, chemically pure, and reproducible compounds. Once production specifications have been defined, products can be manufactured almost indefinitely, in contrast with biological products, which tend to have batch-to-batch variations. Furthermore, synthetic peptides can be easily transported and stored, so that strict control of the properties and use of the final product is possible.

Peptides used in immunoassays typically contain from 7 to 25 amino acids; they may or may not belong to the primary sequence of the native protein, but even so they maintain their reactivity with specific antiprotein antibodies. Basically the length of a peptide should be that of a B epitope—in other words, the number of amino acids recognized in the interaction with an immunoglobulin

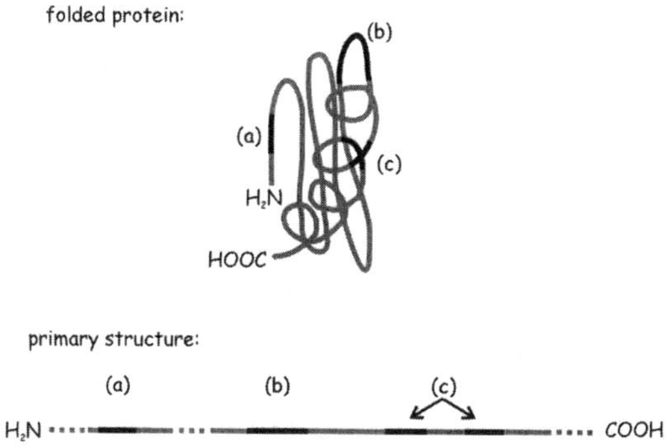

FIG. 4 Synthetic peptides as B epitopes. B epitopes, generally classified as linear (a, b) and conformational (c), are schematically shown in both the folded protein and the corresponding portions on its primary structure. Only linear epitopes are easily reproduced by synthetic peptides. This can be achieved either directly (a) or through modifications of the original amino acidic sequence in order to mimic the structure it adopts within the folded protein (b).

paratope. Crystallographic studies of the antigen–antibody complex have shown that the interactions involve about 15 amino acids [24,25]. In practice, a greater number of amino acids are frequently required; although they do not participate directly in the interaction, they contribute to overall structure and so favor recognition [26].

Two strategies have been developed for searching for peptides to substitute for the reactivity of a given antigen [27]: (1) primary sequence analysis of the target protein [28–30], and (2) random generation of amino acid sequences that are then selected according to their reactivity with antibodies raised against the antigen [31,32]. The first method starts from knowledge of the primary sequence of the protein of diagnostic value; epitopes are predicted by algorithms [33,34] or experimental methods [35]. The algorithms used to analyze the primary sequence select sequences that have high probability of being exposed on the surface of the protein and have certain exposure, flexibility, and conformational characteristics [33]. The experimental methods consist basically of the synthesis of peptides of a given length, 6–10 amino acids, immobilized on derivatized paper or plastic. These peptides cover the entire linear sequence with overlaps of 2 to 3 amino acids. Next, the peptides that react with appropriate antibodies—polyclonal antibodies derived from infection or from immunization with the native protein—are identified by immunochemical techniques [36].

These methods identify linear epitopes. These should be hard to find because most epitopes are conformational (structural), involving regions that are far apart in the primary sequence, but close to each other in the secondary or tertiary structure of the protein [37]. However, this assertion leads to controversy, as in fact it is not difficult to find linear peptides that constitute an epitope; some authors have therefore said that the different methods for finding and/or defining epitopes are suspect [38,39].

Random peptide generation may be carried out by chemical synthesis [40] as well as by molecular biology techniques [41,42]. Peptides are selected using antibodies against the native protein, whether raised by infection or monoclonal. In these peptide epitopes, the residues and the interactions involved with the antibody may not necessarily be present in the protein; rather, they are mimicking the epitope and are therefore called *mimotopes*. These methods offer the advantage of being able to simulate conformational epitopes with mimotopes.

B. Improvement of Molecular Orientation and Bond Stability by Covalent Immobilization

As discussed previously, immobilizing a peptide by physical adsorption is very inefficient, since the peptide structure is wholly or partially "frozen" on the solid surface, making reaction with the antibody improbable (Fig. 5). In order to avoid these problems with proteins as well as with peptides, methods have been developed to synthesize nanoparticles with functional groups on their surface, which make it possible to immobilize molecules more precisely by means of covalent bonding. The aims are: (1) to establish bonds that are more stable with respect to time and (2) to orient the peptide precisely so as to favor recognition by the antibody.

When using covalent bonding, contradictions may arise between the selected peptide's reactivity in solution and its reactivity in the immunoassay, since the immobilization process may restrict the peptide from adopting conformations that are important for its recognition—as frequently happens with physical ad-

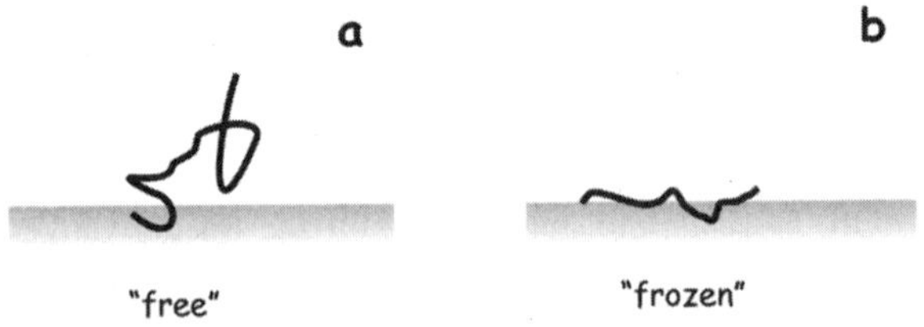

FIG. 5 Physical adsorption is not the method of choice for peptide immobilization. Attachment to the surface should be controlled so as not to impair peptide flexibility and exposure to the aqueous phase (a). Physical adsorption usually involves multiple interaction points that yield constrained structures (b).

sorption. This does not mean that a peptide always loses antigenicity when covalently immobilized; on the contrary, reduction of mobility in a given region of the peptide may enhance recognition. Therefore, the antigenicity of a peptide can only be determined operationally, with explicit reference to the experimental conditions under which it is measured [43]. Flexibility is also important for these small molecules because it determines their ability to adopt appropriate complementarity with the paratope [44]; such flexibility will be possible if the peptides are correctly oriented to the aqueous phase and/or their interactions with the solid phase are minimized.

1. Modification of Peptide Synthesis to Optimize Its Covalent Immobilization

The selected peptide can be modified during synthesis by adding or modifying amino acids (1) in order to achieve efficient immobilization in terms of reaction yield as well as of the final immunoreactivity of the immobilized peptide (Fig. 6); (2) since frequently linear peptides, free from their framework within the native protein, do not adopt the same conformation, it may be necessary to increase conformational similarity between the peptide epitope and the corresponding region of the native protein in order to promote binding of the peptide with the antiprotein antibody.

When designing modifications in order to improve the orientation and exposure of the peptide, the following factors should be taken into consideration: the proper orientation of the peptides on the surface; their separation from the surface and their freedom to rotate around their point of anchorage; a satisfactory immobilization reaction yield; and the possibility of blocking some terminal or

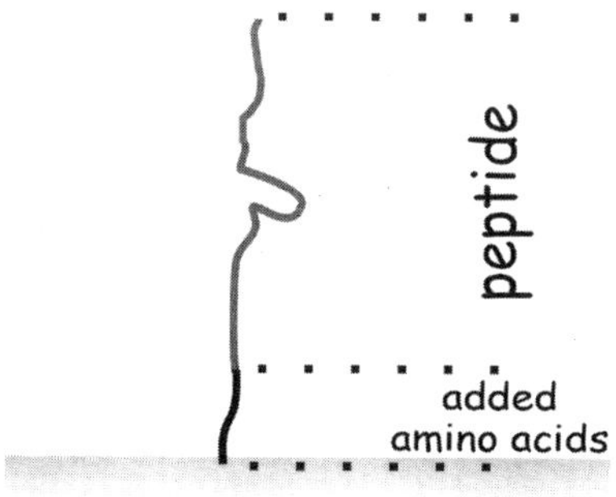

FIG. 6 Sequence modifications of the peptides are usually needed to improve immobilization. A common practice is to add amino acids (black line) to the end where covalent immobilization will take place so as to allow conventional conjugation chemistry and improve exposure as well as flexibility.

lateral functional groups, such as thiol, amine, or carboxyl groups, in order to obtain a charge distribution similar to that of the original protein segment.

With respect to how the peptides should be oriented on the surface, which essentially means which end should be used for attachment, some authors give some guidelines for general practice [45]. If the peptide is from the carboxy terminal or amino terminal of the protein, where flexible linear epitopes are frequently found, it is relatively simple to simulate those situations upon immobilization. If the peptide is from the carboxy terminal end of the protein, it should be joined to the surface by its amino end, and vice versa. If the peptide contains a cysteine residue in midsequence, this will be an appropriate anchorage site, and particularly specific in the case of this unique bond, since it is likely that in the original protein this residue forms a disulfide bridge with another part of the protein, and this information may well be available.

When considering giving the peptide freedom to rotate on its axis (Fig. 7), and distance from the surface, the aim is sterically to facilitate its interaction with the antibody while at the same time avoiding interaction with the surface (Fig. 8). Generally, the addition of various glycine residues at the attachment end of the peptide will achieve these aims, as glycine has no lateral side chains to hinder rotation, and if several glycines are used the rest of the peptide will be distanced from the surface.

As will be seen below, certain strategies for making large peptide complexes require the addition of charged amino acids to the attachment end, so as to create a specific "charge cluster." This helps the peptide to diffuse to the molecule to which it will be attached, which bears opposite charges, and also favors a given

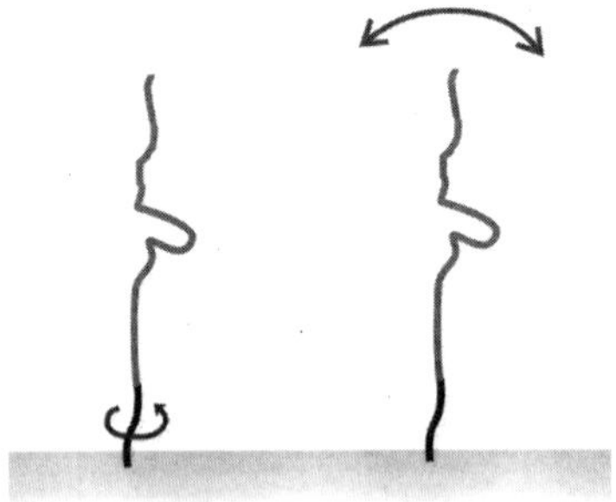

FIG. 7 Sequence modifications of the peptides may favor antibody reaction. Amino acids added to the attachment end (black line) will also provide free rotation and movement of the peptide upon its axis (arrows), facilitating antibody recognition.

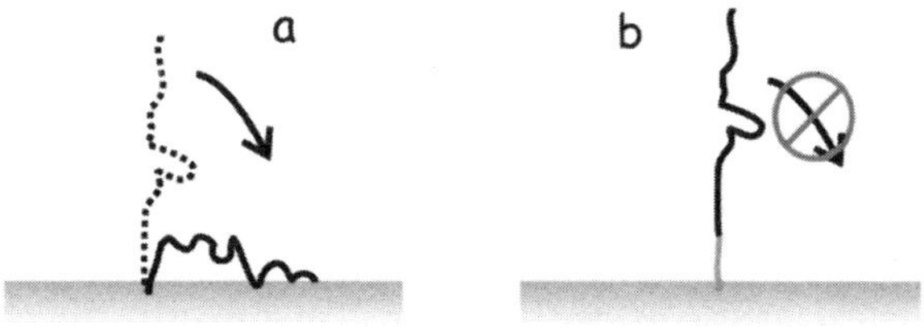

FIG. 8 Glycine tags as spacer sequences. When peptides are attached directly to the surface, additional undesirable interactions might occur (a). The inclusion of several glycine residues at the attachment end is a good strategy to avoid these interactions (b) and usually does not contribute to nonspecific antibody recognition.

orientation upon immobilization (Fig. 9). Such a strategy also increases the reaction efficiency.

Synthesis modifications to simulate the structure of the peptide within the original protein can be carried out as additional chemical reactions in order to give the peptide a cyclic structure [46], or to add on to either side of the epitope amino acids that are not involved in its reactivity but that can induce a particular conformation [47]. This is especially important in the case of "hairpin" structures [48], which are closely associated with protein antigenicity [49].

2. Methods of Covalent Immobilization onto Particles

Covalent immobilization of the peptide can be carried out in three ways: (1) covalent reaction onto reactive surfaces; (2) forming a macromolecular complex

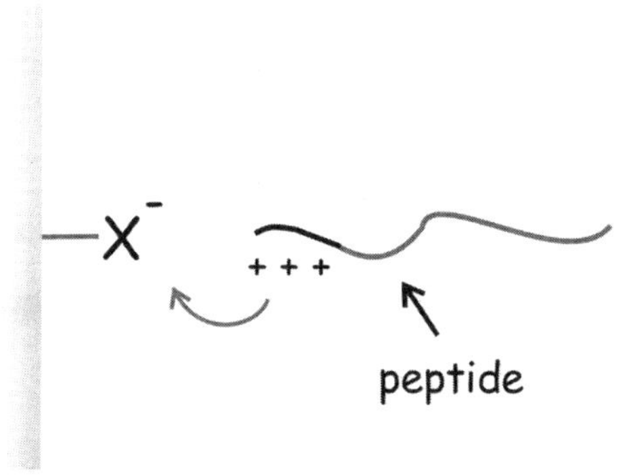

FIG. 9 A charged tag or cluster can also be added to the end of the peptide. This procedure may drive an optimal orientation of the peptide, thus improving a covalent immobilization process.

with the peptide, followed by physical adsorption or covalent linkage to the surface; and (3) covering the surface by physical adsorption of polymers possessing reactive groups, followed by covalent immobilization of the peptide.

During covalent reaction processes, it is difficult to avoid physical adsorption onto the surface; therefore, the characteristics of the nonreactive part of the surface are very important. The more hydrophilic it is, the less physical adsorption will occur. Some authors have included inert detergents in their reactions in order to minimize physical adsorption. Several authors have developed special synthesis methods in order to ensure very hydrophilic surfaces for these reactions [50].

(*a*) *Reactive Surfaces.* Use of reactive surfaces, or latex particles with added functional groups, is an alternative strategy that bypasses the problems of physical adsorption (Fig. 10). Particles with functional groups are obtained by a process of synthesis in which monomers that produce the base polymer are copolymerized with other monomers that provide the functional groups. These groups are on the surface of the particles, either in reactive form or in a protected form if they are unstable upon storage.

The first latexes with added functional groups were designed to form covalent bonds with the ε-amino groups of lysine, present in almost all known proteins. Among the earliest reactive groups used were carboxyl [51], aldehyde [52,53], and chloromethyl groups [54]. At present, other functional groups can be used in latex particle preparation, including amino [55,56], thiol, and epoxy [57] groups, among others (Table 1).

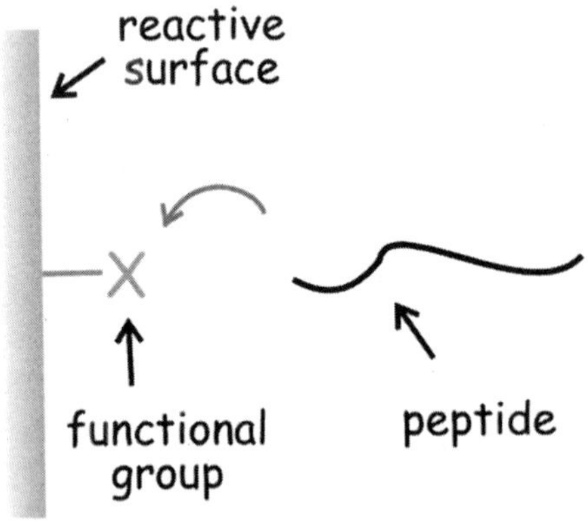

FIG. 10 Schematic representation of the covalent immobilization of a peptide onto a reactive surface. Reactive groups on the peptide and the solid phase are usually ligated via cross-linking reagents.

TABLE 1 Types of Functional Groups and Their Corresponding Ligands Used for Covalent Linkage

Functional group on particle	Ligand group in peptide	Comments
Aldehyde (CHO)	Amino (NH_2)	Nucleophilic addition No activation step required May require deprotection of the aldehyde group
Chloro-activated ($ClCH_2$)	Amino (NH_2)	Nucleophilic substitution No activation step required
Epoxy ($CH-CH_2$, bridged by O)	Amino (NH_2)	Nucleophilic substitution No activation step required Requires pH above the pK_b of the amino group
Thiol (SH)	Thiol (SH)	Reduction No activation step required Susceptible to oxidation and displacement by other SH groups
Thiol (SH)	Maleimide, iodoacetyl	No activation step required Requires prior modification of peptide
Carboxyl (COOH)	Amino (NH_2)	Nucleophilic substitution Requires activation step
Amino (NH_2)	Amino (NH_2) Carboxyl (COOH)	Requires activation step with bifunctional reagent
Hydroxyl (OH)	Amino (NH_2)	Nucleophilic substitution Requires activation step, often with use of organic solvent

Source: Adapted from Ref. 61.

Some of these latexes with added functional groups require activation steps prior to conjugation with peptide, e.g., those with carboxyl or hydroxyl groups; others must be stored with their functional groups protected until the moment of use, e.g., those with aldehyde groups. The variety of functional groups available allows peptides to be attached covalently not only by their primary amino groups but also by thiol, carboxyl, and hydroxyl groups.

When immobilizing a peptide onto a reactive surface, it is possible to add a nonpeptide molecule to act as a "spacer," i.e., as an extension of the peptide toward the aqueous phase, so preventing the peptide from sinking to the particle surface, and allowing it to gain flexibility and improve its recognition (Fig. 11). Adding a nonpeptide spacer has advantages over adding amino acid residues during peptide synthesis for the same purpose. Hydrophilic spacers, like poly-

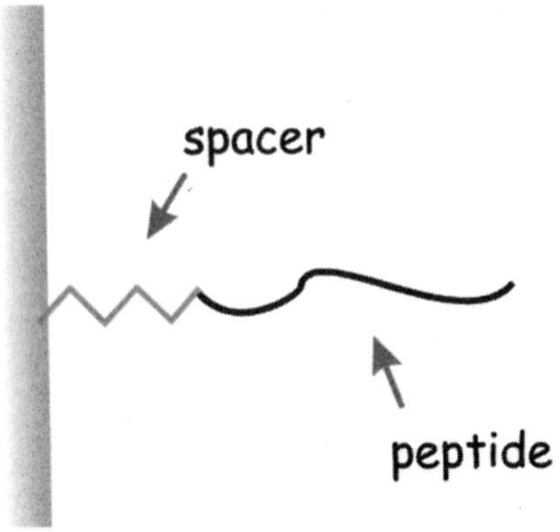

FIG. 11 Use of nonpeptidic spacers for peptide immobilization. Nonpeptidic spacers have advantages over peptidic ones, especially in terms of undesired reactivity with the antibody. Furthermore, the bound spacer may bear a functional group at the free end and thus allow changing of the reactive properties of a given surface.

ethylene glycol [58,59], can improve peptide exposure to aqueous phase, where otherwise the peptide would tend to interact with the surface (Fig. 12). In addition, any spacer can be used to change the functional group available on a surface. For instance, if an aliphatic diamine is made to react with a surface that has been carboxylated with carbodiimide, the resulting surface will be cationic

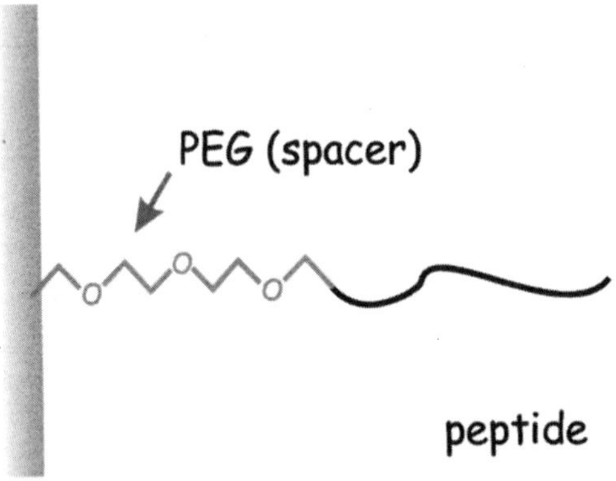

FIG. 12 Use of hydrophilic spacers. Reagents such as PEG avoid additional undesirable interactions between the surface and the peptides, since they "mask" the hydrophobic nature of the surface and, more importantly, improve peptide exposure to the aqueous media.

due to the amino groups. Therefore, the use of spacers can make a given latex reagent extremely versatile, especially since it does not depend on modifications carried out during peptide synthesis. Furthermore, adding many extra amino acids to a peptide of diagnostic value may create a neoepitope in the anchorage zone; this would have a negative effect on specificity. Use of other spacers, especially polyethylene glycol [60], would minimize this problem.

(*b*) *Immobilization of Macromolecular Complexes.* Another procedure for immobilizing peptides is by formation of peptide–macromolecule complexes (Fig. 13). The peptide is covalently linked to a macromolecule and the resulting conjugate is immobilized onto a particle surface. If the macromolecule is a protein, it should be chosen so as not to affect the specificity of the reagent. Macromolecules used for forming complexes with peptide are generally nonprotein polymers possessing functional groups or giving rise to functional groups after a simple activation step. A large multiepitopic complex is formed, with the peptides ideally exposed to the aqueous phase. This approach permits the immobilization of large quantities of peptide or of different kinds of peptides. The conjugate may be covalently linked onto particles bearing functional groups, or passively adsorbed onto hydrophobic particles, such as the traditional polystyrene latex. The number of peptide molecules that are denatured in the adsorption process is greatly reduced (Fig. 14A).

The use of polymers with functional groups has become widespread in the field of biological sciences; examples include their application for immobilizing ligands for affinity chromatography [62,63], for enzyme immobilization [64],

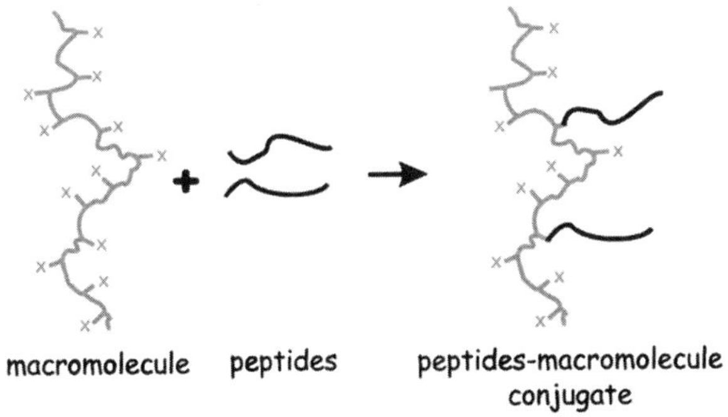

FIG. 13 Formation of the peptide–macromolecule complex. The macromolecule can be a natural or synthetic polymer bearing reactive groups. Inert proteins (from the point of view of the final application) or polylsaccharides are the natural polymers most commonly used. In these cases, coupling reaction requires cross-linking reagents. Synthetic polymers offer the advantage of carrying reactive groups that can be directly used for peptide immobilization.

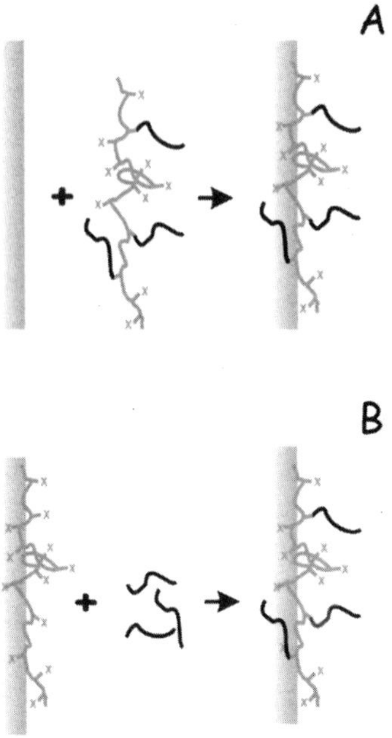

FIG. 14 Two different strategies can be used for peptide immobilization through conjugation to macromolecules. Peptide–macromolecule complexes can be formed in solution and subsequently immobilized, either physically or covalently, onto the surface (A). Peptides can be conjugated or grafted to the macromolecule previously immobilized on the surface (B).

for immobilizing proteins onto polyvinylidene difluoride (PVDF) membranes [65]. Subsequently, in immunochemical applications, Isosaki et al. [66] described protein immobilization onto polystyrene surfaces (ELISA plates) using a copolymer of methyl vinyl ether–maleic anhydride (MAMVE copolymer). More recently, Lavadière et al. [67] described a method for covalently linking peptides by means of their amino terminals to the same copolymer. Our group, together with that of Dr. Pichot (CNRS-bioMerieux, Lyon, France), has carried out the immobilization of peptide-MAMVE conjugates onto hydrophilic particles with functional amino groups. Thus, the use of polymers bearing functional groups has become a very useful approach for immobilizing peptides onto microparticles for diagnostic applications.

A good example of the use of polymers requiring a prior activation step is that of dextran in the preparation of peptide–macromolecule conjugates [68,69]. Hydroxyl groups present in dextran are first activated by reaction with tresyl chloride [68] or $NaIO_4$ [69], and then the peptide is covalently bound through amino groups, in a nucleophilic attack reaction. Recognition by antibody of peptides immobilized by this method is significantly improved with respect to peptides that have been physically adsorbed.

(*c*) *Covalent Immobilization of Peptides onto Surfaces Covered Over by Physically Adsorbed Molecules.* In this strategy the aim is to make the most of the advantages offered by immobilization by physical adsorption. Onto a single type of hydrophobic surface various macromolecules bearing the desired functional groups can be immobilized, and the peptides are covalently linked to those macromolecules. Two main kinds of macromolecules can be used: (1) Proteins, which are generally adsorbed almost irreversibly onto the hydrophobic surface, in very low-cost processes. Another advantage is that the rest of the surface can be made hydrophilic, thus avoiding surface adsorption of peptides when they are covalently immobilized onto the protein. The disadvantage, which may make the technique nonviable, is a loss of specificity owing to the reactivity of the protein macromolecule in the assay. (2) Synthetic polymers bearing functional groups, such as polylysine, polylysine-phenylalanine, polyglutamate, or other nonpeptide synthetic polymers, bearing functional groups and containing hydrophobic segments that enable their adsorption. Use of a synthetic polymer has the great advantage of giving a choice regarding the functional group that we wish to have on the particle surface. Problems of nonspecificity will be much lower than when a protein macromolecule is used; a wider variety of functional groups are available in synthetic polymers, compared to the limited range in proteins; and a greater density of functional groups per unit of surface area can be achieved (Fig. 14B).

To summarize, then, various assays using peptides in latex particle agglutination reactions have been described in the literature, and also some using peptides in enzyme immunoassays. The shared characteristics of these assays are that excellent specificity is achieved but that the diagnostic sensitivity attained is generally not acceptable.

The problem may arise for the following reasons: (1) The immune response is highly complex, and perhaps additional peptides should be included in the assay in order to reflect the immune responses of an entire population. This view is grounded on the fact that the first commercial assays with good diagnostic performance that were based totally on peptides were for the detection of immune responses to virus infections. At a molecular scale, a virus is far less complex than a bacterium, and even more so compared to a helminth; therefore, the immune response elicited by a virus is also much simpler. In the case of infective agents such as parasites, their mechanisms for evading the host immune response increases the complexity, thus emphasizing the need for a variety of peptides on the solid phase to adequately represent that response. (2) Certain problems, such as ensuring proper exposure of the peptide to the aqueous phase for its optimal recognition, have not yet been fully solved. This could result in interactions of reduced affinity between peptides and antibodies. Again, the density of peptide on the particle surface may be too low, resulting in a poor signal.

III. FUTURE PERSPECTIVES

Although there are few reagents for latex particle agglutination reactions that are entirely based on peptides, there is nevertheless an appreciable amount of accumulated information on this topic, which needs to be systematized. This information encourages the development of peptide-based reagents of very high sensitivity and specificity. No systematic study of the immobilization of one or various peptides by means of different methods, and the comparison of the diagnostic performances obtained, has been published in the literature.

It is to be hoped that more research into immobilization strategies will be carried out; simulated reactions using molecular modeling techniques may have much to offer in this exploratory process.

As mentioned at the beginning of this chapter, the use of peptides in particle agglutination assays, as well as in any other kind of immunoassay, has great advantages, as shown by the excellent specificity of the assays that have been developed so far. It is to be hoped that advances will be made in the following areas:

1. Improvements in methods of covalent linkage of peptides to latex or other synthetic surfaces, so as to minimize interactions of the peptide with the solid phase.
2. Development of new methods for peptide–macromolecule complex formation, which can extend a great quantity of epitopes (or multiepitopes) out into the aqueous phase.
3. Development of more versatile immobilization processes involving fewer, much simpler reactions, which can be adapted to production processes.

ACKNOWLEDGMENTS

The authors thank the ECOS-Sud program.

REFERENCES

1. Bradford, E.B.; Vanderhoff, J.W.; Alfrey, T., Jr. The use of monodisperse latexes in an electron microscope investigation of the mechanism of emulsion polymerization. J. Colloid Sci. **1956**, *11*, 135–149.
2. van den Hul, H.J.; Vanderhoff, J.H.W. The characterization of latex particle surfaces by ion exchange and conductometric titration. J. Electroanal. Chem. **1972**, *37*, 161–182.

3. Singer, J.M.; Plotz, C.M. The latex fixation test. I. Application to the serologic diagnosis of rheumatoid arthritis. Am. J. Med. **1956**, *21*, 888–892.
4. Hechemy, K.E.; Michaelson, E.E. Latex particle in laboratory medicine. Part 1. Lab. Manage. **1984**, *22* (6), 27–40.
5. McCarthy, L.R. Latex agglutination test for the rapid diagnosis of infectious disease. In *Rapid Detection and Identification of Infectious Agents*; Kingsbury, D.T., Falkow, S., Eds.; Academic Press: New York, 1985; 165–175.
6. Watillon, A.; Joseph-Petit, A.-M. Interactions between spherical particles of monodisperse polystyrene latices. Discuss. Faraday Soc. **1961**, *42*, 143–153.
7. Hatton, W.; McFadyen, P. Rapid flocculation rates of polystyrene latex dispersions. J. Colloid Interface Sci. **1973**, *37*, 655–660.
8. Ottewill, R.H. Direct measurement of particle-particle interactions. Prog. Colloid Polym. Sci. **1980**, *67*, 71–83.
9. Hidalgo-Alvarez, R.; de las Nieves, F.J.; van der Linde, A.J.; Bijsterbosch, B.H. Electrokinetic studies on positively charged polystyrene latices. Colloids Surf. **1986**, *21*, 259–266.
10. Herrington, T.M.; Midmore, B.R. Determination of rate constants for the rapid coagulation of polystyrene microspheres using photon correlation spectroscopy. J. Chem. Soc. Faraday Trans. 1 **1989**, *85* (10), 3529–3536.
11. Lee, R.E.; Kim, S.W. Adsorption of proteins onto hydrophobic polymer surfaces: adsorption isotherms and kinetics. J. Biomed. Mater. Res. **1974**, *8*, 251–259.
12. Kolthoff, I.M.; Miller, I.K. The chemistry of persulfate. I. The kinetics and mechanism of the decomposition of the persulfate ion in aqueous media. J. Am. Chem. Soc. **1951**, *73*, 3055–3059.
13. Galisteo-Alvarez, F.; Puig, J.; Martin-Rodriguez, A.; Serra-Domenech, J.; Hidalgo-Alvarez, R. Influence of electrostatic forces on IgG adsorption onto polystyrene beads. Colloids Surf. B Biointerfaces **1994**, *2*, 435–441.
14. Kochwa, S.; Brownell, M.; Rosenfeld, R.E.; Wasserman, L.R. Adsorption of proteins by polystyrene particles. I. Molecular unfolding and acquired immunogenicity of IgG. J. Immunol. **1967**, *99* (5), 981–986.
15. Soderquist, M.E.; Walton, A.G. Structural changes in proteins adsorbed on polymer surfaces. J. Colloid Interface Sci. **1980**, *75* (2), 386–397.
16. Kodo, A.; Oku, S.; Higashitani, K. Structural changes in protein molecules on ultrafine silica particles. J. Colloid Interface Sci. **1990**, *43* (1), 214–221.
17. Tan, S.; Martic, P.A. Protein adsorption and conformational change on small polymer particles. J. Colloid Interface Sci. **1990**, *36* (2), 415–430.
18. Arai, T.; Norde, W. The behavior of some model proteins at solid–liquid interfaces. 1. Adsorption from single proteins solutions. Colloids Surf. **1990**, *51*, 1–15.
19. Butler, J.E.; Ni, L.; Nessler, R.; Joshi, K.S.; Suter, M.; Rosemberg, B.; Chang, J.; Brown, W.R.; Cantarero, L.A. The physical and functional behavior of capture antibodies adsorbed on polystyrene. J. Immunol. Meth. **1992**, *150*, 77–90.
20. Maniatis, T.; Fristch, E.F.; Sambrook, J. *Molecular Cloning: A Laboratory Manual.* Cold Spring Harbor Laboratory: Cold Spring Harbor, NY, 1982.
21. Meloen, R.H.; Langedijk, J.P.M.; Langeveld, J.P.M. Synthetic peptides for diagnostic use. Vet. Q. **1997**, *19* (3), 114–118.

22. Walker, B. Solid phase peptide synthesis. In *Peptide Antigens. A Practical Approach*; Wisdom, G.B., Ed.; Oxford University Press: New York, 1994.
23. Sela, M.; Fuchs, S. Preparation of synthetics peptides. In *Handbook of Experimental Immunology. Vol 1, Immunochemistry*, 4th Ed.; Weir, D.M., Ed.; Blackwell Scientific Publications: Oxford, UK, 1986; 2.1–2.12.
24. van Regenmortel, M.H.V. Structural approaches to the study of protein antigenicity. Immunol. Today **1989**, *10* (8), 266–271.
25. Alzari, P.M.; Lascombe, M.B.; Poljak, R.J. Three-dimensional structure of antibodies. Annu. Rev. Immunol. **1988**, *6*, 555–580.
26. Getzoff, E.D.; Tainer, J.A.; Lerner, R.A.; Geysen, H.M. The chemistry and mechanism of antibody binding to protein antigens. Adv. Immunol. **1988**, *43*, 1–97.
27. Horsfall, A.C.; Hay, F.C.; Soltys, A.J.; Jones, M. Epitope mapping. Immunol. Today **1991**, *12* (7), 211–213.
28. Tainer, J.A.; Getzoff, E.D.; Peterson, Y.; Olson, A.; Lerner, R. The atomic mobility component of protein antigenicity. Annu. Rev. Immunol. **1985**, *3*, 501–535.
29. Geysen, H.M.; Rodda, S.J.; Mason, T.J.; Tribbick, G.; Schoofs, P.G. Strategies for epitope analysis using peptide synthesis. J. Immunol. Meth. **1987**, *102*, 259–264.
30. Stren, P.S. Predicting antigenic sites on proteins. Trends Biotechnol. **1991**, *9*, 163–169.
31. Felci, F. Selection of antibody ligands from a large library of oligopeptides expressed on multivalent exposition vector. J. Mol. Biol. **1991**, *220*, 821–827.
32. Lam, K.S.; Salmon, S.E.; Hersh, E.M.; Hruby, V.J.; Kazmierski, W.M.; Knapp, R. A new type of synthetic peptide library for identifying ligand-binding activity. Nature **1991**, *354*, (Nov 7), 82–84.
33. Hopp, T.P.; Wood, K.R. Prediction of protein antigenicity determinants from amino acid sequences. Proc. Natl. Acad. Sci. **1981**, *78*(6), 3824–3828.
34. Pellequer, J.L.; Westhof, E.; Regenmortel, M.H.V. Epitope predictions from primary structure of proteins. In *Peptide Antigens. A Practical Approach*; Wisdom, G.B., Ed.; Oxford University Press: Oxford, UK, 1994.
35. Molina, F; Laune, D.; Gougat, C.; Pau, B.; Granier, C. Improved performances of spot multiple peptide synthesis. Pept. Res. **1996**, *9* (3), 151–155.
36. Morris, G.E., Ed. *Epitope Mapping Protocols*: *Methods in Molecular Biology*. Humana Press: Totowa, NJ, 1996.
37. Blundell, T.L.; Sibanda, B.L.; Strenberg, M.J.; Thornton, J.M. Knowledge-based prediction of protein structures and design of novel molecules. Nature **1987**, *326* (Mar 26–Apr 1), 347–352.
38. Laver, W.G.; Air, G.M.; Webster, R.G.; Smith-Gill, S.J. Epitopes on protein antigens: misconceptions and realities. Cell **1990**, *61*, 553–556.
39. Greenspan, N.S. Epitopes, paratopes and other topes: do immunologists know what they are talking about? Bull. Inst. Pasteur **1992**, *90*, 267–279.
40. Houghten, R.A.; Appel, J.R.; Blondelle, S.E.; Cuervo, J.H.; Dooley, C.T.; Pinillla, C. The use of synthetic peptide combinatorial libraries for the identification of bioactive peptides. Biotechniques **1992**, *13* (3), 413–421.
41. Smith, G.P. Filamentous fusion phage: novel expression vectors that display cloned antigens on the virions surface. Science **1985**, (4705), 1315–1317.

42. Scott, J.K.; Smith, G.P. Searching for peptide ligands with an epitope library. Science **1990**, *249* (4967), 386–390.
43. van Regenmortel, M.H.V. Antigenicity and immunogenicity of synthetic peptides. Biologicals **2001**, *29* , 209–213.
44. Rini, J.M.; Schulze-Gahmen; Wilson, I.A. Structural evidence for induced fit as mechanism for antigen–antibody recognition. Science **1992**, *255*, 959–965.
45. Tam, J.P. Immunization with peptide–carrier complexes: traditional and multiple-antigen peptide systems. In *Peptide Antigens. A Practical Approach*; Wisdom, G.B., Ed.; Oxford University Press: Oxford, UK, 1994.
46. Valero, M.L.; Camarero, J.A.; Haack, T.; Mateu, M.G.; Domigo, E.; Girakt, E.; Andreu, D. Native-like cyclic peptide models of a viral antigenic site: finding a balance between rigidity and flexibility. J. Mol. Recognit. **2000**, *13*, 5–13.
47. van Regenmortel, M.H.V.; Mueller, S. *Synthetic Peptides as Antigens*. Elsevier: Amsterdam, 1999; 1–381.
48. Ramirez-Alvara do, M.; Blanco, B.J.; Niemann, H.; Serrano, L. Role of beta turn residues in beta hairpin formation and tability in designed peptides. J. Mol. Biol. **1997**, *273*, 898–912.
49. Dyson, H.J.; Cross, K.J.; Houghten, R.A.; Wilson, I.A.; Wright, P.E.; Lerner, R.A. The immunodominant site of a synthetic immunogen has a conformational preference in water for a type-II reverse turn. Nature **1985**, *18* (6045), 480–483.
50. Okubo, M.; Azuma, I.; Hattori, H. Preferential adsorption of bovine fibrinogen dimer onto polymer microspheres having heterogeneous surfaces consisting of hydrophobic and hydrophilic parts. J. Appl. Polym. Sci. **1992**, *45*, 245–251.
51. Pelton, R. Chemical reaction at the latex solution interface. In *Scientific Methods for the Study of Polymer Colloids and Their Applications*; Candau, F., Ottewill, Eds.; Kluwer Academic Publishers: Dordrecht, 1990.
52. Bale Oenick, M.D.; Warshawsky, A. Protein immobilization on surface modified lattices bearing aldehyde groups. Colloid Polym. Sci. **1991**, *269*, 139–145.
53. Basinska, T.; Kowalczyk, D.; Miska, B.; Slomkowsky, S. Interaction of proteins with polymeric latexes. Polym. Adv. Technol. **1995**, *6*, 526–533.
54. Miraballes-Martínez, I.; Martín-Rodríguez, A.; Hidalgo-Alvarez, R. Chloroactivated latex for covalent coupling of antibodies. Application to immunoassays. J. Biomater. Sci. Polym. Ed. **1997**, (10), 765–777.
55. Bangs, L.B. *Uniform Latex Particles* ; Seragen Diagnostics Inc., 1984.
56. Miraballes-Martinez, I.; Martín-Molina, A.; Galisteo-González, F.; Forcada, J. Synthesis of amino-functionalized latex particles by a multistep method. J. Polym. Sci. **2001**, *39*, 2929–2936.
57. Inomata, Y.; Wada, T.; Fujimoto, K.; Kawaguchi, H. Preparation of DNA-carrying affinity latex and purification of transcription factors with the latex. J. Biomater. Sci. Polym. Ed. **1994**, *5* , 293–302.
58. Delgado, C.; Francis, G.E.; Derek, F.F. The uses and properties of PG-linked proteins. Crit. Rev. Ther. Drug Carrier Syst. **1992**, *9*, 249–304.
59. Veronese, F.M. Peptide and protein PEGylation: a review of problems and solutions. Biomaterials **2001**, *22*, 405–417.
60. Kodera, Y.; Matsushima, A.; Hiroto, M.; Nishimura, H.; Ishii, A.; Ueno, T.; Inada,

Y. Pegylation of proteins and bioactive substances for medical and technical applications. Prog. Polym. Sci. **1988**, *23*, 1233–1271.
61. Pichot, C.; Delair, T.; Elaïssari, A. Polymer colloids for biomedical and pharmaceutical applications. In *Polymeric Dispersions: Principles and Applications*; Asua, J.M., Ed.; Kluwer Academic Publishers: Netherlands, 1997.
62. Wilchek, M. Affinity chromatography. New approaches for the preparation of spacer containing derivates and for specific isolation of peptides. Adv. Exp. Med. Biol. **1974**, *42*, 15–31.
63. Narayanan, S.R.; Kakodkar, S.V.; Crane, L.J. Glutaraldehyde-P, a stable, reactive aldehyde matrix for affinity chromatography. Anal. Biochem. **1990**, *188*, 278–284.
64. Goldstein, L. Water-insoluble derivatives of proteolytic enzymes. Meth. Enzymol. **1970**, *19*, 935–962.
65. Pappin, D.J.C.; Coull, J.M.; Koster, H. Solid-phase sequence analysis of proteins electroblotted or spotted onto polyvinylidene difluoride membranes. Anal. Biochem. **1990**, *187*, 10–19.
66. Isosaki, K.; Seno, N.; Matsumoto, I. Immobilization of protein ligands with methyl vinyl ether–maleic anhydride copolymer. J. Chromatogr. **1992**, *597* , 123–128.
67. Ladavière, C.; Lorenzo, C.; Elaïssari, A.; Mandrand, B.; Delair, T. Electrostatically driven immobilization of peptides onto (maleic anhydride-*alt*-methyl vinyl ether) copolymers in aqueous media. Bioconj. Chem **2000**, *11*, 146–152.
68. Gregorius, K.; Mouritsen, S.; Elsner, H.I. Hydrocoating: a new method for coupling biomolecules to solid phases. J. Immunol. Meth. **1995**, *181*, 65–73.
69. Böcher, M.; Böldicke, T.; Kieβ, M.; Bilitewski, U. Synthesis of mono- and bifunctional peptide–dextran conjugates for the immobilization of peptide antigens on ELISA plates: properties and application. J. Immunol. Meth. **1997**, *208*, 191–202.

11

Preparation and Applications of Silicone Emulsions Using Biopolymers

MUXIN LIU, AMRO N. RAGHEB, PAUL M. ZELISKO, and MICHAEL A. BROOK McMaster University, Hamilton, Ontario, Canada

I. INTRODUCTION

A. Fundamentals

Dispersions of water in oil, or the inverse, are inherently unstable. Emulsification is thus a nonequilibrium process such that the average droplet size in an emulsion tends to increase over time. However, the characteristic time scales for coarsening of emulsions can span a remarkably wide range, from seconds to several years, that depends on the nature of the oil, the surfactants used to stabilize the emulsion, and the processing history. It is fair to say that not all the parameters affecting emulsion stability are completely understood. Current practices in emulsion formulation thus combine art with science. With silicones, being specialty materials and very unlike their organic counterparts, both theory and art are less well explored than organic surfactants.

B. Silicone Properties

Silicones possess very unusual properties by organic standards. For example, the low torsional force constant of the Si-O-Si-O linkage [1] results in exceptionally flexible molecules: simple dimethylsilicone polymers [polydimethylsiloxane, **1** PDMS $(Me_2SiO)_n$, Scheme 1] have T_g values of approximately −123°C irrespective of molecular weight over a range of 1000–1,000,000 [2]. This backbone flexibility, in combination with the high hydrophobicity of the *gem*-dimethyl groups and relatively high ionic character of the Si-O linkage, results in exceptional surface properties for silicone polymers [3]. These properties are greatly amplified when either nonpolar or polar functional groups are added to the silicone, creating true surfactants [4]. Depending on their structure, silicones can be used as wetting agents, for foaming or defoaming applications, as lubricants,

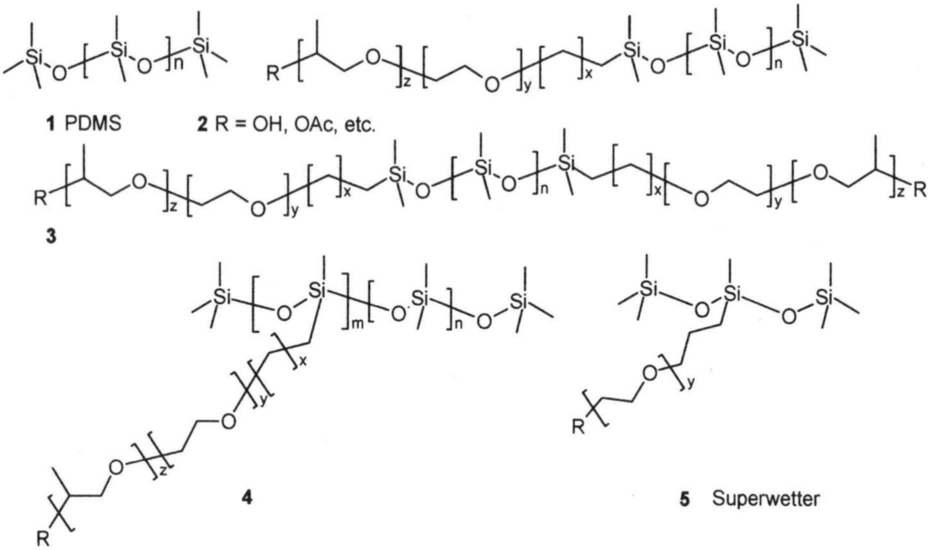

SCHEME 1

and, perhaps most important of all, as compounds that render liquid/air or solid/air surfaces hydrophobic: the methyl groups extend into the air at the interface generating very low-energy surfaces [3]. The surface tension of silicones is a function of molecular weight, increasing from about 16 mN/m for Me_3SiO-$(Me_2SiO)_nSiMe_3$, $n = 0$, to 20–21 mN/m for medium and high molecular weight silicones, $n > 10$ [5].

PDMS is not compatible with aqueous media. If the molecular weight of a given PDMS is high enough, typically starting at six to eight Me_2SiO units, PDMS is also incompatible with mineral oils or more polar oils (e.g., ester oils, natural fats or oils). These hydrophobic and oleophobic properties make it very difficult to form emulsions between silicones and aqueous solutions or organic oils, though high shear can lead temporarily to emulsions that break down easily and rapidly [6].

C. Use and Interest in Silicone Emulsions

With the appropriate surfactant(s), silicones can be formulated into emulsions of a variety of types including W/O, O/W, O/W/O, W/O/W, etc. (W, water; O, oil; the oil may additionally be silicone oil or organic oil). The specific morphology of the emulsion depends on the surfactants used and the processing history.

Silicone emulsions can also arise inadvertently, as a consequence of adventitious surfactants, particularly in biological environments, as shall be discussed below.

D. Impact of Silicone Emulsions in Biological Domains: Purpose of This Chapter

In this chapter, we shall outline some of the basic parameters associated with the formation of colloidal silicone dispersions and then provide some examples of typical silicone emulsions and their application. The remainder of the review will focus on silicone emulsions that form in contact with biological materials. Initially, we shall describe emulsions that spontaneously form in contact with the inner eyeball following retinal repair surgery. Finally, the utilization of proteins and atypical functional silicones to prepare water/silicone emulsions will be described. Both the features necessary for a stable emulsion and the consequences on protein/enzyme tertiary structure will be examined.

II. FUNDAMENTALS OF SILICONE EMULSIONS

A. Classes of Silicone Emulsifiers

A wide variety of surfactants has been used to emulsify and stabilize water/silicone emulsions. As early as 1958, Sato [7] examined the use of many conventional surfactants, including fatty acid esters of polyethylene glycol (nonionic), quaternary ammonium salts (cationic), and alkyl sulfates or alkylarenesulfonates (anionic) to stabilize poly(organoalkylsiloxane) emulsions. It was demonstrated that conventional surfactants based on hydrocarbon chains are generally not very efficient surfactants for silicone emulsions, although sodium dodecyl sulfate (SDS) is useful as a probe for examining the stability of silicone emulsions [6]. The adsorption of the alkyl groups at the silicone interface is not as strong as at an oil interface. In spite of this, because of their relatively low cost, these surfactants are commonly used in commercial emulsions (see discussion of oil-in-water emulsions below). More efficient, but more expensive, silicone-based surfactants were developed both as emulsifiers and stabilizers for silicone emulsions, particularly water-in-oil emulsions [4]. The most common silicone surfactants are described below.

B. Types of Surfactants

Although a variety of phenyl- and trifluoropropyl-modified silicones are sold commercially, there has been little investigation of surfactants based on these compounds. The vast majority of research has focused on modified dimethylsilicones. A wide variety of hydrophils have been combined with silicones to make viable surfactants, including ionic groups such as sulfonates [8], sulfosuccinates

[9], phosphates [10], thiosulfates [11], betaines [12], sulfobetaines [13], and quaternary ammonium salts [14]. The two major classes of silicone-based surfactants are based on amines or polyethers, with the latter holding the lion's share of commercial usage.

Silicone-based emulsifiers involve linear oligomeric or polymeric silicone molecules modified with hydrophilic and, optionally, hydrophobic residues. Both linear block (AB and ABA) and comblike structures are known. In the linear surfactants, functional groups can be located only at the ends of the structures. By contrast, the distribution of functional groups in comb structures is ruled by statistics (Scheme 1).

1. Polyethers

The most important class of emulsifiers for silicones is based on polyethers. Both poly(ethylene oxide) and mixed poly(ethylene oxide)/poly(propylene oxide) polar blocks may be grafted to the silicone backbone (Scheme 1). Although linear block **2**, **3**, and comb **4** structures are available, the comblike silicone polyethers (often known as silicone or dimethicone copolyols) are currently of greatest commercial importance. Even compounds with very low silicone content can be powerful emulsifiers, as exemplified by the trisiloxane **5** [15]. Mixed alkyl/polyether modified silicones are also known: these have applications in emulsifying organic oils [16].

2. Amino-Modified Silicones

Another important class of silicone-based emulsifiers is modified with organoamine groups **6** [1–20] or, in some cases, amino-modified copolyols (polyethers) **7** [21]. Such compounds are widely used as hair softeners and conditioners as well as in cosmetic products (Scheme 2). Since the amine groups on these compounds will be protonated at pHs <10, these compounds are usually ionic surfactants.

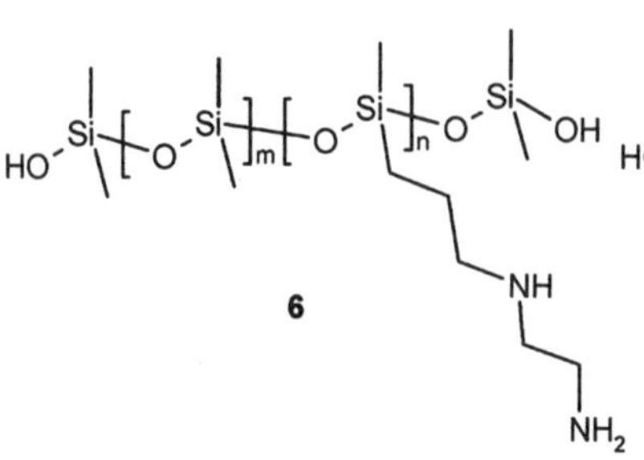

6

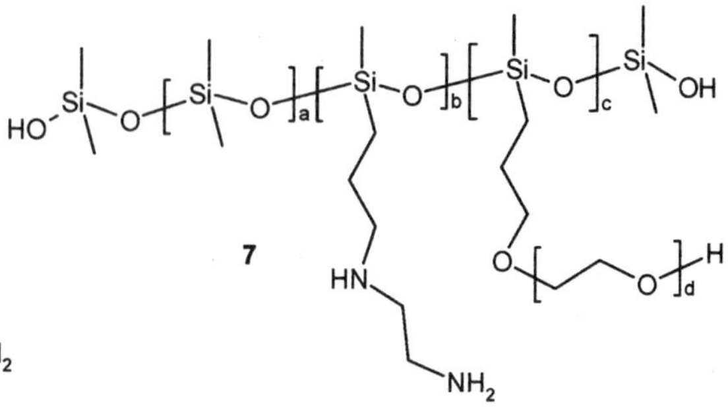

7

SCHEME 2

C. Silicone Emulsion Formulations

Silicones can be formulated into emulsions of a variety of types including W/O, O/W, O/W/O, W/O/W, and microemulsions with the appropriate surfactant(s). The specific morphology of the emulsion depends on the surfactants used and processing procedures. As noted above, organic surfactants are not as efficient as silicone surfactants for stabilizing silicone/water emulsions. However, they are much cheaper and are commonly used in commercial formulations for coating applications and for hair conditioning. Oil-in-water emulsions are generally formed mechanically or through emulsion polymerization of cyclic monomers. In mechanically formed emulsions, the incorporation of the existing materials occurs with specialized agitation/mixing that is beyond the scope of this chapter; no chemical reactions take place. Alternatively, cyclic monomers [typically $D_4(Me_2SiO)_4$] can be polymerized in the presence of emulsifying surfactants and water to form polymeric o/w emulsions. Emulsion polymerization facilitates the incorporation of water-soluble moieties that are not otherwise easily introduced to hydrophobic silicones. In addition, the emulsion composition can be manipulated to adjust the product characteristics as required [21].

While there is extensive knowledge of the stabilization of w/o emulsions by silicone emulsifiers (see next section), examples of o/w emulsions are less well described [22]. Nonionic dimethylsiloxane polyoxyalkylene copolymers are generally used to prepare such dispersions. They must, of necessity, carry a high degree of polyoxyalkylene substituents, which render the surfactant more hydrophilic. The resulting emulsions are sterically stabilized by the polyether chains. Alternatively, amino-modified silicones will form o/w emulsions.

The other important class of silicone emulsion is of the w/o type, for which polydimethylsiloxane-polyoxyalkylene copolymers are preferred as emulsifiers; organic surfactants are generally ineffective. Increasing the molecular weight of the emulsifier is an effective means of preparing emulsions with improved stability. Thus, emulsions containing potentially destabilizing alcohols in the aqueous phase can be successfully emulsified with silicone-copolyols having an approximate molecular weight of 30,000 [23]. The molecular weight of the emulsifier can be further increased by slight cross-linking. To achieve emulsions with even greater stability, organic w/o emulsifiers such as polyglycerol fatty acid esters may also be used.

D. Microemulsions

In both types of emulsions noted above, droplet sizes are typically rather large (>500 nm diameter) such that the emulsions are opaque. It is possible to prepare microemulsions of silicone oils and water with **5** as the emulsifier [17,24]. These optically clear dispersions are isotropic mixtures. Aminosilicone copolyols also form microemulsions spontaneously in water [21], although in this case the dispersion has internal phase particles of 5–50 nm. Because of their smaller particle sizes, microemulsions are more stable than the conventional emulsions [21]. At the time of writing, these emulsions are more of fundamental interest [4,21] than commercial applications, although this situation is expected to change as a result of their intriguing properties.

E. Water-in-Oil-in-Water and Oil-in-Water-in-Oil Emulsions

The fascinating properties of multiple emulsions, which may be of the W/O/W or O/W/O type, have attracted recurring interest, in particular when the protection of sensitive ingredients or controlled release of active substances is required. Two different water–oil interfaces have to be stabilized in both types of multiple emulsions. Like other polymeric surfactants, silicone-based emulsifiers are especially suited to stabilize these emulsions because their polymeric nature permits them to be adsorbed strongly at the oil interface, which prevents the migration of the emulsifiers from one interface to the other leading to destabilization. In one example of this, a w/o/w emulsion was established using two polymeric emulsifiers (see other examples, below) [25]: a hydrophobic polyacrylate copolymer, which carries lipophilic alkyl and hydrophilic polyoxyalkylene groups for stabilization of the oil–water interface, and poly(hexadecylmethylsiloxane)-*co*-poly(ethylene/propylene oxide), which stabilizes the water–oil interface. The hydrophilic–lipophilic balance (HLB) values of the emulsifiers should be above 10 for the hydrophilic emulsifier and below 6 for the hydrophobic emulsifier.

F. Theory

The classical and empirical approach to formulation of organic emulsions uses the HLB [26] surfactant classification system. In general, one uses surfactants that are soluble in the continuous phase to make emulsions successfully. Thus, low HLB surfactants are used for w/o and high HLB for O/W emulsions [27].

The HLB system was developed for alkoxylated nonionic surfactants [27]. The characteristics of silicone surfactants are very different from this class of compound. As a result, it is difficult to apply the HLB system to silicone emulsifiers. Calculations based on critical micelle concentration (CMC) give some idea of the hydrophobicity of the silicone component. Typically, each Me or CH_2 group on a silicone contributes to the hydrophobicity in silicones as much as a CH_2 group does in organic surfactants, while the Si-O does not significantly affect the HLB [28]. More recently, researchers have attempted to predict emulsification behavior of silicone surfactants by use of three-dimensional HLB (3D-HLB) [16]. Thus, an HLB of about 4 is calculated for a polymer of the structure

$[Me(H_{33}C_{16})SiO]_nEO_m$ of molecular weight 10,000–15,000; tests of actual emulsification ability placed the HLB value between 4 to 6, demonstrating reasonable correlation between theory and experiment [29].

Silicone emulsions based on **6** have limited stability as a result of the fairly narrow range of HLB they achieve [20]. Compound **7**, which is made by emulsion polymerization in the absence of water, oils, and surfactants, expands the ability of incorporating silicones in formulations, as the addition of polyether group allows tailoring of the water solubility of these silicones, resulting in an increase in the HLB range [21]. Silanol groups (typically found at the end of linear polymers) also contribute to an increase in the HLB [21].

The origin of the utility of silicone polyethers to stabilize (particularly w/o) emulsions has been the source of significant discourse. Several proposals have been made to explain their notable ability to prevent droplet coalescence. Factors that increase the viscosity of the continuous phase increase emulsion stability, and it is clear that the relatively long silicone spans between each hydrophilic polyether group can serve this purpose. Furthermore, the highly flexible silicone chains, which can extend to the silicone oil continuous phase, may also provide steric stabilization. More careful studies of these systems with well-characterized surfactants are warranted.

III. APPLICATIONS AND SPECIFIC EXAMPLES OF W/O, O/W AND W/O/W, O/W/O EMULSIONS

A. Cosmetics

Silicones are important ingredients in body care, face care, and cosmetic products. Silicone w/o emulsions are used in skin care products such as skin cleansers because they improve spreadability and, more importantly, because they are aesthetically attractive: they impart a smooth and silky feel and reduce greasiness. It is possible to formulate "non-oil" personal care products that have 60% of their composition as silicones and no more than 10% as mineral oil [30]. Dispersions based on low molecular weight cyclomethicones and hexamethyldisiloxane are used in antiperspirant deodorant formulations, again for their aesthetic feel. With a lower heat of evaporation, they do not seem "cool," as do alcohols, the competing materials, and possess an attractive "feel" on the skin.

B. Hair Care

Silicones are extensively used as conditioners in hair care. They impart softness, combing ease, fast drying, and shiny appearance. Two basic classes predominate in this market. In the first, emulsified high molecular weight silicone droplets are deposited on the hair. In the second, cationic amino-modified silicones (the amino group in aqueous solutions/emulsions will be protonated below pH <10)

bind to anionic hair, a protein (see also below), and provide substantivity. Both formulations require extended colloidal stability on the shelf. The latter organofunctional silicones are important constituents of two-in-one shampoos, laminators, conditioners, and so forth. The aminosilicones used in hair treatment are generally used as o/w emulsion–based formulas.

C. Drug Delivery

Drugs are encapsulated, not only for taste and odor masking but also for drug stabilization, gastrointestinal tolerance, and controlled rate of release. Appropriately formed silicone emulsions can be considered as another form of drug encapsulation, as the drug is entrapped in the emulsion droplets, which serve as a carrier as well as protective shell for the drug.

Silicone oil, for instance, was employed as the external phase in an o/w/o multiple phase emulsion, during microsphere formation, using an emulsion/internal gelation technique [31]. In this process, the lipophilic encapsulant (Sudan orange G) was dissolved in the edible oil and then dispersed in alginate sol. This dispersion was dispersed again in silicone oil to form O/W/O emulsion. This was followed by an internal gelation process to give alginate microspheres that contain immobilized oil droplets, which were subjected to a further coating step using chitosan to control the release rate. Silicone is beneficial in this instance not only for its high hydrophobicity, which provides a control element in release kinetics, but also for its regulatory acceptance as an oral excipient in antacid and related applications.

The W/O/W emulsions have potential applications in many areas, such as pharmaceuticals, cosmetics, and agriculture. However, their inherent thermodynamic instability and their fast, uncontrolled release of the entrapped materials limit their use for drug delivery. In order to overcome these problems, Sela et al. [32] studied the release of ephedrine hydrochloride and other compounds from w/o/w emulsions stabilized with commercially available hydrophobic surfactants and hydrophilic silicone surfactants that they synthesized by grafting undecanoic esters of poly(ethylene oxide) (45 ethylene oxide units) to poly(hydromethylsiloxane)-co-(dimethylsiloxane) using hydrosilylation (for hydrosilylation, see Scheme 3). They found that this emulsion exhibited enhanced stability and rate of release. Other silicone emulsion systems for drug delivery have been described [33,34].

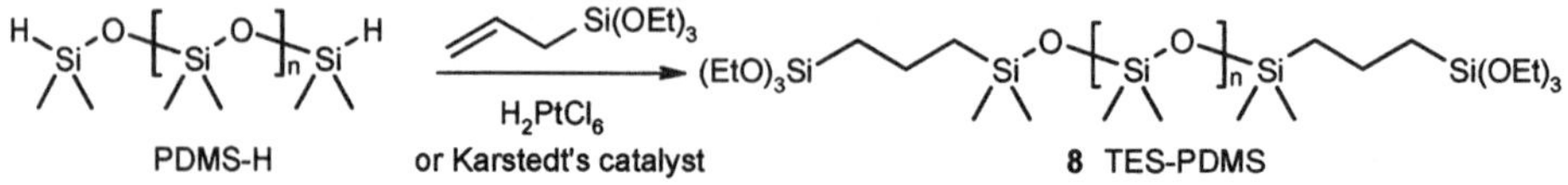

SCHEME 3

IV. EMULSIONS OF SILICONES WITH BIOLOGICAL MATERIALS—ADVENTITIOUS EMULSIFICATION

A. Retinal Repair Fluids and Emulsions

Silicones have an impressive record for biocompatibility and have been used in many applications that require topical (e.g., cosmetics as noted above) or internal use. (At the time of writing, the breast implant controversy appears to have nearly run its course, with most epidemiological studies showing only very weak or no association at all between disease and silicone polymers [35]. Silicone elastomers are still constituents of a variety of medical implants, and silicone oils continue to be approved as oral antacid excipients.) Depending on the application, the surface activity of silicones may be beneficial. For example, silicones, usually in combination with hydrophobed silica, are widely used as defoamers in oral antacid formulations. However, their surface activity is not always desirable.

Silicones have been extensively used as replacement fluids following repair of retinal detachment. Following reattachment of the retina, silicone oils are used to replace the vitreous humor (fluid inside the eyeball) on a temporary basis for up to several months. The viscosity of the PDMS oil used in this process ranges from 1000 to 12,500 centistokes (cSt); while fluorosilicone oils [containing $F_3C(CH_2)_2(Me)SiO$ moieties] are somewhat lower, i.e., 1000 to 10,000 cSt. PDMS has a lower density than the intraocular fluid, and hence is considered useful in dealing with retinal detachment in the superior portion of the eye. Fluorosilicones, on the other hand, are considered more appropriate for repair of inferior detachment, as they possess a higher density than the intraocular fluid [36]. The optical clarity and high permeability to oxygen of silicones are of particular benefit; their ability to form emulsions is not.

It has been frequently observed that after incorporation in the eye, dimethyl-, phenylmethyl-, and fluorosilicones emulsify in vivo. The dispersion droplets migrate to several places including the interior chamber, occlude vision, and, more problematically, change the permeability of the corneal endothelium [37] and the ability of the eye to clear undesirable materials; small silicone oil droplets can cause secondary glaucoma by blocking aqueous outflow [38]. Neither dimethyl- nor fluorosilicones readily form stable emulsions when mixed with water or saline, and several groups have attempted to assess the source of the emulsifier in the eye and understand how the emulsions form. Very informative discussions by Miller clarify that spontaneous emulsification can occur under ideal conditions [39]. These depend on the specific o/w phase diagram, the characteristics of the oil, and the presence of at least a dilute concentration of surfactant(s), which may initially be present or may arise from chemical reaction [40]. What, however, is the active surfactant in the eye?

The relationships between physical and functional characteristics of the silicones and the ease of emulsification were thus assessed in vitro. Some general comments can be made. Lower molecular weight silicones, cyclic and linear oligomers in particular, are associated with emulsification, and great care is now taken by commercial suppliers of these materials to reduce the content of low molecular weight materials (personal communication, Labtician, Oakville, Canada) [41,42]. Phenyl groups were shown not to facilitate emulsion formation whereas, perhaps not surprisingly, the presence of silanol groups did [42]. Fluorosilicones, which have a higher density than the other silicones or of water, were also found to facilitate emulsification in vitro and in vivo [43,44]. In in vitro tests, it was noted that the absence of an air interface (no head space), as is the case in the eye, greatly reduced the degree of emulsification. It was observed that methylsilicones were less emulsified than fluorosilicones of the same viscosity, suggesting that the smaller density difference between silicones and intraocular fluid makes intermixing more difficult compared with fluorosilicones. This, in combination with the observation that reduced emulsification accompanied the use of higher molecular weight, viscous silicones, [38,41], suggests that ease of mixing is an important aspect of emulsion formation.

1. In Vitro Tests

As noted above, emulsions involving silicones are not that easily formed in the absence of surfactants. By performing both in vivo and in vitro tests with plausible biological surfactants, several groups have attempted to determine if bodily fluids could provide surfactants in vivo to stabilize the silicone emulsions.

Emulsification of a medium molecular weight silicone (about 1000 cSt, MW 28,000) [2] was attempted with blood plasma, serum, lipoprotein-deficient serum, and high- or low-density lipoprotein, respectively [45] Lipoprotein-deficient serum did not support the emulsification, but emulsions formed readily in the presence of plasma lipoproteins and constitutents of red blood cell membranes, including phospholipids [46], implicating them in the in vivo emulsification process. Other additives further enhanced emulsification [42]. Emulsions of droplet size about 45 μm were formed in vitro, which compares with 38-μm droplets found in patients. Emulsions could also be made simply by the addition of proteins to the silicone/water system. Emulsifying efficiency followed the order fibrinogen, fibrin, and serum, followed by albumin [44], gamma globulin [38], very low density lipoprotein, and acidic α_1-glycoprotein fibrin. An independent comparison of emulsification in the presence of vitreous humour, blood serum, or collagen showed that the former was the best emulsifier of high molecular weight (about 5000 cSt, 50,000 MW) silicone, although all three produced emulsions in vitro [47]. The presence of balanced salt solution, rather than deionized water, facilitated emulsification in all cases [38].

2. In Vivo Tests

Short implantations (1 week) of silicone preemulsified with bovine serum albumin as the surfactant demonstrated that the combination of silicone and protein, and the ionic strength of the emulsifying liquid, are important factors in physiological effects in the eye (corneal permeability). Inflammation in the eye was observed to facilitate emulsification of the silicone [37]. In longer term implantation studies (6 months) of silicone oil, the surfactant necessary to form a water-in-silicone oil emulsion was judged by infrared spectroscopy to be a protein–silicone complex at the emulsion interface [47].

Two types of mechanisms were proposed to explain silicone oil emulsification in the eye: thermodynamic and hydrodynamic [44]. In the thermodynamic mechanism, emulsification occurs when some surface-active substances populate the interface and the interfacial tension decreases. In the case of the hydrodynamic mechanism, emulsification takes place as a result of oil surface deformation that is induced by external mechanical energy. The above mixing studies suggest the latter mechanism could be important when emulsification is facilitated by an air interface. Based on work with highly phenylated silicone copolymers, Ikeda suggested that the emulsification of the copolymer is more generally facilitated as the interfacial energy decreases, due to the attachment of proteins to the oil surface; that is, in this case the thermodynamic mechanism predominates [44].

3. Speculation About Specific Nature of Silicones and Biological Materials That Leads to Emulsions

In general, the ideal nonemulsifying intraocular tamponade should be an optically clear, nontoxic, hydrophobic liquid that is protein repellant, so there will be no attachment of protein at the oil–aqueous phase interface. However, is there a protein-repellant silicone? These studies suggest that proteins play an important role in stabilizing water-in-silicone oil emulsions. In the remainder of the chapter we shall focus on patents and fundamental studies of silicone emulsions in which proteins are a required constituent.

B. Silicones and Proteins as Cosurfactants

1. Cosmetics

Silicones, as noted above, have a wonderful aesthetic feel and are widely used in consumer products. A current trend in the cosmetics industry is the utilization of natural materials as cosmetic constituents either as "nutriceuticals" or because natural materials are perceived to be beneficial to consumers. In the case of cosmetics, different silicone emulsions containing proteins have been patented. The patents do not claim proteins as required constituents of the water–oil interface, but their presence at the oil–water interface is very likely (see below). Thus, patents have been issued for silicone oil/water emulsions containing nonionic silicone surfactants based on polyethers, with additional protein such as collagen or protein hydrolysate that acts in concert with other agents to gel the emulsion, [48] or that helps to form a film upon application to skin [49]. The latter patent explicitly included the albumin and globulin fraction of soy proteins.

2. Hair Care

The treatment of hair—a protein—was only briefly mentioned in this chapter (see above). Here note is made of a patent describing an emulsion that incorporates a protein, which is added to facilitate silicone deposition on hair by exploiting favorable hair–protein interactions. The inventors do not comment on the role of the protein as a surfactant and add an emulsifier with HLB 8–10, such as Neodol (a C_{12}/C_{13} linear alcohol) or Tergitol (nonylphenylethoxylate) to stabilize the emulsion [50].

3. Biodiagnostics

An interesting patent describes the formation of an O/W emulsion in which the dispersed fluorocarbon or silicone phase is coated with a biodiagnostic protein [51]. Although no claims are made about the stabilization of the emulsion, with droplet sizes about 0.5–5 μm, the protein is critical in this regard. These surface-active proteins serve a supplementary role. In addition to stabilizing the emulsion, they serve as binding sites for biomolecules contained in bodily fluids. Binding to the surface-active proteins reduces the colloidal stability of the emulsion: agglutination is indicative that such binding has occurred. Thus, selective binding of biomolecules to judiciously chosen surface-active proteins provides a convenient biodiagnostic system.

V. PROTEINS INVOLVED IN WATER-IN-SILICONE OIL EMULSIONS

Proteins have been shown to play an integral role in the stabilization of natural emulsion systems, such as milk [52] and, as noted above, silicone oil in the interior of the eye. The same principles that are involved in stabilizing these natural systems should be applicable to engineered emulsions involving a water and silicone oil interface.

A. Enzymes Entrapped within Emulsions—PEO-Modified Silicone

As we have seen, commercial silicone polyethers are excellent surfactants with which to generate stable water-in-silicone oil emulsions [53]. Water-in-D_4

[(Me_2SiO_4)] emulsions, stabilized by a silicone copolyol **4**, do not undergo phase separation over extended periods (in excess of 6 months). From small- (40 mL) to very large-scale emulsions (thousands of liters) may be readily formed using the same experimental protocol; the ability to perform such linear "scale-up" is rather rare. Particles were relatively monodispersed and averaged 2–5 μm in diameter. Much smaller volume emulsions (5 mL or less) required special mixers, which led to broader dispersity emulsions; most particles were submicrometer sized with a few larger droplets of water [52].

It has been shown, perhaps not surprisingly, that neither adding proteins to the emulsions nor changing the ionic strength of the dispersed phase had any significant effect on the ease of emulsification or the stability of the resulting emulsion: differing concentrations of α-chymotrypsin, lysozyme, and alkaline phosphatase have been entrapped within the dispersed phase of water-in-silicone oil emulsions stabilized by comb silicone polyethers [54–58]. This suggested that the protein itself does not play an integral part in stabilizing the oil–water interface. However, confocal microscopy experiments of fluorescently labeled enzyme entrapped within the emulsion droplets clearly revealed that the proteins preferentially adsorb at the emulsion interface, placing them in intimate contact with the functionalized silicone [54]. Thus, although they do not measurably alter the properties of the interface, they are a constituent of it. An added property of the silicone polyether–based emulsion system is its selective permeability. Emulsions formulated using this surfactant allowed the free exchange of neutral species across the oil–water interface, while charged entities did not traverse the interface [58].

Generally, silicone oils and elastomeric surfaces are powerful denaturants for protein tertiary structure [59]. Enzyme assays for the entrapped protein revealed that despite the fact that the enzyme was adsorbed at the water-in-silicone oil emulsion interface and was therefore in contact with the functionalized silicone, the enzymatic activity was equal to or in some cases greater than that of the same enzyme in a buffered solution serving as a control [54–57]. Silicone polyethers are thought to partition at the oil–water interface with the poly(ethylene oxide) chains inserted into the dispersed aqueous phase anchoring the silicone spacers to the water–oil interface [60]. The poly(ethylene oxide) chains must act to passivate the water–oil interface for the proteins that reside there, a phenomenon that is well known on solid surfaces [61–64].

B. Enzymes Entrapped within Emulsions—TES-PDMS

It is possible to prepare functional silicone polymers that possess alkoxysilane groups (RO-Si). Such species are often used for elastomer formation catalyzed by water, among other things (RTV, room temperature vulcanization). One such polymer, triethoxysilyl-modified polydimethylsiloxane **8** (TES-PDMS), is prepared by the hydrosilylation of commercially available H-Me_2Si-terminated silicone (Scheme 3). This material, irrespective of molecular weight (MW 500–60,000) does not function as a surfactant that will stabilize w/o emulsions. Similarly, albumin (human serum albumin, HSA), a protein well known for its high surface activity, will not stabilize a water-in-silicone oil emulsion for extended periods [65,66]. However, the combination of TES-PDMS and albumin leads to the formation of stable water-in-D_4 emulsions. At about 2–5 μm, particle sizes are comparable to those observed in emulsions stabilized with a silicone copolyol surfactant. These emulsions can be formulated on 5-mL (micromixer), 40-mL, and larger scales, and are stable on the order of 25–45 days, although greater stability has been observed in some cases [53].

Confocal microscopy experiments on the emulsions utilizing fluorescein-modified HSA (FITC-HSA) clearly demonstrate that the labeled protein adsorbs at the oil–water interface. HSA has been shown to preferentially adsorb at the interface even in the presence of a second protein [67]. Fluorescence spectroscopy experiments have revealed that although HSA is in intimate contact with the TES-PDMS and is subjected to a great deal of shear stress during the emulsification process, it retains its native conformation. These results suggest that the TES- PDMS serves as a physical buffer against the mixing stress being imparted on the system during emulsification, and further that the alkoxysilyl groups present at the termini of the TES-PDMS are hydrophilic enough to passivate the protein against spontaneous denaturation [67].

Model studies that examined the desorption of HSA from well-defined HSA/TES-PDMS films demonstrated that protein desorption occurred at a much slower rate with functional silicone, compared to normal PDMS of the same molecular weight [68]. In addition, angle-dependent X-ray photoelectron spectroscopy and contact angle measurements were consistent with the interpretation that very high affinity exists between HSA and TES-PDMS, similar perhaps to the case at a water–silicone oil emulsion interface [69,70].

HSA entrapped in silicone-modified starch microparticles displays very interesting immunological properties. The microparticles are easily prepared by precipitating a water-in-vegetable oil emulsion containing the HSA in acetone containing the silicone [71,72]. When silicone-modified starch microparticles containing HSA are orally or nasally administered to animal models, a Th2 antibody response is generated [73]. Enhanced antibody titers (IgG) were observed with the TES-PDMS-modified microparticles when compared to animal models immunized intraperitoneally with unmodified starch microparticles, PDMS-coated particles, or HSA alone. Following oral delivery of the microparticles, HSA-specific IgA antibodies were isolated from the gut washings [74]. Both sets of results indicate that the protein-containing microparticles are a viable means by which to nasally, orally, or via the peritoneum delivery entrapped antigens for vaccination as a result of the protective nature for the protein by the functional silicone.

VI. NATURE OF THE INTERACTIONS NECESSARY FOR THE STABILIZATION

It is apparent from the experiments described above that proteins can act in concert with silicones of various types to stabilize water-in-silicone emulsions. The active surfactant varies from the silicone (e.g., added silicone polyether or amine-modified silicone), the protein (e.g., the emulsifiers in retinal repair), or a combination of the two as in the TES-PDMS albumin emulsions. While there is no clear evidence for the specific nature of the silicone–protein interactions, particularly in the latter case, the commercial amino-modified silicones provide some clues. These compounds, as with the proteins, are ionized at physiological pHs, such that they possess charged end groups, hydrophilic (PEO) linkers, and the hydrophobic silicone. It is conceivable that the TES-PDMS surfactant is building up a similar structure once combined with the protein, except that the protein must act as both the hydrophilic linker *and* the charged hydrophil. There is little evidence in our experiments that there is a direct covalent bond between the trialkoxysilane terminus and the protein. However, an ionic interaction between a hydrolyzed alkoxysilane (silanolate-SiO^{-}-H_3N^{+}) and a protein-based quaternary ammonium ion cannot be ruled out. Research into exact nature of the protein–silicone interaction is ongoing in our laboratory.

VII. CONCLUSIONS

Silicone/water emulsions of a variety of structural morphologies are readily formed in the presence of appropriate surfactants: normally, these surfactants are silicone based. Water-soluble/dispersible proteins favor sitting at silicone–water interfaces and can facilitate emulsification on their own or in combination with other excipients. Many applications for such emulsions can be envisaged, including as immobilized enzymes, for biodiagnostics, and for drug delivery (or personal care products). In order to realize these goals, it is necessary to first understand more clearly the nature of the protein–silicone interaction and to optimize their characteristics.

REFERENCES

1. Grigoras, S.; Lane, T.H. Conformation analysis of substituted polysiloxanes polymers. In *Silicon-Based Polymer Science: A Comprehensive Resource*; Zeigler, J.M.,

Fearon, F.W.G., Eds.; ACS Adv. Chem. Ser. 224; American Chemical Society: Washington, DC, 1990; 125.

2. Arkles, B., Ed. *Silicon, Germanium, Tin and Lead Compounds, Metal Alkoxides, Diketonate and Carboxylates: A Survey of Properties and Chemistry*; Gelest Inc. Catalog, 612 William Leigh Dr., Tullytown, PA 19007–6308 USA; 386–428.
3. (a) Owen, M. Siloxane surface activity. In *Silicon-Based Polymer Science: A Comprehensive Resource*; Zeigler, J.M., Fearon, F.W.G., Eds.; ACS Adv. Chem. Ser. 224; American Chemical Society: Washington, DC, 1990; 705. (b) Owen, M.J. Surface chemistry and applications. In *Siloxane Polymers*; Clarson, S.J., Semlyen, J.A., Eds.; Prentice Hall: Englewood Cliffs, NJ, 1993; 309. (c) Owen, M.J. Ind. Eng. Chem. Prod. Res. Dev. **1980**, *19*, 97.
4. Hill, R.M., Ed. *Silicone Surfactants*; Surfactant Science Series, Vol. 86; Marcel Dekker: New York, 1999.
5. Grüning, B.; Bungard, A. Silicone surfactants: emulsification. In *Silicone Surfactants*; Hill, R.M., Ed.; Surfactant Science Series, Vol. 86; Marcel Dekker: New York, 1999; 209–240.
6. Bibette, J.; Morse, D.C.; Witten, T.A.; Weits, D.A. Phys. Rev. Lett. **1992**, *69*, 2439.
7. Sato, K. Japanese Patent Application 33,002,782, Matsushita Electric Industrial Co., 1958; Chem. Abstr. **1959**, *53*, 70381.
8. Kanner, B.; Pike, R.A. US Patent 3,507,897, Union Carbide Corp., 1987.
9. (a) Colas, A.R.L.; Renauld, F.A.D. US Patent 4,477,377, Dow Corning, 1988. (b) Maxon, B.D. US Patent 4,717,498, Mclntyre Chemical, 1987.
10. O'Lenick, A.J., Jr. US Patent 5,070,171, Siltech, 1990.
11. Gruning, B.; Holtschmidt, U.; Koerner, G. German Patent 3,323,881, T. Goldschmidt, 1983.
12. Hoffmann, K.; Kollmeier, H.J.; Langenhagen, R.D. German Patent 3,417,912, T. Goldschmidt, 1985.
13. Fenton, W.N.; Owen, M.J.; Snow, S.A. US Patent 4,918,210, Dow Corning, 1990.
14. Schaefer, D.; Kradenberg, M. German Patent 3,719,086, T. Goldschmidt, 1987.
15. Hill, R.M. Silicone surfactants: emulsification. In *Silicone Surfactants*; Hill, R.M., Ed.; Surfactant Science Series, Vol. 86; Marcel Dekker: New York, 1999, 1–48.
16. McKellar, R.L. US Patent 3,427,271, Dow Corning, 1996.
17. Hill, R.M. Ternary phase behavior of mixtures of siloxane surfactants, silicone oils, and water. In *Silicone Surfactants*; Hill, R.M., Ed.; Surfactant Science Series, Vol. 86; Marcel Dekker: New York, 1999, 313–348.
18. (a) Merrifield, J.H.; Thimineur, R.J.; Traver, F.J. US Patent 5,244,598, General Electric, 1993. (b) Berthiaume, M.D.; Merrifield, J.H. US Patent 5,683,625, General Electric, 1997. (c) Gee, R.P. US Patent 5,852,110, Dow Corning, 1996. (d) Gee, R.P. European Patent 138192 B1, Dow Corning, 1988. (e) Katayama, H.; Tagawa, T.; Kunieda, H. J. Colloid Interface Sci. **1992**, *153*, 429.
19. Cheng, J.; Wu, Q.;Wang, X. Youjigui Cailiao **2001**, *15*, 9.
20. Halloran, D.; Hoag, C. Cosmet. Toiletries **1998**, *113*, 61.
21. O'Lenick, A.J.,Jr.; Sitbon, C.S. Cosmet. Toiletries **1998**, *113*, 63.
22. Grüning, B.; Bungard, A. Silicone surfactants: emulsification. In *Silicone Surfac-*

tants; Hill, R.M., Ed.; Surfactant Science Series, Vol. 86; Marcel Dekker: New York, 1999; 232.

23. Starck, M.S. US Patent 4,311,695, Dow Corning, 1979.
24. Hill, R.M. US Patent 5,705,562, Dow Corning, 1998.
25. Grüning, B.; Hameyer, P.; Weitemeyer, C. Tenside Surf. Detergents **1992**, *29*, 78.
26. HLB = (molar% hydrophilic group in a surfactant)/5. Thus maximum HLB = 20 (100/5). Surfactants with HLB = 0 are completely water insoluble.
27. Porter, M.R. *Handbook of Surfactants*; Blackie: Glasgow, 1991; 42.
28. Grüning, B.; Bungard, A. Silicone surfactants: emulsification. In *Silicone Surfactants*; Hill, R.M., Ed.; Surfactant Science Series, Vol. 86; Marcel Dekker: New York, 1999; 217.
29. (a) Grüning, B.; Bungard, A. Silicone surfactants: emulsification. In *Silicone Surfactants*; Hill, R.M., Ed.; Surfactant Science Series, Vol. 86; Marcel Dekker: New York, 1999; 218. (b) Griffin, W.C. J. Soc. Cosmet. Chem. **1954**, *5*, 249. (c) Hameyer, P. Seifen-Oele-Fette-Waechse **1990**, *116*, 392.
30. De Baker, G.; Ghirade, D. Parfumes Cosmet. Arom. **1993**, *114*, 61.
31. Ribeiro, A.J.; Neufeld, R.J.; Arnaud, P.; Chaumeil, J.C. Int. J. Pharm. **1999**, *187*, 115.
32. Sela, Y. ; Magdassi, S. ; Garti, N. J. Controlled Rel. **1995**, *33*, 1.
33. Clément, P.; Laugel, C.; Marty, J. J. Controlled Rel. **2000**, *66*, 243.
34. Dahms, G.H. SÖFW J. **1999**, *125*, 6.
35. (a) For the Institute of Medicine report, see: Bondurant, S.; Ernster, V.; Herdman, R.; Eds. *Safety of Silicone Breast Implants*; National Academy of Sciences: Washington, DC, 2000; summarized in Rouhi, M. Chem. Eng. News **1999**, (June 28), 10. (b) Gabriel, S.E.; O'Fallon, M.W.; Kurland, L.T.; Beard, C.M.; Woods, J.E.; Melton, L.J., III. N. Engl. J. Med. **1994**, *330*, 1697. (c) Sánchez-Guerrero, J.; Colditz, G.A.; Karlson, E.W.; Hunter, D.J.; Speizer, F.E.; Liang, M.H. N. Engl. J. Med. **1995**, *332*, 1666. (d) For the review by the panel appointed by Judge Pointer, see: www.fjc.gov/BREIMLIT/SCIENCE/summary.htm.
36. Miyamoto, K.; Refojo, M.F.; Tolentino, F.I.; Fournier, G.A.; Albert, D.M. Arch. Opthalmol. **1986**, *104*, 1053.
37. Green, K.; Tsai, J.; Kearse, E.C.; Trask, D.K. J. Toxic Cutan. Ocul. Toxicol. **1996**, *15*, 325.
38. Heidenkummer, H.P.; Kampik, A.; Thierfelder, S. Retina **1992**, *12* (3 Suppl), S28.
39. For a review, see: (a) Miller, C.A.; Rang, M.J.; Mittal, K.L.; Kumar, P., Eds. *Emulsions, Foams, and Thin Films*; Marcel Dekker: New York, 2000. (b) Nishimi, T.; Miller, C.A. J. Colloid Interface Sci. **2001**, *237*, 259.
40. We thank a referee for this helpful suggestion.
41. Crisp, A.;de Juan, E., Jr.; Tiedeman, J. Arch. Ophthalmol. **1987**, *105*, 546.
42. Heidenkummer, H.P.; Kampik, A.; Thierfelder, S. Graefe's Arch. Clin. Exp. Ophthalmol. **1991**, *229*, 88.
43. Nakamura, K.; Refojo, M.F.; Crabtree, D.V. Inv. Ophthalmol. Vis. Sci. **1990**, *31*, 647.
44. Ikeda, T.; Nakamura, K.; akagami, K.; Iwahashi, H.; Sugimoto, K.; Matsuda, T.; Tano, Y. Japan J. Ophthalmol. **2001**, *45*, 53.

45. Savion, N.; Alhalel, A.; Treister, G.; Bartov, E. Inv. Ophthalmol. Vis. Sci. **1996**, *37*, 694.
46. Bartov, E.; Pennarola, F.; Savion, N.; Naveh, N.; Treister, G. Retina **1992**, *12* (3 Suppl), S23.
47. Lukasiak, J.; Dorosz, A.; rokopowicz, M.; Raczynska, K.; Falkiewicz, B. Polimery (Warsaw, Poland) **2001**, *46*, 428; Chem. Abstr. **2001**, *135*, 335071.
48. Bara, I.; Mellul, M. US Patent 5,942,213, l'Oreal, 1999.
49. Russ, J.G.; Sandewicz, I.M.; Zamyatin, T. US Patent 6,299,890, Revlon, 2001.
50. O'Lenick, A.J., Jr.; Buffa, C.W. US Patent 5,854,319, Lambent Technologies, Biosil Technologies, 1998.
51. Giavere, I.;. Keese, C.R. US Patent 4,619,904, General Electric, 1986.
52. Dickinson, E. J. Chem. Soc. Faraday Trans. **1998**, *94*, 1657.
53. Brook, M.A.; Zelisko, P.M.; Walsh, M.J.; Crowley, J.N. *Silicon Chem 1*; **2002**, *1*, 99.
54. Zelisko, P.M.; Brook, M.A. Langmuir **2002, *18*, 8982.**
55. Zelisko, P.M.; Brook, M.A. Polym. Prepr. (Am. Chem. Soc. Div. Polym. Chem. **2001**, *42* (2), 115.
56. Brook, M.A.; Zelisko, P.M. Polym. Prepr. (Am. Chem. Soc. Div. Polym. Chem.) **2001**, *42* (1), 97.
57. Zelisko, P.M.; Bartzoka, V.; Brook, M.A. In *Synthesis and Properties of Silicones and Silicone-Modified Materials*; Clarson, S. J., Fitzgerald J. J., Owen, M. J., Smith, S. D., van Dyke, M. E., Eds.; ACS Symposium Series, 2003, Ch. 19.
58. Brook, M.A.; Zelisko, P.M.; Walsh, M.J. In *Organosilicon Chemistry: From Molecules to Materials*; Auner, N., Weis, J., Eds.; VCH: Weinheim, 2003; vol. 5, in press.
59. (a) Darst, .A.; Roberston, C.R.; Berzofsky, J.A. J. Colloid Interface Sci. **1986**, *111*, 466. (b) Anderson, A.B.; Roberston, C.R. Biophys. J. **1995**, *68*, 2091. (c) Sun, L.; Alexander, H.; Lattarulo, N. Biomaterials **1997**, *18*, 1593.
60. Floyd, D. Silicone surfactants: applications in the personal care industry. In *Silicone Surfactants*; Hill, R.M., Ed.; Surfactant Science Series, Vol. 86; Marcel Dekker: New York, 1999; 181–207.
61. Holmlin, R.E.; Chen, X.; Chapman, R.G.; Takayama, S.; Whitesides, G.M. Langmuir **2001**, *17*, 2841.
62. Ostuni, E.; Chapman, R.G.; Liang, M.N.; Meluleni, G.; Pier, G.; Ingber, D.E.; Whitesides, G.M. Langmuir **2001**, *17*, 6336.
63. Malmsten, M.; Emoto, K.; Van Alstine, J.M. J. Colloid Interface Sci. **1998**, *202*, 507.
64. Van Alstine, J.M.; Malmsten, M. Langmuir **1997**, *13*, 4044.
65. Bartzoka, V.; Chan, G.; Brook, M.A. Langmuir **2000**, *16*, 4589.
66. Bartzoka, V.; McDermott, M.R.; Brook, M.A. Protein-silicone interactions at liquid/liquid interfaces. In *Emulsions, Foams and Thin Films*; Mittal, K.L., Kumar, P., Eds.; Marcel Dekker: New York, 2000; 371–380.
67. Zelisko, P.M.; Flora, K.; Brook, M.A.; Brennan. J.D. Langmuir submitted.
68. Bartzoka, V.; Brook, M.A.; McDermott, M.R. Langmuir **1998**, *14*, 1892.
69. Bartzoka, V.; Brook, M.A.; McDermott, M.R. Langmuir **1998**, *14*, 1887.

70. Bartzoka, V.; McDermott, M.R.; Brook, M.A. Adv. Mater. **1999**, *11*, 257.
71. McDermott, M.R.; Brook, M.A.; Heritage, P.L.; Underdown, B.J.; Loomes, L.M.; Jiang, J. US Patent 5,571,531, McMaster University, 1995.
72. Brook, M.A.; Jiang, J.; Heritage, P.; Bartzoka, V.; Underdown, B.; McDermott, M.R. Langmuir **1997**, *13*, 6279.
73. McDermott, M.R.; Heritage, P.L.; Bartzoka, V.; Brook, M.A. Immunol. Cell Biol. **1998**, *76*, 256.
74. Heritage, P.L.; Loomes, L.M.; Jiang, J.; Brook, M.A.; Underdown, B.J.; McDermott, M.R. Immunology **1996**, *88*, 162.

12

Colloidal Particles

Elaboration from Preformed Polymers

THIERRY DELAIR CNRS-bioMérieux, Lyon, France

I. INTRODUCTION

Colloidal particles can be obtained by a wide variety of methods relying on the polymerization of at least one monomer. Whether the processes are precipitation, dispersion, suspension, or emulsification, they allow the efficient elaboration of a great variety of colloids used in paints, paper coatings, and other materials for which the cost of the process is of paramount importance for the final product. Besides these products, other applications may be less sensitive to the cost of the process as in pharmaceutical sciences, for instance. For the pharmaceutical industry the priority for the use of a polymer in vivo will be the lack of toxicity of the synthetic macromolecule itself and of components used during the elaboration and/or formulation processes (as solvents, for example). Thus, formulation approaches, though less cost effective than the most popular polymerization processes, can be of great value.

The use of preformed polymers in the elaboration of colloidal particles allows a wide choice of structures and modes of synthesis of the macromolecules. Polymers can be obtained by polycondensation reactions or ionic polymerization (e.g., ring-opening polymerization), and the chemical nature of the monomers can be of a great variety. The architecture of the polymer entering the composition of the colloids can be homopolymers, diblocks, triblocks, grafted structures, or a mix of them. Finally, the structure of the colloid itself can be modified; hard particles, capsules composed of an aqueous core and a hard shell, and lipoparticles, in which the main components of the colloid are natural lipids, can be obtained. The various formulation techniques that can be used for that purpose will be presented in the following sections.

The elaboration of a colloidal dispersion of particles in a continuous water phase from a preexisting polymer requires, at some point, that water–polymer interactions be minimized to allow the formation of the desired nanospheres.

This can be achieved either by using non–water-soluble macromolecules or by favoring intra- or interpolymer chain interactions rather than those with the molecules of the continuous phase. Depending on this strategic choice, will be the use of an organic solvent and thus of its nature and method of elimination before any in vivo applications.

This chapter is organized into three sections. The first deals with the uses of water-insoluble polymers and of organic solvents. The second section addresses the methods using only an aqueous medium. The third section is devoted to the physical methods for obtaining colloids.

II. METHODS USING AN ORGANIC SOLVENT

Two main formulation approaches are used for obtaining particle dispersions from an organic solution of preformed polymer. Formation of the particles may depend on an emulsification step of the organic phase in an aqueous medium, the elimination of the solvent from the stabilized emulsion being responsible for the precipitation of the polymer to form particles. In the second approach, no emulsification is required. The solvent is either miscible in the water phase or dialyzed through a membrane and replaced by water. In both cases, the particle formation is concomitant with the modification of the chemical nature of the continuous phase.

A. Methods Based on an Emulsification Step

Obtaining an emulsion of an organic phase in water requires a high-energy source such as a homogenizer or an ultrasonic device. In order to favor the emulsion formation step, formulations have been improved using salts or saturated liquid phases.

1. Emulsification/Evaporation: Simple and Double Emulsions

This method is based on the procedure developed by Vanderhoff et al. [1] for the preparation of pseudolatexes. The preformed polymer is dissolved in a water-immiscible organic solvent that is emulsified in an aqueous solution containing a stabilizer (Fig. 1a). The organic solvent used needs to be volatile, such as chloroform, acetone, and dichloromethane, and is quite often removed by evaporation under reduced pressure. Hard spheres have been obtained from polyesters such as poly(lactic acid) (PLA), poly(ε-caprolactone), or poly(hydroxybutyric acid) [2] or nanogels of cross-linked poly(ethylene oxide) (PEO) and polyethyleneimine (PEI), PEO-*cl*-PEI [3].

Diblock copolymers can also be used to obtain particles. For instance, monomethoxypolyoxyethylene (MPOE)-*b*-PLA [4], or POE-*b*-PLA [5], or even

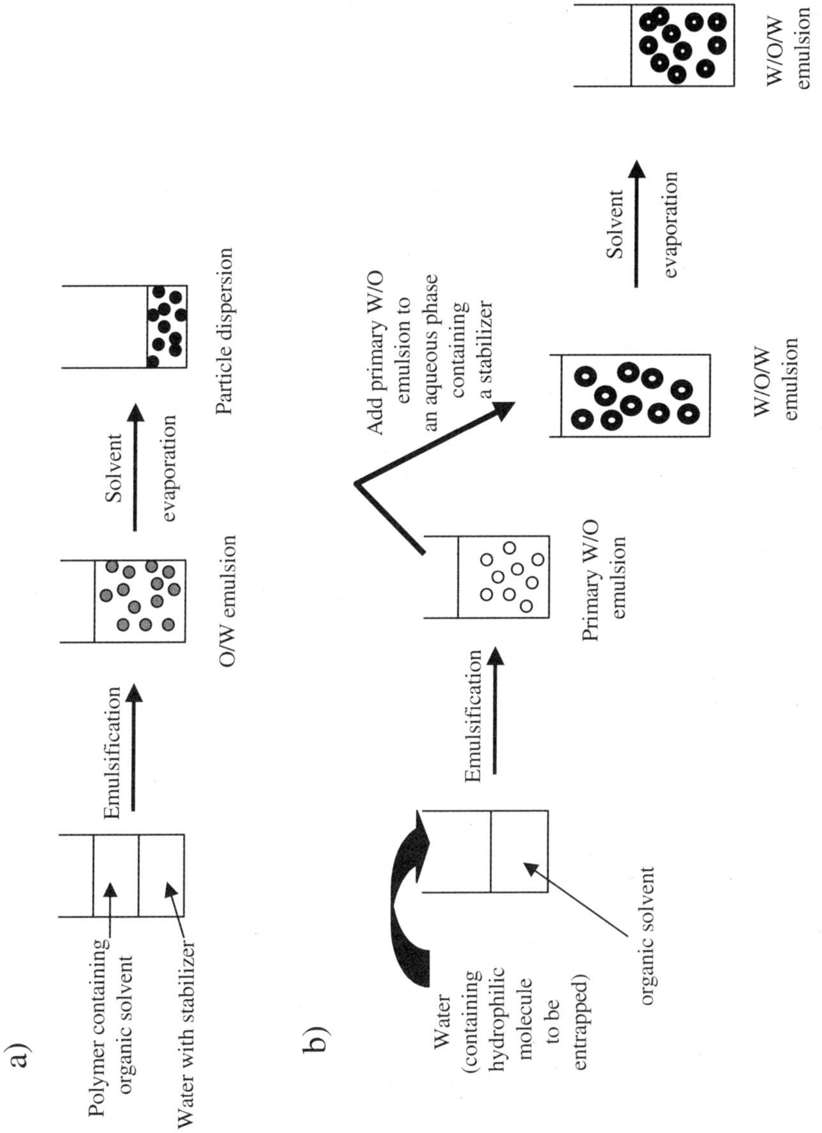

FIG. 1 Emulsification–solvent evaporation. (a) Simple emulsion process; (b) double-emulsion process.

triblock PLA-*b*-POE-PLA [6] was used in the elaboration of POE-covered particles that would display low interactions with plasma proteins.

The nature of the stabilizer in the water phase can be a surfactant such as sodium dodecyl sulfate, potassium oleate, or a macromolecule such as poly(vinyl alcohol) (PVA), POE-PPO-POE triblocks (PPO being the polypropylene oxide block that constitutes the hydrophobic segment of the macromolecular stabilizer), a protein like albumin [2] or a hydrophobized dextran [7]. On the nature of the stabilizer depend the size and the colloidal stability of the particles, as well as the nature of the resulting solid–liquid interface. It has been shown that PVA could bind in an irreversible manner to poly(DL-lactide-*co*-glycolide) (PLGA), and that the influence of the surface PVA layer would be larger in smaller particles but be independent of the PVA concentration of the continuous phase during the manufactoring process [8].

The solvent evaporation method, based on the formation of an o/w emulsion, allows entrapment within the core of the particles of a lipophilic substance by dissolution of the latter in the organic phase. To encapsulate hydrophilic compounds such as proteic antigenes, for example, a double-emulsion w/o/w is required (Fig. 1b). First the protein solution is emulsified in dichloromethane containing PLGA by homogenization at 5000 rpm. Thereafter, this first emulsion is poured into a PVA-containing aqueous solution and homogenized at 10,000 rpm. Then, the solvent is removed under reduced pressure. Using bovine serum albumin (BSA) as a model protein, 64% efficiency was attained with PLGA and only 49% with PEO-PLGA diblock copolymer [9]. The use of the diblock copolymer allowed an increase of the half-life of BSA in rats from 14 min to 4.5 h. The use of a PLGA-POE-PLGA triblock copolymer improved the release profile of BSA as compared to PLGA alone [10].

Lipoparticles composed of triglycerides can be obtained by either the o/w or the w/o/w methods [11]. In this approach, copolymers such as PVA, PEO derivatives, or polysaccharides are used as stabilizers or surface modifiers [11].

2. Salting Out

The classical emulsification–solvent evaporation method described above relies on the use of water-immiscible solvents, most of them being incompatible for further in vivo applications. To allow the separation from an aqueous phase of water-miscible solvents, more prone to be accepted for in vivo use, a method was developed using a salting-out effect. Acetone is often used as a water-miscible solvent because it is easily separated from aqueous solutions by salting-out with electrolyte [2]. A polymer solution in acetone, eventually containing a drug to be entrapped, is emulsified under mechanical stirring in an aqueous solution containing the salting-out agent and a colloidal stabilizer. The resulting emulsion is then diluted with a large amount of water to allow the diffusion of acetone into the aqueous phase, which induces the formation of the particles. Solvents and electrolytes can be removed by cross-flow filtration [2].

3. Emulsification–Diffusion

Recently, a new method for manufacturing nanoparticles from preformed polymers was developed to reduce the level of energy of the emulsification step, to allow the use of pharmaceutically acceptable organic solvents, and to produce particles in high yields with high reproducibility [12]. Named emulsification–diffusion, the process involves the formation of an o/w emulsion of a water-saturated organic solvent containing the polymer in a solvent-saturated aqueous phase containing a colloidal stabilizer. The subsequent dilution of the system in water causes diffusion of the solvent into the aqueous phase responsible for particle formation (Fig. 2).

In this process, besides increasing the PLA concentration in the organic phase, the critical step in the control of the particle size of the resulting colloidal dispersion is the size of the droplets formed during the emulsification step. A decrease in particle size (from 1.66 μm to 170 nm) and of polydispersity was obtained with an increase of the steering rate [13] for the emulsion step but not for the dilution step. Having shown as well that there was no particle size dependency on the solvent diffusion rate, the authors suggested that the particle formation arose from a chemical instability due to the presence of polymer-supersaturated regions rather than mechanical instability during the diffusion step [12]. Nevertheless, the nature of the stabilizer also has to be taken into consideration. Using propylene carbonate (PC) as an organic solvent, large aggregates were formed independently of the polymer concentration and process conditions, with polysorbate 80 or polyvinylpyrrolidone and dextran. But in the presence of poloxamer 188, stable nanoparticles were obtained [13].

Nanocapsules containing an oily core surrounded by a hard polymeric shell have also been obtained by the emulsification–diffusion method. Using ethyl acetate as a solvent, capsules containing medium-chain triglycerides have been obtained. Density gradient centrifugation allowed confirmation of the formation of nanocapsules whose densities were intermediate between those of nanoemulsions and nanospheres [14].

B. Solvent Displacement Method

This technique does not require the prior formation of an emulsion to obtain a dispersion of colloidal particles in an aqueous phase. It relies on the spontaneous emulsification of a polymer solution in a semipolar solvent such as acetone or ethanol. This solution is poured or injected, under moderate magnetic stirring, into an aqueous medium containing a stabilizer, and particles are formed instantaneously by rapid diffusion of the solvent in water. Then the organic solvent is

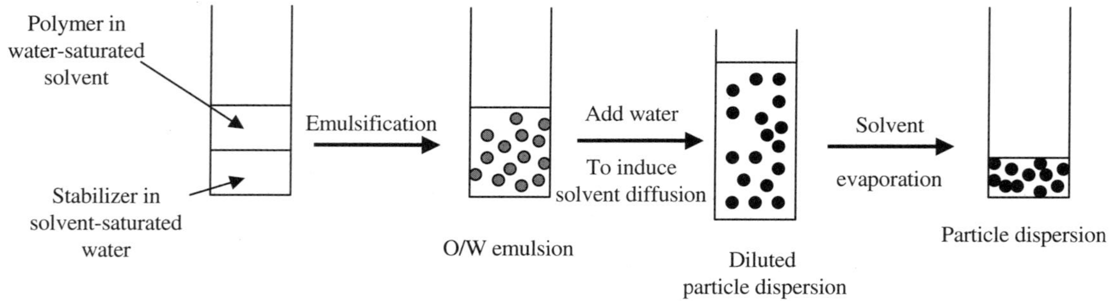

FIG. 2 Emulsification–diffusion process.

removed under reduced pressure (Fig. 3) [15]. Lipophilic drugs and, eventually, phospholipids, acting as stabilizers, can be added to the organic phase. Poly(DL-lactic acid) or poly(ε-caprolactone) can be used to obtain particles and the mostly used stabilizers in the water phase are pluronics or PVA. In this system, superficial instabilities of interfaces are sufficient to allow emulsification.

Nanocapsules can be obtained by the solvent displacement method. An acetonic solution of benzyl benzoate and poly(ε-caprolactone) was added to an aqueous solution of pluronic F68. Nanocapsules composed of a benzyl benzoate core and poly(ε-caprolactone) hard shell were obtained. The incorporation of benzyl benzoate was evaluated by a gas chromatography–mass spectrometry procedure. Particles of mean diameter 200 nm were obtained with an incorporation higher than 86% [16].

Finally, nanoemulsions can be obtained with this procedure. Elbaz et al. have described a positively charged submicrometer emulsion based on lecithin, medium-chain triglycerides, poloxamer 188, and stearylamine using the solvent displacement method [17]. With a slightly modified approach, Trimaille et al. showed that it was possible to modify the interface of these emulsions by depos-

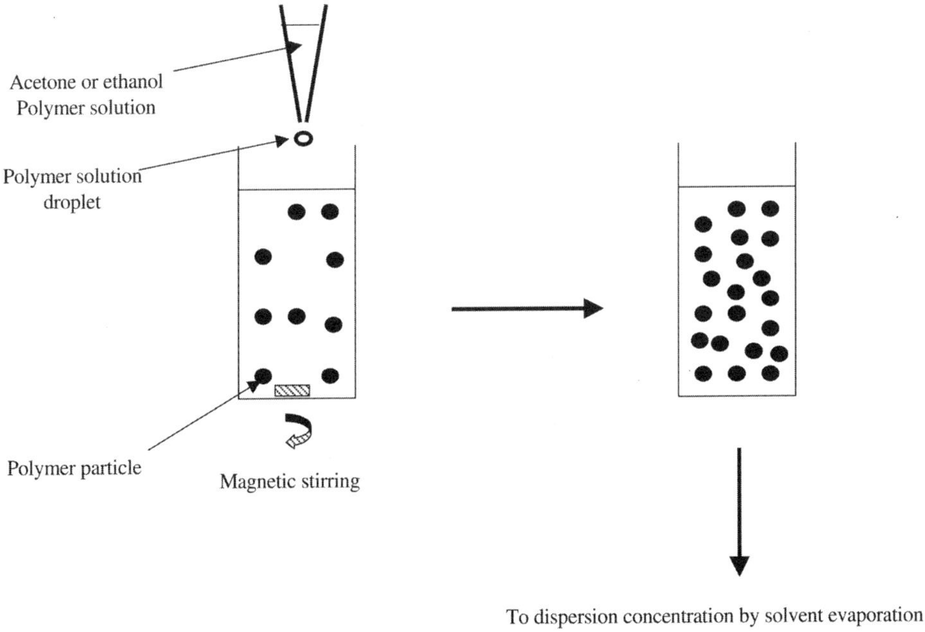

FIG. 3 Solvent displacement process.

iting, concomitantly with the particle formation step, an adequately functionalized copolymer [18].

Homopolymers, as well as block copolymers (i.e., PEO-PLA), have been used to obtain particles [19] and nanocapsules [20]. For both systems, the mean diameters and colloidal stability were dependent on the ratio of the two chains in the block copolymers. Increasing the hydrophobic segment resulted in decreased stability.

C. Dialysis-Based Methods

To avoid the emulsification step, the dialysis method can be used. It consists of filling a dialysis tube with a polymer solution in a water-miscible solvent such as dimethylsulfoxide (DMSO), dimethylformamide (DMF), or acetone. This solution is put to dialysis against water and, as the organic solvent diffuses away from the dialysis tube and is substituted by water, the polymer precipitates under the form of spherical particles. PLA particles coated with poly(L-lysine) (PLL) were obtained by dialyzing a DMSO mixture of PLA and derivatized PLL against water [21]. At a fixed PLA concentration, increasing the polysaccharide-PLL copolymer resulted in a reduction of the average particle size from 300 nm to 80 nm. With this approach, PLA particles were surface functionalized with a galactose-carrying polystyrene to target hepatocytes via their galactose receptors [22].

Block copolymers of PLA [23] and poly(ε-caprolactone) [24] with amphiphilic properties have led to nanoparticles with a hydrophobic core–hydrophilic shell structure whose particle size was less than 200 nm in both cases. Both amphiphilic copolymers were obtained by ring-opening polymerization of glycolide in the presence of MPEO [23] or of poly(ε-caprolactone) in the presence of PEO-PPO-PEO triblock copolymer [24]. Interestingly, since PEO-PPO-PEO triblock copolymers are known to be thermoresponsive, the authors investigated the role of the temperature on particle size and reported a marked decrease in particle diameter with increasing temperature. Recently, the elaboration, characterization, and application of block copolymer micelles for drug delivery was reviewed by Kataoka et al., who cited the dialysis method for the preparation of reactive polymeric micelles [25].

Logically, if amphiphilic block copolymers could be formulated under the form of particles by the dialysis method, this method should apply as well to hydrophobized water-soluble polymers. Pullulan acetates, as hydrophobized pullulans, were synthezised in a DMF suspension in the presence of acetic anhydride and pyridine. After purification, the white powder was dissolved in DMSO and dialyzed against water to obtain particles of size of less than 80 nm [26].

D. Supercritical Fluid Technology

Supercritical fluid technology is an environment-friendly alternative to process particles in high yield and exempt of traces of solvent that can be toxic in many pharmaceutical applications. The two main technologies for obtaining particles from preformed polymers will be quickly overviewed in this section.

In the rapid expansion of supercritical solution (RESS), the polymer is solubilized in a supercritical fluid. On rapid expansion of the solution, the solvent power of the critical fluid is drastically reduced, provoking the precipitation of the polymer under the form of particles [27]. This technique is very attractive because the precipitated particles are completely solvent free; however, unfortunately, most polymers exhibit low to no solubility in supercritical fluids. This method is, in fact, limited to low molecular weight polymers.

To address this solubility issue, the supercritical antisolvent method was developed. The idea was to dissolve the polymer in a solvent and use a nonsolvent supercritical fluid to induce the precipitation of the polymer and thus the particle formation. Practically, the polymer solution is introduced in the precipitation chamber, which is then charged with the supercritical fluid. At high pressure, enough nonsolvent can enter the liquid phase and induce formation of the particles. When the particle-forming step is complete, the supercritical fluid is used to remove the solvent (Fig. 4). In a modified version of this technique the polymer solution can be directly introduced into the supercritical fluid [28].

III. METHODS USING AN AQUEOUS MEDIUM

The use of organic solvents for the preparation of colloids for in vivo biomedical applications is an important issue because one has to use pharmaceutically ac-

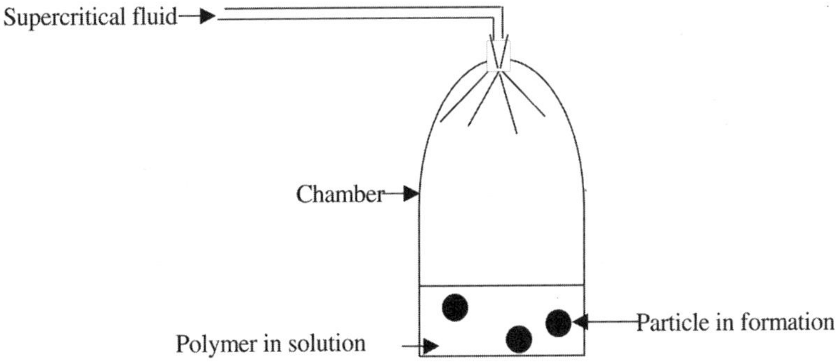

FIG. 4 Supercritical fluid antisolvent process.

ceptable solvents and quantify the trace amounts in the final preparation. Thus, to address this safety issue, various formulation procedures have been, or are being, designed, avoiding the use of organic media.

A. Ionic Gelation

Polyelectrolytes can be formulated under the form particles by polyion-mediated gelation due to the formation of inter- and intramolecular cross-links.

Chitosan (poly 1–4 glucosamine obtained by deacetylation of the natural chitin from crab shell) as a polycation was formulated under the form of beads by tripolyphosphate (TPP) [29]. A modification of the original procedure allowed the formation of nanoparticles of chitosan alone or in combination with other hydrophilic polymers. Thus, a mild and simple procedure was developed based on the mixture of two aqueous phases, one containing the chitosan and PEO or a PEO-PPO diblock copolymer, and the second containing the TPP [30]. Size (100–200 nm) and ζ potential could be modulated by the chitosan-to-PEO polymer ratio. Porous chitosan beads have been obtained by using an aqueous solution of TPP at pH 8.9 [31]. The morphology of the particles was evidenced by scanning electron microscopy and compared to the dense structure of the colloids obtained at pH values lower than 6. The particles were stabilized by cross-linking using ethylene glycol diglycidyl ether and then could be further modified with reagents such as succinic anhydride and benzoic anhydride. Glutaraldehyde-cross-linked chitosan nanospheres have been reported [32], but the dialdehyde had too many negative effects on cell viability to allow a general application of these particles as drug carriers. Sodium sulfate has been used as well to obtain nanoparticles by dropwise addition of sodium sulfate solution and a chitosan solution containing a nonionic stabilizer [33].

Alginates are sodium salts of a linear glycuronan composed of a mixture of β-mannosyluronic acid and α-gulosyluronic acid residures. They are often used in the elaboration of capsules for oral delivery of bioactive molecules. Upon interaction with calcium, alginates cross-link under the form of microcapsules by ionic gelation. An alginate solution (1–2 wt %) is slowly added under stirring to a calcium chloride solution in water. The resulting capsules can be collected and washed with water [34] before further use.

B. Polyelectrolytic Complexes and LbL Approach

Colloidal dispersions of interpolyelectrolyte complexes can be obtained in an aqueous medium by mixing dilute solutions of polycations and anions. An aqueous solution of the sodium salt of the alternating copolymer of maleic acid and propene (MAPE) was added to a dilute solution of poly(dimethyldiallylammonium chloride) (DMDAA). A threefold excess of cationic groups vs. anionic group was used to obtain positively charged colloids [35]. Particles in the 100-

to 200- nm range were obtained depending on the ionic strength of the complexation medium. On increasing the ionic strength, the particle size increased. The particle formed via the aggregation of the polyelectrolyte complexes due to hydrophobic interactions. Electrostatic stabilization was brought by the excess of positive charges used in the formulation.

By slowly mixing equal volumes of a solution of negatively charged calf thymus DNA and an aqueous solution of polycation poly(L-lysine) (PLL), loaded in different syringes and driven to a T-mixer, DNA-PLL particles were obtained. The particles had a surface charge density depending on the +/– charge ratio. Here again, the colloidal stability depended on ionic strength. At low pH and salt concentration, monodispersed DNA-PLL particles of 200 nm could be obtained [36]. A similar approach was used for obtaining isodisperse chitosan/DNA nanospheres; sodium sulfate was added to the DNA solution, and the mixing speed and temperature of the blending of the two solutions were controlled as well [37].

A template-assisted synthesis of hollow capsules was developed by Caruso et al. [38]. On a preformed charged colloid, alternate deposition of polyions of opposite charges was achieved constituting thus a shell of polyelectrolytic complexes (Fig. 5). The control of the deposition of each layer of opposite charge polymers was be achieved by measuring the surface charge inversion of the ζ potential of the colloid. In a next step, the template particle was degraded either by chemical or by thermal treatment for, respectively, a melamine or a polystyrene core so as to give rise to the formation of hollow capsule [38].

Recently, the colloidal template was replaced by enzyme crystals. This process led to the reversible encapsulation of the enzyme with retention of its bioactivity, thus offering new perspectives in the production of tailored and optimized systems in biotech applications [39].

C. Self-Assemblies

Particle formation can arise from the self-association of a polymeric structure designed to develop segment–segment interactions rather than solvent–segment interactions that are responsible for the solubility of the macromolecules. The most common structures used for self-assembies are amphiphilic macromolecules, bearing hydrophobic and hydrophilic segments. For instance, there are reports on the use of diblocks [25] or comblike polymers (a hydrophilic backbone grafted with hydrophobic pendant groups) [40]. In some cases, the polymeric structure can acquire the amphiphilic character after (1) interactions with an other counterpart, such as the formation of neutral polyelectrolyte complexes onto a ionic segment of a diblock copolymer [25]; (2) after stimulation by an external factor such as heat [41] or pH [41,42]. Though hydrophobic interactions between macromolecules are widely described to ensure the autoassocia-

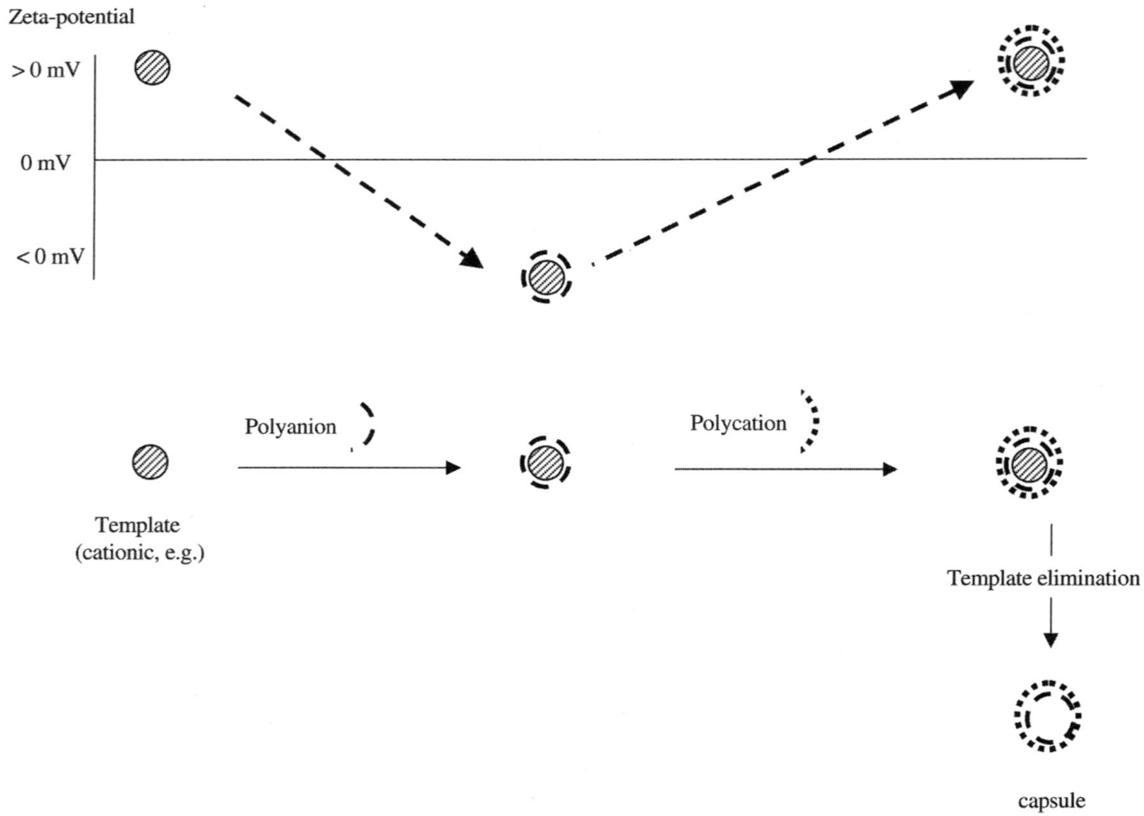

FIG. 5 Template-assisted capsule synthesis.

tion of polymers, just as phospholipids associate in uni- or multi-lamellar vesicles, hydrogen bonding has also been involved in the formation of colloids [43].

To come back to the amphiphilic structures described above, the role of the hydrophilic segments of the polymer is to ensure the colloidal stability during and after formation of the particles. The chemical nature of these segments can be ionic [25,41], when, for instance, a polyanion is grafted with hydrophobic groups [40], or nonionic but hydrophilic, such as in the case of PEO chains [25].

The self-assembly method is fairly versatile in terms of structures that can be obtained. The formation of micelles [25], particles [43], and vesicles [44] has been reported. From a practical standpoint, colloids can be obtained by redispersion of the dried amphiphilic polymer in an aqueous mixture [40,43] with or without a sonication step to improve the colloid size dispersion. In the presence of cholesterol, vesicles can be elaborated with the amphiphilic copolymer being at the oil–water interface [45]. The dissolution–sonication procedure was used for the elaboration of particles via intermolecular hydrogen bonding of PEO-grafted chitosan molecules [43] and the obtained particles were in the 90- to 150-nm range, depending on the mole percent of PEO chains per sugar unit. A pH-induced modification of the solubility of the polymer has been used to obtain particles from modified polyamino acids [42] or polyacrylamide [41], the latter also exhibiting thermosensitive swelling properties.

IV. PHYSICAL METHODS

Following is a short overview of physical methods that are continuous processes for the production of microsphere whose main applications are found in the pharmaceutical industry or food industry. Two main processes will be developed—extrusion and spray-drying—though these procedures lead mainly to the formation of micrometer-sized particles.

Extrusion consists in the production of droplets from a polymer solution that will be further hardened to solid microspheres. This latter step can be achieved by melt solidification, gelation, chelation, solvent extraction, or evaporation. For instance, alginate microparticles can be obtained by extrusion into an aqueous solution of calcium chloride, with the electrolyte serving as the gelation agent to yield the hardened particles [46].

Spray-drying relies on the atomization of a polymer solution, which is then spray-dried to yield the particles to be collected (Fig. 6).

The spray-drying step requires the use of a hot chamber in which the solvent can evaporate. This is fairly easy to achieve when the solvent is volatile such as methylene chloride for PLGA particles [47], but requires higher temperature (between 110°C and 180°C) when polysaccharides are used [48]. For these

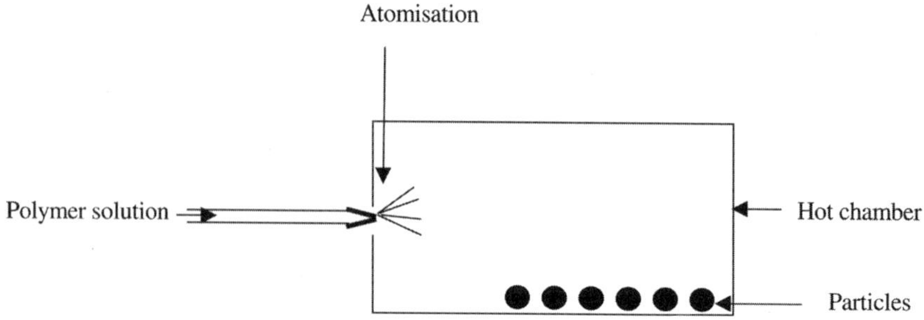

FIG. 6 Spray-drying method.

polymers requiring aqueous solutions, an alternate method was developed that consisted of spraying the polymer solution into a liquid coagulating agent [48]. The separation of the microspheres from the coagulation medium was achieved by centrifugation. The mass ratio of air to polymer liquid, the relative velocity of air to polymer liquid and the viscosity of the polymer solution affected the particle size distributions.

V. CONCLUSIONS AND PROSPECTIVES

Most of the methods described in this chapter barely meet the requirements for large-scale production. Thus, applications for these procedures are mainly found in the pharmaceutical domain, for which the added value is very high and the production scale lower than for some other industries. Therefore, micelles [25], nanocapsules [49], and nanoparticles [50] obtained from polymeric materials have been used as drug delivery devices. Molecules larger than drugs, such as proteins [51] or DNA [52], have been loaded into or onto colloidal vectors. The variety in chemical nature of the polymers for the manufacture of these colloids, though limited due to toxicity constraints, includes polyesters [50], polycyanoacrylates [50], and polysaccharides [53]. For the sustained delivery of bioactive molecules, tuning the degradation kinetics of the carrier to reach the expected goals or performances is still a challenging issue, as much for small drugs as for larger antigens used for vaccination.

Since capsules with either lipophilic or hydophilic compartments can be made, as we have seen in this chapter, one can envision new colloids that would contain various types of cavities. For instance, one could think of the presence of a hydrophilic compartment to contain water-soluble drugs and a lipophilic one filled with hydrophobic drugs. These colloids could be well suited in multitherapy approaches.

Another direction in the improvement of bioactive molecule delivery systems will probably be the stimuli-responsive carriers. We can imagine potential synthetic vectors which, once injected in a patient, could be driven from the outside, with a magnet, for instance, to the targeted organ. Once the target is reached, a second stimulus would burst the carrier and thus release the active compound.

The ultimate carrier should be capable of recognizing the cellular target, binding and/or entering the target, and delivering the bioactive molecule to a specific site of the target. This could be achieved by building multilayered colloids that could bear a variety of successive information. These colloids would mimic viruses during the infection process. Viruses specifically bind to the targeted cells and enter them. Fusion of the cell membrane and the viral membrane induces the unwrapping of the viral genome associated with proteins essential for viral replication. Then, intracellular trafficking takes place to drive the viral genome to the nucleus, which is then entered. Integrases allow the intercalation of the viral genome into the host cell and a complete takeover of the cell functions is achieved by the virus. Thus, to achieve efficient gene delivery to cells, multifunctional colloids would be needed, bearing a variety of information essential for each stage of the process but that should be removed to allow progression to the next step.

Most applications dealt with so far were focused on the delivery of bioactive molecule. A new developing field is the detoxification of the body by specific removal of defined substances. It could be cancer cells, as already achieved ex vivo, as well as small molecules like cholesterol, hormones, sugar, or larger assemblies such as circulating viruses in chronic infections. These colloids should circulate long enough in the body to adsorb the specific toxin and then to be eliminated in the natural ways.

Applications other than in the pharmaceutical domain can be envisioned for "sophisticated colloids." For instance, smart garments have appeared on the market, made of fabrics containing encapsulated perfumes or antiperspirants or antiwrinkle creams. Surely human imagination will provide numerous perspectives to colloids and new challenges for the chemistry of materials.

REFERENCES

1. Vanderhoff, J.M.; El-Aasser, M.S.; Ugelstad, J. US Patent 4,177,177, 1979.
2. Quintanar-Guerrero, D.; Alléman, E.; Fessi, H.; Doelker, E. Preparation techniques and mechanism of formation of biodegradable nanoparticles from preformed polymers. Drug Dev. Ind. Pharm. **1998**, *24*, 1113–1128.
3. Vinogradov, S.V.; Bronich, T.K.; Kabanov, A.V. Nanosized cationic hydrogels for drug delivery: preparation, properties and interactions with cells. Adv. Drug Deliv. Rev. **2001**, *57*, 135–147.
4. Bouillot, P.; Petit, A.; Dellacherie, E. Protein encapsulation in biodegradable am-

phiphilic microspheres. I. Polymer synthesis and characterization and microsphere elaboration. J. Appl. Polym. Sci. **1998**, *68*, 1695–1702.

5. Deng, X.M.; Li, X.H.; Yuan, M.L.; Xiang, C.D.; Huang, Z.T.; Jiu, W.X.; Zhang, Y.H. Optimisation of the preparation conditions for poly-(DL-lactide)-poly (ethylene glycol) microspheres with entrapped vibrio cholera antigens. J. Controlled Rel. **1999**, *58*, 123–131.
6. Matsumoto, J.; Nakoda, Y.; Sakurai, K.; Nakamuna, T.; Takahashi, Y. Preparation of nanoparticles consisted of poly(L-lactide)-poly(ethylene glycol)-poly(L-lactide) and their evaluation in vitro. Int. J. Pharmacol. **1999**, *185*, 93–101.
7. Rouzes, C.; Gref, R.; Leonard, M.; De Sousa Delgado, A.; Dellacherie, E. Surface modification of poly(lactic acid) nanospheres using hydrophobically modified dextrans as stabilizers in an o/w emulsion/evaporation technique. Biomed. Mater. Res. **2000**, *50*, 557–564.
8. Lee, S.C.; Oh, J.T.; Jang, M.H.; Chung, S.I. Quantitative analysis of polyvinylalcohol on the surface of poly(D,L-lactide-co-glycolide) microparticles prepared by solvent evaporation method; effect of particle size and PVA concentration. J. Controlled Rel. **1999**, *59*, 123–132.
9. Li, Y.P.; Pei, Y.P.; Zhang, X.Y.; Gu, Z.H.; Zhou, Z.H.; Yuan, W.F.; Zhou, J.J.; Zhu, J.H.; Gao, X.J. Pegylated PLGA nanoparticles as protein carriers: synthesis, preparation and biodistribution in rats. J. Controlled Rel. **2001**, *71*, 203–211.
10. Bittner, B.; Witt, C.; Möher, K.; Kissel, T. Degradation and protein release properties of microspheres prepared from biodegradable poly(lactide-co-glycolide) and ABA triblock copolymers: influence of buffer media on polymer erosion and bovine serum albumin release. J. Controlled Rel. **1999**, *60*, 297–309.
11. Cortesi, R.; Esposito, E.; Luca, G.; Nastruzzi, C. Production of lipospheres as carriers for bioactive compounds. Biomaterials **2002**, *23*, 2283–2294.
12. Quintanar-Guerrero, D.; Alléman, E.; Doelker, E.; Fessi, H. A mechanistic study of the formation of polymer nanoparticles by the emulsification-diffusion technique. Colloid Polym. Sci. **1997**, *275*, 640–647.
13. Quintanar-Guerrero, D.; Fessi, H.; Alléman, E.; Doelker, E. Influence of stabilizing agents and preparation variables on the formation of poly(D, L-lactic acid) nanoparticles by an emulsification technique. Int. J. Pharmacol. **1999**, *143*, 133–141.
14. Quintanar-Guerrero, D.; Alléman, E.; Doelker, E.; Fessi, H. Preparation and characterization of nanocapsules from preformed polymers by a new process based on emulsification-diffusion technique. Pharm. Res. **1998**, *15*, 1056–1061.
15. Fessi, H.; Puisieux, F.; Devisaguet, J.P. Eur. Patent 274,961, 1987.
16. Benabah, M.; Zahidi, A. Preparation of poly-ε–caprolactone nanocapsules by nanoprecipitation and evaluation of benzyl benzoate by GC-MS. J. Pharm. Clin. **1999**, *18*, 313–317.
17. Elbaz, E.; Zeevi, A.; Klang, S.; Benita, S. Positively charged submicronemulsions: a new type of colloidal drug carrier. Int. J. Pharmacol. **1993**, *96*, 121–126.
18. Trimaille, T.; Chaix, C.; Delair, T.; Pichot, C.; Teixeria, H.; Dubernet, C.; Couvreur, P. Interfacial deposition of functionalized copolymers onto nanoemulsions produced by the solvent displacement method. Colloid Polym. Sci. **2001**, *279*, 784–792.

19. Riley, T.; Govender, T.; Stolnik, S.; Xiong, C.D.; Garnett, M.C.; Illum, L.; Davis, S.S. Colloidal stability and drug incorporation aspects of micellar-like PLA-PEG nanoparticles. Colloids Surf. B Biointerfaces **1999**, *16*, 147–159.
20. Ma, J.; Feng, P.; Yo, C.; Wang, Y.; Fan, Y. An improved interfacial coacervation technique to fabricate biodegradable nanocapsules of an aqueous peptide solution from polylactide and its block copolymer with poly(ethylene glycol). Colloid Polym. Sci. **2001**, *279*, 387–392.
21. Muruyama, A.; Ishihara, T.; Kim, J.S.; Kim, S.W.; Akaike, T. Nanoparticle DNA carrier with poly(L-lysine) grafted polysaccharide copolymer and poly(D, L lactic acid). Bioconj. Chem. **1997**, *8*, 735–742.
22. Cho, C.S.; Cho, K.Y.; Park, I.K.; Kim, S.H.; Sasagawa, T.; Uchiyamam, M.; Akaike, T. Receptor-mediated delivery of all trans-retinoic acid to hepatocyte using poly(L-lactic acid) nanoparticles coated with galactose carrying polystyrene. J. Controlled Rel. **2001**, *77*, 7–15.
23. Kim, S.Y.; Shin, I.G.; Lee, Y.M. Amphiphilic diblock copolymeric nanospheres composed of methoxy poly(ethylene glycol) and glycolide: properties, cytotoxicity and drug release behaviour. Biomaterials **1999**, *20*, 1033–1042.
24. Kim, S.Y.; Hu, J.C.; Lee, Y.M. Poly(ethylene oxide)-poly(propylene oxide)-poly-(ethylene oxide)/poly(ε-caprolactone) (PCL) amphiphilic block copolymeric nanospheres. II. Thermo-responsive drug release behaviors. J. Controlled Rel. **2000**, *65*, 345–358.
25. Kataoka, K.; Harada, A.; Nagasaki, Y. Block copolymer micelles for drug delivery: design, characterization and biological significance. Adv. Drug Deliv. Rev. **2001**, *47*, 113–131.
26. Jeong, Y.I.; Nah, J.W.; Nah, H.K.; Na, H.K.; Na, K.; Kim, I.S.; Cho, C.S.; Kim, S.H. Self-assembling nanospheres of hydrophobized pullulans in water. Drug Dev. Ind. Pharm. **1999**, *25*, 917–927.
27. Tom, J.W.; Debenedetti, P.G. Particle formation with supercritical fluids—a review. J. Aerosol Sci. **1991**, *22*, 555–584.
28. Randolph, T.W.; Randolph, A.D.; Mebes, M.; Yeung, S. Sub-micron-sized biodegradable particles of poly(L-lactic acid) via the gas antisolvent spray precipitation process. Biotechnol. Prog. **1993**, *9*, 429–435.
29. Bodmeier, R.; Chen, H.; Paeratakul, O. A novel approach to the delivery of microparticles or nanoparticles. Pharm. Res. **1989**, *6*, 413–417.
30. Calvo, P.; Remunan-Lopez, C.; Vila-Jato, J.L.; Alonso, M.J. Novel hydrophilic chitosan-polyethylene oxide nanoparticles as protein carriers. J. Appl. Polym. Sci. **1997**, *63*, 125–132.
31. Mi, F.L.; Shyu, S.S.; Chan, C.T.; Schoung, J.Y. Porous chitosan microspheres for controlling the antigen release of Newcastle disease vaccine: preparation of antigen-adsorbed microsphere and in vitro release. Biomaterials **1999**, *20*, 1603–1612.
32. Ohya, Y.; Shirotani, M.; Kobayashi, H.; Ouchi, T. Release behavior of 5 fluorouracil from chitosan gel nanospheres immobilizing 5-fluorouracil coated with polysaccharides and their cell specific cytotoxicity. Pure Appl. Chem. **1994**, *A31*, 629–642.
33. Berthold, A.; Cremer, K.; Kreuder, J. Preparation and characterization of chitosan microspheres. J. Controlled Rel. **1996**, *39*, 17–25.

34. Baverstock, T.L.; Hogenesch, H.; Suckow, M.; Porter, R.E.; Jackson, R.; Park, H.; Park, K. Oral vaccination with alginate microsphere systems. J. Controlled Rel. **1996**, *39*, 209–220.
35. Pergushov, D.V.; Bucchammer, H.M.; Lunkwitz, K. Effect of a low-molecular-weight salt on colloidal dispersions of interpolyelectrolyte complexes. Colloid Polym. Sci. **1999**, *277*, 101–107.
36. Lee, L.K.; Mount, C.N.; Shamlou, P.A. Characterization of the physical stability of colloidal polycation-DNA complexes for gene therapy and DNA vaccines. Chem. Eng. Sci. **2001**, *56*, 3163–3172.
37. Mao, H.Q.; Roy, K.; Walsh, S.M.; August, J.T.; Leong, K.W. DNA-chitosan nanospheres for gene delivery. Proc. Int. Symp. Controlled Rel. Bioact. Mater. **1996**, *23*, 401–402.
38. Caruso, F. Hollow capsule processing through colloidal templating and self-assembly. Chem. Eur. J. **2000**, *6*, 413–419.
39. Caruso, F.; Trau, D.; Mohwald, H.; Renneberg, R. Enzyme encapsulation in layer-by-layer engineered polymer multilayer capsules. Langmuir **2000**, *16*, 1485–1488.
40. Lee, K.Y.; Kwon, I.C.; Kim, Y.H.; Jo, W.H.; Jeong, S.Y. Preparation of chitosan self-aggregates as a gene delivery system. J. Controlled Rel. **1998**, *51*, 213–220.
41. Kim, E.J.; Cho, S.H.; Yuk, S.H. Polymeric microspheres composed of pH/temperature polymer complex. Biomaterials **2001**, *22*, 2495–2499.
42. Haas, S.; Miura-Fraboni, J.; Zavala, F.; Murata, K.; Leone-Bay, A.; Santiago, N. Oral immunization with a model protein entrapped in microspheres prepared from derivatized α-amino acids. Vaccine **1996**, *14*, 785–791.
43. Ohya, Y.; Nishizawa, H.; Hora, K.; Ouchi, T. Preparation of PEG-grafted chitosan nanoparticles as peptide drug carriers. S.T.P. Pharma. **2000**, *10*, 77–82.
44. Uchegu, I.F.; Schötzlein, A.G.; Tetley, L.; Gray, A.I.; Sludden, J.; Siddique, S.; Mosha, E. Polymeric chitosan-based vesicles for drug delivery. J. Pharm. Pharmacol. **1998**, *50*, 453–458.
45. Brown, M.D.; Schötzlein, A.G.; Browlie, A.; Jack, V.; Wang, W.; Tetley, L.; Gray, A.I.; Uchegu, I.F. Preliminary characterization of novel aminoacid based polymeric vesicles as gene and drug delivery agents. Bioconj. Chem. **2000**, *11*, 880–891.
46. Arshady, R. Manufacturing methodology of microcapsules In *Micropsheres, Microcapsules and Liposomes*; Arshady, R., Ed.; Citus Books: London, 1999; 279–322.
47. Tracy, M.A.; Ward, K.L.; Firouzabedian, L.; Wang, Y.; Doug, N.; Qian, R.; Zhang, Y. Factors affecting the degradation rate of poly(lactide-co-glycolide) microspheres in vivo and in vitro. Biomaterials **1999**, *20*, 1057–1062.
48. Mi, F.L.; Wong, T.B.; Shyu, S.S.; Chang, S.F. Chitosan microspheres: modification of polymeric chemico-physical properties of spray-dried microspheres to control the release of antibiotic drug. J. Appl. Polym. Sci. **1999**, *71*, 747–759.
49. Legrand, P.; Barratt, G.; Mosqueira, V.; Fessi, H.; Devissaguet, J.P. Polymeric nanocapsules as drug delivery systems—a review. S.T.P. Pharma. **1999**, *9*, 411–418.
50. Soppimath, K.S.; Aminabhavi, T.M.; Kulkarni, A.R.; Rudeinski, W.E. Biodegradable polymeric particles as drug delivery devices. J. Controlled Rel. **2001**, *70*, 1–20.

51. Singh, M.; O'Hagan, D. The preparation and characterization of polymeric antigen delivery systems for oral administration. Adv. Drug Deliv. Rev. **1998**, *34*, 285–304.
52. Singh, M.; Briones, M.; Ott, G.; O'Hagan, D. Cationic microparticles: a potent delivery system for DNA vaccines. Proc. Natl. Acad. Sci. USA **2000**, *97*, 811–816.
53. Janes, K.A.; Calvo, P.; Alonso, M.J. Polysaccharide colloidal particles as delivery systems for macromolecules. Adv. Drug Deliv. Rev. **2001**, *47*, 83–97.

13

Poly(alkylcyanoacrylates)

From Preparation to Real Applications as Drug Delivery Systems

CHRISTINE VAUTHIER, PATRICK COUVREUR, and CATHERINE DUBERNET Université de Paris Sud, Chatenay-Malabry, France

I. INTRODUCTION

Poly(alkylcyanoacrylates) (PACA) have not been much employed as polymers. On the contrary, their corresponding monomers, alkylcyanoacrylates, have received increased interest since their synthesis in 1947 [1] because of their remarkable polymerization ability and strong reactivity. Indeed, cyanoacrylate monomers can polymerize extremely rapidly in the presence of moisture or basic component traces. Their excellent adhesive properties result from bonds of high strength they are able to form with most polar substrates. Therefore, they were extensively used as adhesives [2].

A very active field of investigation for poly(alkylcyanoacrylate)-based materials is biomedical research. Alkylcyanoacrylates have been used as surgical glue because of their excellent adhesion to living tissues, including skin. They were also developed as tissue adhesive for the closure of skin wounds [3] and as embolitic material for endovascular surgery [4]. At the moment, an exciting application is the use of poly(alkylcyanoacrylates) as drug particulate carriers. This is an area of research that emerged in the 1980s [5–7] for cancer treatments, which generally involve highly toxic molecules on healthy tissue. Other molecules of interest consist of poorly stable compounds like peptides and nucleic acids [8–10].

Today, PACA nanoparticles are considered to be the most promising polymer colloidal drug delivery system in clinical development for cancer therapy [11–15]. As described in this chapter, PACA nanoparticles can be prepared by emulsion and interfacial polymerization of alkylcyanoacrylates and also from copolymers of poly(alkylcyanoacrylate)-*co*-poly(ethylene glycol cyanoacrylate). The main developments pertaining to use of PACA nanoparticles as drug carriers for cancer treatments will be described and discussed in another section of this chapter.

II. PREPARATION OF POLY(ALKYLCYANOACRYLATE) NANOPARTICLES

Poly(alkylcyanoacrylate) nanoparticles can be prepared either by polymerization of alkylcyanoacrylates or directly from the polymer. In this case, nanoparticles can be prepared by nanoprecipitation and emulsification–solvent evaporation methods.

A. Preparation of Nanoparticles by Polymerization

For polymerization, monomers are mainly prepared by a method based on a Knoevenagel condensation of formaldehyde with the corresponding alkylcyanoacetate synthesized in turn by Fisher esterification of cyanoacetic acid with the glycolate ester. This method provides a high yield of monomer (≥80%) [1] and a purity grade over 99% [16–18]. Monomers generally occurring as clear and colorless liquid with a low viscosity are highly reactive compounds and extremely difficult to handle in the pure form [1,16,19–21]. Inhibitors are essential to maintain their stability. The most common anionic inhibitors used are acidic gases (sulfur dioxide) and strong acids (aliphatic and aromatic sulfonic acids, mineral acids). Free-radical polymerization inhibitors are usually added to a level of 100 to several thousand ppm (hydroquinone, other quinones, methyl ether of hydroquinone, methoxyhydroquinone) [18,19,21].

The polymerization of alkylcyanoacrylates can theoretically occur according to three different mechanisms: free-radical, anionic, and zwitterionic mechanisms (Fig. 1). In practice, the anionic and zwitterionic routes are strongly favored because they are rapidly initiated at ambient temperature. Classical initiators of the anionic polymerization are anions (i.e. I^-, CH_3COO^-, Br^-, OH^-, etc.), weak bases such as alcohols, water, and amino acids encountered in living tissues [16]. Tertiary bases such as phosphine and pyridine derivatives were described to initiate zwitterionic polymerization [19,22]. In both cases, the growing end is a carbanion and the reaction is difficult to control. Indeed, in the case of the anionic polymerization carried out in an organic solvent, the polymer chain growth could only be interrupted by the addition of a very strong mineral acid [20].

In contrast to anionic and zwitterionic polymerization processes, radical polymerization is slower and requires much higher activation energy (125 kJ/mol).

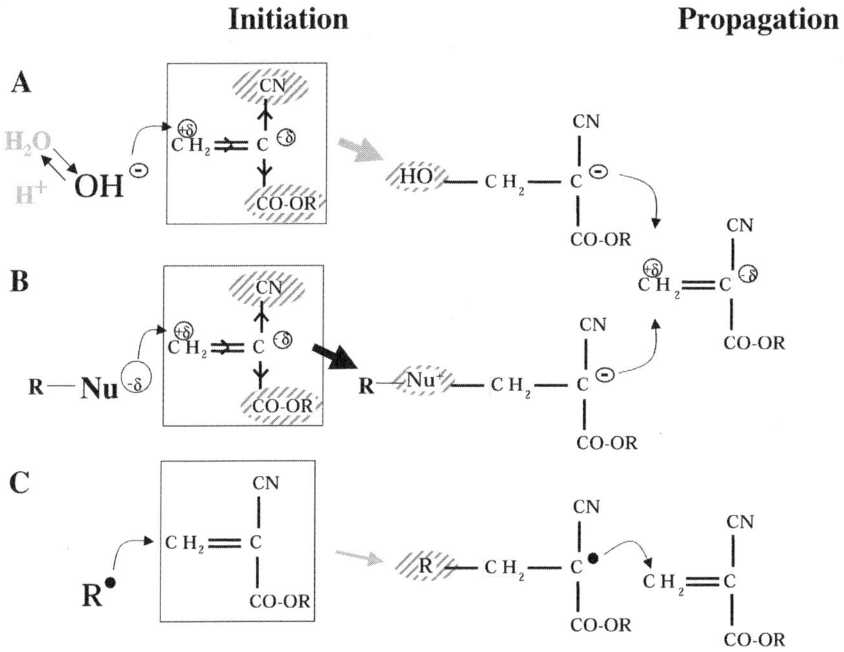

FIG. 1 Different routes of the polymerization initiation of alkylcyanoacrylates. (A) Anionic polymerization; (B) zwitterionic polymerization; (C) radical polymerization.

In addition, the reaction rate greatly depends on temperature and quantity of radicals. This polymerization has been described in bulk in such a condition that the anionic polymerization was mainly under control [23–27].

1. Preparation of Matricial Poly(alkylcyanoacrylate) Nanoparticles (i.e., Nanospheres) by Emulsion Polymerization

Emulsion polymerization of alkylcyanoacrylates was introduced by Couvreur et al. in 1979 [5] to design biodegradable polymer particles suitable for in vivo delivery of drugs. Polymerization media formulated to prepare poly(alkylcyanoacrylate) nanoparticles to be used as drug carriers are usually very complex. For example, the monomer (100 μL) is dispersed in acidified water containing a surfactant or a stabilizing agent (10 mL of a 0.5–1% solution of Pluronic F68 or dextran 70 at pH 2.5 with HCl) and let polymerization occur spontaneously for a few hours (3–4 h). This unusual mode of polymerization for such reactive monomers leads to the formation of colloidal polymer particles with a diameter ranging of 50–300 nm. These particles presenting a matricial structure were named "nanospheres" [5]. At pH higher than 3, the polymerization is too fast and polymer aggregates are formed. It is noteworthy that the polymerization can even be initiated in the presence of acids normally capable of its inhibition in organic solvents at pH lower than 1 because of the presence of additives, including surfactants or stabilizing agents dissolved in the reaction medium [28–31].

The molecular weight of the polymers forming the nanoparticles is usually low as evaluated by size exclusion chromatography. It is affected by the pH of the polymerization medium, by the presence or the absence of surface-active agents, and by the presence or the absence of a drug [32–35]. Recently, Nehan et al. [36] investigated in detail the polymerization of alkylcyanoacrylates at different pH in the presence of 0.1% dextran. They suggested a rather complex mechanism for the formation of the polymer nanoparticles involving first oligomers that are allowed to polymerize further through a reinitiation, repolymerization process. This complex mechanism was based on the assumption that only the hydroxyl groups from water were responsible for the polymerization initiation. It totally omitted the fact that hydroxyl groups from dextran could also initiate the polymerization of alkylcyanoacrylate under the same conditions, as previously demonstrated by Douglas et al. [29].

The size of the nanoparticle formed can be controlled by the amount of surface-active agent or by the molecular weight of colloidal stabilizer like dextran [29,37,38]. Another important parameter to consider for controlling particle size is the concentration of sulfur dioxide dissolved in the monomer as the inhibitor of anionic polymerization. Contradictory results have been reported considering either the effect of the pH or the effect of the monomer sulfur dioxyde concentration [32,39]. Since the pH of the polymerization medium controls the partition and ionization of sulfur dioxide in water, it affects in turn the sulfur dioxide concentration in the monomer once it is dispersed in the aqueous polymerization medium. Thus, both the pH and the sulfur dioxide concentration are affecting the polymerization process and controlling the size of the nanoparticles that form [40].

Many drugs were entrapped with success in polyalkylcyanoacrylate nanoparticles [41]. However, in rare cases, drugs were shown to initiate the polymerization reaction and lost their biological activity [33–35]. For certain drugs, cyclodextrins can be added to the polymerization medium to promote their association with the carrier [42,43]. Side reactions occurring during polymerization can advantageously be used to associate compounds by covalent binding with nanoparticles. This has been applied to naphthalocyanines, photo-sensitizers used in phototherapy of tumors [44], and a series of molecules containing diethyltriaminepentaacetic acid (DTPA) capable of complexing radioactive metals for radiolabeling of nanoparticles in medical imaging [45]. These reactions were also used to produce nanoparticles with modified surface properties, allowing the covalent coupling of macromolecules on the nanoparticle surface [29–31, 46,47].

2. Preparation of Reservoir-Type Poly(alkylcyanoacrylate) Nanoparticles (i.e., Nanocapsules) by Interfacial Polymerization

The interfacial polymerization of alkylcyanoacrylates at an oil–water interface occurring either in oil-in-water or in water-in-oil emulsions or microemulsions leads to particles with liquid core surrounded by a polymeric shell. Such particles having a diameter of a few hundred nanometers were named "nanocapsules." Their core-shell structure can be highlighted by transmission electron microscopy after cryofracture [48–50]. They were developed as drug delivery systems [48,51–53].

Oil-containing nanocapsules are obtained by the polymerization of alkylcyanoacrylates at the oil–water interface of a very fine oil-in-water emulsion [48]. In practice, the oil (Miglyol 1 mL), the monomer (isobutyl-2-cyanoacrylate 0.125 mL), and the drug are dissolved in a water-miscible organic solvent (ethanol 25 mL) to prepare the organic phase. This organic phase is injected in the aqueous phase (50 mL) containing a hydrophilic surfactant (Pluronic F68 0.25%) via a canula and under strong magnetic stirring. The outlet of the cannula is placed inside the aqueous phase. The nanocapsules formed immediately to give a milky suspension. The organic phase is then removed under reduced pressure using a rotoevaporator. The polymerization of alkylcyanoacrylates is initiated when monomers encounter water molecules. In such a system, the organic solvent, which is totally miscible with water, serves as a monomer vehicle and the polymerization is believed to occur at the surface of the oil droplets that form during emulsification [10,54–56].

To promote nanocapsule formation, an ideal oil/ethanol ratio of 2% in the organic phase has been suggested [54], and the use of aprotic solvents such as acetone and acetonitrile was recommended [57]. Protic solvents like ethanol, *n*-butanol, and isopropanol appeared to enhance the formation of nanospheres.

Nanocapsules prepared by this method contain more than 90% oil in weight [56]. Thus, this method is mainly adapted for the encapsulation of oil-soluble substances [50]. However, because of the extremely fast formation of the polymer shell around the oil droplets, highly water-soluble molecules such as insulin could be encapsulated with high encapsulation yields (up to 97%) [55,58,59].

Water-containing nanocapsules may be obtained by interfacial polymerization of alkylcyanoacrylate in water-in-oil microemulsions. In these systems, water-swollen surfactant micelles of small and uniform size are dispersed in an organic phase. The monomer is added to the oily phase once the microemulsion is formed and the anionic polymerization is initiated at the surface of the water-swollen micelles. The polymer, which forms locally at the water–oil interface, precipitates allowing the formation of the nanocapsule shell [51–53]. The nanocapsules obtained by this method are especially useful for the encapsulation of water-soluble molecules such as peptides [53] and nucleic acids, including antisense oligonucleotides [52]. However, since they are dispersed in an oily continuous phase, they can mainly be developed as drug delivery systems for the oral route. For intravenous administrations, nanocapsules must be transferred to an aqueous continuous phase. This can be achieved by ultracentrifugation of the suspension over a layer of pure water [52].

B. Preparation of Nanoparticles by Nanoprecipitation and Emulsification–Solvent Evaporation

Nanoparticles can also be prepared from a polymer that is prepared by a totally independent method. This approach presents the major advantage that the polymers entering the composition of the nanoparticles are well characterized and their intrinsic physicochemical characteristics will not depend on the conditions encountered during the preparation of the nanoparticles. To prepare nanoparticles from a poly(alkylcyanoacrylate) two methods were applied.

The first one, *nanoprecipitation*, is based on the formation of colloidal polymer particles during phase separation induced by the addition of a nonsolvent of the polymer to a rather diluted polymer solution [60]. At that stage, the particles form spontaneously and quasi-instantaneously. The solvent of the polymer is then removed from the suspension by rotoevaporation. To facilitate formation of the colloidal polymer particles during the first step of the procedure, the phase separation is performed with totally miscible solvent and nonsolvent of the polymer.

The second method, *emulsification–solvent evaporation*, includes two steps. First is the emulsification of the polymer solution in an aqueous phase with the help of a high-pressure homogenizer or microfluidizer to produce very small emulsion droplets. In the second step, the solvent of the polymer is allowed to diffuse through the aqueous continuous phase to be evaporated at the air–water interface. The decrease of the amount of solvent in the emulsion droplet induces the precipitation of the polymer as nanoparticles.

These methods of nanoparticle preparation were applied on poly(alkylcyanoacrylates) to prepare poly(ethylene glycol)-coated nanoparticles from copolymers of hexadecylcyanoacrylate (HDCA) and monomethoxypoly(ethylene glycol cyanoacrylate) (PEGCA) [12–15,61–64]. Synthesis of these copolymers was achieved according to an original route of synthesis based on Knoevenagel condensation of the corresponding cyanoacetate derivatives with formaldehyde [61]. This method for polymer synthesis was seldom applied until now because most applications of cyanoacrylates were based on the properties of the monomers. However, it could attract much interest in the development of new polymers to promote the conception of original biomaterials. For example, preparing the copolymer from an aminopoly(ethylene glycol cyanoacetate) and hexadecylcyanoacetate allowed preparation of very small nanoparticles that represented an excellent support for coupling ligands at the nanoparticle surface, and facilitated development of nanoparticules for cell-selective targeting of drugs [12].

III. STABILITY AND TOXICITY OF POLY(ALKYLCYANOACRYLATE) NANOPARTICLES

Degradation and toxicity of poly(alkylcyanoacrylates) is often discussed in the literature especially for the in vivo applications of nanoparticles as drug delivery systems.

Concerning the stability of the nanoparticles, different pathways for their degradation have been described. The predominent mechanism greatly depends on the experimental conditions. Among the known degradation pathways, the most likely to occur in vivo are of two kinds. One mechanism consists of hydrolysis of the ester bond of the alkyl side chain of the polymer [32,65] (Fig. 2A). Degradation products consist of an alkylalcohol and poly(cyanoacrylic acid), which can dissolve in water and be eliminated in vivo through kidney filtration. This degradation has been shown to be catalyzed by esterases from serum, lysosomes, and pancreatic juice [66,67] and is believed to occur as the major degradation pathway in vivo. According to this mechanisms, nanoparticles are usually degraded within in a couple of hours depending on the alkyl side chain length of the poly(alkylcyanoacrylates).

The other mechanism that may occur in biological systems consists of an unzipping depolymerization of the parent polymer with immediate repolymerization to give a new polymer of much lower molecular weight. The whole phenomenom occurs within a few seconds and is generally induced by a base (Fig. 2B) [21].

The well-known inverse Knoevenagel reaction resulting from water hydrolysis reaches an equilibrium of formaldehyde and cyanoacetic ester production that is limited at physiological pH to only 5% after 24 h. This degradation pathway is much slower to compete with the other mechanisms occurring much faster in vivo and being further catalyzed by enzymes [68–70].

Toxicity of poly(alkylcyanoacrylates) is still a subject of debate. However, as applied in various biomedical applications, poly(alkylcyanoacrylates) did not reveal major problems of toxicity [17,71–75]. This is in agreement with the suspected in vivo degradation mechanism leading to the release of alcohol and poly(cyanoacrylic acid) as major metabolization products. The transitory minor

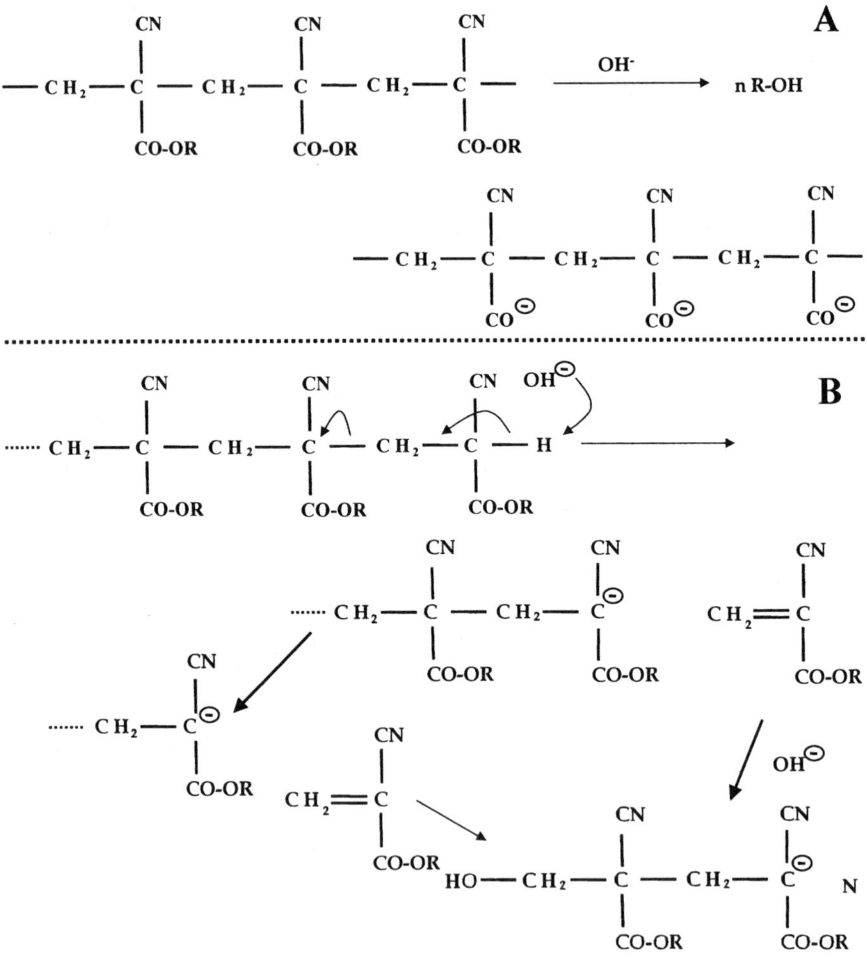

FIG. 2 More probable degradation routes of poly(alkylcyanoacrylate) encountered in vivo. (A) Hydrolysis of the ester bond of the alkyl side chain [32,65]; (B) unzipping depolymerization–repolymerization mechanism [21].

toxic effects or inflammatory responses that may be observed with these polymers were a function of the rate of the polymer degradation [17,73–75]. Polymers expected to degrade slowly showed the best compatibility with living tissues. Regarding the toxicity, the LD_{50} values found for poly(isobutylcynoacrylate) and poly(isohexylcyanoacrylate) nanospheres given intravenously to mice were 200 mg/kg and 700 mg/kg, respectively [71,72]. The low toxicity found with PACA nanoparticles has inspired clinical trials for human cancer [11,72]. A phase I trial confirmed the good tolerance of the drug carrier because only secondary effects due to the associated drug could be highlighted.

IV. PACA NANOPARTICLES IN CANCER THERAPY

It is well known that cancer treatment requires the use of very potent chemotherapeutic agents, that induce severe secondary toxic effects due to their wide distribution in the body. Because drug targeting can modulate drug distribution, the use of colloidal carriers has been proposed for a long time as a promising way to increase the efficacy of chemotherapy while reducing adverse effects. Because of its very limiting cardiotoxicity, doxorubicin was likely the most studied molecule in liposomal formulations as well as in nanoparticulate carriers. In addition, the emergence of multidrug resistance led some to explore the ability of nanoparticles to maintain doxorubicin efficacy in resistant tumor cells, as will be explained below. Besides conventional chemotherapy, oligonucleotides have been proposed in the last decade or so as a new strategy to specifically treat cancers. Here again, PACA nanoparticles have attracted interest as carriers for these large polyanionic molecules.

A. Efficacy of Doxorubicin-Loaded Nanoparticles Against Liver Metastasis

After intravenous administration, nanoparticles concentrate rapidly in the liver, which makes them particularly convenient for the treatment of hepatic tumors and even more for hepatic metastasis. The antitumor efficacy and superiority of doxorubicin-loaded nanoparticles over free doxorubicin in solution were clearly demonstrated in the M5076 reticulosarcoma hepatic metastasis model in mice [76]. The nanoparticles were found to be effective in reducing the number of metastases even when administered very late, when a great deal of metastasis has already occurred and when free doxorubicin has no effect, even at high doses. Such a result has been explained by the tissue targeting leading to a 2.5-fold increase of doxorubicin concentration in the tumor tissue compared to free drug. More precisely, nanoparticles are actually taken up by the Küpffer cells, playing the role of drug reservoir, allowing doxorubicin to diffuse toward the neighboring malignant cells [77].

Such tumor targeting allows nanoparticles to dramatically decrease doxorubicin cardiotoxicity [78,79]. Nevertheless, the high concentrations of doxorubicin and polymer in the liver raise the question of long-term toxicity of both compounds, as pointed out in rats treated with doxorubicin-loaded liposomes [80,81]. In this respect, only minor and mostly reversible effects were described with PACA nanospheres [82,83]. Finally, it must be noted that PACA nanoparticles also accumulate in the bone marrow, especially PIHCA because of its low bioerosion rate [84]. This tropism of PIHCA nanoparticles has been found useful to deliver myelostimulating compounds to reverse the suppressive effects of intense chemotherapy [85].

B. PACA Nanoparticles as a Mean to Overcome Multidrug Resistance

One of the main mechanisms involved in multidrug resistance (MDR) is the overexpression of a transmembrane glycoprotein, Pgp, acting as a pump and decreasing the intracellular concentration of a variety of lipophilic drugs. To remain cytotoxic against resistant cells, the doses of chemotherapeutic agents have to be considerably increased, far above the tolerated doses in clinics.

One strategy for bypassing this difficulty is the use of reversing agents such as cyclosporine or verapamil, inhibiting Pgp and allowing the intracellular concentrations of anticancer drug to be restored. This strategy has a limited interest in clinics because of the toxicity of the reversing agents. An alternative is to use drug targeting, the hypothesis being that drugs would be protected against the efflux from Pgp by the colloidal carrier, entering the cells probably by endocytosis.

The first interesting results obtained with nanoparticles encapsulating doxorubicin are dated from 1989 [86] and were confirmed on a variety of resistant cell lines in 1992 [87] and 1994 [88]. It was then found that PACA nanoparticles were able to completely restore the sensitivity of various cell lines even in the case of a resistance factor of more than 200. Surprisingly, further studies [89] showed that only PACA nanoparticles made it possible to overcome multidrug resistance. For example, polylactide and alginate nanoparticles exhibited the same cytotoxicity as the free drug, probably because of a too rapid release of doxorubicin in the culture medium.

Searching for a mechanism of action, extensive work has been done on P388 leukemia cell line, sensitive and resistant to doxorubicin (P388/ADR). First, it has been found that PACA nanoparticles were not endocytosed by the tumor cells [90] but that direct contact between the nanoparticles and the tumor cells was essential to observe the reversion of the resistance [91]. The most likely hypothesis was that nanoparticles adsorb onto the cell surface, leading to a locally increased drug concentration gradient favoring in turn cellular penetration. PACA nanoparticles release doxorubicin along with their bioerosion due to esterases, and lead to the formation of soluble poly(cyanoacrylic) acid. Colin de Verdière et al. [91] showed that the presence of this soluble polymer coming from the bioerosion of nanoparticles was sufficient to increase both doxorubicin cytotoxicity and drug uptake by the resistant cells. In the same time, Hu et al. [92] found that PACA did not inhibit Pgp. The role of poly(cyanoacrylic acid) had then to be sought elsewhere: actually, both doxorubicin and poly(cyano-acrylic acid) can form a complex by ion pairing [93], leading to an uncharged and more hydrophobic compound than the original drug. Diffusion through cell membranes is then favored and intracellular concentrations can be increased.

Hence, PACA nanoparticles were shown to restore cell sensitivity to doxorubicin by a mechanism completely independent of the presence of the Pgp. Actually, resistant cells, including P388/ADR, present an increased membrane potential compared to sensitive cells, inhibiting the diffusion of cationic drugs, which remain sequestered in the phospholipidic bilayer by electrostatic interactions. Bypassing these interactions through charge masking is probably the way doxorubicin/polycyanoacrylic acid ion pairs succeeded in overcoming MDR.

Encouraging results were also obtained in vivo with a P388 model growing as ascites [87], after intraperitoneal administration in mice. Furthermore, results obtained in a new in vitro model consisting of a coculture between macrophages and tumor cells suggest that doxorubicin-loaded nanoparticles will also have interest after intravenous administration in the case of resistant hepatic metastases, for example [94,95]. A complete reversion has even been obtained by combining doxorubicin and cyclosporin A, a reversing agent, in the same PACA nanoparticles. These "mixed" carriers will allow delivery of both drugs in the same cells and at the same time, increasing their synergistic effect while suppressing their general side effects [96].

C. PACA Nanoparticles and Oligonucleotide Therapy of Cancer

Oligonucleotides with base sequences complementary (antisense) to a specific RNA offer the exciting potential of selectively modulating the expression of an individual gene. However, crucial problems, such as the stability of oligonucleotides in relation to nuclease activity in vitro and in vivo and the low penetration into cells, have to be solved. Thus, the development of PACA particulate carriers [97,98] for oligonucleotide delivery may be considered as being an interesting approach to improve the in vivo efficacy of these molecules by protecting them against degradation and by increasing their delivery to the cell interior.

The association of oligonucleotides with PACA nanospheres was found possible in the presence of a hydrophobic cation such as cethyltrimethylammonium bromide (CTAB) [97]. Indeed, the hydrophilic character of the nucleic acid chains doesn't allow interaction with the polymer through hydrophobic interactions. In the proposed method, oligonucleotide adsorption on the nanospheres was mediated by the formation of ions pairs between the negatively charged phosphate groups of the nucleic acid chain and the positive charge of the hydrophobic cations. Oligonucleotides could also associate with DEAE-dextran-coated PACA nanospheres according to the same mechanism [99].

When adsorbed onto PACA nanospheres, oligonucleotides were totally protected against enzymatic degradation even after 5 h incubation with phosphodiesterase [100]. It was observed that the uptake of a 15mer oligonucleotide by U937 cells was dramatically increased when associated with nanospheres [100], and intact oligonucleotides could be found in both the nuclear and extranuclear fractions even after 24 h of incubation, whereas no more intact molecules could be found in the extranuclear fractions after 6 h of incubation in the case of free oligonucleotides. This finding suggests that nanoparticles greatly contribute to the protection of oligonucleotides in the cells, even if this type of colloidal carrier is known to be finally digested by lysosomes where oligonucleotides will be degraded too. It is not out of the question that nanoparticles could escape from the endosomal compartment due to the presence of the adsorbed CTAB which is susceptible to exert a lytic action against the endosomal membranes.

Nakada et al. [101] performed pharmacokinetic studies of the ^{33}P-labeled oligonucleotide, free or associated to PACA nanoparticles, after intravenous administration in mice. Free oligonucleotides rapidly disappear from the blood and concentrate in the kidney where they are eliminated. Nanoparticles did not markedly increase the blood half-life of the ^{33}P oligonucleotide but significantly modified its tissue distribution. Briefly, liver uptake was increased in association with a subsequent decrease in the other organs, especially the kidney. Furthermore, control of the integrity of the oligonucleotides showed that 5 min after intravenous administration, a significant number of molecules were found intact in both the blood and the liver when associated with nanoparticles, whereas no traces of intact oligonucleotide could be detected when injected free. However, some oligonucleotide was still degraded probably because these molecules were located at the surface of the particles, which allowed a certain accessibility for the nucleases.

Finally, this system has been proven efficient in vivo after intratumoral injection when applied to an anti-*ras* oligonucleotide: this system markedly inhibited Ha-*ras*-dependent tumor growth in nude mice in a highly specific manner [102].

As discussed above, the main drawback of PACA nanospheres for oligonucleotide delivery is that oligonucleotides are adsorbed onto the surface of the particles, which is not the optimal localization to avoid degradation by nucleases in vivo. In order to completely mask oligonucleotides from nuclease attack, water-containing PACA nanocapsules have been proposed [52]. Fluorescence quenching assays have shown that fluorescent oligonucleotides were located in the aqueous core of the nanocapsules, surrounded by a polymeric wall [52]. As expected, nanoencapsulation led to a much more efficient protection of oligonucleotides against degradation by serum nucleases than that obtained with CTAB-coated nanospheres.

Antisense oligonucleotides against a fusion protein (EWS FLI-1) were successfully encapsulated in PACA nanocapsules and injected intratumorally in rats grafted with an experimental Ewing sarcoma. Impressive results were obtained in terms of tumor inhibition [103].

D. Toward a Real Targeting of Solid Tumors

After intravenous administration, "conventional" PACA nanoparticles show particular potential for therapeutic applications in which macrophages and organs of the mononuclear phagocyte system (MPS) including the liver, the spleen, and the bone marrow are concerned [77,84,104,105]. This is due to the very specific pattern of distribution of PACA nanoparticles which are normally recognized as foreign body by the MPS and rapidly cleared from the bloodstream. The main mechanism governing this recognition involves adsorption of blood proteins (opsonins) and complement activation [85,106,107]. The first challenge has been to develop colloidal systems capable of avoiding such recognition and lasting longer in the bloodstream. Another major challenge was to develop drug delivery systems having those properties together with the additional capacity to recognize a biological target with a high specificity at a molecular level.

In this view, different approaches were proposed to modify the surface properties of PACA nanospheres [41,108–111]. The synthesis of a new PEG-containing PACA copolymer [poly(PEGCA-*co*-HDCA)] allowing the preparation of nanospheres by methods avoiding in situ polymerization (i.e., nanoprecipitation or emulsification–solvent evaporation) can be considered as the more promizing approach [62,112]. Indeed, in these nanospheres, PEG chains are covalently grafted to the surface by the means of PACA anchors forming the core of the particles. The composition of the copolymer and the PEG structure at the nanosphere surface could be perfectly controlled [112,113]. A reduction of protein adsorption on the nanoparticle surface was clearly demonstrated by two-dimensional polyacrylamide gel electrophoresis and by transmission electron microscopy [113]. Furthermore, after intravenous administration, poly(PEGCA-*co*-HDCA) nanospheres showed an increase of their half-life in the blood [13, 114]. Due to their prolonged circulating time, they are supposed to be capable of extravasating through the damaged endothelium encountered in certain types of tumors, thus allowing the selective delivery of drugs to these tissues.

Recombinant human tumor necrosis factor–α (rHuTNF-α) has been successfully associated with these poly(PEGCA-*co*-HDCA) nanospheres [13,64]. The pharmacokinetic and antitumor effects were evaluated in vivo in mice with sarcoma 180 cells implanted intradermally. A higher accumulation of rHuTNF-α in the tumor and an increased antitumoral activity were measured when this compound was injected intravenously as poly(PEGCA-*co*-HDCA) nanospheres [13] compared to the free drug.

To develop cell-selective targeting, folic acid has been conjugated to poly-(PEGCA-*co*-HDCA) nanospheres [12], the rationale behind this construction being that folic acid–binding protein is frequently overexpressed on the surface of human cancer cells. The nanospheres were prepared using a copolymer containing PEG with a terminal amino group. This amino group was exposed at the particle surface, and subsequent coupling with activated folic acid was successful. Based on surface plasmon resonance assays, it was found the folate-conjugated nanospheres interacted much more with folate-binding protein than nanospheres coated only with PEG on their surface. Folate-decorated nanospheres also showed a greater affinity for the receptor than a single molecule of folic acid. This could be explained by the cooperative interactions obtained with the nanospheres [12]. These folic acid–decorated nanospheres open interesting perspectives for the selective application of anticancer compounds in tumoral cells and tissues.

V. CONCLUSION

As shown in this chapter, progress in the chemistry of alkylcyanoacrylate polymerization and, more recently, in the chemistry of PEG-containing copolymers leads to the development of interesting materials for the conception of biodegradable nanosystems with high potential for the delivery of drugs in vivo to various organs, tissues, and cells. By exploring different approaches of cancer therapy, including chemotherapy and antisense therapy, PACA nanoparticles showed very promising results. For example, they were the only ones among polymer particles to be able to restore the sensitivity of multidrug-resistant cells to chemotherapy. They were also shown to allow the delivery of active antisense oligonucleotides to implanted tumor in vivo. PEG-coated nanospheres will be preferred when long circulating properties are needed for tumor extravasation, for example. Finally, the development of PACA nanoparticles for biomedical applications is still in progress considering the attachment of molecular targeting moieties to the nanoparticle surface with potential for cell-selective targeting.

REFERENCES

1. Coover, H.W.; Dreifus, D.W.; O'Connor, J.T. Cyanoacrylate adhesives. In *Handbook of Adhesives*, 3rd Ed.; Skeist, I, Ed.; Van Nortrand Reinhold: New York, 1990; 463–477.
2. Skeist, I.; Miron, J. Adhesive composition. In *Encyclopedia of Polymer Science and Engineering*; Bikales, N., Ed.; Wiley Interscience: New York, 1977; Supp. Vol 2., 1–19.
3. Montanaro, L.; Arciola, C.R.; Cenni, E.; Ciapetti, G.; Savioli, F Filippini, F.; Barsanti, L.A. Cytotoxicity, blood compatibility and antimicrobial activity of two cyanoacrylate glues for surgical use. Biomaterials **2001**, *22*, 59–66.
4. Oowaki, H.; Matsuda, S.; Sakai, N.; Ohta, T.; Iwata, H.; Sadato, A.; Taki, W.;

Hashimoto, N.; Ikada, Y. Non-adhesive cyanoacrylate as an embolic material for endovascular neurosurgery. Biomaterials **2000**, *21*, 1039–1046.

5. Couvreur, P.; Kante, B.; Roland, M.; Guiot, P.; Bauduin, P.; Speiser, P. Polycyanoacrylate nanocapsules as potential lysosomotropic carriers: preparation, morphological and sorptive properties. J. Pharm. Pharmacol. **1979**, *31*, 331–332.
6. Couvreur, P.; Kante, B.; Roland, M. Les vecteurs lysosomotropes. J. Pharm. Belg. **1980**, *35*, 51–60.
7. Couvreur, P.; Kante, B.; Grislain, L.; Roland, M.; Speiser, P. Toxicity of polyalkycyanoacrylate nanoparticles. II. Doxorubicin-loaded nanoparticles. J. Pharm. Sci. **1982**, *71*, 790–792.
8. Barratt, G.; Couarraze, G.; Couvreur, P.; Dubernet, C.; Fattal, E.; Gref, R.; Labarre, D.; Legrand, P.; Ponchel, G.; Vauthier, C. Polymeric micro and nanoparticles as drug carriers. In *Polymeric Biomaterials*, 2nd Ed.; Dumitriu, S., Ed.; Marcel Dekker: New York, 2001; 753–782.
9. Fattal, E.; Vauthier, C. Nanoparticles as drug delivery systems. In *Encyclopedia of Pharmaceutical Technology*; Swarbrick, J., Boylan, J.C., Eds.; Marcel Dekker: New York, 2002; 1874–1892.
10. Couvreur, P.; Barratt, G.; Fattal, E.; Legrand, P.; Vauthier, C. Nanocapsule technology: a review. Crit. Rev. Drug Del. Syst. **2002**, *19*, 99–134.
11. Kattan, J.; Droz, J.P.; Couvreur, P.; Marino, J.P.; Boutan-Laroze, A.; Rougier, P.; Brault, P.; Vranks, H.; Grognet, J.M.; Morge, X.; Sancho-Garnier, H. Phase I clinical trial and pharmacokinetic evaluation of doxorubicin carried by polyisohexylcyanoacrylate nanoparticles. Invest. New Drugs **1992**, *10*, 191–199.
12. Stella, B.; Arpicco, S.; Peracchia, M.T.; Desmaële, D.; Hoebeke, J.; Renoir, M.; D'Angelo, J.; Cattel, L.; Couvreur, P. Design of folic acid–conjugated nanoparticles for drug targeting. J. Pharm. Sci. **2000**, *89*, 1452–1464.
13. Li, Y.P.; Pei, Y.Y.; Zhou, Z.H.; Zhang, X.Y.; Gu, Z.H.; Ding, J.; Zhou, J.J.; Gao, X.J.; Zhu, J.H. Stealth polycyanoacrylate nanoparticles as tumor necrosis factor-α carriers: pharmacokinetics and anti-tumor effects. Biol. Pharm. Bull. **2001**, *24*, 662–665.
14. Brigger, I.; Chaminade, P.; Marsaud, V.; Appel, M.; Besnard, M.; Gurny, R.; Renoir, M.; Couvreur, P. Tamoxifen encapsulation within polyethylene glycol-coated nanospheres. A new antiestrogen formulation. Int. J. Pharmacol. **2001**, *214*, 37–42.
15. Calvo, P.; Gouritin, B.; Chacun, H.; Desmaële, D.; D'Angelo, J.; Noël, J.P.; Georgin, D.; Fattal, E.; Andreux, J.P.; Couvreur, P. Long-circulating polycyanoacrylate nanoparticles as new drug carrier for brain delivery. Pharm. Res. **2001**, *18*, 1157–1166.
16. Leonard, F.; Kulkarni, R.K.; Brandes, G.; Nelson, J.; Cameron, J.J. Synthesis and degradation of poly(alkyl α-cyanoacrylates). J. Appl. Polym. Sci. **1966**, *10*, 259–272.
17. Jaffre, H.; Wade, C.W.R.; Hegyeli, A.F.; Rice, R.; Hodge, J. Synthesis and bioevaluation of alkyl-2-cyanoacryloyl glycolate as potential soft tissue adhesives. J. Biomed. Mater. Res. **1986**, *20*, 205–212.
18. Tseng, Y.C.; Hyon, S.H.; Ikada, Y. Modification of the synthesis and investigation of properties for 2-cyanoacrylates. Biomaterials **1990**, *11*, 73–79.

19. Donnelly, E.F.; Johnston, D.S.; Pepper, D.C. Ionic and zwitterionic polymerization of n-alkyl 2-cyanoacrylates. Polym. Lett. Ed. **1977**, *15*, 399–405.
20. Pepper, D.C. Anionic and zwitterionic polymerization of α-cyanoacrylates. J. Polym. Sci. Polym. Symp. **1978**, *62*, 65–77.
21. Ryan, B.; McCann, G. Novel sub-ceiling temperature rapid depolymerization-repolymerization reactions of cyanoacrylate polymers. Macromol. Rapid Commun. **1996**, *17*, 217–227.
22. Cronin, J.P.; Pepper, D.C. Zwitterionic polymerization of butyl cyanoacrylate by triphenylphosphine and pyridine. Makromol. Chem. **1988**, *189*, 85–102.
23. Canale, A.J.; Goode, W.E.; Kinsinger, J.B.; Panchak, J.R.; Kelso, R.L.; Graham, R.K. Methyl α-cyanoacrylate. I. Free-radical homopolymerization. J. Appl. Polym. Sci. **1960**, *4*, 231–236.
24. Bevington, J.C.; Jemmett, J.A.L. Polymerization of methyl α-cyanoacrylate. I. Initiation by benzoyl peroxide. J. Chem. Soc. Faraday Trans. I. **1973**, *69*, 1866–1871.
25. Bevington, J.C.; Jemmett, J.A.L.; Onyon, P.F. Polymerization of methyl α-cyanoacrylate. II. Conditions for radical polymerization. Eur. Polym. J. **1976**, *12*, 255–257.
26. Otsu, T.; Yamada, B.; Kusayama, S.; Nagao, S. Radical copolymerization of ethyl α-cyanoacrylate in the presence of effective inhibitors against anionic polymerization. Kobunshi Ronbunshu **1979**, *36*, 797–802.
27. Yamada, B.; Yoshioka, M.; Otsu, T. Determination of absolute rate constants for radical polymerization and copolymerization of ethyl α-cyanoacrylate in the presence of effective inhibitors against anionic polymerization. Makromol. Chem. **1983**, *184*, 1025–1033.
28. Costa, G.; Cronin, J.P.; Pepper, D.C.; Loonan, C. Termination and transfer by acids in the pyridine-initiated polymerization of butyl cyanoacrylate. Eur. Polym. J. **1983**, *19*, 939–945.
29. Douglas, S.J.; Illum, L.; Davis, S.S. Particle size and size distribution of poly(butyl 2-cyanoacrylate) nanoparticles. II. Influence of stabilizers. J. Colloid Interface Sci. **1985**, *103*, 154–163.
30. Peracchia, M.T.; Vauthier, C.; Puisieux, F.; Couvreur, P. Development of sterically stabilized poly(isobutyl 2-cyanoacrylate) nanoparticles by chemical coupling of poly(ethylene glycol). J. Biomed. Mater. Res. **1997**, *34*, 317–326.
31. Peracchia, M.T.; Vauthier, C.; Popa, M.I.; Puisieux, F.; Couvreur, P. Investigation of the formation of sterically stabilized poly(ethylene glycol/isobutylcyanoacrylate) nanoparticles by chemical grafting of polyethylene glycol during the polymerization of isobutyl cyanoacrylate. S.T.P. Pharma Sci. **1997**, *7*, 513–520.
32. Vansnick, L.; Couvreur, P.; Christiaens-Ley, D.; Roland, M. Molecular weights of free and drug-loaded nanoparticles. Pharm. Res. **1985**, *1*, 36–41.
33. Gallardo, M.M.; Roblot-Treupel, L.; Mahuteau, J.; Genin, I.; Couvreur, P.; Plat, M.; Puisieux, F. Nanocapsules et nanospheres d'alkylcyanoacrylate. Interaction principe actif-polymère. Proceedings of the 5th International Congress of Pharmaceutical Technology *4*, 36–45, 1989.
34. Guize, V.; Drouin, J.Y.; Benoit, J.; Mahuteau, J.; Dumont, P.; Couvreur, P. Vidar-

abine-loaded nanoparticles: a physicochemical study. Pharm. Res. **1990**, *7*, 736–741.

35. Grangier, J.L.; Puygrenier, M.; Gautier, J.C.; Couvreur, P. Nanoparticles as carriers for growth hormone releasing factor. J. Controlled Rel. **1991**, *15*, 3–13.
36. Nehan, N.; Birkinshaw, C.; Clarke, C. Poly n-butyl cyanoacrylate nanoparticles: a mechanistic study of polymerization and particle formation. Biomaterials **2001**, *22*, 1335–1344.
37. Seijo, B.; Fattal, E.; Roblot-Treupel, L.; Couvreur, P. Design of nanoparticles of less than 50 nm diameter: preparation, characterization and drug loading. Int. J. Pharmacol. **1990**, *62*, 1–7.
38. Alonso, M.J.; Sanchez, A.; Torres, D.; Seijo, B.; Vila-Jato, J.L. Joint effect of monomer and stabilizer concentrations on the physico-chemical characteristics of poly(butyl-2-cyanoacrylate) nanoparticles. J. Microencapsulation **1990**, *7*, 517–526.
39. Lenaerts, V.; Raymond, P.; Juhasz, J.; Simard, M.A.; Jolicoeur, C. New method for the preparation of cyanoacrylic nanoparticles with improved colloidal properties. J. Pharm. Sci. **1989**, *78*, 1051–1052.
40. Lescure, F.; Zimmer, C.; Roy, D.; Couvreur, P. Optimization of polycyanoacrylate nanoparticle preparation: influence of sulfur dioxide and pH on nanoparticle characteristics. J. Colloid Interface Sci. **1992**, *154*, 77–86.
41. Vauthier, C.; Couvreur, P. Poly(alkylcyanoacrylates). In *Handbook of Biopolymers, Vol 9. Miscellanous Biopolymers and Biodegradation of Synthetic Polymers*; Matsumara, J.P., Steinbuchel, A., Eds.; Wiley-VHC: New York, 2003.
42. Monza da Silveira, A.; Ponchel, G.; Puisieux, F.; Duchène, D. Combined poly-(isobutylcyanoacrylate) and cyclodextrins nanoparticles for enhancing the encapsulation of lipophilic drugs. Pharm. Res. **1990**, *15*, 1051–1055.
43. Duchene, D.; Wouessidjewe, D.; Ponchel, G. Cyclodextrins and carrier systems. J. Controlled Rel. **1999**, *62*, 263–268.
44. Labib, A.; Lenaerts, V.; Chouinard, F.; Leroux, J.C.; Ouellet, R.; Van Lier, J.E. Biodegradable nanospheres containing phthalocyanines and naphthalocyanines for targeted photodynamic tumor therapy. Pharm. Res. **1991**, *8*, 1027–1031.
45. Ghanem, G.E.; Joubran, C.; Arnould, R.; Lejeune, F.; Fruhling. Labelledpolycyanoacrylate nanoparticles for human in vivo use. Appl. Radiat. Isot. **1993**, *44*, 1219–1224.
46. Douglas, S.J.; Illum, L.; Davis, S.S. Poly(butyl-2-cyanoacrylate) nanoparticles with differing surface charges. J. Controlled Rel. **1986**, *3*, 15–23.
47. Yang, S.C.; Ge, H.X.; Hu, Y.; Jiang, X.Q.; Yang, C.Z. Formation of positively charged poly(butyl cyanoacrylate) nanoparticles stabilized by chitosan. Colloid Polym. Sci. **2000**, *278*, 285–292.
48. Al Khoury-Fallouh, N.; Roblot-Treupel, L.; Fessi, H.; Devissaguet, J.P.; Puisieux, F. Development of a new process for the manufacture of poly(isobutylcyanoacrylate) nanocapsules. Int. J. Pharmacol. **1986**, *28*, 125–136.
49. Chouinard, F.; Kan, F.W.F.; Leroux, J.C.; Foucher, C.; Lenaerts, V. Preparation and purification of polyisohexylcyanoacrylate nanocapsules. Int. J. Pharmacol. **1991**, *71*, 211–217.

50. Fresta, M.; Cavallaro, G.; Giammona, G.; Wehrli, E.; Puglisi, G. Preparation and characterization of polyethyl-2-cyanoacrylate nanocapsules containing antiepileptic drugs. Biomaterials **1996**, *17*, 751–758.
51. Gasco, M.; Trotta, M. Nanoparticles from microemulsions. Int. J. Pharmacol. **1986**, *29*, 267–268.
52. Lambert, G.; Fattal, E.; Pinto-Alphandary, H.; Gulik, A.; Couvreur, P. Polyisobutylcyanoacrylate nanocapsules containing an aqueous core as a novel colloidal carrier for the delivery of oligonucleotides. Pharm. Res. **2000**, *17*, 707–714.
53. Watnasirichaikul, S.; Davies, N.M.; Rades, R.; Tucker, I.G. Preparation of biodegradable insulin nanocapsules from biocompatible microemulsions. Pharm. Res. **2000**, *17*, 684–689.
54. Gallardo, M.M.; Couarraze, G.; Denizot, B.; Treupel, L.; Couvreur, P.; Puisieux, F. Preparation and purification of isohexylcyanoacrylate nanocapsules. Int. J. Pharmacol. **1993**, *100*, 55–64.
55. Aboubakar, M.; Puisieux, F.; Couvreur, P.; Deyme, M.; Vauthier, C. Study of the mechanism of insulin encapsulation in poly(isobutylcyanoacrylate) nanocapsules obtained by interfacial polymerization. J. Biomed. Mater. Res. **1999**, *47*, 568–576.
56. Wohlgemuth, M.; Mächtle, W.; Mayer, C. Improved preparation and physical studies of polybutylcyanoacrylate nanocapsules. J. Microencapsulation **2000**, *17*, 437–448.
57. Puglisi, G.; Fresta, M.; Giammona, G.; Venture, C.A. Influence of the preparation conditions in poly(ethylcyanoacrylate) nanocapsules formation. Int. J. Pharm. **1995**, *125*, 283–287.
58. Damgé, C.; Michel, C.; Aprahamiam, M.; Couvreur, P. New approach for oral administration of insulin with polyalkylcyanoacrylate nanocapsules as drug carrier. Diabetes **1988**, *37*, 246–251.
59. Damgé, C.; Vonderscher, J.; Marbach, P.; Pinget, M. Poly(alkylcyanoacrylate) nanocapsules as a delivery system in the rat for octreotide, a long-acting somatostatin analogue. J. Pharm. Pharmacol. **1997**, *49*, 949–954.
60. Fessi, H.; Devissaguet, J.P.; Puisieux, F. Procédé de préparation de systèmes colloïdaux dispersibles d'une substance sous forme de nanoparticules. French Patent 2,608,988, 1986.
61. Peracchia, M.T.; Desmaële, D.; Couvreur, P.; D'Angelo, J. Synthesis of a novel poly(MePEG cyanoacrylate-co-alkylcyanoacrylate) amphiphilic copolymer for nanoparticle technology. Macromolecules **1997**, *30*, 846–851.
62. Peracchia, M.T.; Vauthier, C.; Desmaële, D.; Gulik, A.; Dedieu, J.C.; Demoy, M.; D'Angelo, J.; Couvreur, P. Pegylated nanoparticles from a novel methoxypolyethylene glycol cyanoacrylate-hexadecyl cyanoacrylate amphiphilic copolymer. Pharm. Res. **1998**, *15*, 548–554.
63. Calvo, P.; Gouritin, B.; Brigger, I.; Lasmezas, C.; Deslys, J.P.; Williams, A.; Andreux, J.P.; Dormont, D.; Couvreur, P. PEGylated polycyanoacrylate nanoparticles as vector for drug delivery in prion diseases. J. Neurosci. Meth. **2001**, *111*, 151–155.
64. Li, Y.P.; Pei, Y.Y.; Zhou, Z.H.; Zhang, X.Y.; Gu, Z.H.; Ding, J.; Zhou, J.J.; Gao, X.J. PEGylated polycyanoacrylate nanoparticles as tumor necrosis factor-α carrier. J. Controlled Rel. **2001**, *71*, 287–296.

65. Langer, K.; Seegmüller, E.; Zimmer, A.; Kreuter, J. Characterization of polybutylcyanoacrylate nanoparticles: quantification of PBCA polymer and dextran. Int. J. Pharmacol. **1994**, *110*, 21–27.
66. Scherer, D.; Robinson, J.R.; Kreuter, J. Influence of enzymes on the stability of polybutylcyanoacrylate nanoparticles. Int. J. Pharmacol. **1994**, *101*, 165–168.
67. Müller, R.H.; Lherm, C.; Herbort, J.; Couvreur, P. In vitro model for the degradation of alkylcyanoacrylate nanoparticles. Biomaterials **1990**, *11*, 590–595.
68. Vezin, W.R.; Florence, A.T. In vitro heretogenous degradation of poly(n-alkyl α-cyanoacrylates). J. Biomed. Mater. Res. **1980**, *14*, 93–106.
69. Ley, D.; Couvreur, P.; Lenaerts, V.; Roland, M.; Speiser, P. Etude du mécanisme de dégradation des nanoparticules de polycyanoacrylate d'alkyle. Labo Pharma-Probl. Tech. **1984**, *32*, 100–104.
70. Cooper, A.W.; Harris, P.J.; Kumar, G.K.; Tebby, J.C. Hydrolysis of alkylcyanoacrylate and ethoxycarbonylpropenoylphosphonate polymers. J. Polym. Sci. A Polym. Chem. **1989**, *27*, 1967–1974.
71. Kante, B.; Couvreur, P.; Dubois-Krack, G.; De Meester, C.; Guiot, P.; Roland, M.; Mercier, M.; Speiser, P. Toxicity of polyalkylcyanoacrylate nanoparticles I. free nanoparticles. J. Pharm. Sci. **1982**, *71*, 786–790.
72. Couvreur, P.; Vrancks, H.; Brasseur, F.; Roland, M. Toxicité des nanoparticles à base de polycyanoacrylates d'alkyle. S.T.P. Pharma Sci. **1989**, *5*, 31–37.
73. Henderson, A.M.; Stenphenson, M. 3-Methoxybutylcyanoacrylate: evaluation of biocompatibility and bioresorption. Biomaterials **1992**, *13*, 1077–1084.
74. Fernandez-Urrusuno, R.; Fattal, E.; Féger, J.; Couvreur, P.; Théron, P. Evaluation of hepatic antioxidant sytems after intravenous administration of polymeric nanoparticles. Biomaterials **1997**, *18*, 511–517.
75. Simeonova, M.; Chorbadjiev, K.; Antchva, M. Study of the effect of polybutylcyanoacrylate nanoparticles and their metabolites on primary immune response in mice to sheep red blood cells. Biomaterials **1998**, *19*, 2187–2193.
76. Chiannikulchai, N.; Driouich, Z.; Benoit, J.P.; Parodi, A.L.; Couvreur, P. Doxorubicin-loaded nanoparticles: increased efficiency in murine hepatic metastases. Select. Cancer Ther. **1989**, *5*, 1–11.
77. Chiannikulchai, N.; Ammoury, N.; Caillou, B.; Devissaguet, J.P.; Couvreur, P. Hepatic tissue distribution of doxorubicin-loaded nanoparticles after i.v. administration in M5076 metastasis-bearing mice. Cancer Chemother. Pharmacol. **1990**, *26*, 122–126.
78. Couvreur, P.; Grislain, L.; Lenaerts, V.; Brasseur, F.; Guiot, P.; Biornacki, A. Biodegradable polymeric nanoparticles as drug carrier for antitumor agents. In *Polymeric Nanoparticles and Microspheres*; Guiot, P., Couvreur, P., Eds.; CRC Press: Boca Raton, 1986; 27–93.
79. Verdun, C.; Couvreur, P.; Vranckx, H.; Lenaerts, V.; Roland, M. Development of a nanoparticle controlled release formulation for human use. J. Controlled Rel. **1986**, *3*, 205–210.
80. Storm, G.; Steerenberg, P.A.; Emmen, F.; van Borssum Waalkes, M.; Crommelin, D.J.A. Release of doxorubicin from peritoneal macrophages exposed in vivo to doxorubicin-containing liposomes. Biochim. Biophys. Acta **1988**, *965*, 136–145.
81. Daeman, T.; Regts, J.; Meesters, M.; Ten Kate, M.T.; Bakker-Woudenberg,

I.A.J.M.; Scherphof, G. Toxicity of doxorubicin entrapped within long-circulating liposomes. J. Controlled Rel. **1997**, *44*, 1–9.
82. Fernandez-Urrusuno, R.; Fattal, E.; Porquet, D.; Feger, J.; Couvreur, P. Evaluation of liver toxicological effects induced by polyalkylcyanoacrylate nanoparticles. Toxicol. Appl. Pharmacol. **1995**, *130*, 272–279.
83. Fernandez-Urrusuno, R.; Fattal, E.; Rodrigues, J.M., Jr.; Féger, J.; Bedossa, P.; Couvreur, P. Effect of polymeric nanoparticle administration on the clearance activity of the mononuclear phagocyte system in mice. J. Biomed. Mater. Res. **1996**, *31*, 401–408.
84. Gibaud, S.; Andreux, J.P.; Weingarten, C.; Renard, M.; Couvreur, P. Increased bone marrow toxicity of doxorubicin bound to nanoparticles. Eur. J. Cancer **1994**, *30A*, 820–826.
85. Gibaud, S.; Rousseau, C.; Weingarten, C.; Favier, R.; Douay, L.; Andreux, J.P.; Couvreur, P. Polyalkylcyanoacrylate nanoparticles as carriers for granulocyte-colony stimulating factor (g-csf). J. Controlled Rel. **1998**, *52*, 131–139.
86. Kubiak, C.; Couvreur, P.; Manil, L.; Clausse, B. Increased cytotoxicity of nanoparticle carried adriamycin in vitro and potentiation by verapamil and amiodarone. Biomaterials **1989**, *10* (10), 553–556.
87. Cuvier, C.; Roblot Treupel, L.; Millot, J.M.; Lizard, G.; Chevillard, S.; Manfait, M.; Couvreur, P.; Poupon, M.F. Doxorubicin loaded nanospheres bypass tumor cell multidrug resistance. Biochem. Pharmacol. **1992**, *44* (3), 509–517.
88. Bennis, S.; Chapey, C.; Couvreur, P.; Robert, J. Enhanced cytotoxicity of doxorubicin encapsulated in polyosohexylcyanoacrylate nanospheres against multidrug resistant tumor cells in culture. Eur. J. Cancer **1994**, *30A*, 89–93.
89. Némati, F.; Dubernet, C.; Fessi, H.; Co lin de Verdière, A.; Puisieux, F.; Couvreur, P. Reversion of the multidrug resistance using nanoparticles in vitro: influence of the nature of the polymer. Int. J. Pharmacol. **1996**, *138*, 237–246.
90. Colin de Verdière, A.; Dubernet, C.; Némati, F.; Poupon, M.F.; Puisieux, F.; Couvreur, P.Uptake of doxorubicin from loaded nanoparticles in multidrug resistant leukemic murine cells. Cancer Chemother. Pharmacol. **1994**, *33*, 504–508.
91. Colin de Verdière, A.; Dubernet, C.; Némati, F.; Soma, E.; Appel, M.; Ferté, J.; Bernard, S.; Puisieux, F.; Couvreur, P. Reversion of the multidrug resistance with polyalkylcyanoacrylate nanoparticles: towards a mechanism of action. Br. J. Cancer **1997**, *76* (2), 198–205.
92. Hu, Y.P.; Jarillon, S.; Dubernet, C.; Couvreur, P.; Robert, J. On the mechanism of action of doxorubicin encapsulation in nanospheres for the reversal of multidrug resistance. Cancer Chemother. Pharmacol. **1996**, *37*, 556–560.
93. Dubernet, C.; Soma, E.; Couvreur, P.; Pépin, X.; Attali, L.; Dombrault, C.; Gallet, S.; Metreau, J.M.; Renault, Y.; Imalalen, M.; Cardot, P. On the use of chromatographic ion pairing elution mode to elucidate doxorubicin release mechanism from polyalkylcyanoacrylate nanoparticles at the cellular level. J. Chromatogr. Biol. Appl. **1997**, *702*, 181–191.
94. Soma, C.E.; Dubernet, C.; Barratt, G.; Nemati, F.; Appel, M.; Benita, S.; Couvreur, P. Ability of doxorubicin-loaded nanoparticles to overcome multidrug resistance of tumor cells after their capture by macrophages. Pharm. Res. **1999**, *16* (11), 1710–1716.

95. Soma, C.E.; Dubernet, C.; Barratt, G.; Benita, S.; Couvreur, P. Investigation of the role of macrophages on the cytotoxicity of doxorubicin and doxorubicin-loaded nanoparticles on M5076 cells in vitro. J. Controlled Rel. **2000**, *68*, 283–289.
96. Soma, C.E.; Dubernet, C.; Bentolila, D.; Benita, S.; Couvreur, P. Reversion of multidrug resistance by co-encapsulation of doxorubicin and cyclosporin A in polyalkylcyanoacrylate nanoparticles. Biomaterials **2000**, *21*, 1–7.
97. Chavany, C.; Le Doan, T.; Couvreur, P.; Puisieux, F.; Helene, C. Polyalkylcyanoacrylate nanoparticles as polymeric carriers for antisense oligonucleotides. Pharm. Res. **1992**, *9*, 441–449.
98. Fattal, E.; Vauthier, C.; Aynié, I.; Nakada, Y.; Lambert, G.; Malvy, C.; Couvreur, P. Biodegradable polyalkylcyanoacrylate nanoparticles for the delivery of oligonucleotides. J. Controlled Rel. **1998**, *53*, 137–143.
99. Zimmer, A. Antisense oligonucleotide delivery with polyhexylcyanoacrylate nanoparticles as carriers. Meth. Companion Meth. Enzymol. **1999**, *18* 286–295.
100. Chavany, C.; Saison-Behmoaras, T.; Le Doan, T.; Puisieux, F.; Couvreur, P.; Helene, C. Adsorption of oligonucleotides onto polyisohexylcyanoacrylate nanoparticles protects them against nucleases and increases their cellular uptake. Pharm. Res. **1994**, *11*, 1370–1378.
101. Nakada, Y.; Fattal, E.; Foulquier, M.; Couvreur, P. Pharmacokinetics and biodistribution of oligonucleotide adsorbed onto poly(alkylcyanoacrylate) nanoparticles after intravenous administration to mice. Pharm. Res. **1996**, *13*, 38–43.
102. Schwab, G.; Chavany, C.; Duroux, I.; Goubin, G.; Lebeau, J.; Helene, C.; Saison-Behmoaras, T. Antisense oligonucleotides adsorbed to polyalkylcyanoacrylate nanoparticles specifically inhibit mutated Ha-ras-mediated cell proliferation and tumorigenicity in nude mice. Proc. Natl. Acad. Sci. USA **1994**, *91*, 10460–10464.
103. Lambert, G.; Bertrand, J.R.; Fattal, E.; Subra, F.; Pinto-Alphandary, H.; Malvy, C.; Auclair, C.; Couvreur, P. EWS Fli-1 antisense nanocapsules inhibits Ewing sarcoma–related tumor in mice. Biochim. Biophys. Res. Commun. **2001**, *279*, 401–406.
104. Youssef, M.; Fattal, E.; Alonso, M.J.; Roblot-Treupel, L.; Sauzière, C.; Omnes, A.; Couvreur, P.; Andremont, A. Effectiveness of nanoparticle-bound ampicillin in the treatment of *Listeria monocytogenes* infection in athymic nude mice. Antimicrob. Agents Chemother. **1988**, *32*, 1204–1207.
105. Fattal, E.; Youssef, M.; Couvreur, P.; Andremont, A. Treatment of experimental salmonellosis in mice with ampicillin-bound nanoparticles. Antimicrob. Agents Chemother. **1989**, *33*, 1540–1543.
106. Verdun, C.; Brasseur, F.; Vranckx, H.; Couvreur, P.; Roland, M. Tissue distribution of doxorubicin associated with polyisohexyl-cyanoacrylate nanoparticles. Cancer Chemother. Pharmacol. **1990**, *26*, 13–18.
107. Grislain, L.; Couvreur, P.; Lenaerts, V.; Roland, M. Pharmacokinetics and biodistribution of a biodegradable drug-carrier. Int. J. Pharmacol. **1983**, *15*, 335–345.
108. Peracchia, M.T.; Desmaële, D.; Vauthier, C.; Labarre, D.; Fattal, E.; D'Angelo, J.; Couvreur, P. Development of novel technologies for the synthesis of biodegradable pegylated nanoparticles. In *Targeting of Drugs 6: Strategies for Stealth Therapeutic Systems*; Gregoriadis, G., McCormack, B., Eds.; Plenum Press: New York, 1998; 225–239.

109. Illum, L.; Davis, S.S.; Muller, R.H.; Mak, E.;. West, P. The organ distribution and circulation of intravenously injected colloidal carriers sterically stabilized with a block copolymer–poloxamin 908. Life Sci. **1987**, *40*, 356–374.
110. Olivier, J.C.; Vauthier, C.; Taverna, M.; Puisieux, F.; Ferrier, D.; Couvreur, P. Stability of orosomucoid-coated poly(isobutylcyanoacrylate) nanoparticles in the presence of serum. J. Controlled Rel. **1996**, *40*, 157–168.
111. Peracchia, M.T.; Vauthier, C.; Passirani, C.; Couvreur, P.; Labarre, D. Complement consumption by poly(ethylene glycol) in different conformations chemically coupled to poly(isobutyl 2-cyanoacrylate) nanoparticles. Life Sci. **1997**, *61*, 749–761.
112. Brigger, I.; Chaminade, P.; Desmaële, D.; Peracchia, M.T.; D'Angelo, J.; Gurny, R.; Renoir, M.; Couvreur, P. Near infrared with principal component analysis as a novel analytical approach for nanoparticle technology. Pharm. Res. **2000**, *17*, 1124–1132.
113. Peracchia, M.T.; Harnisch, S.; Pinto-Alphandary, H.; Gulik, A.; Dedieu, J.C.; Desmaële, D.; D'Angelo, J.; Muller, R.H.; Couvreur, P. Visualization of in vitro protein-rejecting properties of PEGylated stealth polycyanoacrylate nanoparticles. Biomaterials **1999**, *20*, 1269–1275.
114. Peracchia, M.T.; Fattal, E.; Desmaële, D.; Besnard, M.; Noël, J.P.; Gomis, J.M.; Appel, M.; D'Angelo, J.; Couvreur, P. Stealth PEGylated polycyanoacrylate nanoparticles for intravenous administration and splenic targeting. J. Controlled Rel. **1999**, *60*, 121–128.

14
Preparation of Biodegradable Particles by Polymerization Processes

S. SLOMKOWSKI Center of Molecular and Macromolecular Studies, Lodz, Poland

I. INTRODUCTION

Standard procedures that have been used for many years for fabrication of polymeric objects usually consist of the following sequence of stages: polymerization, isolation and/or purification of synthesized polymer, followed by shaping obtained polymeric material into the desired form. In spite of many advantages of such approach, which has not only proved practical but in addition allowed relatively easy determination of fundamental reactions and physicochemical processes leading to final products, there is a growing interest in processes with all of the mentioned above steps proceeding in parallel. It is worth noting that all kinds of living organisms use the latter approach to build their bodies. Apparently, parallel synthesis of macromolecules and their self-assembly into desired objects could be less time consuming and more energy efficient. Moreover, recent studies of macromolecular self-assembly prove that this approach could be used for fabrication of nano- and micro-objects with control and accuracy not attainable in any other way (e.g., in miniemulsion, emulsion, and dispersion polymerizations [1]).

For many years biodegradable nanoparticles, microparticles, and microcapsules have been considered useful candidates for various medical applications, particularly in the area of drug delivery as injectable carriers or for pulmonary drug administration by inhalation. The unique properties of small-particle carriers are related to their large specific surface area and to the fact that their content is close to the interface. Thus, the release from small particles could be steadier than from the macroscopic objects (e.g., plates and slabs). Microparticles and microcapsules are obtained today from the earlier synthesized polymers either by any of the polymer solution emulsification processes followed by solvent evaporation/extraction or by spray drying of polymer solutions [2–8]. The above-mentioned processes were successfully used for formation of particles loaded with bioactive compounds. Proper selection of starting biodegradable polymer used as a matrix (its chemical nature and molecular weight) helped in controlling the rates of biodegradation.

When planning administration of microparticles into living organisms one has to take into account, among others, simple geometrical relations between dimensions of particles and dimensions of various tissue elements. On the basis of published data it is possible to make the following rough prediction for localization of microspheres introduced into the human body (cf. Table 1). The data in Table 1 indicate that selection of microparticles with proper size and size distribution is crucial. Microspheres with size in the wrong size window either would be removed by the mononuclear phagocyte system or would end at an undesirable location. In the case of microspheres with very broad size distribution, regardless of their average diameter, only a fraction would have appropriate dimensions. Therefore, one could expect that administration of drugs loaded into such microspheres would be less effective and might lead to some side effects (e.g., drugs delivered to the wrong tissue or organs). Unfortunately, control of diameters, and in particular of diameter distribution, of microspheres and microcapsules formed by the above-mentioned standard procedures from earlier synthesized polymers is difficult. On the other hand, in the past polymer chemists elaborated emulsion and dispersion polymerization techniques allowing synthesis of microspheres with predictable diameters and very narrow diameter distributions. Unfortunately, these processes were developed mainly for polymerizations of vinyl monomers carried on in protic (mainly water) media and yielding nonbiodegradable polymers. Nevertheless, several years ago we decided to explore the possibility of controlled synthesis of polyesters [poly(ε-

TABLE 1 Size of Particles and Their Possible Localization in Human Body

Diameter of microspheres, D (μm)	Localization in human body
$D > 10$	Particles too large to pass capillary blood vessels of many organs; localized activity
$5 < D < 10$	Particles captured in lung capillary blood vessels
$1 < D < 5$	Particles captured by the mononuclear phagocyte system
$1 < D < 3$	Particles captured in spleen
$0.1 < D < 1$	Particles captured in liver
$D < 0.1$	Particles captured in bone marrow

Source: Data from Ref. 6.

caprolactone) and poly(lactide)] yielding product in the form of microspheres. This chapter presents a summary of our studies.

II. BASIC CONCEPTS

Lactides and ε-caprolactone can be polymerized according to cationic, anionic, and pseudoanionic mechanisms. The mechanisms and kinetics of these processes have been investigated in detail [9–30]. All of these processes have to be carried out in organic media at anhydrous conditions because in ionic and pseudoionic polymerizations compounds with labile protons (e.g., water and alcohols) act as termination and/or chain transfer agents. Thus, all procedures developed for more than a century for emulsion and/or dispersion polymerizations carried out in water cannot be used for the analogous polymerizations of the above-mentioned cyclic esters.

There are two ways of stabilizing suspensions of colloidal particles: electrostatic and steric [31]. Repulsive forces between particles with the same electric charge are responsible for electrostatic stabilization. However, this type of interaction is possible only in media with very high dielectric constant (like water). In a majority of organic media the relatively (in relation to water) low dielectric constant reduces dissociation of ions, and in effect the ionic double layer at the surface of microspheres is too thin to induce strong repulsive forces at sufficiently long interparticle distances. Thus, steric stabilization is the only method suitable for synthesis of polyester microparticles. Steric stabilization results from desolvation and reduction of conformational freedom of molecular, oligomer, or polymer segments in overlapping interfacial layers of microspheres coming in close contact. Positive changes of free energy related to decreased distance between particles manifest themselves as repulsive forces.

When we begun our studies on dispersion polymerization of ε-caprolactone and lactides, very little was known about these processes. The literature contained a report on a patent to Union Carbide (from 1972) for the synthesis of polyester microspheres by ring-opening polymerization and many important issues were not clear [32]. For our studies we selected anionic and pseudoanionic polymerizations as the methods of choice. In recent years both types of polymerization of cyclic esters in solution and in bulk were investigated comprehensively in many (including ours) laboratories and main features of these processes were established [14–29]. Briefly, anionic polymerizations of ε-caprolactone could be initiated with alkali metal alcoholates. Monomer addition proceeds by nucleophilic attack of alcoholate anion onto carbonyl carbon atom in monomer molecule followed with acyl-oxygen bond scission and regeneration of alcoholate anion (cf. Scheme 1). However, it is necessary to remember that alcoholate active centers may also participate in side inter- and intramolecular transesterification reactions shown in Scheme 2. Intermolecular transesterification results in

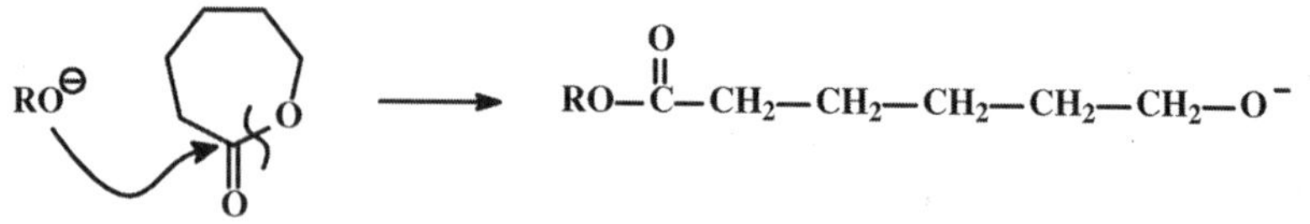

SCHEME 1

broadening of molecular weight distribution of synthesized polymer whereas intermolecular transesterification results in formation of cyclics [18,33–37].

Detailed kinetic studies revealed that in polymerizations with less reactive active centers (e.g., in pseudoanionic polymerization with aluminum alkoxide active centers) rate constants of the side transesterification reactions are much more strongly reduced than rate constants of propagation [18,37–40]. Propagation in pseudoanionic polymerization proceeds with coordination of monomer molecule with aluminum alkoxide active center followed with monomer insertion (cf. Scheme 3 illustrating propagation in pseudoanionic polymerization of ε-caprolactone initiated with diethylaluminum ethoxide). Pseudoanionic polymerizations are especially interesting because reduction of transesterification reactions often allows for practically complete monomer conversion before side reactions can have any effect on molecular weight and molecular weight distribution of synthesized polymer. Therefore, this type of polymerization has been chosen to begin our discussion on dispersion polymerization of ε-caprolactone.

Polymerizations of lactides were usually initiated with stannous octoate [tin(II) 2-ethoxyhexanoate]. For many years the mechanism of initiation and propagation in this system remained obscure. Recently, Duda and Penczek found that in polymerizations initiated with stannous octoate, protic impurities (e.g., traces of water or alcohols) act as coinitiators and active centers contain tin-alkoxide groups [28,41,42]. Studies of dispersion polymerization of lactides were carried out using tin octoate and tin alkoxides as initiators.

For dispersion polymerizations of ε-caprolactone and lactides it was necessary to choose a medium that would be a solvent for monomer and initiator and a nonsolvent for the corresponding polymer. Unfortunately, because no pure solvent suitable for anionic and/or pseudoanionic polymerization fulfilled the above-mentioned requirements, mixed solvents were used as reaction media. Polymerizations of lactides were carried out in heptane/1,4-dioxane 4:1 (v/v) mixtures, whereas for polymerizations of ε-caprolactone the most suitable mixture was heptane/1,4-dioxane 9:1 (v/v) [43].

Surface-active compound providing steric stabilization of poly(ε-caprolactone) and poly(lactide) microspheres was designed taking into account both chemical composition of polymers in microspheres and properties of the reaction media. It was assumed that the necessary properties could have poly(dode-

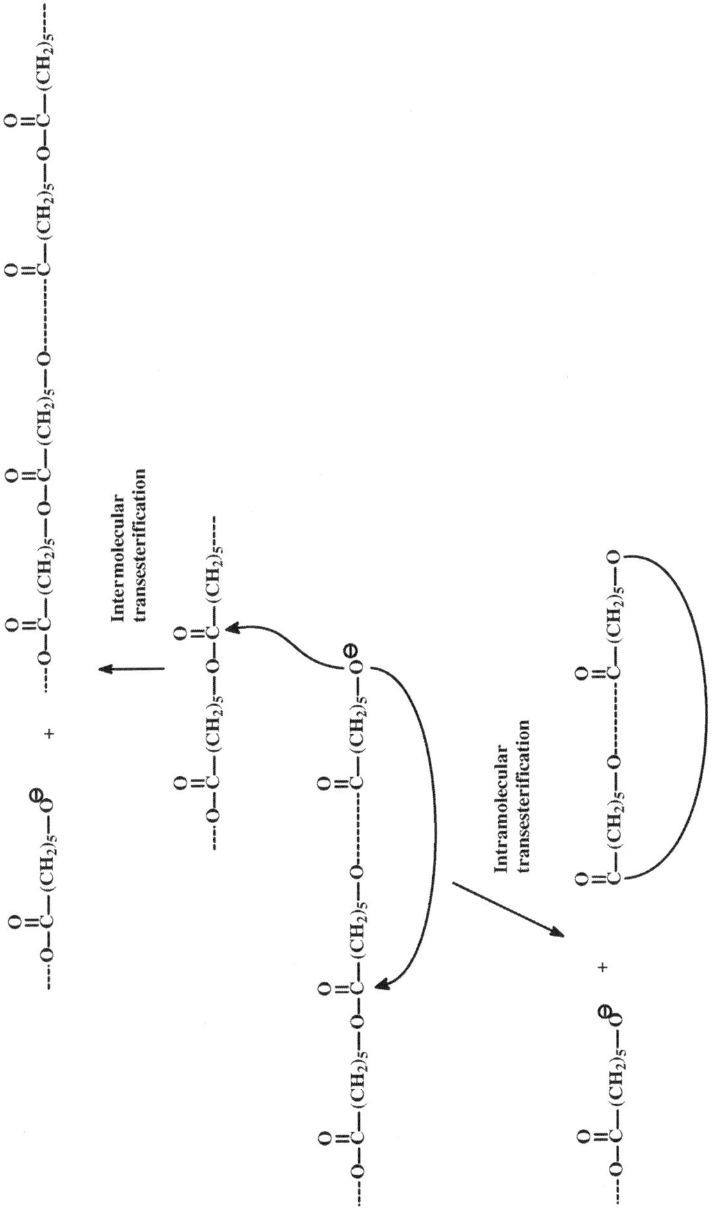
Intermolecular transesterification
Intramolecular transesterification

SCHEME 2

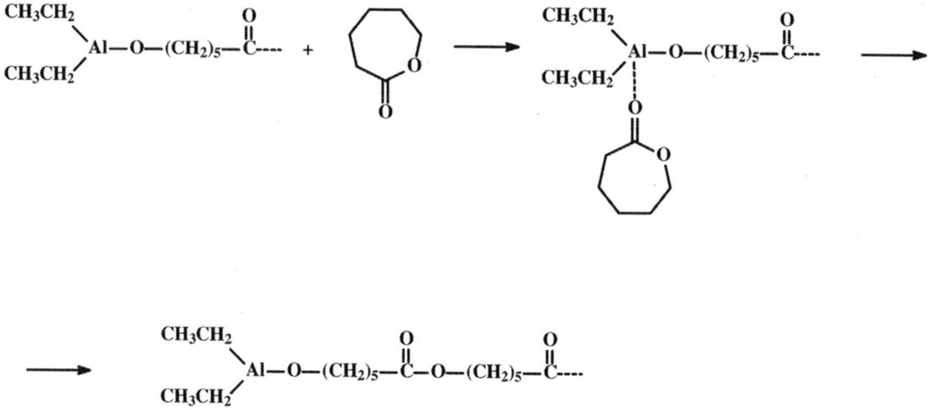

SCHEME 3

cyl acrylate) with poly(ε-caprolactone) grafts [43]. Poly(dodecyl acrylate) is highly soluble in heptane and heptane/1,4-dioxane mixtures, and thus should form the sterically stabilizing layer, whereas poly(ε-caprolactone) could act as an anchor binding poly(dodecyl acrylate) onto the surface of microspheres (cf. Scheme 4).

III. SYNTHESIS OF POLY(DODECYL ACRYLATE)-*g*-POLY(ε-CAPROLACTONE), A SURFACE-ACTIVE COPOLYMER FOR STABILIZATION OF POLY(ε-CAPROLACTONE) AND POLY(LACTIDE) MICROSPHERES

Poly(dodecyl acrylate)-*g*-poly(ε-caprolactone) was synthesized by radical copolymerization of dodecyl acrylate and poly(ε-caprolactone) methacrylate macromonomer. Synthesis of macromonomer and its copolymerization with dodecyl acrylate is illustrated in Scheme 5. A brief description (based on Ref. 43) is given below. Hydroxyl-terminated poly(ε-caprolactone) was synthesized by polymerization of ε-caprolactone (14.2 g, 0.125 mmol) initiated with $(CH_3CH_2)_2$-$AlOCH_2CH_3$ (0.785 g, 6 mmol). Polymerization was carried out for 8 h at 20°C in 145 mL of dry tetrahydrofuran (THF) in a sealed ampoule into which (before sealing) all reagents were introduced under vacuum conditions. Thereafter, ampoule was opened and active centers were killed by addition of acetic acid (fourfold excess with respect to initiator). Polymer was precipitated into cold (−30°C methanol); after isolation it was dissolved in chloroform and purified from aluminum acetate by column chromatography. Eventually, the hydroxyl-

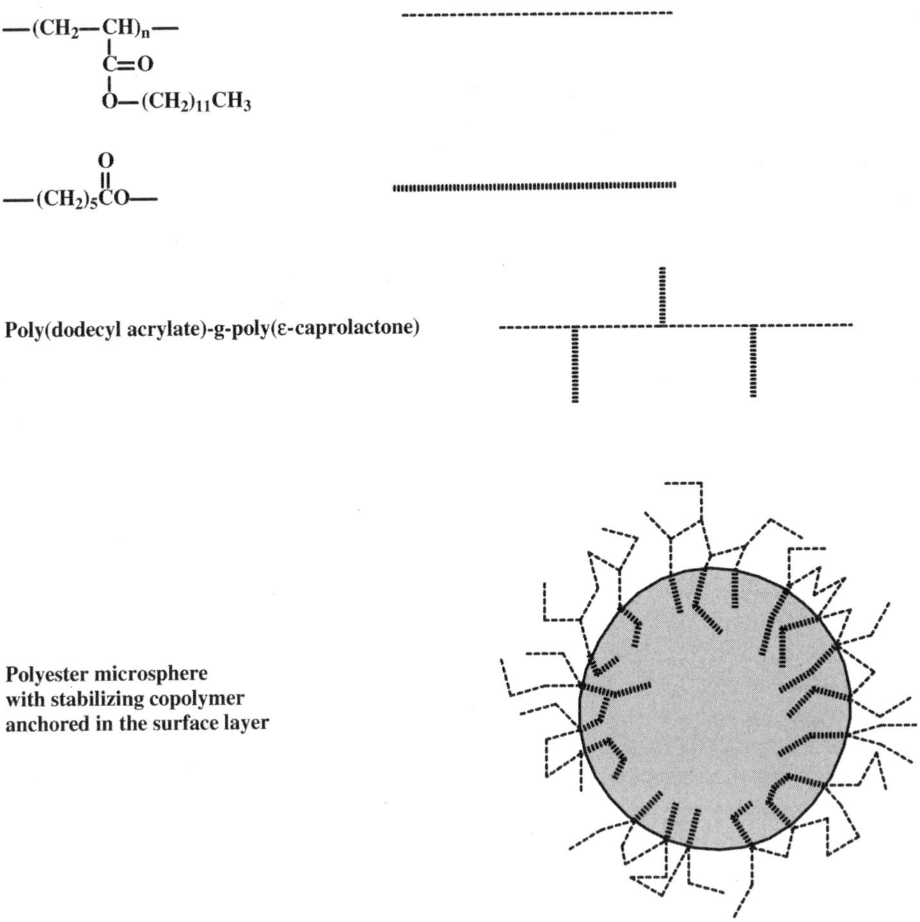

SCHEME 4

terminated poly(ε-caprolactone) was reprecipitated into cold methanol, isolated, and dried to the constant weight. The molecular weight of obtained polymer [gel permeation chromatography (GPC), calibration with poly(ε-caprolactone) samples with narrow molecular weight distribution] was $\overline{M}_n = 3090$. A sample (4.7 g) of hydroxyl-terminated poly(ε-caprolactone) (1.5 mmol of hydroxyl groups), 0.21 g of methacryloil chloride (2 mmol) reacted in the presence of 0.41 g of triethylamine in 5 mL of dry toluene for 48 h. After filtration of triethylamine hydrochloride, poly(ε-caprolactone) methacrylate macromonomer was precipitated into cold methanol and dried at room temperature under high

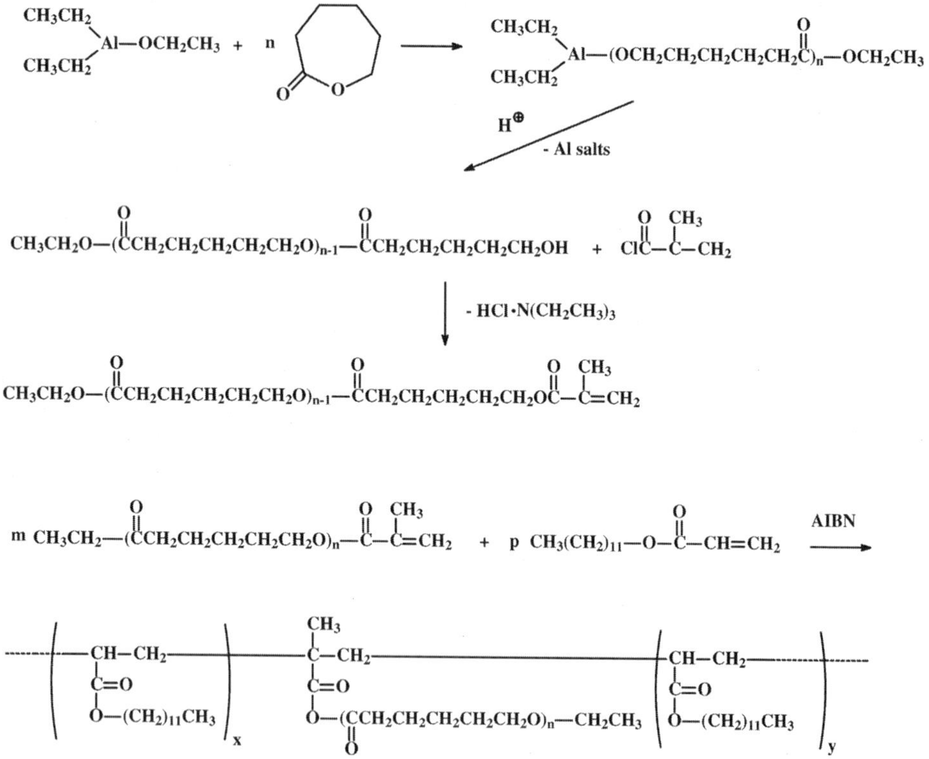

SCHEME 5

vacuum, with a yield of 3 g (64%), $\overline{M}_n = 3000$ ($\overline{M}_w/\overline{M}_n = 1.19$). The structure of poly(ε-caprolactone) methacrylate macromonomer was determined by ^{1}H NMR [$-CH_2CH_2CH_2CH_2\underline{CH_2}OC(O)-$(t) 4.06, $-\underline{CH_2}CH_2CH_2CH_2CH_2OC(O)-$(t) 2.33, $-CH_2\underline{CH_2}CH_2\underline{CH_2}CH_2OC(O)-$(m) 1.64, $-CH_2CH_2\underline{CH_2}CH_2CH_2OC(O)-$(m) 1.38, $-CH_2CH_2CH_2CH_2CH_2OC(O)-C(CH_3)=\underline{CH_2}$ 5.55 and 6.08].

Poly(ε-caprolactone) methacrylate (2.5 g), dodecyl acrylate (10 g), and azoisobutyronitrile (0.135 g) were dissolved in 15 mL of toluene and placed in an ampoule. After removal of air from the mixture by the freeze–thaw method, the ampoule was sealed off. Polymerization was carried out at 70°C for 72 h. Thereafter, polymer solution was diluted with 100 mL of toluene, precipitated into methanol, and isolated poly(dodecyl acrylate)-*g*-poly(ε-caprolactone) was dried under high vacuum, with a yield of 9.5 g (76%). In ^{1}H NMR spectrum of copolymer the signals of $-CH_3$ groups of dodecyl acrylate (t) were at 0.87, but signals of $-CH_2$ groups of dodecyl acrylate overlapped with signals of $-CH_2$

groups of poly(ε-caprolactone) grafts. The molecular weight of copolymer, determined by osmometry, was 49,000. Integration of a signal of CH_3 groups of poly(dodecyl acrylate) units and signal of $-CH_2\underline{CH_2}COO$ groups of poly(ε-caprolactone) allowed determination of molar fraction of poly(ε-caprolactone) monomeric units that for synthesized copolymer was f(ε-caprolactone) = 0.25. Knowledge of the molecular weight of poly(ε-caprolactone) macromonomer ($\overline{M}_n = 3000$), of poly(ε-caprolactone)-*g*-poly(dodecyl acrylate) ($\overline{M}_n = 49{,}000$), and the molar fraction of poly(ε-caprolactone) allowed calculation of the average number of poly(ε-caprolactone) grafts per copolymer macromolecule [NG = $\overline{M}_n$(copolymer) f(ε-caprolactone)/$\overline{M}_n$(poly(ε-caprolactone], which for described copolymer was 4.1.

Proper selection of initial concentrations of initiator [$(CH_3CH_2)_2AlOCH_2CH_3$] and monomer (ε-caprolactone) allowed one to obtain hydroxyl-terminated poly(ε-caprolactone) (block for grafts in copolymer) with $\overline{M}_n$ ranging from 1200 to 8800 [44]. Poly(ε-caprolactone)-*g*-poly(dodecyl acrylate) copolymers with molar fraction of poly(ε-caprolactone) kept in the relatively narrow range from 0.19 to 0.24 were obtained using the above-mentioned macromonomers. The molecular weight of copolymers varied from 22,000 to 50,000 and the average number of poly(ε-caprolactone) grafts from 0.6 to 6.9. It has to be noted that for these copolymers critical concentrations of micellization were strongly dependent on their composition. The lowest ones from 2 to 3 g/L were observed for copolymers with the ratio $\overline{M}_n$[poly(ε-caprolactone)]/$\overline{M}_n$ [poly(ε-caprolactone)-*g*-poly(dodecyl acrylate)] in the range 1.5–2.6.

IV. SYNTHESIS OF POLY(ε-CAPROLACTONE) AND POLY(LACTIDE) MICROSPHERES

A. Poly(ε-caprolactone) Microspheres

An example of a recipe for synthesis of poly(ε-caprolactone) microspheres by dispersion ring-opening pseudoanionic polymerization of ε-caprolactone (based on description in Ref. 43) is given below. Polymerization of 5.55 g (50 mmol) of ε-caprolactone initiated with 0.1 g (0.77 mmol) of $(CH_3CH_2)_2AlOCH_2CH_3$ was carried out in a mixture (100 mL) of dry (dried over sodium-potassium alloy) heptane and 1,4-dioxane (9:1; v:v) containing 0.22 g of poly(dodecyl acrylate)-*g*-poly(ε-caprolactone) (molecular weight of copolymer $\overline{M}_n$[poly(dodecyl acrylate)-*g*-poly(ε-caprolactone)] = 49,000, molecular weight of poly(ε-caprolactone) grafts $\overline{M}_n$[poly(ε-caprolactone)] = 3000, average number of poly(ε-caprolactone) grafts per copolymer molecule NG = 4.1). All reagents were charged into reaction vessel under vacuum conditions. Polymerization was carried out under argon at room temperature for 1 h. Thereafter, propagation was stopped by addition of 2.3 g (30 mmol, about tenfold molar excess with respect to substituents on Al) of CH_3CH_2COOH in 100 mL of heptane. After sedimentation under gravitational forces, microspheres were washed (resuspended in fresh portions of heptane and isolated by sedimentation) 10 times. Diameter and diameter polydispersity index [determined from scanning electron microscopy (SEM) images] of these microspheres were $\overline{D}_n = 630$ nm and $\overline{D}_w/\overline{D}_n = 1.038$, respectively. GPC measurements [columns calibrated using poly(ε-caprolactone) samples with narrow molecular weight distribution] revealed that $\overline{M}_n$ of polyester in microspheres was 8200 and polydispersity parameter $\overline{M}_w/\overline{M}_n$ was 1.11. Tonometric studies confirmed these findings, yielding $\overline{M}_n = 10{,}000$.

B. Poly(lactide) Microspheres

A typical procedure for the synthesis of poly(lactide) (regardless of whether from optically active, racemic, or meso monomers) is described in the following example based on reports in Refs. 43 and 44. Racemic mixture of L,L- and D,D-lactides (4 g) was dissolved in 72 mL of a mixture of dry (dried over sodium-potassium alloy) 1,4-dioxane and heptane (1:4, v/v) containing 0.16 g of dissolved poly(dodecyl acrylate)-*g*-poly(ε-caprolactone) [the same that was used in the previously described synthesis of poly(ε-caprolactone) microspheres]. This solution was saturated with argon, heated to boiling under reflux, and stirred at a rate of 60 revolutions/min. Separately a solution of stannous octoate [tin(II) 2-ethylhexanoate] (0.19 g) in 10 mL of a mixture of dry 1,4-dioxane and heptane (1:4, v/v) has been prepared under vacuum in an ampoule. The ampoule was attached to the flask with monomer solution and whole initiator was added rapidly. Stirring was maintained at 60 revolutions/min and polymerization was carried out at 95°C for 2 h. After this time the reaction mixture was rapidly cooled to about 55°C by rapid addition of 100 mL of cold heptane. After 3 h the microspheres were washed with small portions of heptane cooled to 5°C allowing for crystallization of unreacted monomer. Monomer crystals were removed by fractional sedimentations. In cold heptane the crystals collected at the flask bottom after about 2 min. Sedimentation of microspheres began after about 15 min. The above-described procedure used for purification of microspheres and, in particular, for removal of traces of unreacted monomer was repeated three times. The effective yield of poly(D,L-lactide) shaped into microspheres, after all purification steps, was 67.5%. The number averaged diameter ($\overline{D}_n$) of poly(D,L-lactide) microspheres determined by analysis of SEM images was found equal 2.50 μm. Diameter polydispersity was rather narrow, i.e., $\overline{D}_w/\overline{D}_n =$ 1.15.

Polymerization of lactides is an equilibrium process and there is always some amount of monomer present in the system [43–47]. GPC analysis of microspheres obtained in the synthesis described above revealed that the unreacted monomer content in particles was close to 8 wt %. GPC was also used for determination of molecular weight and molecular weight polydispersity index of poly(D,L-lactide) in microspheres [calibration based on poly(ε-caprolactone) samples with known molecular weight and narrow molecular weight distribution]. In microspheres synthesized according to the recipe given above $\overline{M}_n$ = 9400 and $\overline{M}_w/\overline{M}_n$ = 1.05. The narrow molecular weight distribution indicated that at the moment when polymerization has been stopped the whole system was still far from the full thermodynamic equilibrium.

^{1}H NMR spectra of dissolved microspheres indicated that the content of surface-active compound [poly(dodecyl acrylate)-*g*-poly(ε-caprolactone)] was lower than 1 wt % [43].

It must be noted that proper thermal treatment of poly(L,L-lactide) microspheres suspended in hydrocarbons allows one to control the degree of crystallinity from 0 to 60% [48]. Heating of microspheres suspended in dodecane for 20 min at 175°C followed by rapid cooling of this suspension to –30°C gave microspheres that were fully amorphous. Annealing microspheres suspended in heptane for required time periods at 80°C yielded particles with various degrees of crystallinity up to 60%.

V. DIAMETER POLYDISPERSITY OF MICROSPHERES

Control of diameter polydispersity is a key factor in all dispersion and emulsion polymerizations. Usually it depends on such parameters as monomer and initiator concentrations, and interfacial properties of surface-active agent (e.g., its critical concentration of micellization). For a given composition of the liquid phase, interfacial properties of surfactant usually strongly depend on its chemical structure. There are reports on using poly(dodecyl acrylate)-*g*-poly(ε-caprolactone)s with various ratios of molecular weights of poly(ε-caprolactone) grafts and total molecular weights of copolymer molecules in syntheses of poly(lactide) microspheres [44,49].

Figure 1 illustrates dependence of the optical density at 320 nm of samples poly(dodecyl acrylate)-*g*-poly(ε-caprolactone) [P(DA-CL)] with various ratios of $\overline{M}_n$(CL)/$\overline{M}_n$(DA-CL) dissolved in 1,4-dioxane-heptane (1:4 v/v) on P(DA-CL) concentration. The rapid increase of optical density indicates that copolymer concentration exceeded the critical concentration of micellization. Values of critical concentrations of micellization (ccm) for these copolymers are listed in Table 2.

Data in Table 2 indicate that ccm for copolymers with essentially the same total content of poly(ε-caprolactone) and with similar molecular weights (from 22,000 to 32,000) strongly depends on the $\overline{M}_n$(CL)/$\overline{M}_n$(DA-CL) ratio. Initially, when $\overline{M}_n$(CL)/$\overline{M}_n$(DA-CL) decreases ccm decreases as well to about 5 g/L for copolymers with $\overline{M}_n$(CL)/$\overline{M}_n$(DA-CL) close to 0.2. However, when $\overline{M}_n$(CL)/

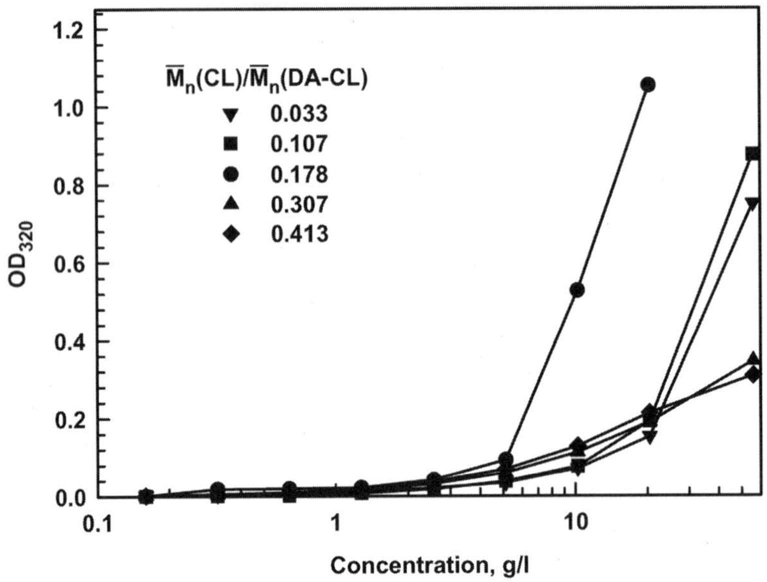

FIG. 1 Dependence of the optical density at 320 nm (OD_{320}) on the concentration of poly(dodecyl acrylate)-*g*-poly(ε-caprolactone); samples with ratio $\overline{M}_n$(CL)/$\overline{M}_n$(DA-CL) equal 0.033, 0.107, 0.178, 0.307, and 0.413. Conditions: 1,4-dioxane/heptane (1:4 v/v) solvent, 1 cm cell. (From Ref. 44.)

TABLE 2 Critical Concentrations of Micellization (ccm) of P(DA-CL) in 1,4-dioxane-heptane (1:4 v/v) for Copolymers with Various Ratios of $\overline{M}_n$(CL)/$\overline{M}_n$(DA-CL)[a]

$\overline{M}_n$(CL)/$\overline{M}_n$(DA-CL)	Weight fraction of poly(ε-caprolactone) in P(DA-CL)	ccm (g/L)
0.033	0.23	21
0.107	0.20	16
0.178	0.24	5
0.307	0.19	> 60 if any
0.413	0.24	> 60 if any

[a]Weight fraction of poly(ε-caprolactone) in these copolymers was in the narrow range from 0.19 to 0.24.
Source: Data from Ref. 44.

$\overline{M}_n$(DA-CL) becomes higher than about 0.3 the copolymers do not form micelles even at concentrations of 60 g/L. Apparently, copolymers with many short poly-(ε-caprolactone) chains are more soluble in a 1,4-dioxane/heptane mixture than those with one chain and thus form micelles inefficiently. On the other hand, copolymers with only one long poly(ε-caprolactone) chain per macromolecule do not form micelles. Probably in these molecules the long poly(ε-caprolactone) chains exceed a certain critical length, fold onto themselves, and become embedded in poly(dodecyl acrylate) long segments from the same copolymer macromolecule. Such behavior is characteristic, for example, for proteins in water solutions. Proteins in water are usually folded so as to hide their hydrophobic fragments inside and expose their hydrophilic segments to exterior. These hydrophilic envelopes limit their tendencies for aggregation. Schematically, the two above-discussed situations are illustrated in Fig. 2. In the figure, A corresponds to poly(DA-CL) having many short and thus soluble poly(ε-caprolactone) side chains and B corresponds to copolymer molecule with one long insoluble poly(ε-caprolactone) chain.

Poly(dodecyl acrylate)-*g*-poly(ε-caprolactone) copolymers with various $\overline{M}_n$(CL)/$\overline{M}_n$(DA-CL) ratios were used as stabilizers in dispersion polymerizations of racemic mixture of D,D- and L,L-lactide and in polymerization of the optically pure L,L-lactide enantiomer [43,44,49]. Dependence of the uniformity parameter [1/($\overline{D}_w/\overline{D}_n - 1$)] on $\overline{M}_n$(CL)/$\overline{M}_n$(DA-CL) ratio is shown in Fig. 3.

Figure 3 indicates that the lowest polydispersity of diameters of poly(lactide) [poly(D,L-lactide) and poly(L,L-lactide)] microspheres could be obtained for $\overline{M}_n$(CL)/$\overline{M}_n$(DA-CL) ratio close to 0.23. Indeed, in polymerization of racemic mixture of D,D-lactide and L,L-lactide carried on in the presence of P(DA-CL) with $\overline{M}_n$(CL)/$\overline{M}_n$(DA-CL) = 0.25, microspheres with very narrow diameter distributions were obtained ($\overline{D}_w/\overline{D}_n = 1.03$). The number average diameter of these particles was 2.71 μm. The SEM image of these uniform microspheres is shown in Fig. 4.

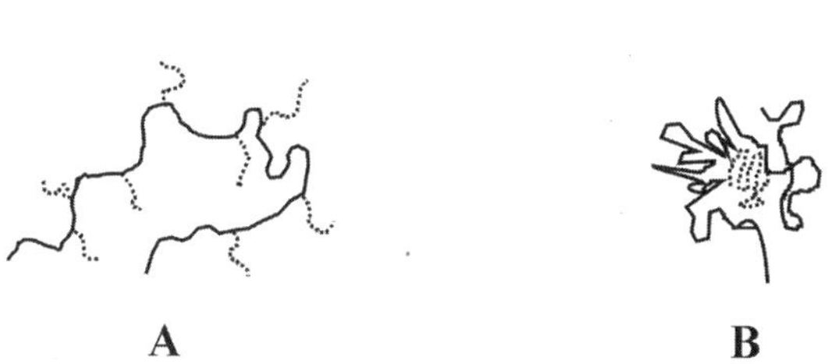

FIG. 2 Schematic illustration of poly(dodecyl acrylate)-*g*-poly(ε-caprolactone) copolymer chains in 1,4-dioxane/heptane mixture: A, a copolymer chain with many short poly-(ε-caprolactone) grafts; B, a copolymer with one long and insoluble poly(ε-caprolactone) side chain.

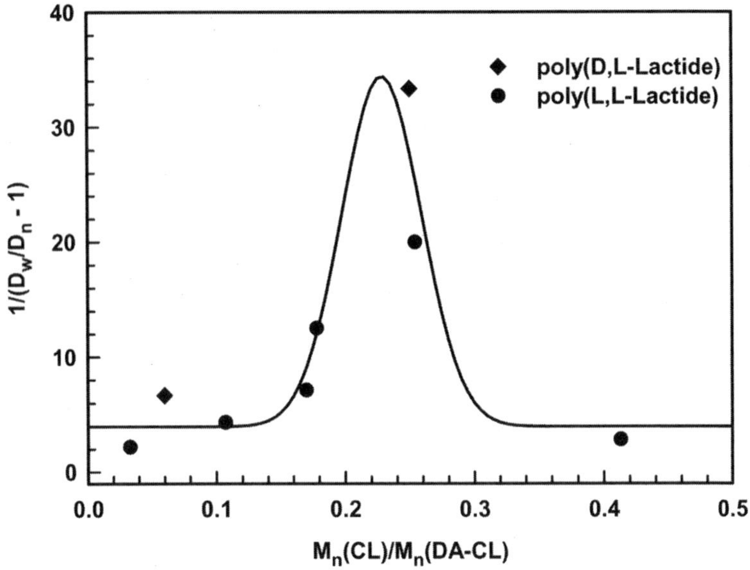

FIG. 3 Dispersion polymerizations of lactides (L,L-lactide and racemic mixture of D,D-lactide and L,L-lactide). Dependence of the diameter monodispersity parameter $(1/(\overline{D}_v/\overline{D}_n - 1))$ on the ratio of molecular weight of poly(ε-caprolactone) grafts and molecular weight of poly(dodecyl acrylate)-*g*-poly(ε-caprolactone) copolymers. (From Ref. 58.)

It was also important to find out to what extent diameter distribution of microspheres would be affected by concentration of P(DA-CL) surfactant. In a series of polymerizations of L,L-lactide initiated with stannous octoate and carried on in the 1,4-dioxane/heptane (1:4 v/v) medium, the concentration of P(DA-CL) $[\overline{M}_n(\mathrm{CL})/\overline{M}_n(\mathrm{DA\text{-}CL}) = 0.178]$ has been varied. Dependence of diameter polydispersity parameter $(\overline{D}_w/\overline{D}_n)$ and yield of polymer in the form of microspheres on concentration of P(DA-CL) is shown in Fig. 5. According to Fig. 5, $\overline{D}_w/\overline{D}_n$ was essentially independent of concentration of P(DA-CL) and varied in the range 1.06–1.08. However, it is important to stress that for P(DA-CL) concentrations lower than about 0.6 g/L the fraction of poly(L,L-lactide) in the form of microspheres decreased significantly. The remaining portion of polymer was in the form of a shapeless precipitate.

VI. CONTROL OF DIAMETERS OF MICROSPHERES

According to Table 1, for any medical applications tailoring of diameters of biodegradable particles would be of great importance. We noticed that the size

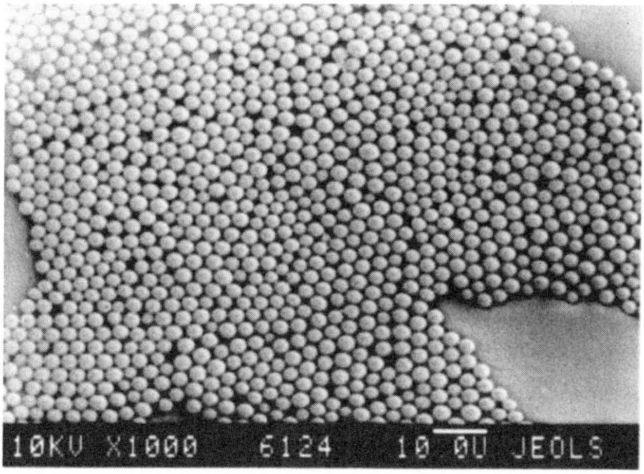

FIG. 4 SEM image of poly(D,D + L,L-lactide) microspheres. $\overline{D}_n = 2.71$ μm, $\overline{D}_w/\overline{D}_n = 1.03$. Conditions of polymerization: $[\text{D,D + L,L-lactide}]_0 = 0.28$ g/L, $[\text{stannous octoate}]_0 = 5.0 \times 10^{-3}$ mol/L, P(DA-CL) 1.6 g/L $[\overline{M}_n(\text{CL})/\overline{M}_n(\text{DA-CL}) = 0.25]$, 1,4-dioxane/heptane (1:4 v/v) medium, polymerization temperature 95°C, time of polymerization 2 h.

of microspheres is related to the initial monomer and initiator concentrations. Smaller microspheres could be obtained by stopping dispersion polymerization of lactide at the earlier stages. For example, for polymerizations L,L-lactide with initial monomer concentration equal 0.29 mol/L, initiated with 2,2-dibutyl-2-stanna-1,3-dioxepane ($[\text{2,2-dibutyl-2-stanna-1,3-dioxepane}]_0 = 8.0 \times 10^{-3}$ mol/L), diameters of microspheres formed at monomer conversion 90%, 71%, and 31% were 2.55, 2.37, and 1.76 μm, respectively. The increased initial initiator concentration also leads to particles with smaller diameters. Polymerization of lactide with the same initial monomer concentration as in experiments described above ($[\text{L,L-lactide}]_0 = 0.29$ mol/L) but with much higher concentration of initiator ($[\text{2,2-dibutyl-2-stanna-1,3-dioxepane}]_0 = 1.02 \times 10^{-2}$ mol/L) yielded particles with $\overline{D}_n = 1.14$ μm even for monomer conversion exceeding 99% [51].

Whereas synthesis of poly(lactide) particles with diameters ranging from about 1 to 4.0 μm was rather easy and could be accomplished by proper adjustment of monomer and initiator concentrations, obtaining larger particles during polymerization posed considerable difficulties. It is well known that radical dispersion and/or emulsion polymerizations with lower concentrations of surfactants (added or produced in situ) often yield microspheres with larger diameters. This is due to aggregation of the less stabilized primary particles at the earlier stages of the polymerization. A similar phenomenon has also been observed for dispersion polymerization of L,L-lactide ($[\text{L,L-lactide}]_0 = 0.278$ mol/L) initiated

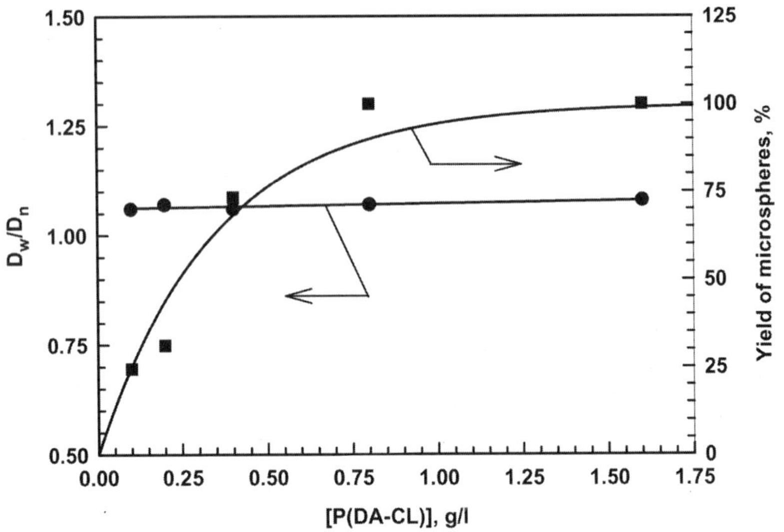

FIG. 5 Dependence of diameter polydispersity of poly(L,L-lactide) microspheres and percent of poly(L,L-lactide) in form of microspheres on concentration of poly(dodecyl acrylate)-*g*-poly(ε-caprolactone) (P(DA-CL)) surfactant. Conditions of polymerization: $[\text{L,L-lactide}]_0 = 0.277$ mol/L, $[\text{stannous octoate}]_0 = 4.9 \times 10^{-3}$ mol/L, 1,4-dioxane/heptane (1:4 v/v) medium, polymerization temperature 95°C, time of polymerization 2 h. (Based on data from Refs. 44 and 50.)

with stannous octoate ($[\text{stannous octoate}]_0 = 4.9 \times 10^{-3}$ mol/L). Figure 6 shows that for polymerizations with concentration of P(DA-CL) decreasing from 1.6 g/L to 0.1 g/L the diameters of microspheres ($\overline{D}_n$) increased to 4.8 μm. Unfortunately, as follows from Fig. 5, in polymerizations with lower concentrations of surfactants the fractions of polymer in the form of microspheres significantly decreases, making such approach somewhat impractical.

It has been noted that with increased monomer conversion diameters of microspheres increased gradually [51]. Thus, it was reasonable to expect that for higher initial monomer concentrations it might be possible to obtain larger particles. Unfortunately, limited solubility of L,L-lactide in 1,4-dioxane/heptane (1:4 v/v) mixture (maximal concentration about 0.35 mol/L) did not allow for this apparently straightforward approach. The above-mentioned problem has been solved by a stepwise monomer addition to the polymerizing mixture [52]. The first step was carried out with initial L,L-lactide concentration equal 0.35 mol/L ($[\text{stannous octoate}]_0 = 5.70 \times 10^{-3}$ mol/L, [poly(DA-CL)] = 1.67 g/L, (P(DA-CL) with $\overline{M}_n(\text{CL})/\overline{M}_n(\text{DA-CL}) = 0.18$). After 1.5 h a new monomer portion was added, making the overall concentration of introduced lactide equal to 0.624

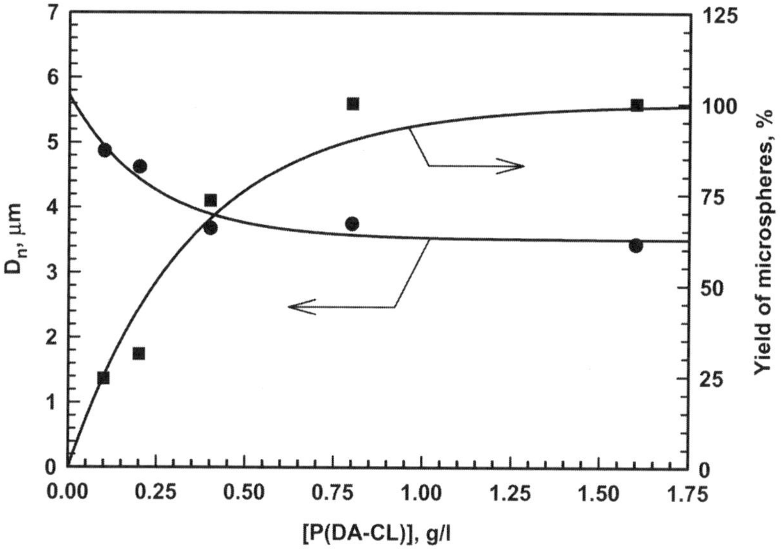

FIG. 6 Dependence of diameters of poly(L,L-lactide) microspheres and percent of poly-(L,L-lactide) in form of microspheres on concentration of poly(dodecyl acrylate)-*g*-poly(ε-caprolactone) [P(DA-CL)] surfactant. Conditions of polymerization: [L,L-lactide]$_0$ = 0.277 mol/L, [stannous octoate]$_0$ = 4.9 × 10^{-3} mol/L, 1,4-dioxane/heptane (1:4 v/v) medium, polymerization temperature 95°C, time of polymerization 2 h. (Based on data from Refs. 44 and 50.)

mol/L. A third portion of L,L-lactide was added, raising the concentration of introduced monomer to 0.90 mol/L. After each step a sample of polymerizing mixture was withdrawn and diameters of microspheres were determined from the SEM images. Figure 7 shows relations between the number averaged diameters and volumes of microspheres and the total monomer concentration after each step. The average volume of microsphere was calculated using the following equation:

$$\overline{V}_{\mathrm{n}} = \frac{\pi \Sigma_{i=1}^{N} D_i^3}{6N} \tag{1}$$

In Eq. (1), N denotes number of microspheres in the analyzed SEM picture. Indeed, according to Fig. 7, $\overline{D}_{\mathrm{n}}$ increased following each monomer addition, reaching a value 6.36 μm. It is also worth noting that the dependence of the average volume of microspheres ($\overline{V}_{\mathrm{n}}$) on total monomer concentration could be represented by a straight line. This suggests that each new added monomer portion was converted to polymer in the already existing particles, without lead-

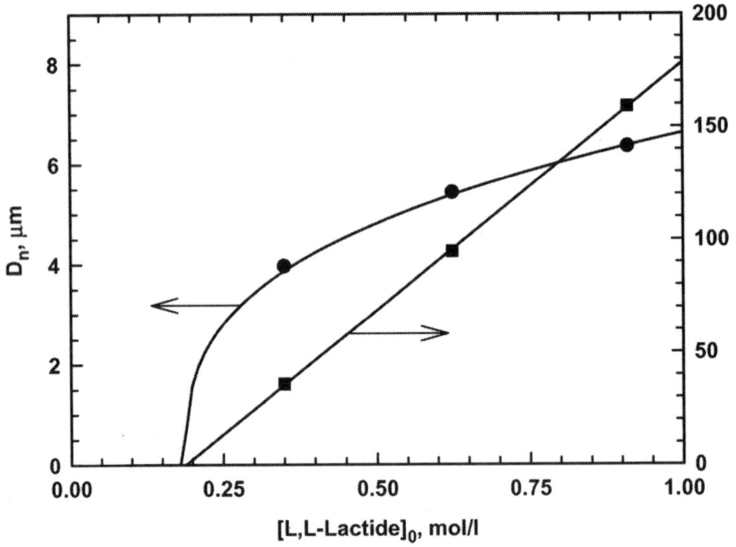

FIG. 7 Dependence of the number averaged diameters and volumes of poly(L,L-lactide) microspheres and total concentration of added monomer. Conditions of polymerization: [stannous octoate]$_0$ = 5.70×10^{-3} mol/L, [poly(DA-CL)] = 1.67 g/L, [P(DA-CL) with $\overline{M}_n$(CL)/$\overline{M}_n$(DA-CL) = 0.18]. (Based on data from Ref. 52.)

ing to formation of new ones. Taking into account this assumption, the dependence of $\overline{D}_n$ on total concentration of L,L-lactide introduced into polymerizing mixture [(L,L-lactide)$_0$] was fitted with the function $\overline{D}_n = a\sqrt[3]{[\text{L,L-lactide}]_0 - b}$ reflecting a linear dependence of $\overline{V}_n$ on [L,L-lactide]$_0$. Both lines, for $\overline{D}_n$ and $\overline{V}_n$, intersected abscissa at a concentration of L,L-lactide equal to 0.18 mol/L. Therefore, below this monomer concentration microspheres apparently could not be formed. Polymerization of lactides is a reversible process; even after long polymerization times residual monomer was always detected in polymerizing systems (regardless of whether polymerizations were carried out in solution or in dispersion). It is worth noting that the equilibrium monomer concentration evaluated from dependencies of $\overline{D}_n$ and $\overline{V}_n$ on [L,L-lactide]$_0$ was close to the value found in separate studies in which the remaining monomer concentration was measured by GPC (0.14 mol/L) [43,44].

From the slope of the dependence of $\overline{V}_n$ on [L,L-lactide]$_0$ it was also possible to estimate concentration of microspheres in polymerizing mixture expressed as number of particles per liter of suspension. From mass balance it follows that

$$Nd(\overline{V}_n)d_{\text{poly(L,L-lactide)}} = \text{FW}_{\text{L,L-lactide}}\, d([\text{L,L-lactide}]_0)10^{12} \quad (2)$$

where N denotes number of microspheres, $d(\overline{V}_n)$ change of the average volume of microsphere (in μm^3), $d_{poly(L,L\text{-lactide})}$ density of polymer (1.25 and 1.28 g/cm^3 for amorphous and crystalline poly(L,L-lactide), respectively), $FW_{L,L\text{-lactide}}$ formula weight for monomer (144.13), $d([L,L\text{-lactide}]_0)$ change of monomer conversion, and coefficient 10^{12} is due to unification of units. Rearrangement of Eq. (2) gives

$$N = \frac{FW_{L,L\text{-lactide}} 10^{12}}{d_{poly(L,L\text{-lactide})} \dfrac{d(\overline{V}_n)}{d([L,L\text{-lactide}]_0)}} \tag{3}$$

in which $d(\overline{V}_n)/d([L,L\text{-lactide}]_0)$ is a slope in the $\overline{V}_n = f([L,L\text{-lactide}]_0)$ plot.

This slope was equal to 220 $\mu m^3 \cdot$ L/mol (cf. Fig. 7), and taking the density of poly(L,L-lactide) at 1.28 g/cm^3 the concentration of microspheres in the above-described three-step polymerization was estimated at 5.24×10^{11} particles/L.

VII. MECHANISM OF FORMATION OF MICROSPHERES IN DISPERSION RING-OPENING POLYMERIZATION OF ε-CAPROLACTONE AND LACTIDES

Understanding how particles are formed and how they grow is essential for comprehensive description of dispersion polymerization of cyclic esters. In principle, the following basic mechanisms of particle formation and growth should be taken into account.

1. According to the first mechanism, initiation is slow. New chains are formed in solution throughout the polymerization process. In solution they grow fast, nucleating new particles or becoming adsorbed onto the already existing ones. Coalescence of particles is absent. In polymer particles due to the low local monomer concentrations propagation stops or proceeds very slowly. Because according to this model the new polymer chains and new particles are continuously formed in solution, particle diameters and polyester molecular weight distributions should be broad. For polymerizations proceeding according to this model, the concentration of initiator and/or propagating centers in the continuous phase should decrease slowly throughout the polymerization process.
2. The second mechanism is similar to the first, but with efficient coalescence of small, not properly stabilized particles until their total surface becomes small enough to be sufficiently saturated with surfactant, providing the necessary stabilization.

3. According to the third mechanism, all chains are initiated in the initial very short period. When growing chains reach the critical length they aggregate into nuclei of microspheres. The microspheres are swollen with monomer and further polymerization proceeds within these particles. Propagation stops when equilibrium monomer concentration is reached. In this type of polymerization all species suitable for reaction with monomer disappear from the liquid phase at a very early stage of polymerization. In an absence of particle coalescence all formed microspheres grow in the same manner. Their number should be constant and one could expect a narrow diameter distribution.
4. The last conceivable mechanism would be similar to the one described in point 3 but allowing for coalescence of particles. In the case of such polymerization, the diameter distribution should be broad.

The mechanism of particle formation and growth not only should affect particle diameter distribution but also, due to the different local concentrations of monomer and active species, should have an influence on the kinetics of polymerization and on molecular weights and molecular weight distributions in microspheres.

The question of whether initiator and propagating active centers are present for a long time in the continuous phase (as for polymerizations according to mechanisms 1 and 2) or are quickly transferred to microspheres has been answered on the basis of studies of the dispersion polymerization of ε-caprolactone initiated with diethylaluminum ethoxide [53]. During polymerization (initial monomer and initiator concentrations = 4.1×10^{-1} and 1.66×10^{-1} mol/L, respectively; reaction medium 1,4-dioxane/heptane 1:9 v/v, room temperature), samples of reaction mixture (known volume) were withdrawn at various moments. Each sample was added to heptane-containing acetic acid terminating the propagation. Microspheres were isolated by centrifugation and content of aluminum-containing compound in supernatant was determined by 8-hydroxyquinoline method [53]. Isolated microspheres were dissolved in THF and molecular weight of polymer that constituted these particles was determined by GPC [calibration with poly(ε-caprolactone) samples with narrow molecular weight distribution].

Figure 8 illustrates the dependence of concentration of aluminum-containing centers in supernatant on time of polymerization. The relation between molecular weight of poly(ε-caprolactone) in microspheres and fraction of active centers in microspheres (assuming that fraction of active centers in microspheres equals $([AC(S)]_0 - [AC(S)])/[AC(S)]_0$, where $[AC(S)]_0$ and $[AC(S)]$ denote the initial and actual concentration of active centers in liquid phase, respectively, is shown in Fig. 9.

Following are discussed characteristic features of relations illustrated by Figs. 8 and 9. After 150 s from the beginning of initiation, the fraction of active

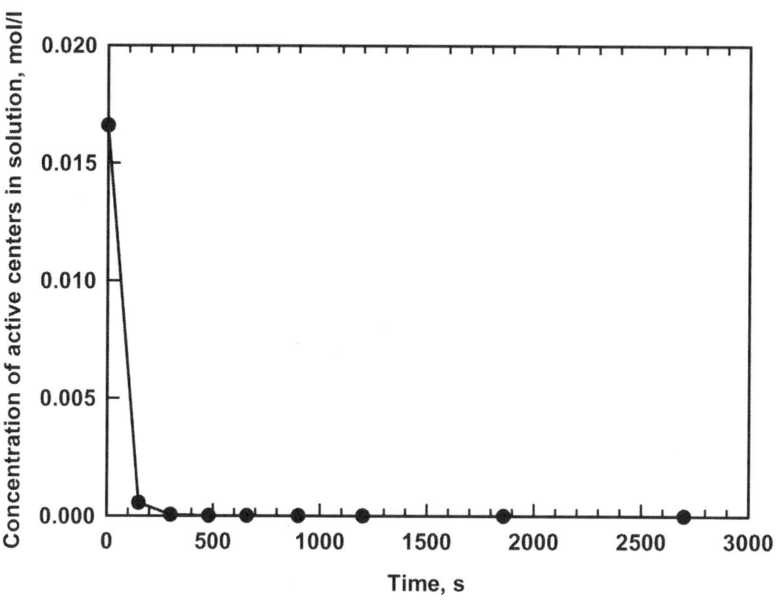

FIG. 8 Polymerization of ε-caprolactone. Dependence of concentration of propagating species in solution and time of polymerization. Polymerization conditions: [ε-caprolactone]$_0$ = 4.1×10^{-1} mol/L, [diethylaluminumalkoxide]$_0$ = 1.66×10^{-1}, mol/L, P(DA-CL) 1.6 g/L [$\overline{M}_n$(CL)/$\overline{M}_n$(DA-CL) = 0.125]. (From Ref. 53.)

centers in microspheres exceeds 96%. Moreover, number average molecular weight of poly(ε-caprolactone) at this stage constitutes only 18% of molecular weight of polymer in microspheres after complete monomer conversion. These findings indicate that in dispersion polymerization of ε-caprolactone propagation of all chains is initiated very early and that essentially all growing chains shortly after initiation, when their molecular weight is in the range 500–1000, become incorporated in microspheres. Subsequently, further propagation proceeds in the already formed microspheres. Thus, the above-described experiments conform to mechanisms 3 and/or 4, describing formation and growth of microspheres, and definitely exclude mechanisms 1 and 2.

Discrimination between mechanisms 3 and 4 was possible after monitoring changes of the concentration of microspheres with time. For dispersion polymerizations of ε-caprolactone and L,L-lactide initiated with anionic and pseudoanionic initiators, the number of microspheres in a given volume of polymerizing mixtures was determined at various monomer conversions [54–57].

Microspheres were counted on Bürker's plate using an optical microscope. Bürker's plate is a device commonly used in medicine and biology for counting blood cells and other cells suspended in a liquid environment. It consists of a

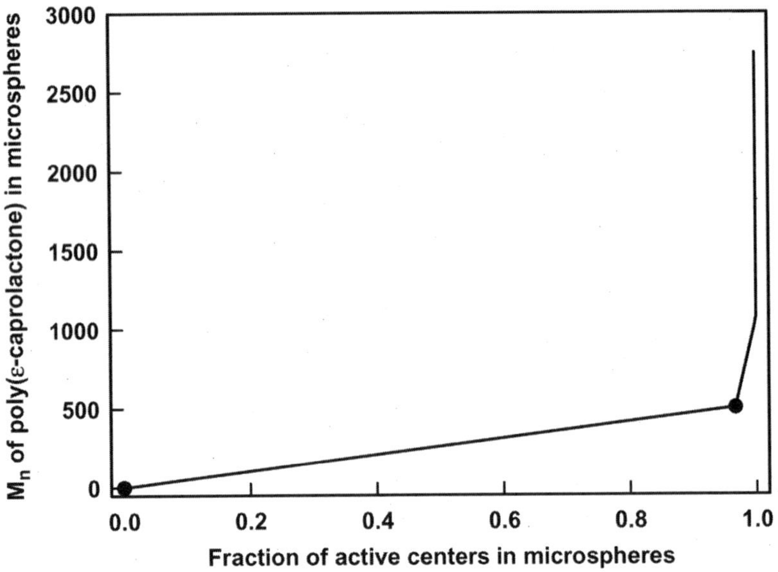

FIG. 9 Polymerization of ε-caprolactone. Relation between molecular weight of poly-(ε-caprolactone) in microspheres and fraction of active centers in microspheres. Polymerization conditions the same as for Fig. 8. (From Ref. 53.)

quartz or glass plate with wells with perpendicular walls and of a cover slide that when placed on the plate closes tops of the wells, thus limiting their volumes. A drop of polymerizing mixture placed on the plate and covered with the cover glass is divided into several wells, allowing for several parallel determinations and thus for determination of the average particle numbers with better precision.

In spite of limitations in using optical microscopy for studies of microspheres with diameters smaller than a few micrometers (due to diffraction of light at the edges of light-scattering objects, spatial dimensions of particles with diameters comparable or only slightly larger than the light wavelength cannot be determined with a reasonable precision), this method can be used for detection of particles with diameters as small as 0.1 μm. Thus, analysis of optical microscopy pictures of wells in the Bürker's plate filled with samples of polymerizing mixture allowed determinations of particle concentration. Examples of results of such determinations for polymerizations of ε-caprolactone and L,L-lactide are shown in Fig. 10. Plots in the figure revealed that for monomer conversions higher than about 20% of monomer conversion the number of particles did not change. Thus, any coalescence of particles that occurred in dispersion polymerization of ε-caprolactone and/or lactide would have happened only at the initial

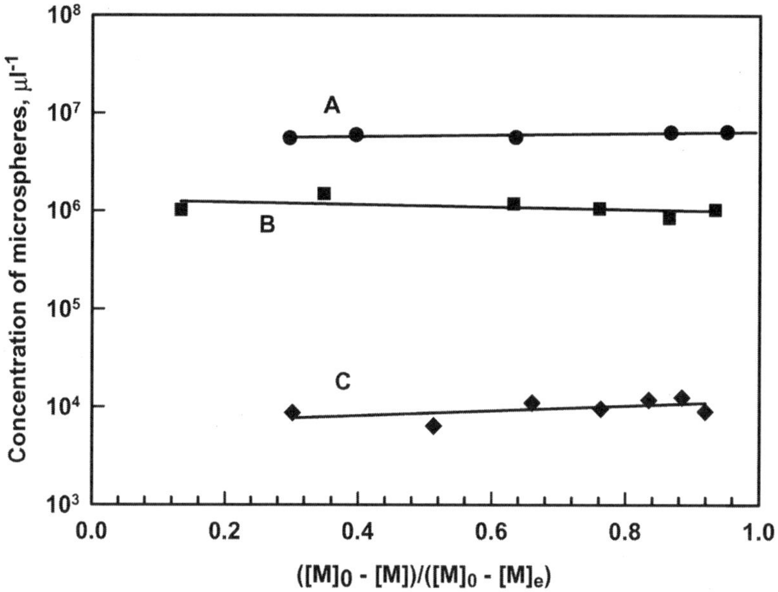

FIG. 10 Dependence of concentration of microspheres on normalized monomer conversion. A, [ε-caprolactone]$_0$ = 4.3×10^{-1} mol/L, $[(CH_3CH_2)_2AlOCH_2CH_3]_0 = 5.60 \times 10^{-3}$ mol/L; B, [ε-caprolactone]$_0$ = 4.2×10^{-1} mol/L, $[(CH_3)_3SiONa]_0 = 1.83 \times 10^{-3}$ mol/L; C, [L,L-lactide]$_0$ = 4.4×10^{-1} mol/L, [Tin(II) 2-ethylhexanoate]$_0$ = 7.16×10^{-4} mol/L. (From Ref. 56.)

stage. Therefore, all results of the above-mentioned studies conformed to the third mechanism listed at the beginning of this section.

Comprehensive description of particle formation and growth requires finding a law to determine concentrations of microspheres in a polymerizing system. For polymerizations of L,L-lactide carried out at various ratios of monomer and initiator concentrations, a linear relationship has been observed between the average mass of microsphere ($\overline{Mm}_n$) and mass of monomer molecules per mole of propagating chains [57]. An example of such dependence for polymerization of L,L-lactide initiated with 2,2-dibutyl-2-stanna-1,3-dioxepane is shown in Fig. 11 in which $\overline{Mm}_n$ has been plotted as a function of ([L,L-Lc]$_0$ − [L,L-Lc]$_e$)144.13/$[I]_0$. It must be remembered that ([L,L-Lc]$_0$ and [L,L-Lc]$_e$ denote the initial and equilibrium monomer concentrations, $[I]_0$ denotes the initial concentration of initiator which for quantitative initiation is equal to concentration of growing chains, and 144.13 is the molar mass of L,L-lactide. The ratio of the average mass of microsphere and mass of momomer molecules per mole of growing

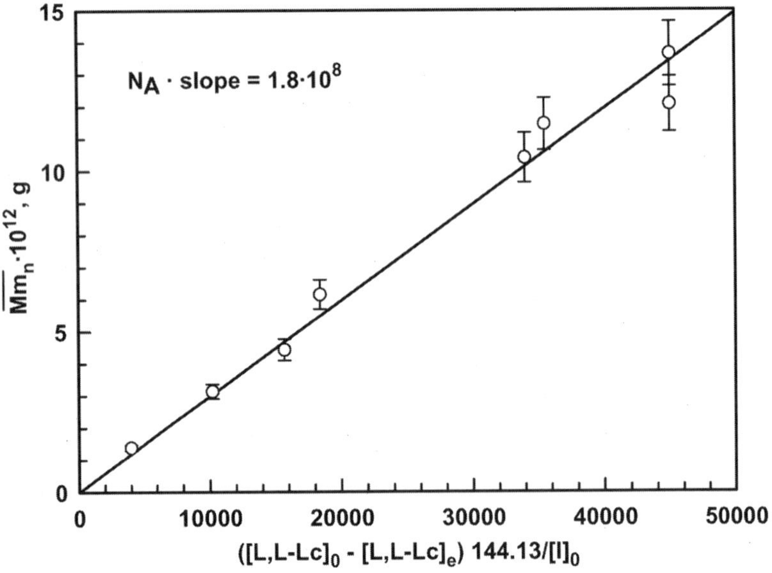

FIG. 11 Number-averaged mass of microsphere as a function of mass of monomer converted to polymer per mole of growing chains $(([\text{L,L-Lc}]_0 - [\text{L,L-Lc}]_e)144.13/[I]_0)$. Dispersion polymerization of L,L-lactide initiated with 2,2-dibutyl-2-stanna-1,3-dioxepane. (From Ref. 57.)

chains equals the number of moles of growing chains per microsphere. Thus, the average number of growing chains per microsphere (N_C) equals:

$$N_C = N_A 144.13\ [I]_0 \frac{\overline{Mm_n}}{[\text{L,L-Lc}]_0 - [\text{L,L-Lc}]_e} \tag{4}$$

where N_A is Avogadro's number.

From the plot in Fig. 11 in which slope equals 3×10^{-16} $(144.13[I]_0\overline{Mm_n}/([\text{L,L-Lc}]_0 - [\text{L,L-Lc}]_e))$ the average number of growing chains per microsphere in all polymerizations was 1.8×10^8. The linear dependence of $\overline{Mm_n}$ on $([\text{L,L-Lc}]_0 - [\text{L,L-Lc}]_e)144.13/[I]_0$ suggests that regardless of concentration of growing chains (i.e., initiator concentration) the number of propagating chains per microsphere was similar. Indeed, a plot of N_C as a function of the initial initiator concentration indicated that for initial concentrations of 2,2-dibutyl-2-stanna-1,3-dioxepane varied from 8.0×10^{-4} mol/L to 1.02×10^{-2} mol/L the average number of growing chains in microsphere varied in a narrow range from 1.61×10^8 to 1.99×10^8 (cf. Fig. 12, [57]).

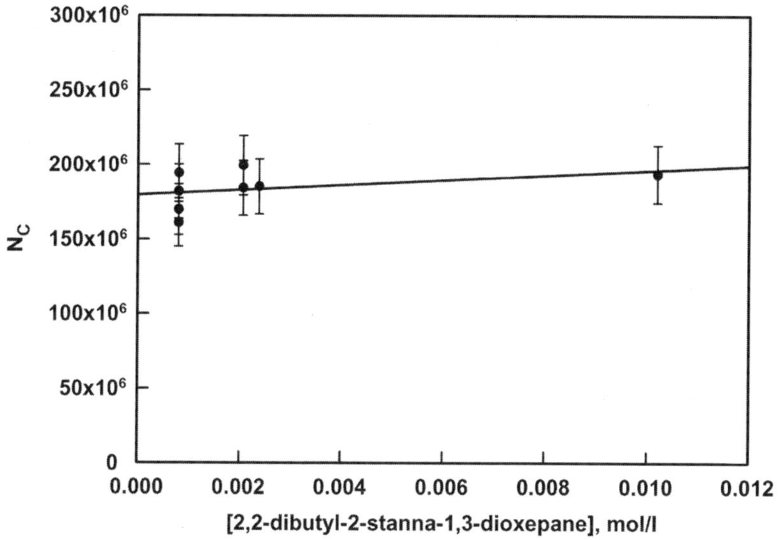

FIG. 12 Relation between anumber of propagating chains per microsphere (N_C) and concentration of initiator (2,2-dibutyl-2-stanna-1,3-dioxepane) in dispersion polymerization of L,L-lactide. (From Ref. 57.)

At first it seems strange that at the beginning of polymerization, in spite of various concentrations of initiator, always the same very large number of propagating chains (about 2×10^8) form the nucleus of a microsphere. However, it could be shown that such a characteristic feature conforms to a rather simple model describing nucleation of microspheres. According to this model, at the stage at which growing chains reach their critical length (for polymerization of ε-caprolactone when the molecular weight of macromolecules is between 500 and 1000) they disappear from solution in one of the following two processes: nucleation of new microspheres or adsorption onto existing ones. Nucleation of new microspheres could be described by a differential equation corresponding to a process in which collision between two macromolecules creates nuclei of a microsphere:

$$\frac{dN}{dt} = \frac{1}{2} k_1 n^2 \tag{5}$$

In Eq. (5) dN/dt denotes the rate of formation of new microspheres (N represents the concentration of microspheres expressed as number of microspheres in a unit volume), n the concentration of growing chains (also expressed as number of chains in a unit volume), and k_1 the rate constant of microsphere nucleation. The coefficient ½ reflects creation of a new microsphere by collision of two chains.

The rate at which growing chains become incorporated into microspheres (dn/dt) could be described as a sum of the rate at which chains disappear from solution due to nucleation of microspheres (R_1) and the rate at which chains are removed from solution due to adsorption onto the existing particles (R_2).

$$R_1 = k_1 n^2 \tag{6}$$

$$R_2 = k_A N S n \tag{7}$$

In Eq. (7), k_A denotes rate of adsorption and S the average surface of one microsphere (i.e., R_2 is proportional to the total surface of microspheres in a unit volume of propagating mixture). The relation between surface and volume of spherical particle is as follows:

$$S = \sqrt[3]{36\pi V^2} \tag{8}$$

Based on the simplifying assumption that formation of particles is so fast that during completion of this process propagating chains do not considerably change in length (molecular weight of aggregating chains was found to be in the range from 500 to 1000; in further estimations it was taken equal 750), it was possible to express the average volume of nucleated microsphere as:

$$V = \frac{750}{N_A d} \frac{n - n_0}{N} \tag{9}$$

In Eq. (9), N_A denoted Avogadro's number, d density of polymer in microspheres [e.g., for poly(lactide) 1.25 g/cm^3], N concentration of microspheres, n and n_0 concentration of aggregating chains at time t and at the beginning of aggregation, respectively.

Combining Eqs. (7) and (9), one could write:

$$R_2 = 400 k_A \frac{1}{\sqrt[3]{N_A^2 d^2}} \frac{n\sqrt[3]{(n - n_0)^2}}{\sqrt{N}} \tag{10}$$

Introducing $k_2 = 400k_A/\sqrt[3]{N_A^2 \mathrm{d}^2}$, R_2 could be expressed as:

$$R_2 = k_2 \frac{n\sqrt[3]{(n - n_0)^2}}{\sqrt{N}} \tag{11}$$

and the total rate with which polymer chains disappear from solution ($dn/dt = R_1 + R_2$) would be described by an equation:

$$\frac{dn}{dt} = -k_1 n^2 - k_2 \frac{n\sqrt[3]{(n - n_0)^2}}{\sqrt{N}} \tag{12}$$

Numerical solution of a set of Eqs. (5) and (12), for any given initial concentration of chains (n_2) and values of k_1 and k_2, allows calculations of changes of a number of microspheres during a process of particle formation. Results of such calculations (with $k_1 = 300$ L/mol·s and $k_2/k_1 = 7.36 \times 10^4$) for initial concentration of chains (n_0) varied from 10^{-3} mol/L to 5×10^{-2} mol/L are shown in Fig. 13. Plots in the figure indicate that in the chosen range of propagating chains concentrations (corresponding to a typical range of initiator concentrations) the final concentrations of microspheres are the same and equal 1.8×10^8 (as in experimental studies of the above-described polymerization of L,L-lactide).

VIII. SPECIFIC FEATURES OF KINETICS OF DISPERSION RING-OPENING POLYMERIZATION OF ε-CAPROLACTONE AND LACTIDES

Since in ionic and pseudoionic dispersion polymerization of cyclic esters all propagating chains are located in microspheres that occupy only a relatively small fraction of the total volume of reacting mixture (after a short induction

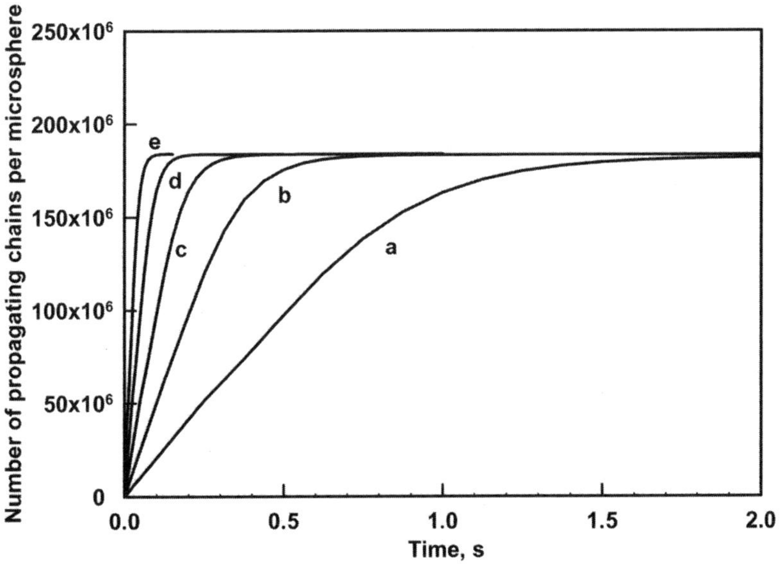

FIG. 13 Number of chains per microsphere as a function of time. Calculations based on Eqs. (5) and (12), values of kinetic parameters: $k_1 = 300$ L/mol × s and $k_2/k_1 = 7.36 \times 10^4$. Initial concentrations of chains: a, 1.0×10^{-3} mol/L; b, 2.5×10^{-3} mol/L; c, 5.01×10^{-3} mol/L; d, 1.0×10^{-2} mol/L; e, 2.0×10^{-2} mol/L.

period when particles are nucleated), kinetics of this process should differ from kinetics of homogeneous polymerizations in solution and/or in bulk in which active centers are uniformly distributed over the whole volume. Kinetics of ionic and pseudoionic dispersion polymerizations should also be different from the standard radical emulsion and dispersion polymerizations because in radical polymerizations propagating species are produced continuously throughout the polymerization process, often in a liquid phase, and primary radicals enter into the growing particles during polymerization. (In some systems there is a certain probability of their escape from microspheres.)

Formal kinetics of polymerization in solution has been well developed in recent decades and has been presented in many textbooks. The kinetics of radical dispersion polymerizations in heterogeneous and "miniheterogeneous" systems (emulsion, miniemulsion, dispersion polymerizations), albeit more complicated, has been discussed in many textbooks and monographs as well (e.g., [1]). However, the kinetics of ionic and/or pseudoionic polymerizations still awaits thorough analysis. Below is given a basic description of kinetic relations characterizing dispersion ionic and/or pseudoionic processes from the moment when microspheres have been nucleated. This description is based on analysis discussed in Ref. 54.

In dispersion polymerization the total volume of reaction mixture (V) is a sum of the volumes of microspheres (V_m) and of a liquid phase (V_s):

$$V = V_m + V_s \tag{13}$$

Similarly, the total number of moles of monomer (m) partitioned between the microspheres (m_m) and the liquid phase (m_s) is derived as follows:

$$m = m_m + m_s \tag{14}$$

Monomer concentration averaged over the whole volume of polymerizing mixture ($[M]$) is described by:

$$[M] = \frac{m}{V} = \frac{V_m}{V}[M_m] + \frac{V_s}{V}[M_s] \tag{15}$$

In Eq. (15) $[M_m]$ and $[M_s]$ describe monomer concentrations in particles and in the liquid phase, respectively.

The concentration of active centers averaged over the whole volume of polymerizing mixture ($[I]_0$) is derived as follows:

$$[I]_0 = \frac{i}{V} = \frac{V_m}{V}[I_m] \tag{16}$$

where i and $[I_m]$ ($[I_m] = i/V_m$) denote number of moles of propagating species (in case of quantitative initiation with initiator with functionality 1 equal to the initial initiator concentration) and their concentration in microspheres.

Differentiation of Eq. (15) with respect to time (the simplifying assumption has been made that contraction of the volume of polymerizing mixture equals 0 and thus, V is constant) yields:

$$\frac{d[M]}{dt} = \frac{1}{V}\frac{d(V_{\mathrm{m}}[M_{\mathrm{m}}])}{dt} + \frac{1}{V}\frac{d(V_{\mathrm{s}}[M_{\mathrm{s}}])}{dt} \tag{17}$$

When all propagating chains are already inside of microspheres, changes of monomer concentration in solution are due only to monomer diffusion from the liquid phase to microspheres and from microspheres to the liquid phase. Thus, for $d(V_{\mathrm{s}}[M_{\mathrm{s}}])/dt$, one could write a formula:

$$\frac{d(V_{\mathrm{s}}[M_{\mathrm{s}}])}{dt} = -S_{\mathrm{m}}F_{\mathrm{s,m}}\,[M_{\mathrm{s}}] + S_{\mathrm{m}}F_{\mathrm{m,s}}\,[M_{\mathrm{m}}] \tag{18}$$

in which S_{m} denotes total surface of microspheres and $F_{\mathrm{s,m}}$ and $F_{\mathrm{m,s}}$ coefficients determining flux of monomer from liquid phase to microspheres and from microspheres to liquid phase, respectively.

The differential equation describing changes of monomer concentration inside of microspheres could be written as follows:

$$\frac{d(V_{\mathrm{m}}[M_{\mathrm{m}}])}{dt} = -V_{\mathrm{m}}k_{\mathrm{p,m}}^{\mathrm{app}}[I_{\mathrm{m}}][M_{\mathrm{m}}] + S_{\mathrm{m}}\;F_{\mathrm{s,m}}[M_{\mathrm{s}}] - S_{\mathrm{m}}F_{\mathrm{m,s}}\,[M_{\mathrm{m}}] \tag{19}$$

The first term in the right side of this equation corresponds to monomer consumption due to polymerization inside of microspheres; the second term characterizes flux of monomer from liquid phase into particles; and the third term describes a flux of monomer from microspheres to the liquid phase. In Eq. (19), $k_{\mathrm{p,m}}^{\mathrm{app}}$ denotes apparent propagation rate constant in particles. This constant is called apparent because it is a function of rate constants of propagation involving various physical forms of active centers (e.g., ions, ion pairs, and ionic aggregates in ionic polymerization, and monomeric and aggregated propagating species in the pseudoionic process) and of the equilibrium constant between these species.

Substituting expressions (18) and (19) in Eq. (17) and taking into account Eq. (16) yields:

$$\frac{d[M]}{dt} = -k_{\mathrm{p,m}}^{\mathrm{app}}[I]_0[M_{\mathrm{m}}] \tag{20}$$

According to Eq. (20), the rate of monomer consumption averaged over the whole volume of reaction mixture depends on the average concentration of active centers ($[I]_0$) and on the local monomer concentration in microspheres ($[M_{\mathrm{m}}]$).

At stationary state $d(V_{\mathrm{m}}[M_{\mathrm{m}}])/dt = 0$ and, thus, from Eq. (19) and formula (16), the following relation follows:

$$[M_{\mathrm{m}}] = \frac{S_{\mathrm{m}}\ F_{\mathrm{s,m}}}{k_{\mathrm{p,m}}^{\mathrm{app}} V[I]_0 + S_{\mathrm{m}}\ F_{\mathrm{m,s}}}\ [M_{\mathrm{s}}] \tag{21}$$

Combination of Eqs. (13), (15), and (21) gives:

$$[M_{\mathrm{m}}] = \frac{S_{\mathrm{m}}\ F_{\mathrm{s,m}}}{S_{\mathrm{m}}[F_{\mathrm{m,s}} + \alpha(F_{\mathrm{s,m}} - F_{\mathrm{m,s}})] + (1 - \alpha)k_{\mathrm{p,m}}^{\mathrm{app}}\ [I]_0}\ [M] \tag{22}$$

In Eq. (22), $\alpha = V_{\mathrm{m}}/V$ denotes volume fraction occupied by microspheres. Substituting the right-hand side of Eq. (22) in Eq. (20) gives:

$$\frac{d[M]}{dt} = -\frac{S_{\mathrm{m}}\ F_{\mathrm{s,m}}\ k_{\mathrm{p,m}}^{\mathrm{app}}}{S_{\mathrm{m}}[F_{\mathrm{m,s}} + \alpha(F_{\mathrm{s,m}} - F_{\mathrm{m,s}})] + (1 - \alpha)k_{\mathrm{p,m}}^{\mathrm{app}}[I]_0}\ [I]_0[M] \tag{23}$$

Introducing a parameter $k_{\mathrm{p,d}}^{\mathrm{app}}$ that could be called the apparent propagation rate constant in dispersion,

$$k_{\mathrm{p,d}}^{\mathrm{app}} = \frac{S_{\mathrm{m}}\ F_{\mathrm{s,m}}\ k_{\mathrm{p,m}}^{\mathrm{app}}}{S_{\mathrm{m}}[F_{\mathrm{m,s}} + \alpha(F_{\mathrm{s,m}} - F_{\mathrm{m,s}})] + (1 - \alpha)k_{\mathrm{p,m}}^{\mathrm{app}}[I]_0} \tag{24}$$

allows simplification of Eq. (23) to the following:

$$\frac{d[M]}{dt} = -k_{\mathrm{p,d}}^{\mathrm{app}}[I]_0[M] \tag{25}$$

In principle $k_{\mathrm{p,d}}^{\mathrm{app}}$ depends to some extent on monomer conversion (it depends on $\alpha = V_{\mathrm{m}}/V$, i.e., on the volume of microspheres V_{m} that might change during polymerization). However, there are several limiting cases when Eq. (25) could be used to obtain interesting and important information.

1. For polymerizations with propagation in microspheres much faster than diffusion of monomer into and outside of particles $(1 - \alpha)k_{\mathrm{p,m}}^{\mathrm{app}}[I]_0 >> S_{\mathrm{m}}[F_{\mathrm{m,s}} + \alpha(F_{\mathrm{s,m}} - F_{\mathrm{m,s}})]$ the rate monomer conversion becomes diffusion controlled and independent of $k_{\mathrm{p,m}}^{\mathrm{app}}$ and of the concentration of propagating species:

$$\frac{d[M]}{dt} = -\frac{S_{\mathrm{m}}\ F_{\mathrm{s,m}}}{(1 - \alpha)}[M] \tag{26}$$

   When microspheres occupy a small fraction of the total volume of reaction mixture ($\alpha << 1$), Eq. (25) could be simplified further to $d[M]/dt = -S_{\mathrm{m}}F_{\mathrm{s,m}}[M]$.
2. For polymerizations with very slow propagation in microspheres $(1 - \alpha)$ $k_{\mathrm{p,m}}^{\mathrm{app}} << S_{\mathrm{m}}[F_{\mathrm{m,s}} + \alpha(F_{\mathrm{s,m}} - F_{\mathrm{m,s}})]$, $k_{\mathrm{p,d}}^{\mathrm{app}}$ could be expressed as:

$$k_{p,d}^{app} = \frac{\dfrac{F_{s,m}}{F_{m,s}}}{(1-\alpha)+\alpha\dfrac{F_{s,m}}{F_{m,s}}} k_{p,m}^{app} \tag{27}$$

In the case when coefficients characterizing flux of monomer into and outside of microspheres are equal ($F_{s,m} = F_{m,s}$), the propagation rate constant in dispersion should be equal to the propagation rate constant in microspheres ($k_{p,d}^{app} = k_{p,m}^{app}$).

For polymerizations with very slow propagation in microspheres and with coefficients characterizing flux of monomer into particles that is much higher than the corresponding coefficient for monomer flux from microspheres into solution (so high that relation $\alpha F_{s,m} >> F_{m,s}$ is fulfilled, i.e., particles are highly swollen with monomer) Eq. (27) is reduced to:

$$k_{p,d}^{app} = \frac{1}{\alpha} k_{p,m}^{app} \tag{28}$$

In this case, the apparent propagation rate constant strongly depends on the volume fraction occupied by growing microspheres. Generally, α depends on monomer conversion; however, for particles highly swollen with monomer (i.e., when almost all monomer becomes located inside of particles) the volume fraction of microspheres may change very little during polymerization. In a subsequent section some experimental results for dispersion polymerization of ε-caprolactone will be discussed and analyzed.

The kinetics of dispersion polymerization of ε-caprolactone was investigated for systems with initiators of pseudoanionic [$(CH_3CH_2)_2AlOCH_2CH_3$] [53,54,58] and anionic [$(CH_3)_3SiONa$] [54,58] polymerization. The initial monomer concentration in these studies was kept in a narrow range (0.41 ± 0.02 mol/L) and initiator concentrations were varied from 1.4×10^{-3} to 5.4×10^{-3} mol/L for $(CH_3)_3$-SiONa and from 3.4×10^{-3} to 5.4×10^{-2} mol/L for $(CH_3CH_2)_2AlOCH_2CH_3$. Polymerizations were carried out at room temperature in 1,4-dioxane/heptane (1:9 v/v) mixtures in a presence of poly(dodecyl acrylate)-*g*-poly(ε-caprolactone) surface-active copolymer. Concentration of poly(dodecyl acrylate)-*g*-poly(ε-caprolactone) in polymerizing mixture was 2.2 g/L. Poly(dodecyl acrylate)-*g*-poly(ε-caprolactone) with $\overline{M}_n = 28{,}800$ contained 15 wt% of poly(ε-caprolactone) grafts with $\overline{M}_n = 3600$ and $\overline{M}_w/\overline{M}_n = 1.1$. Dodecane (0.6 vol%) used as an internal standard for GPC determination of residual monomer was also present in the mixture. Polymerizations were carried out under argon with stirring at 60 revolutions/min. At chosen times small samples of polymerizing mixture were withdrawn and added to THF containing acetic acid (concentration 4×10^{-4}

mol/L) that terminated propagation. Samples with dissolved polymer were frozen in liquid nitrogen and evacuated for 10 min. Thereafter, they were melted and the whole liquid contents distilled to evacuated vials. The amount of unreacted monomer was determined using GPC (from ratio of signal integrals of ε-caprolactone and dodecane using the separately prepared calibration curve). Obtained data were formally analyzed using Eq. (25) in which $k_{p,d}^{app}$ was considered to be a time-independent parameter.

In Eq. (25) no monomer equilibrium concentration was taken into account. For polymerization of ε-caprolactone values of thermodynamic parameters (in THF, $\Delta H_p = -28.8$ kJ/mol, $\Delta S_p = -53.9$ J/mol·K [58,59]) yield equilibrium monomer concentration equal to 5.9×10^{-3} mol/L. Thus, for initial concentration of ε-caprolactone equal to 0.4 mol/L such simplification could be justified for monomer conversion up to about 90%.

Solution of Equation (25) leads to the relation:

$$\ln\frac{[M]_0}{[M]} = k_{p,d}^{app}[I]_0 t \tag{29}$$

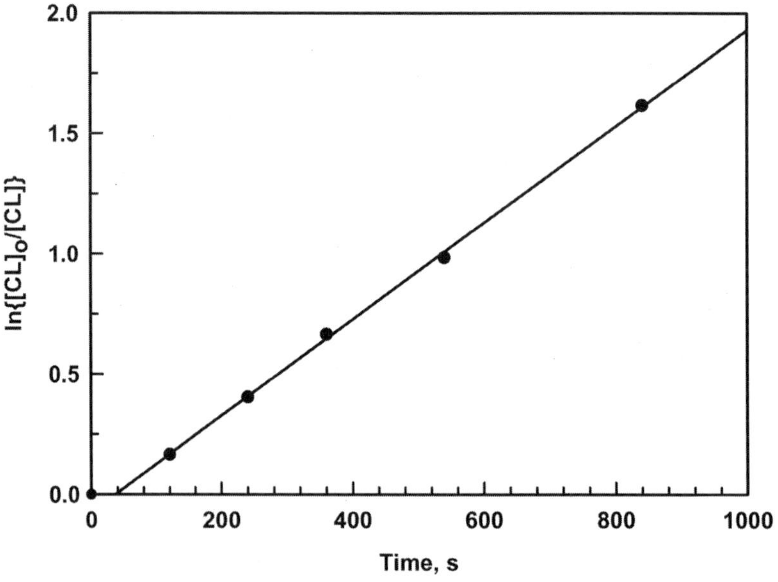

FIG. 14 Kinetics of the dispersion polymerization of ε-caprolactone. Conditions of the polymerization: [ε-caprolactone]$_0$ = 4.3×10^{-1} mol/L, $[(CH_3CH_2)_2AlOCH_2CH_3]_0 = 1.6 \times 10^{-2}$ mol/L. (From Ref. 53.)

in which $[M]_0$ denotes the initial monomer concentration and t denotes time. Thus, from plots of $\ln([M]_0/[M])$ vs. time, $k_{p,d}^{app}$ was evaluated dividing slope by $[I]_0$.

A typical kinetic plot is shown in Fig. 14. Experimental data in the kinetic plot could be perfectly fitted with a straight line with the slope 2.02×10^{-3} L/s yielding the corresponding $k_{p,d}^{app} = 1.25 \times 10^{-1}$ L/(mol·s). The short induction period has been attributed to the stage when microspheres were nucleated.

Rate constants for dispersion polymerizations of ε-caprolactone initiated with $(CH_3CH_2)_2AlOCH_2CH_3$ and $(CH_3)_3SiONa$ initiators are given in Table 3. Data in the table indicate that rates of dispersion polymerization depend on the chemical nature of propagating species and on their concentration. Thus, the possibility that propagation rates are controlled by monomer diffusion into growing particles should be ruled out. It has to be noted, however, that $k_{p,d}^{app}$ depends on concentration of active centers. Plots of $k_{p,d}^{app}$ as a function of active center concentrations (assumed to be equal to initial concentrations of initiators) are shown in Fig. 15. Dependencies of the apparent propagation rate constants on initiator concentration have been observed earlier for many ionic polymerizations in solution, including solution polymerizations of lactones [12,18,30,38,60–63]. They reflect the presence of equilibria between various physical forms of ionic

TABLE 3 Rates and Apparent Propagation Rate Constants of Dispersion Pseudoanionic [Initiated with $(CH_3CH_2)_2AlOCH_2CH_3$] and Anionic [Initiated with $(CH_3)_3SiONa$] Polymerizations of ε-Caprolactone

$[\text{Initiator}]_0$ (mol/L)	[ε-caprolactone]$_0$ (mol/L)	Rate of polymerization (L/s)	$k_{p,d}^{app}$ [1/(mol × s)]
	$(CH_3CH_2)_2AlOCH_2CH_3$		
3.4×10^{-3}	3.9×10^{-1}	1.63×10^{-3}	4.97×10^{-1}
4.8×10^{-3}	3.9×10^{-1}	1.50×10^{-3}	3.13×10^{-1}
1.0×10^{-2}	4.2×10^{-1}	1.98×10^{-3}	1.98×10^{-1}
1.6×10^{-2}	4.3×10^{-1}	2.02×10^{-3}	1.25×10^{-1}
2.6×10^{-2}	4.1×10^{-1}	1.69×10^{-3}	6.50×10^{-2}
	$(CH_3)_2SiONa$		
1.4×10^{-3}	4.0×10^{-1}	3.00×10^{-1}	2.14
1.4×10^{-3}	4.0×10^{-1}	4.51×10^{-1}	3.22
2.0×10^{-3}	4.0×10^{-1}	2.56×10^{-1}	1.28
2.0×10^{-3}	4.0×10^{-1}	2.62×10^{-1}	1.31
2.4×10^{-3}	4.0×10^{-1}	2.86×10^{-1}	1.19
3.5×10^{-3}	4.0×10^{-1}	7.20×10^{-1}	1.80
5.4×10^{-3}	4.0×10^{-1}	6.12×10^{-1}	1.53

Source: Data from Ref. 55.

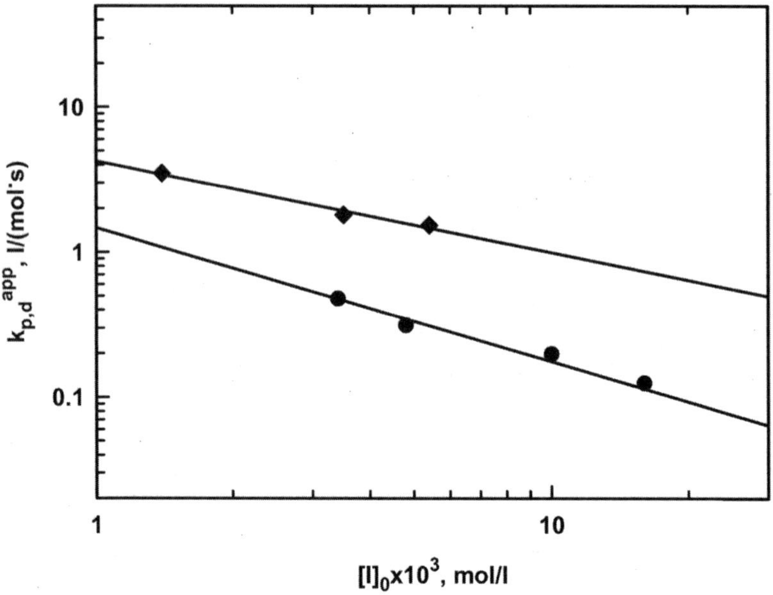

FIG. 15 Dependence of the apparent propagation rate constants ($k_{p,d}^{app}$) in pseudoanionic [initiated with $(CH_3CH_2)_2AlOCH_2CH_3$] and anionic [initiated with $(CH_3)_3SiONa$] dispersion polymerizations of ε-caprolactone on initiator concentration. Initial monomer concentration 0.41 ± 0.02 mol/L, reaction medium 1,4-dioxane/heptane 1:9 v/v mixture, room temperature. (Plots based on data from Ref. 55.)

and pseudoionic active species (ions, ion pairs, ionic aggregates, covalent species, and covalent species aggregates) propagating with different rate constants. The positions of these equilibria (molar fractions of various physical forms) strongly depend on the overall concentration of active centers.

It is interesting to compare the apparent propagation rate constants for polymerization of ε-caprolactone in dispersed systems ($k_{p,d}^{app}$) with the corresponding propagation rate constants (on active centers with the same chemical structure) for polymerizations in solution ($k_{p,s}^{app}$). Figure 16 illustrates dependence of the $k_{p,d}^{app}/k_{p,s}^{app}$ ratio on concentration of initiator. The plot indicates that for anionic polymerization $k_{p,d}^{app}/k_{p,s}^{app}$ is essentially independent from concentration of active centers and that the apparent propagation rate constant for polymerization in dispersion is about 10 times higher than for the corresponding propagation in solution. In the case of pseudoanionic polymerization initiated with $(CH_3CH_2)_2AlOCH_2CH_3$ $k_{p,d}^{app}/k_{p,s}^{app}$ decreases from 21 to 12 when concentration of active centers increases from 3.4×10^{-3} to 2.6×10^{-2} mol/L. Apparently, pseudoanionic

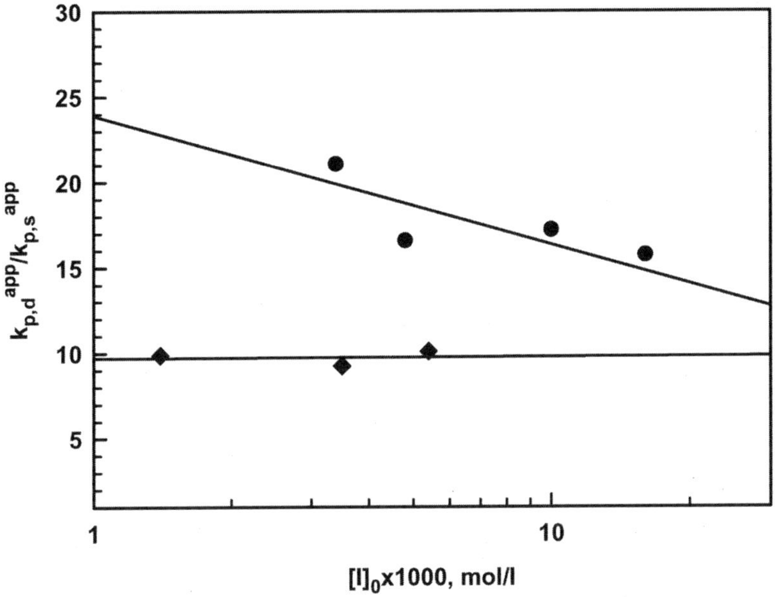

FIG. 16 Dependence of the ratio of apparent propagation rate constants for pseudoanionic and anionic polymerizations in dispersion and in solution ($k^{app}_{p,d}/k^{app}_{p,s}$) on initiator concentration. Conditions of polymerization: initiators $(CH_3CH_2)_2AlOCH_2CH_3$, and $(CH_3)_3$-SiONa for pseudoanionic and anionic polymerizations, respectively, initial monomer concentration 0.41 ± 0.02 mol/L, reaction media 1,4-dioxane/heptane (1:9 v/v mixture) and THF for dispersion and solution polymerizations, respectively, room temperature. (Plots based on data from Refs. 54 and 55.)

active centers confined in microspheres aggregated efficiently with increased overall (averaged over the while reaction volume) concentration.

One could expect that high values of $k^{app}_{p,d}$ are due to high local monomer concentration within microsheres. Indeed, in investigated dispersed system microspheres occupied a small fraction of the total volume ($\alpha \approx 0.04$). Thus, assuming that monomer diffusion is fast and that Eq. (28) can be used, it was possible to estimate $k^{app}_{p,m}$ ($k^{app}_{p,m} = \alpha\, k^{app}_{p,d}$). For example, for pseudoanionic polymerization initiated with $(CH_3CH_2)_2AlOCH_2CH_3$ (initial initiator concentration 3.4×10^{-3} mol/L), $k^{app}_{p,m}$ was estimated to be 1.9×10^{-2} L/(mol·s), whereas the corresponding propagation rate constant in for polymerization solution was equal to 2.3×10^{-2} L/(mol·s). Similar estimation for anionic dispersion polymerization and for the polymerization in solution ($[(CH_3)_3SiONa]_0 = 3.5 \times 10^{-3}$ mol/L) gave $k^{app}_{p,m} = 7.2 \times 10^{-2}$ L/(mol·s) and $k^{app}_{p,s} = 2.0 \times 10^{-1}$ L/(mol·s). Thus, the propagation rate constants inside microspheres are lower that the corresponding rate constants for propagation in solution, probably due to higher local viscosity and more efficient aggregation of active centers due to their confinement in small particles.

Direct evidence that poly(ε-caprolactone) microspheres in 1,4-dioxane/heptane (1:9 v/v) mixed medium are efficiently swollen with a monomer has been obtained by determination of monomer partition between microspheres and liquid phase [55]. Figure 17 shows dependence of ε-caprolactone concentration in liquid phase and in particles plotted as a function of the monomer concentration ($[\varepsilon\text{-caprolactone}]_{av}$) averaged over the whole volume (volume of liquid phase plus volume of microspheres). The volume fraction of microspheres in this system was 0.037.

Some investigators [12,38,61,64] indicate that in the anionic and pseudoanionic polymerizations of ε-caprolactone in solution concentration of active centers does not change during the polymerization process (absence of termination). In dispersion polymerizations of this monomer linear dependencies of $\ln([M]_0/[M])$ on time (cf. Fig. 14, slope equal to the product of $k_{p,d}^{app}$ and concentration

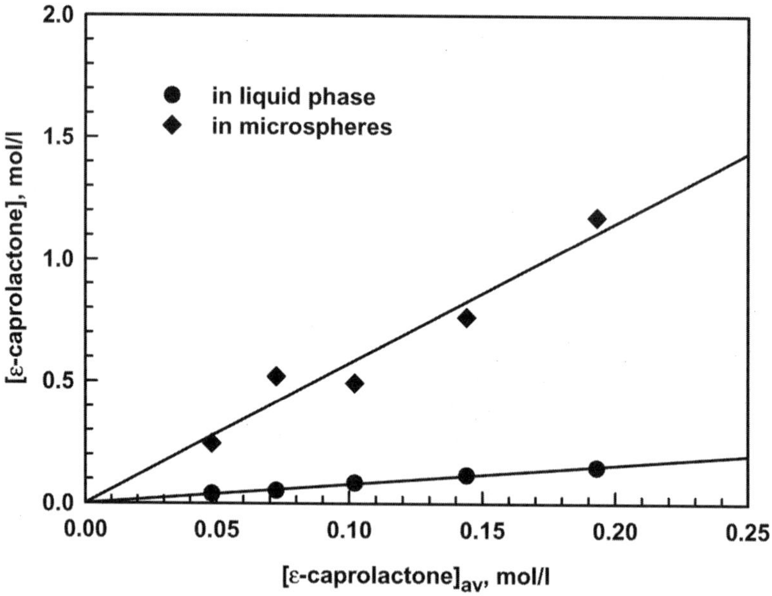

FIG. 17 Partition of ε-caprolactone between poly(ε-caprolactone) microspheres and 1,4-dioxane/heptane (1:9 v/v) liquid phase. Volume fraction of poly(ε-caprolactone) microspheres 2.06%. (From Ref. 55.)

of propagating species) indicate that also in these processes significant termination is absent, at least for monomer conversion up to about 90%. In pseudoanionic polymerization of ε-caprolactone in solution initiated with $CH_3CH_2)_2AlOCH_2$-CH_3 it was also found that chain transfer reactions (inter- and intramolecular transesterification) are reduced to such an extent that during time needed for monomer conversion up to 95% these side reactions could be practically neglected [30,38,62,65]. On the contrary, in the anionic polymerization of ε-caprolactone in solution with alkali metal counterions, transesterification reactions (including back biting leading to macrocycles) play a significant role and products contain substantial fraction of cyclic oligomers [12,33,34]. For anionic and pseudoanionic dispersion polymerizations of ε-caprolactone [initiated with $(CH_3)_3$-SiONa and $(CH_3CH_2)_2AlOCH_2CH_3$, respectively], good agreement between the measured and calculated (assuming absence of chain transfer and quantitative initiation) values of the average molecular weight for obtained polymers [$\overline{M}_n$ and $\overline{M}_n$(calcd)] has been observed (cf. Fig. 18). The plot also indicates that samples with higher molecular weight that were synthesized with lower initiator concentration had lower molecular weight polydispersity down to $\overline{M}_w/\overline{M}_n = 1.15$. Apparently, at lower initiator concentrations when formation of microspheres is completed at lower monomer conversions (microspheres are formed when $\overline{M}_n$ is

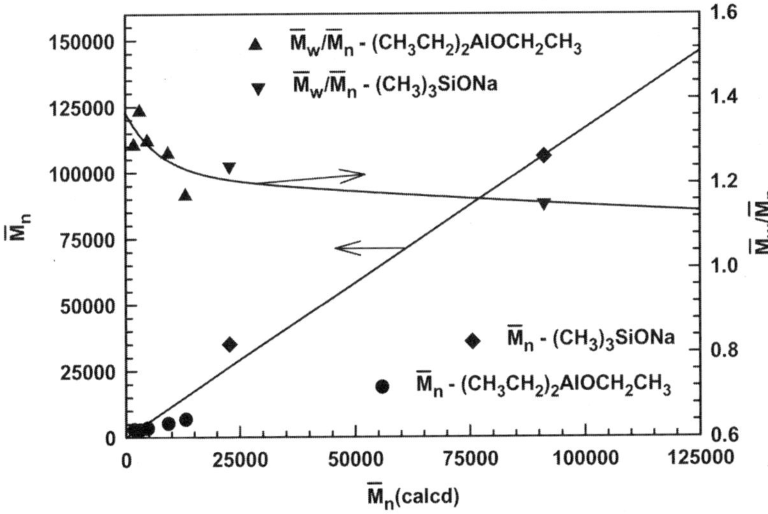

FIG. 18 Relation between measured $\overline{M}_n$, $\overline{M}_w/\overline{M}_n$, and calculated molecular weight [$\overline{M}_n$ (calcd)] of poly(ε-caprolactone) for pseudoanionic and anionic dispersion polymerization. (From Ref. 56.)

between 500 and 1000) the subsequent propagation proceeds without significant contribution of side transesterification reactions.

Absence of transesterification reactions, in particlular intramolecular back biting leading to formation of cyclics, in dispersion polymerization of ε-caprolactone with Na^+ cations was clearly manifested in GPC trace of synthesized polymers. GPC traces of poly(ε-caprolactone) synthesized in anionic solution polymerization in which back biting is important display a very broad signal of linear polymer and signals of cyclic oligomers at long elution times (Fig. 19). On the contrary, GPC trace of poly(ε-caprolactone) synthesized in dispersion polymerization with Na^+ counterion contains only a narrow signal of a linear polymer. Signals of cyclics are absent in these chromatograms (cf. Fig. 20, Ref. 55).

The following reasons might contribute to this difference. It is known that due to thermodymanic relations the fraction of cyclic oligomers is lower for polymerizations with higher initial monomer concentration [12,33,34,37]. Since microspheres are swollen with monomer the initial concentration in particles could be close to that in bulk. At room temperature the total concentration of monomeric units in cyclics is close to 0.25 mol/L. Thus, for polymerization in

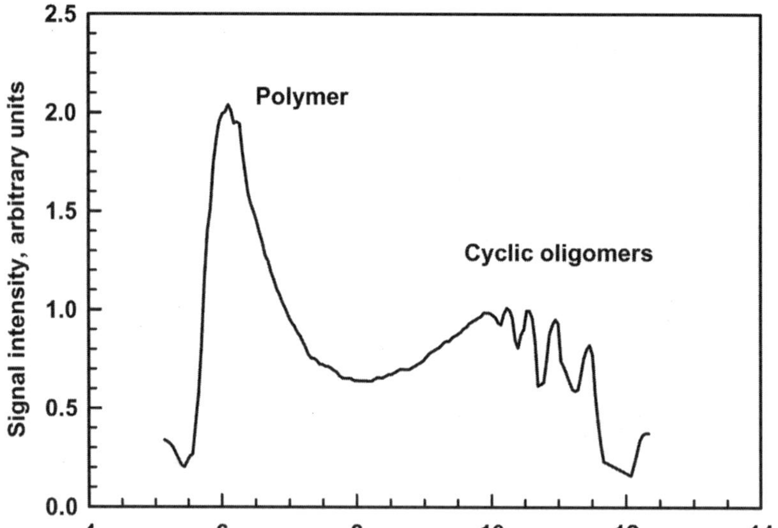

FIG. 19 GPC tracing of the product of anionic polymerization of ε-caprolactone in THF solution. Polymerization conditions: $[\varepsilon\text{-caprolactone}]_0 = 6.1 \times 10^{-1}$ mol/L, $[(CH_3)_3\text{-}SiONa]_0 = 6.1 \times 10^{-3}$ mol/L, temperature 20°C.

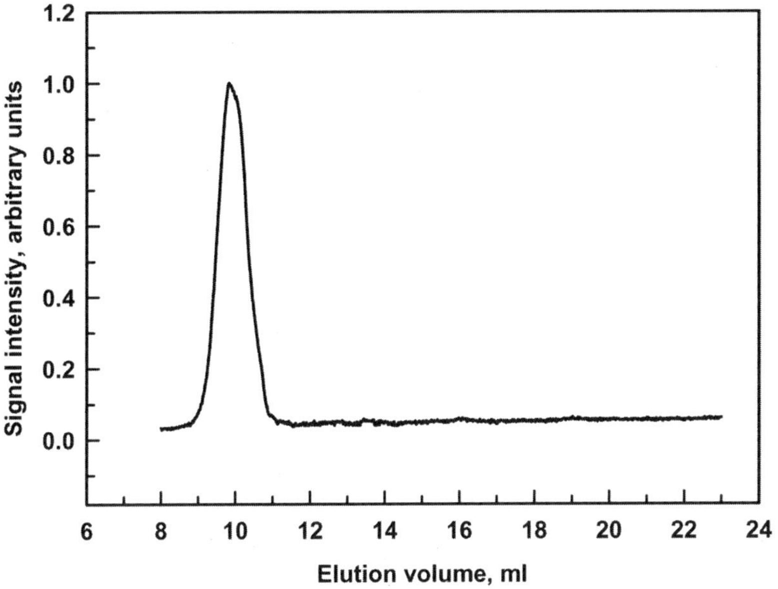

FIG. 20 GPC tracing of the product of anionic dispersion polymerization of ε-caprolactone. Polymerization conditions: $[\varepsilon\text{-caprolactone}]_0 = 4.0 \times 10^{-1}$ mol/L, $[(CH_3)_3SiONa]_0 = 5.1 \times 10^{-4}$ mol/L, room temperature. From a calibration curve obtained using poly(ε-caprolactones) with narrow molecular weight distribution $\overline{M}_n = 106,600$, $\overline{M}_w/\overline{M}_n = 1.15$. (From Ref. 55.)

bulk (initial monomer concentration equal to 9.0 mol/L) the fraction of cyclics would amount to only as high as 2.8%. It is also possible that relatively low mobility of polymer chains inside of microspheres (in comparison with chain mobility in solution) decreases rates of transesterification reactions. Thus, in anionic dispersion polymerization of ε-caprolactone, both thermodynamic and kinetic factors could contribute to practical elimination of cyclics up to the moment of full monomer conversion.

IX. PHASE TRANSFER OF POLY(ε-CAPROLACTONE) AND POLY(LACTIDE) MICROSPHERES FROM HYDROCARBONS TO THE WATER-BASED MEDIA

Poly(lactide) and poly(ε-caprolactone) microspheres were synthesized in 1,4-dioxane/heptane mixed media. After transfer to pure heptane these particles form stable suspensions (if they settle down they can be resuspended by gentle shaking). However, for any biological or medical applications such particles must be transferred to water or water-based media. It was reported (cf. Ref. 65) that such transfer was accomplished in the following way. A sample of suspension of poly(ε-caprolactone) and/or poly(lactide) microspheres (volume 1–1.7 mL) was added to the ethanol solution of a given surfactant [Triton X-405, ammonium sulfobetaine-2 (ASB), and/or sodium dodecyl sulfate (SDS); chemical structures of these surfactants are given in Fig. 21]. Total volume of obtained suspension was 7.5 mL. Concentrations of surfactants down to about 0.5% (wt/vol) could be used.

Thereafter, to the suspension 0.6 mL of 1 M KOH was added. Poly(ε-caprolactone) microspheres were incubated in this medium for 10 min, poly(lactide) particles for 1 h. Thereafter, the microspheres were isolated by centrifugation and redispersed in ethanol solution of any of the above-mentioned surfactants (1% wt/vol). This procedure was repeated at least once more and then particles isolated by centrifugation were transferred to 1 mL of the aqueous surfactant solution. Eventually, pH of suspending medium was adjusted to 11 by addition of the appropriate amount of KOH solution. Suspensions of microspheres were

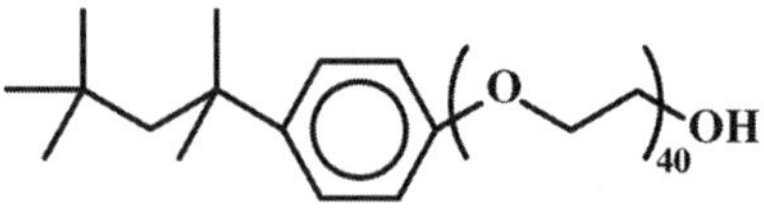

Triton X-405, cmc 2.0 10^{-1} % wt/v

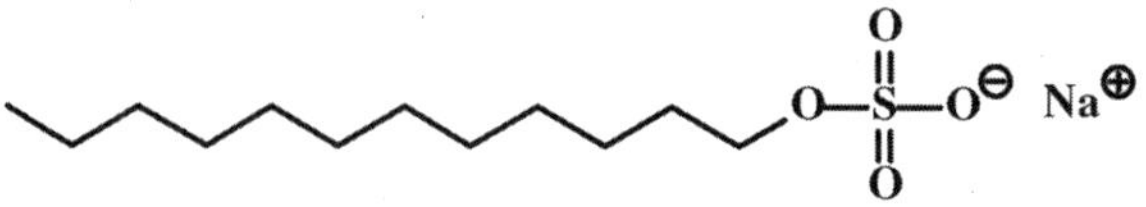

SDS, cmc 2.2 10^{-1} % wt/v

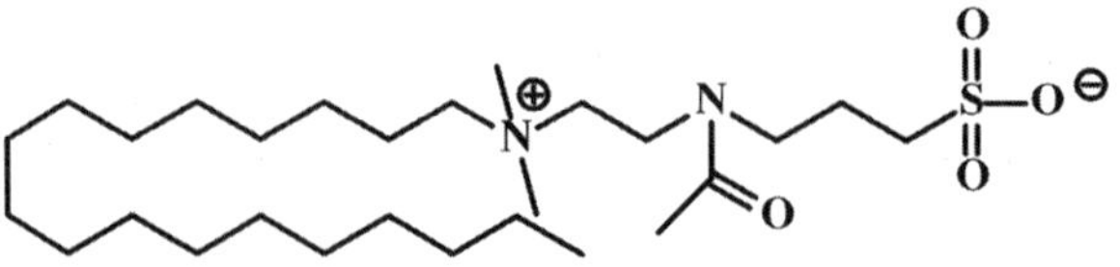

ASB, cmc 5.0 10^{-3} % wt/v

FIG. 21 Chemical structures of Triton X-405, sodium dodecyl sulfate (SDS), and ammonium sulfobetaine-2 (ASB) surfactants.

kept at 4°C. The lowest concentrations of surfactants in water providing sufficient stabilization of suspensions of microspheres were 1.5×10^{-1}, 1.7×10^{-1}, and 4.7×10^{-1}% (wt/vol) for ABS, Triton X-405, and SDS, respectively. Stability of suspensions was monitored by quasi-elastic light scattering. When microspheres begun to coalesce the averaged diameter of light scattering objects in suspension increased [65].

Particles in the above-described buffered water suspensions were stabilized electrostatically and sterically. Electrostatic repulsion between particles resulted from presence of carboxylic anions in their surface layers. These groups were formed by basic hydrolysis of poly(ε-caprolactone) induced by KOH. Steric stabilization was due to adsorption of surfactants (ABS, Triton X-405, and/or SDS). It has to be noted that alone neither electrostatic stabilization (due to KOH treatment) nor the steric one (using the above-listed surfactants) were sufficient for stabilization of poly(lactide) and/or poly(ε-lactide) microspheres.

Hydrolysis of polyester molecules in surface layer of poly(ε-caprolactone) and/or poly(lactide) microspheres obviously should lead to changes in the overall molecular weight of polymer in particles. One should also take into account changes in microsphere diameters resulting from hydrolysis. In Table 4 are given values of averaged diameters ($\overline{D}_n$) and diameter polydispersity factors ($\overline{D}_w/\overline{D}_n$) for particles that were stored in heptane and after their transfer to the buffer. The molecular weights of polymers before and after transfer to water media are given in Table 5.

Diameters of particles from suspension in heptane and from the buffer were determined from SEM images. In addition, diameters of microspheres suspended in the buffer were determined by quasi-elastic light scattering (this method was not suitable for measurements in heptane since particles from suspensions in hydrocarbons were adsorbed onto the cell's surface).

TABLE 4 Diameters and Polydispersity Indexes of Poly(ε-caprolactone) and Poly(L,L-lactide) Microspheres Before and After Transfer from Heptane to Water/ABS System

	Before transfer to water/ASB		After transfer to water/ASB		
	Measurements by SEM		Measurements by SEM		Measurements by QELS
Microspheres	$\overline{D}_n$ (μm)	$\overline{D}_w/\overline{D}_n$	$\overline{D}_n$ (μm)	$\overline{D}_w/\overline{D}_n$	$\overline{D}_h$ (μm)
Poly(ε-caprolactone)	0.65	1.47	0.62	1.55	0.7
Poly(L,L-lactide)	2.74	1.13	2.73	1.13	2.9

Source: Ref. 65.

TABLE 5 Molecular Weight and Molecular Weight Distribution of Poly(ε-caprolactone) and Poly(L,L-lactide) Microspheres Before and After Their Transfer from Heptane to the Water/ASB System

	Before transfer from heptane		After transfer to water/ASB	
Microspheres	$\overline{M}_n$	$\overline{M}_w/\overline{M}_n$	$\overline{M}_n$	$\overline{M}_w/\overline{M}_n$
Poly(ε-caprolactone)	42,600	1.52	8,350	1.57
Poly(L,L-lactide)	38,400	1.12	19,300	1.53

Source: Ref. 65.

Data in Table 4 revealed that the procedure used for transferring microspheres from heptane to buffers did not change the diameters of microspheres. However, hydrolysis constituting an important step in transfer of particles to water substantially decreased the molecular weight of microspheres and, in the case of particles composed of polymer with low $\overline{M}_w/\overline{M}_n$, also significantly increased the molecular weight polydispersity (cf. Table 5).

Hydrolysis led to formation of carboxylic groups in microsphere surface layer. In media with sufficiently basic pH these groups would be ionized providing particles that are negatively charged. Ionic surfactants (SDS and ASB) also modify the charge at surfaces of microspheres. Surface charge densities (combining charge of polyester carboxyl end groups, ionic groups of adsorbed surfactants, and counterions embedded into loose surface layer) for both kinds of microspheres and for all kinds of surfactants were determined on the basis of the measurements of microsphere electrophoretic mobility [65]. The following relation holds for surface charge density of microspheres (σ) and their electrophoretic mobility (μ_e)

$$\sigma = \mu_e \eta \kappa \tag{30}$$

In Eq. (30), η denotes viscosity of the medium and κ the reciprocal of the Debye length given by the following expression:

$$\kappa = \sqrt{\left(\frac{8\pi e^2 n \gamma^2}{\varepsilon k T}\right)} \tag{31}$$

in which e denotes charge of electron, n number of ions in 1 mL, γ valence of ions, ε dielectric constant of the medium, k Boltzman constant, and T temperature in K.

The dependence of electrophoretic mobility of poly(ε-caprolactone) and poly(L,L-lactide) microspheres on pH of suspending medium is shown in Fig. 22.

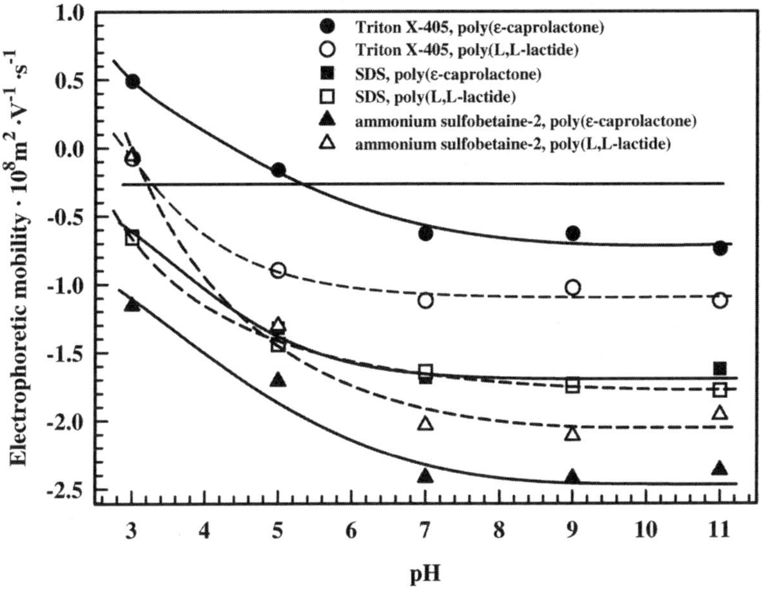

FIG. 22 Dependence of electrophretic mobilities of poly(ε-caprolactone) and poly(L,L-lactide) microspheres on pH. (From Ref. 65.)

Stabilization of microspheres with Triton X-405 does not introduce any additional charge (molecules of this surfactant do not contain any ionic or ionophoric groups). Thus, electrophoretic mobility of microspheres in this system should depend solely on electrostatic charge due to the presence of carboxylic groups resulting from polyester chain scission during hydrolysis. Indeed, according to plots in Fig. 22, the isoelectric point for poly(ε-caprolactone) and poly(L,L-lactide) microspheres is at pH 2.9 and 4.4, respectively. It is also known that typical carboxylic acids have pK_a ranging from 4 to 6 [67] and therefore in a range of pH from 2 to 4 these acids are ionized in less than 1%.

The electrophoretic mobility of microspheres stabilized with ionic surfactants (SDS and ASB) was negative in the pH range used in the studies (from pH 3 to 11). This was due to $-SO_3^-$ and $-OSO_3^-$ groups (more acidic than carboxyl groups) introduced into surface layer by surfactants.

It is worth noting that for pH > 7 for all types of poly(ε-caprolactone) and poly(L,L-lactide) microsphere suspensions the electrostatic mobility reached plateau. This means that under these conditions all carboxyl, sulfate, and sulfite groups in particle surface layers were fully ionized. Values of electrostatic charge densities calculated from these maximal (with respect to their absolute

TABLE 6 Surface Charge Density for Poly(ε-caprolactone) and Poly(L,L-lactide) Microspheres (Stabilized with Triton X-405, SDS, and ASB Surfactants) Calculated from Microsphere Electrophoretic Mobilities at Their Maximal Absolute Values

Microspheres	Stabilizing surfactant	$\mu_e \cdot 10^8$ ($m^2\ V^{-1}\ s^{-1}$)	σ ($\mu C\ cm^{-2}$)	$\sigma \cdot 10^7$ ($mol\ m^{-2}$)
Poly(ε-caprolactone)	Triton X-405	−0.72 ± 0.04	−2.5 ± 0.1	−2.6 ± 0.1
	SDS	−1.70 ± 0.07	−5.9 ± 0.2	−6.1 ± 0.2
	ASB	−2.5 ± 0.1	−8.7 ± 0.4	−8.9 ± 0.4
Poly(L,L-lactide)	Triton X-405	−1.09 ±	−3.8 ± 0.2	−3.9 ± 0.2
	SDS	−1.77 ± 0.09	−6.2 ± 0.3	−6.4 ± 0.3
	ASB	−2.05 ± 0.08	−7.2 ± 0.3	−7.4 ± 0.3

Source: Ref. 63.

values) electrophoretic mobilities are given in Table 6. The stability of microsphere suspensions stabilized not only due to steric but also due to ionic interactions could be affected not only by pH but by ionic strength as well. Destabilization of microsphere suspensions was monitored visually by observation of a drop of suspension on a black plate used for the diagnostic latex tests. Drops of stable suspension look homogeneous whereas loss of stability is manifested by formation of particle aggregates. It has been found [65] that poly(ε-caprolactone) microspheres formed stable suspensions in buffers within pH range from 3 to 11 and for ionic strength from 10^{-3} to 1 mol/L, regardless of the nature of the surfactant (Triton X-405, SDS, ASB). In the case of the larger poly(L,L-lactide) microspheres the stability of suspensions was dependent on pH and on the ionic strength of the medium as well on the nature of surfactant. ASB stabilized suspension of poly(L,L-lactide) microspheres effectively in the full pH range (from 3 to 11) and ionic strength (from 10^{-3} to 1 mol/L). Stabilization of poly(L,L-lactide) microspheres with Triton X-405 and SDS was dependent on pH and ionic strength in the same manner (cf. Fig. 23; destabilization occurred for higher values of ionic strength and lower values of pH).

X. MICROSPHERES LOADED WITH BIOACTIVE COMPOUNDS

There are four strategies that can be used for incorporation of bioactive compounds into microspheres obtained by dispersion ring-opening polymerization:

1. Precipitation of bioactive compound during polymerization onto growing microspheres and its penetration into particles

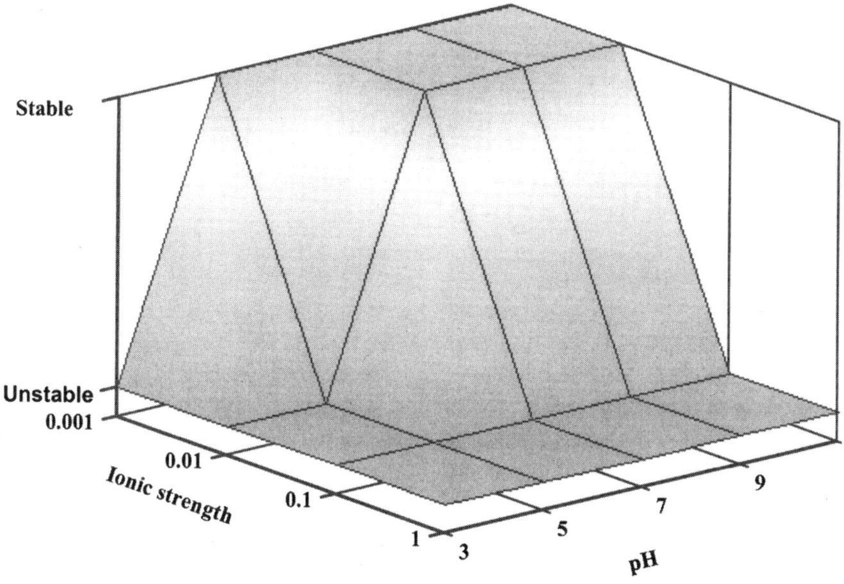

FIG. 23 Stability of poly(L,L-lactide) microspheres atbilized with SDS and Triton X-405 surfactants as a function of ionic strength and pH.

2. Chemical binding of bioactive compound molecules to growing polymer chains (bioactive compound acting as chain transfer agent)
3. Swelling of synthesized microspheres with bioactive compound
4. Adsorption of bioactive compounds onto surface of microspheres

Examples of incorporation of bioactive compounds into poly(L,L-lactide) and poly(ε-caprolactone) microspheres according to the listed strategies are given below.

A. Strategy A—Synthesis of Poly(L,L-lactide) Microspheres Loaded with Omeprasol [50]

Omeprasol (5-methoxy-2-{[(4-methoxy-3,5-dimethyl-2-pirydyl)-methyl]-sulfinyl}-1H-benzimidazole) (structure shown in Fig. 24), known as an inhibitor of an enzyme $(H^+$-$K^+)$ATPase [66,67], has been used as a model drug. Polymerization of L,L-lactide was prepared and initiated according to description given in Section IV.B. Initial monomer, initiator (stannous octoate), and poly(dodecyl acrylate)-*g*-poly(ε-caprolactone) were equal to 2.5×10^{-1} mol/L, 4.6×10^{-3} mol/L and 1.6×10^{-1} g/L, respectively. Polymerization was carried out at 95°C. When polymerizing mixture became turbid, i.e., when microspheres were nucle-

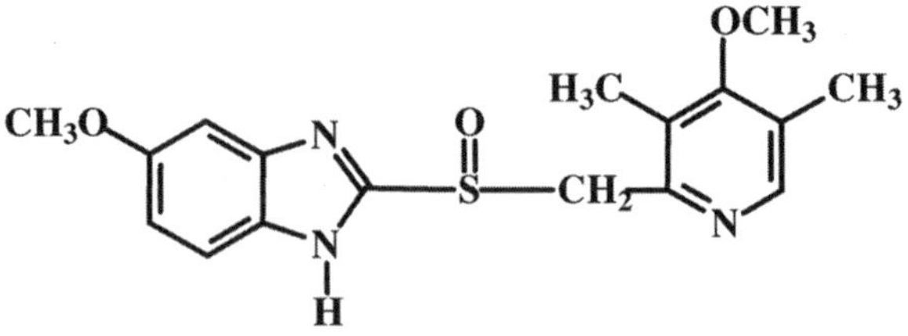

FIG. 24 Chemical structure of omeprasol.

ated, solution of omeprasol in heptane was added dropwise. Final concentration of omeprasol in polymerizing mixture was 5×10^{-3} mol/L. Polymerization was carried out for 2 h and afterward microspheres were isolated in the usual way (cf. Section IV.B). The SEM image of these microspheres is shown in Fig. 25. The number average diameter ($\overline{D}_n$) and diameter polydispersity index ($\overline{D}_w/\overline{D}_n$) were 1.73 μm and 1.17, respectively.

It has been found that omeprasol was simply entrapped in microspheres without being bound chemically. In GPC traces of dissolved microspheres the signals of poly(L,L-lactide) and omeprasol gave isolated peaks (cf. Fig. 26). It is important to note also that in the ^{1}H NMR spectrum of poly(L,L-lactide)/omeprasol microspheres only signals of polyester and drug were present, suggesting

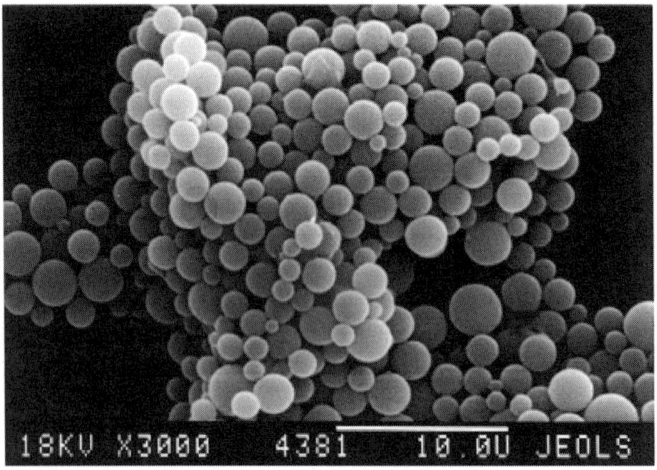

FIG. 25 SEM image of poly(L,L-lactide/omeprasol) microspheres. (From Ref. 50.)

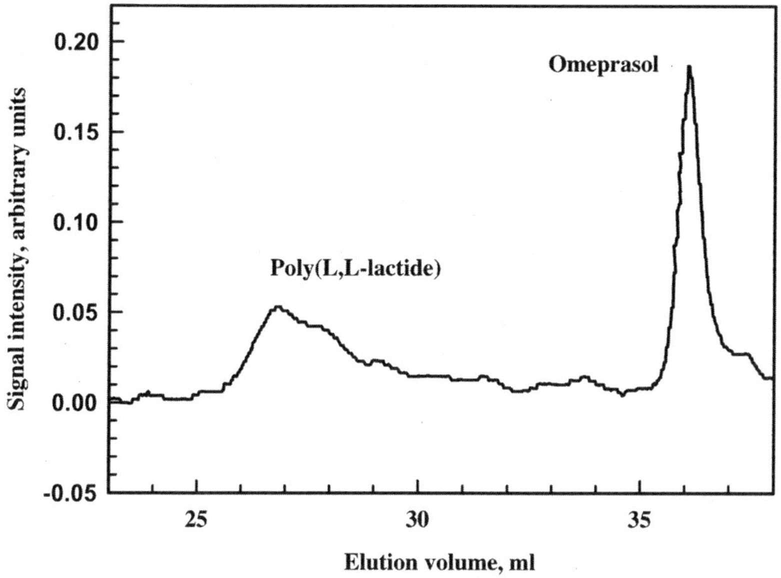

FIG. 26 GPC trace of poly(L,L-lactide)/omeprasol microspheres in THF. (From Ref. 50.)

absence of any side products (cf. Fig. 27). On the basis of UV spectra of poly(L,L-lactide)/omeprasol microspheres dissolved in THF, loading was found equal to 11 wt %.

B. Strategy B—Synthesis of Poly(ε-caprolactone) Microspheres Loaded with *N,N*-Bis(hydroxyethyl)isonicotinamide

N,N-Bis(hydroxyethyl)isonicotinamide being equipped with two hydroxyl groups could be used as a chain transfer agent and in effect become incorporated into poly(ε-caprolactone) chains forming poly(ε-caprolactone)/*N,N*-bis(poly(ε-caprolactone)isonicotinamide microspheres (cf. Scheme 6). Synthesis of poly-(ε-caprolactone) microspheres with chemically bound *N,N*-bis(hydroxyethyl) isonicotinamide was performed similarly to that described in Section IV.B. 1,4-Dioxane/heptane 1:9 (v/v) mixture was used as reaction medium. Monomer, initiator [$(CH_3)_3SiONa$], poly(dodecyl acrylate)-*g*-poly(ε-caprolactone), and *N,N*-bis(hydroxyethyl)isonicotinamide were used in the following concentrations: [ε-caprolactone]$_0$ = 3.8×10^{-1} mol/L, [$(CH_3)_3SiONa$]$_0$ = 1.76×10^{-3} mol/L, [*N,N*-bis(hydroxyethyl)isonicotinamide]$_0$ = 4.0×10^{-3} mol/L, [poly(dodecyl

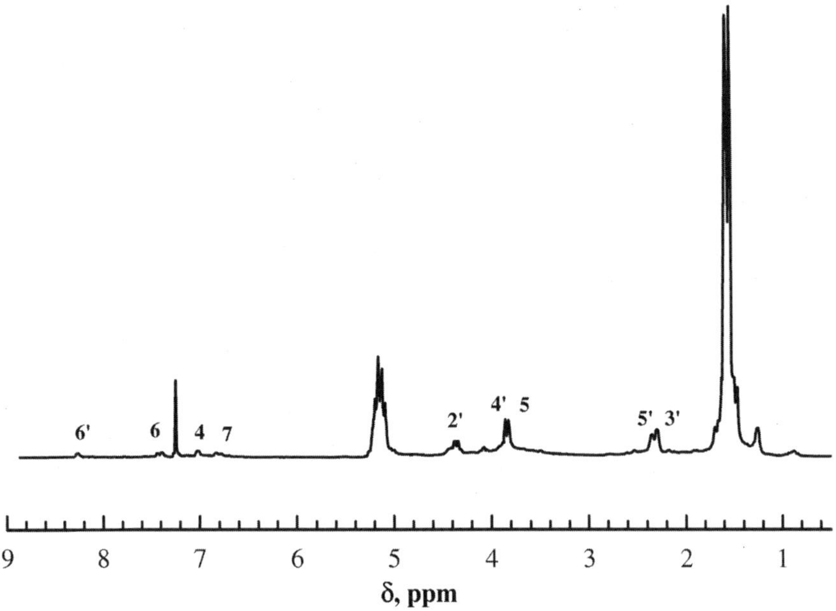

FIG. 27 ^{1}H NMR spectrum of poly(L,L-lactide)/omeprasol microspheres in $CDCl_3$. (From Ref. 50.)

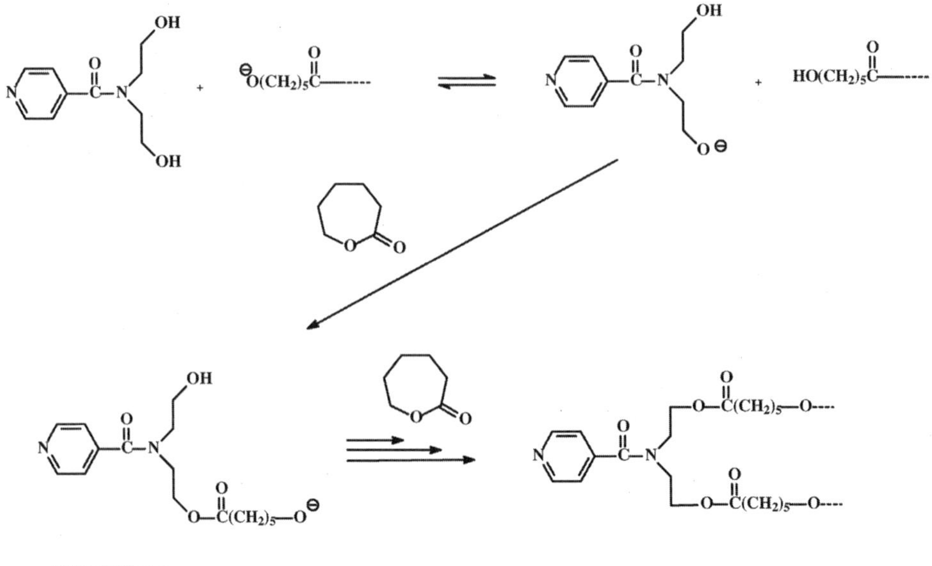

SCHEME 6

acrylate)-*g*-poly(ε-caprolactone)] = 1.6 g/L. Polymerization was carried out at room temperature for 20 min. The yield of microspheres was 72%. Drug loading determined from UV spectra of microsopheres dissolved in 1,4-dioxane (cf. Fig. 28) was 0.45 wt %.

It must be noted that higher concentrations of *N,N*-bis(hydroxyethyl)isonicotinamide in polymerizing mixture (2.0×10^{-2} mol/L) in polymerizing mixture allowed synthesis of microspheres with drug loading equal 6.4 wt %.

C. Strategy C—Poly(ε-caprolactone) Microspheres Loaded with Ethyl Salicylate by Swelling of Microspheres [67]

Ethyl salicylate has been used as a model liquid lipophilic drug for studies of loading of poly(ε-caprolactone) microspheres by swelling. Loading has been performed in the following way [65]. Poly(ε-microspheres) (5.4 mg, $\overline{D}_n = 0.62$ μm, $\overline{D}_w/\overline{D}_n = 1.5$) were suspended in a 7:3 vol/vol ethanol/water mixture containing various amounts of ethyl salicylate (from 1.5 to 50 μL). Total volume

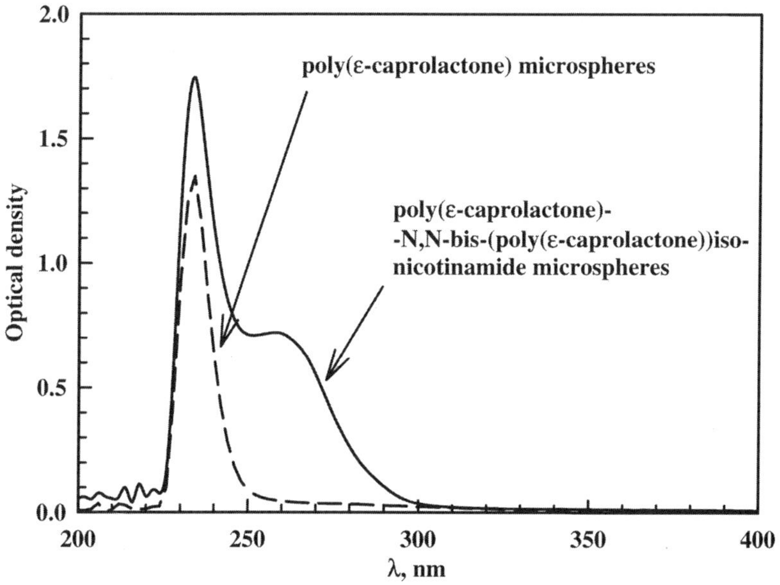

FIG. 28 UV spectra of poly(ε-caprolactone) and poly(ε-caprolactone)/*N,N*-bis-[poly(ε-caprolactone]isonicotinamide microspheres. Solvent 1,4-dioxane, concentration of microspheres 12 g/L.

of the mixture was 1 mL. Microspheres were incubated in this mixture for 48 h. Thereafter, they were isolated by centrifugation. Drug loading was determined from the difference in ethyl salicylate concentration in solution before and after incubation with microspheres. The content of ethyl salicylate in poly(ε-caprolactone) microspheres as a function of the initial drug concentration in solution is shown in Fig. 29. It is worth noting that in the case of investigated drug with relatively low water solubility the loading of microspheres approached 30 wt %.

D. Strategy D—Poly(ε-caprolactone) and Poly(D,L-lactide) Microspheres with Adsorbed Human Serum Albumin and Gamma Globulin [43]

Biodegradable polyester microspheres are considered to be good candidates for protein and oligopeptide transportation. Examples of human serum albumin

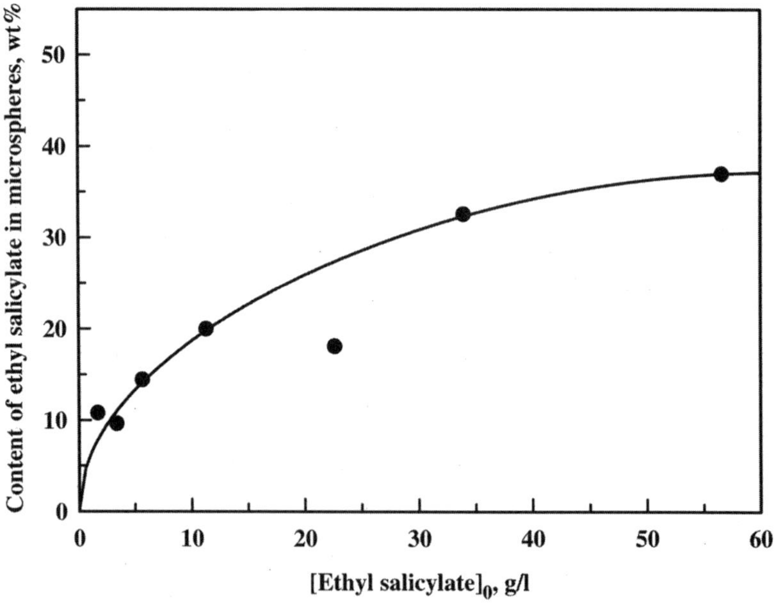

FIG. 29 Loading of ethyl salicylate into poly(ε-caprolactone) microspheres by swelling. Concentration of ethyl salicylate in microspheres as a function the initial drug concentration. (From Ref. 65.)

(HSA) and human gamma globulin (γG) adsorption onto poly(ε-caprolactone) and poly(D,L-lactide) microspheres are given below.

Suspensions of poly(ε-caprolactone) microspheres ($\overline{D}_n = 0.63$ μm, particle concentration 3.12 mg/mL) and/or poly(D,L-lactide) microspheres (2.50 μm, 16.6 mg/mL) in phosphate buffer (pH 7.4) were incubated with HSA and/or γG at varied protein concentrations. Incubation was carried out at room temperature for 12 h. Surface concentrations of attached proteins (Γ_{HSA} and $\Gamma_{\gamma G}$) as a function of protein concentration in solution are shown in Figs. 30 and 31. According to these figures, the maximal surface concentrations of adsorbed proteins at plateaus corresponding to saturation of surfaces of microspheres with proteins are as follows: For poly(ε-caprolactone) microspheres $\Gamma_{HSA}(\max) = 1.0$ mg/m^2 and $\Gamma_{\gamma G}(\max) = 2.6$ mg/m^2, whereas for poly(D,L-lactide) microspheres $\Gamma_{HSA}(\max) = 0.9$ mg/m^2 and $\Gamma_{\gamma G}(\max) = 1.7$ mg/m^2. It is worth noting that adsorption of HSA and γG onto polyester microspheres is only about 30% lower than for polystyrene microspheres used for many applications (e.g., in medical diagnostics) as convenient supports for protein adsorption [43].

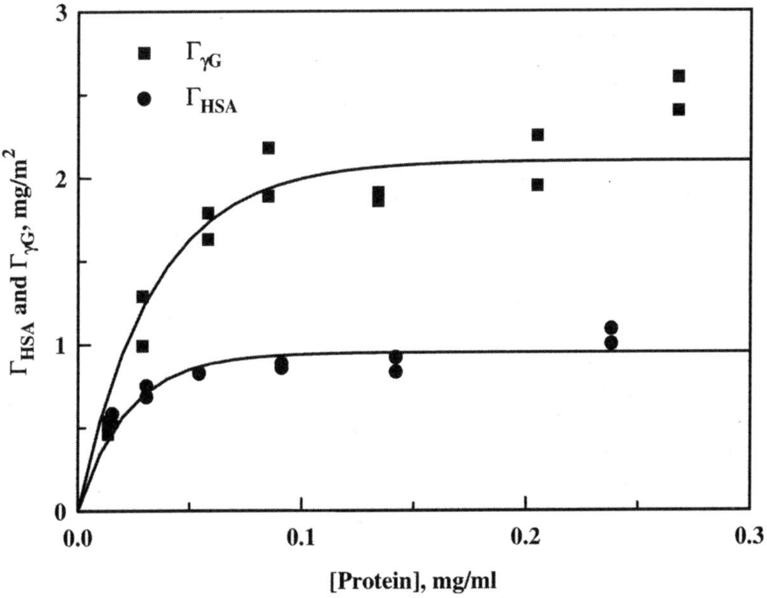

FIG. 30 Surface concentrations of proteins (Γ_{HSA} and $\Gamma_{\gamma G}$) adsorbed onto poly(ε-caprolactone) microspheres as functions of protein concentrations in solution. (From Ref. 43.)

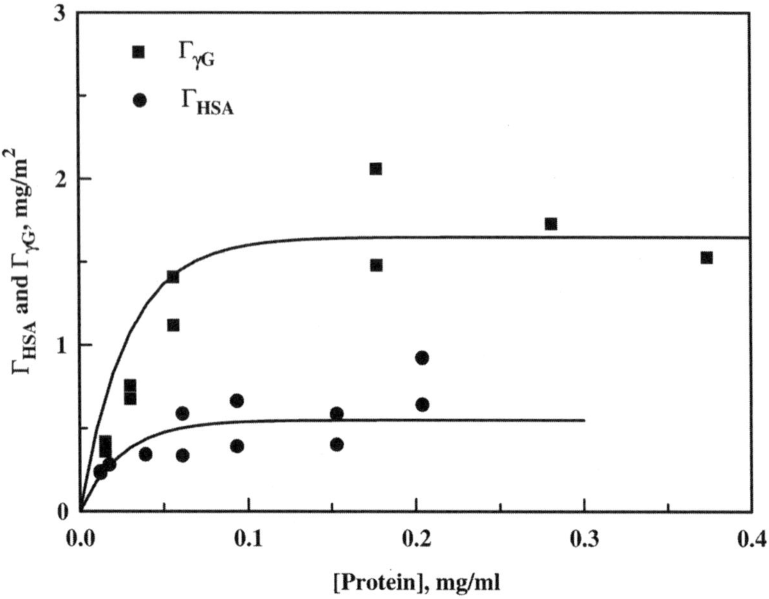

FIG. 31 Surface concentrations of proteins (Γ_{HSA} and $\Gamma_{\gamma G}$) adsorbed onto poly(D,L-lactide) microspheres as functions of protein concentrations in solution. (From Ref. 43.)

XI. MICROSPHERES WITH FLUORESCENT TAGS

Carboxylic groups at the surface of poly(L,L-lactide) and poly(ε-caprolactone) microspheres could be used for covalent immobilization of fluorescent tags with amino and hydrazine reactive groups. Activation of microspheres and immobilization of 6-aminoquinoline and Lucifer yellow is illustrated in Scheme 7. A typical recipe for attachment of fluorophores (e.g., Lucifer yellow) follows. To a suspension of poly(L,L-lactide) microspheres (9.2 mg, $\overline{D}_n = 2.73$ μm) in phosphate buffer (0.325 mL, pH 7.4) containing 0.17% (wt/vol) of SDS the following solutions was added 50 μL of 1-ethyl-3-(3-dimehylaminopropyl)carbodiimine (concentration 20 mg/mL), 50 μl of water solution of *N*-hydroxysulfosuccinimide (concentration 50 mg/mL), and 0.53 mL of water solution of Lucifer yellow (concentration 1 mg/mL). The suspension was gently shaken for 24 h at room temperature. Thereafter, the unbound label was removed by several repeated isolations of microspheres by centrifugation and resuspension in pure water. Figure 32 shows a microscopic picture of fluorescent poly(L,L-lactide) microspheres labeled with Lucifer yellow.

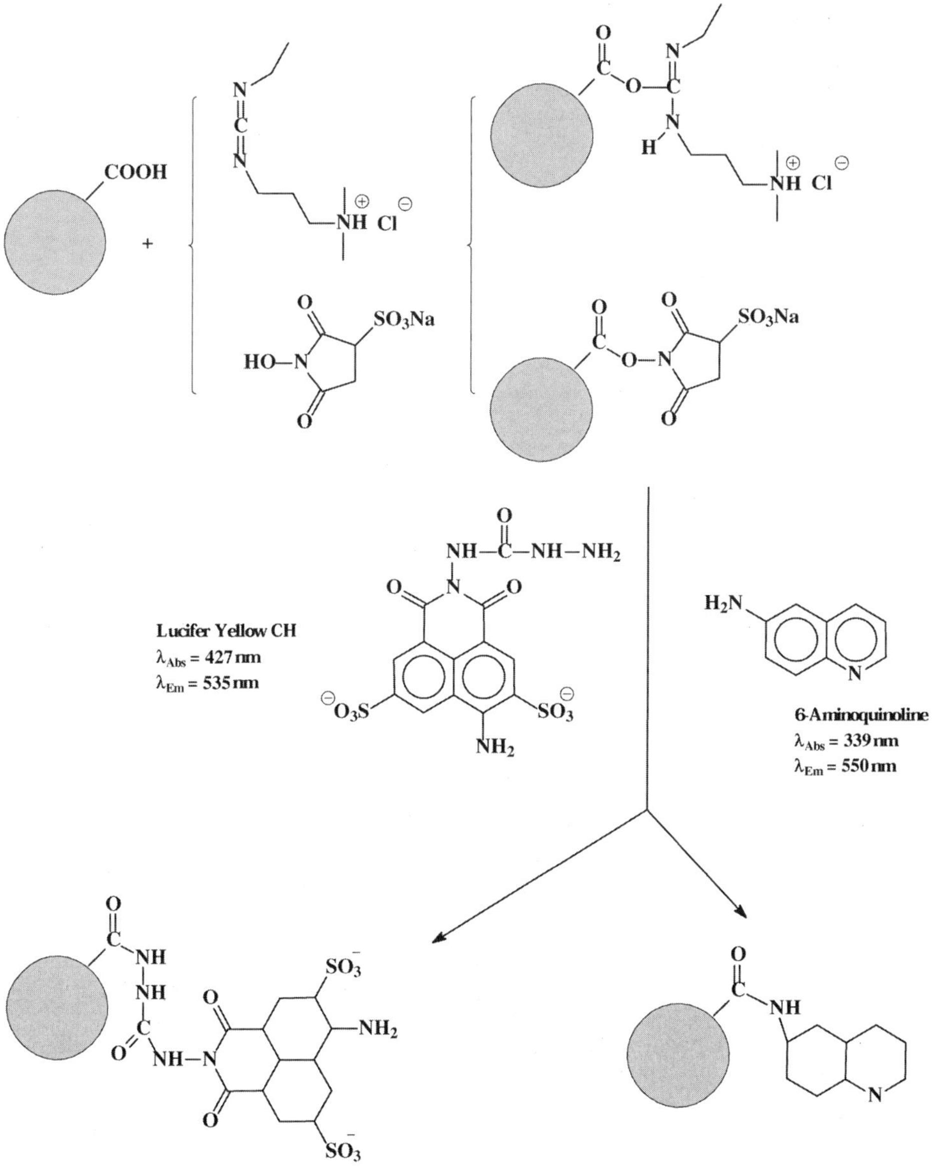
COOH
SO3Na
HO—N
Lucifer Yellow CH
λAbs = 427 nm
λEm = 535 nm
NH—C—NH—NH2
NH2
6-Aminoquinoline
λAbs = 339 nm
λEm = 550 nm
NH
NH—N
SO3
NH2

SCHEME 7

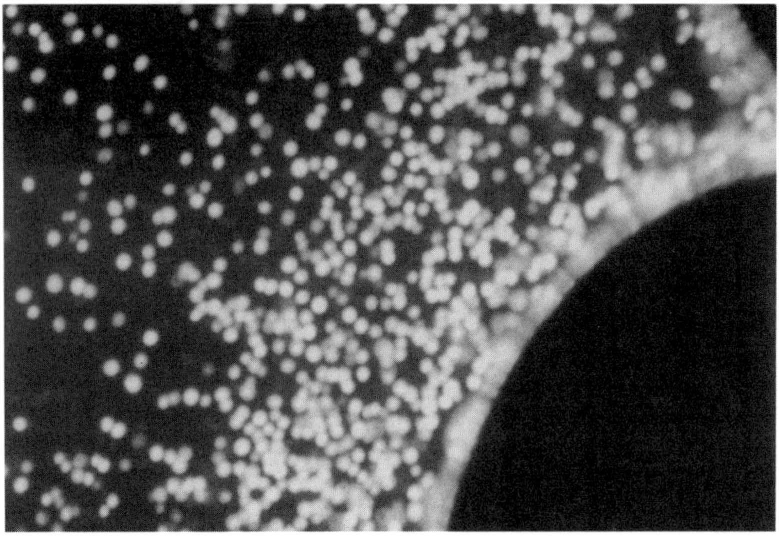

FIG. 32 Fluorescent poly(L,L-lactide) microspheres labeled with Lucifer yellow.

XII. SUMMARY

Biodegradable polyester microspheres have found many applications in fundamental studies in biology and in pharmacy as drug delivery carriers. Often it is essential to use particles with well-determined diameter, diameter distribution, molecular weight, and crystallinity of constituting polymers. This chapter has described fundamental features of the pseudoanionic and anionic polymerizations of ε-caprolactone and lactides leading to product in form of microspheres. Recipes allowing synthesis of microspheres were given. Also discussed were models of microsphere nucleation and growth and experimental results indicating formation of microsphere nuclei at the very beginning of polymerization followed by propagation within microspheres. Formal kinetics of ionic and/or pseudoionic dispersion polymerizations of cyclic esters were analyzed. On the basis of these analyses differences between rates of polymerization in solution and in dispersed systems were explained. Methods suitable for control of degree of crystallinity of poly(L,L-lactide) microspheres in a broad range from 0 to 60% were described. Also described was a method allowing transfer of microspheres from organic liquids (in which they are synthesized) into water-based media in which they form stable suspensions. Strategies of entrapment/attachment of bioactive compounds (including proteins) to poly(ε-caprolactone) and poly(lactide) microspheres were discussed and the corresponding examples given.

REFERENCES

1. Lovell, P.A., El-Aasser, M.S., Eds. *Emulsion Polymerization and Emulsion Polymers*; Wiley: Chichester, 1997.
2. Rosoff, M., Ed. *Controlled Release of Drugs: Polymers and Aggregate Systems*; VCH: New York, 1989.
3. Whateley, T.L., Ed. *Microencapsulation of Drugs*; Harwood Academic Publishers: Reading, 1992.
4. Park, K.; Shalaby, W.S.W.; Park, H. Eds. *Biodegradable Hydrogels for Drug Delivery*; Technomic: Lancaster, 1993.
5. Benita, S., Ed. *Microencapsulation, Methods and Industral Applications*; Marcel Dekker: New York, 1996.
6. Domb, A.J., Ed. *Polymeric Site-Specific Pharmacotherapy*; Wiley: New York, 1994.
7. Clarke, N.; O'Connor, K.; Ramtoola, Z. Drug Dev. Ind. Pharm. **1998**, *24*, 169.
8. Giunchedi, P.; Alpar, H.O.; Conte, U. J. Microencapsulation, **1998**, *15*, 185.
9. Tsuruta, T.; Kawakami, Y. Anionic ring-opening polymerization: stereospecificty for epoxides, episulfides and lactones. In *Comprehensive Polymer Science*; Allen, G., Bevington, J.C., Eds.; Pergamon Press: Oxford, 1989; 489.
10. Jerome, R.; Teyssie, P. Anionic ring-opening polymerization: lactones. In *Comprehensive Polymer Science*, Allen, G., Bevington, J.C., Eds.; Pergamon Press: Oxford, 1989; 501.
11. Penczek, S.; Slomkowski, S. Cationic ring-opening polymerization: cyclic esters. In *Comprehensive Polymer Science*; Allen, G., Bevington, J.C., Eds.; Pergamon Press: Oxford, 1989; 831.
12. Sosnowski, S.; Slomkowski, S.; Penczek, S.; Reibel, L. Makromol. Chem. **1983**, *184*, 2159.
13. Sosnowski, S.; Slomkowski, S.; Penczek, S. J. Macromol. Sci. Chem. **1983**, *A20*, 979.
14. Hofman, A.; Szymanski, R.; Slomkowski, S.; Penczek, S. Makromol. Chem. **1984**, *185*, 655.
15. Hofman, A.; Slomkowski, S.; Penczek, S. Makromol. Chem. Rapid Commun. **1987**, *8*, 387.
16. Jedlinski, Z.; Kowalczuk, M. Macromolecules **1989**, *22*, 3242.
17. Jedlinski, Z.; Walach, W.; Kurcok, P.; Adamus, G. Makromol. Chem. **1991**, *192*, 2051.
18. Slomkowski, S.; Duda, A. Anionic ring-opening polymerization. In *Ring-Opening Polymerization*; Brunelle, D.J., Ed.; Carl Hanser Verlag: Munich, 1993; 87.
19. Kurcok, P.; Penczek, J.; Franek, J.; Jedlinski, Z. Macromolecules **1992**, *25*, 2285.
20. Kricheldorf, H.R.; Lee, S.-R. Macromolecules **1995**, *28*, 6718.
21. Kricheldorf, H.R.; Lee, S.-R.; Bush, S. Macromolecules **1996**, *29*, 1375.
22. Kricheldorf, H.R.; Eggerstedt, S. Macromol. Chem. Phys. **1998**, *199*, 283.
23. Kricheldorf, H.R.; Lee, S.-R.; Schittenhelm, N. Macromol. Chem. Phys. **1998**, *199*, 273.
24. Kowalski, A.; Duda, A.; Penczek, S. Macromolecules **2000**, *33*, 689.
25. Kowalski, A.; Duda, A.; Penczek, S. Macromolecules **2000**, *33*, 7539.

26. Kricheldorf, H.R.; Kreiser-Saunders, I.; Stricker, A. Macromolecules **2000**, *33*, 702.
27. Kowalski, A.; Libiszowski, J.; Duda, A.; Penczek, S. Macromolecules **2000**, *33*, 1964.
28. Majerska, K.; Duda, A.; Penczek, S. Macromol. Rapid Commun. **2000**, *21*, 1327.
29. Penczek, S.; Duda, A.; Kowalski, A.; Libiszowski, J.; Majerska, K.; Biela, T. Macromol. Symp. **2000**, *157*, 61.
30. Duda, A.; Penczek, S. Mechanisms of aliphatic polyester formation. In *Biopolymers, Vol. 3b*: *Polyesters II--Properties and Chemical Synthesis*; Steinbüchel, A., Doi, Y., Eds.; Wiley-VCH: Weinheim, 2001; 371.
31. Ottewill, R.H. Stabilization of polymer colloid dispersions. In *Emulsion Polymerizations and Emulsion Polymers*; Lovell, P.A., El-Aasser, M.S., Eds.; Wiley: Chichester, 1997; 59.
32. Union Carbide. US Patent 3,632,669, 1972.
33. Ito, K.; Hashizuka, Y.; Yamashita, Y. Macromolecules **1997**, *6*, 624.
34. Ito, K.; Yamashita, Y. Macromolecules **1978**, *11*, 68.
35. Sosnowski, S.; Slomkowski, S.; Penczek, S.; Reibel, L. Makromol. Chem. **1983**, *184*, 2159.
36. Slomkowski, S. J. Macromol. Sci. Chem. **1984**, *A21*, 1383.
37. Slomkowski, S. Makromol. Chem. **1985**, *186*, 2581.
38. Penczek, S.; Duda, A.; Slomkowski, S. Makromol. Chem. Macromol. Symp. **1992**, *54/55*, 31.
39. Penczek, S.; Duda, A. Makromol. Chem. Macromol. Symp. **1993**, *67*, 15.
40. Penczek, S.; Duda, A.; Szymanski, R. Macromol. Symp. **1998**, *132*, 441.
41. Kowalski, A.; Duda, S.; Penczek, S. Macromolecules **2000**, *33*, 689.
42. Kowalski, A.; Duda, A.; Penczek, S. Macromolecules **2000**, *33*, 7359.
43. Sosnowski, S.; Gadzinowski, M.; Slomkowski, S.; Penczek, S. J. Bioact. Compatible Polym. **1994**, *9*, 345.
44. Sosnowski, S.; Gadzinowski, M.; Slomkowski, S. Macromolecules **1996**, *29*, 4556.
45. Tunc, D.C.; Rohovsky, W.M.; Jadhav, B.; Lehman, W.B.; Strongwater, A.; Kummer, F. In *Advances in Biomedical Polymers*; Gebelein, C.G., Ed.; Polymer Science Technology, Vol. 35; Plenum Press: New York, 1987; 87.
46. Leenslag, J.W.; Pennings, A.J. J. Makromol. Chem. **1987**, *188*, 1809.
47. Duda, A.; Penczek, S. Macromolecules **1990**, *23*, 1636.
48. Sosnowski, S. Polymer **2002**, *42*, 637.
49. Slomkowski, S.; Sosnowski, S.; Gadzinowski, M. Macromol. Symp. **1997**, *123*, 45.
50. Slomkowski, S.; Sosnowski, S.; Gadzinowski, M.; Pichot, C.; Elaissari, A. Direct synthesis of polyester microspheres, potential carriers of bioactive compounds. In *Tailored Polymeric Materials for Controlled Delivery Systems*; McCulloch, I., Shalaby, W., Eds.; ACS Symp. Ser., 709; ACS: Washington, DC, 1998; 143.
51. Sosnowski, S.; Slomkowski, S.; Lorenc, A.; Kricheldorf, H.R. Colloid Polym. Sci. **2002**, *280*, 107–115.
52. Slomkowski, S.; Sosnowski, S. ACS Polym. Prep. **1998**, *39* (2), 212.
53. Gadzinowski, M.; Sosnowski, S.; Slomkowski, S. Macromolecules **1996**, *29*, 6404.
54. Slomkowski, S.; Gadzinowski, M.; Sosnowski, S. Macromol. Symp. **1998**, *132*, 451.

55. Slomkowski, S.; Sosnowski, S.; Gadzinowski, M. Colloids Surf. A Physicochem. Eng. Asp. **1999**, *153*, 111.
56. Slomkowski, S.; Sosnowski, S.; Gadzinowski, M.; Pichot, C.; Elaissari, A. Macromol. Symp. **2000**, *150*, 259.
57. Sosnowski, S.; Slomkowski, S.; Lorenc, A.; Kricheldorf, H.R. Colloid Polym. Sci. **2002**, *280*, 107.
58. Slomkowski, S.; Sosnowski, S.; Gadzinowski, M. Polym. Degr. Stabil. **1998**, *59*, 153–160.
59. Bonetskaya, A.K.; Sukuratov, S.M. Vysokomol. Soedin. Ser. A **1969**, *11*, 532.
60. Slømkowski, S. Polymer **1986**, *27*, 71–75.
61. Sosnowski, S.; Slomkowski, S.; Penczek, S. Makromol. Chem. **1991**, *192*, 735–744.
62. Duda, A.; Penczek, S. Chem. Rapid Commun. **1994**, *15*, 559.
63. Duda, A.; Penczek, S. Makromol. Chem. Macromol. Symp. **1991**, *42/43*, 135.
64. Duda, A.; Penczek, S.; Makromol. Chem. Macromol. Symp. **1991**, *47*, 127.
65. Gadzinowski, M.; Slomkowski, S.; Elaissari, A.; Pichot, C. J. Biomater. Sci. Polym. Ed. **2000**, *11*, 459.
66. Sturm, E.; Kruger, U.; Senn-Bilfinger, J.; Figala, V.; Klemm, K.; Kohl, B.; Rainer, G.; Schaefer, H.; Blake, T.J.; Darkin, D.W.; Ife, R.J.; Leach, C.A.; Mitchell, R.C.; Pepper, E.S.; Salter, C.J.; Viney, N.J.; Huttner, G.; Zsolnai, L. J. Org. Chem. **1987**, *52*, 4573.
67. Senn-Bilfinger, J.; Kruger, U.; Sturm, E.; Figala, V.; Klemm, K.; Kohl, B.; Rainer, G.; Schaefer, H.; Blake, T.J.; Darhin, D.W.; Ife, R.J.; Leach, C.A.; Mitchell, R.C.; Pepper, E.S.; Salter, C.J.; Viney, N.J. J. Org. Chem. **1990**, *55*, 4163.

15

Supercritical Fluid Processes for Polymer Particle Engineering

Applications in the Therapeutic Area

JOEL RICHARD Ethypharm S.A., Saint-Cloud, France

FRANTZ S. DESCHAMPS Mainelab S.A., Angers, France

I. INTRODUCTION

The production of polymer particles with specific physicochemical and size properties is an increasing challenge for drug delivery in the pharmaceutical area. These particles, whose size typically lies within the micrometer range, consist of synthetic biocompatible polymers that contain active materials and can be applied to living systems (in vivo) without inducing undesirable effects. For this purpose, biodegradable polymers are commonly used as materials for sustained-release delivery systems because the control of their physicochemical features makes it possible to design a specific release profile for the drug [1]. Particulate drug delivery systems are useful for several routes of administration. They can be administered as dry powders for inhalation applications or as aqueous suspensions by parenteral, topical, or oral routes. Parenterally, polymer particles containing a drug whose size is within the 1- to 100-μm range can be administered either subcutaneously or intramuscularly to obtain sustained-release depots [2], thus minimizing the frequency of drug administration. Intravenously, suspensions of polymer nanoparticles containing a drug, whose diameter is less than 1 μm, can be administered to increase the systemic circulating time of the drug, reduce toxicity, and target solid tumors [3]. Thus, polymer micro- and nanoparticles are of great importance for drug delivery and targeting.

Current methods for preparation of drug-containing polymer microparticles are mainly based on emulsion–solvent extraction or phase separation techniques. The former technique requires dissolution of the polymer in an organic solvent, dispersion or dissolution of the drug in this solution, emulsification of this organic phase in an aqueous solution, and subsequent extraction of the solvent by evaporation or dilution in a large excess of water [4]. The latter technique involves dispersion of drug in an organic solution of a polymer and addition of a nonsolvent, either a polymer or an organic solvent, to induce the phase separation of a polymer-rich phase containing the drug; an additional organic solvent is then used in a large excess to harden the microspheres [5].

A serious drawback of these techniques is the extensive use of organic solvents to either dissolve the polymer or induce phase separation and hardening. Most of these organic solvents are toxic; moreover, they are suspected to be responsible for biological inactivation of large molecules such as proteins [6]. Due to the large amounts of solvents used in these processes, it is not possible to completely remove them from the particles, which is required prior to their in vivo use. The removal of residual solvents currently involves heating, which is not appropriate for heat-labile drugs. In addition, these methods display other major drawbacks. For instance, in the emulsification phase, the emulsion–solvent extraction method requires the use of surfactants that have then to be removed from the end product. This process also leads to a wide particle size distribution, and to a low drug loading in the case of water-soluble drugs due to drug partitioning between the organic and aqueous phase.

Other methods reported in the literature for the preparation of particulate polymer drug delivery systems do not use any organic solvent. They are based on hot-melt procedures, which involve shear and heat that can have a detrimental effect on fragile molecules [7].

Accordingly, there is a strong interest in new methods for production of polymer particles, which can be carried out without any organic solvent and under conditions that have minimal detrimental effect on the drug encapsulated, especially in the case of fragile molecules such as peptides and proteins.

The use of supercritical fluids (SCFs) as vehicles for the production of polymer particles with therapeutic applications is a recent development. SCFs appear to be one of the most promising factors in the development of new production techniques that completely eliminate organic solvents or minimize their use, and are carried out under mild temperature conditions. An SCF is a fluid whose pressure and temperature are simultaneously higher than those at the critical point, i.e., the end point of the liquid–gas phase transition line (Fig. 1). Several features of SCFs make them versatile and appropriate for production of polymer particles. They display some liquidlike properties, such as a high density, and other gaslike ones such as low viscosity and high diffusivity. The most important property of an SCF is its large compressibility near the critical point. The combination of liquidlike density and large compressibility is of major interest, since it leads to a fluid whose solvent power can be continuously tuned from that of a liquid to that of a gas, with small variations of pressure. Because of its low critical temperature ($T_c = 31.1°C$) and environmentally acceptable nature, CO_2 is the most widely used SCF in pharmaceutical development and process-

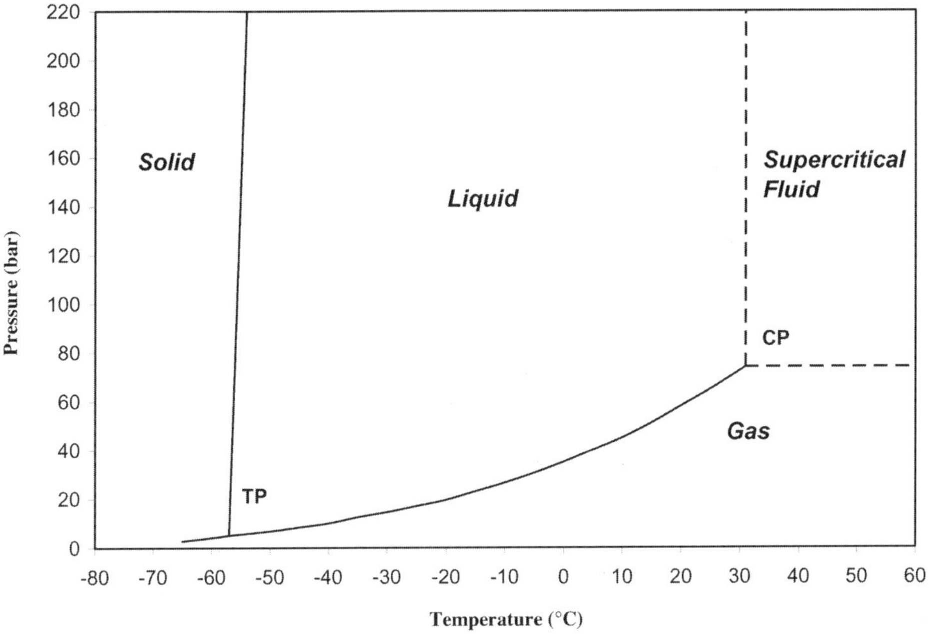

FIG. 1 Schematic projection of the pressure–temperature phase diagram for a substance. CO_2 has been chosen as an illustrative compound (critical temperature, T_c = 31.1°C, critical pressure, P_c = 73.8 bar; TP, triple point; CP, critical point).

ing. It actually offers many advantages. CO_2 is a readily available, nontoxic, and nonflammable agent that is generally recognized as safe and has a low cost. It allows one to work at moderate temperatures and leaves no toxic residues since it turns back to a gas phase at ambient conditions. Due to its unique properties, CO_2 is used routinely in large-scale operations for decaffeination of coffee beans and extraction of hops [8].

Several SC CO_2-based processes have been reported in the literature for the production of drug-loaded polymer micoparticles. These processes currently use CO_2 either as:

A solvent of the polymer and the drug from which particles are precipitated; this is known as the rapid expansion from supercritical solutions (RESS) process [9].

A swelling and plasticizing agent that is dissolved in the polymer until a gas-saturated solution is obtained, which is later expanded to cause supersaturation and particle precipitation. This is known as the particles from gas-saturated solutions (PGSS) process [10].

An antisolvent that causes precipitation of the polymer initially dissolved in an organic solvent. This is known as the gas antisolvent (GAS) [11] or supercritical antisolvent (SAS) process, and has many related modifications, e.g., aerosol solvent extraction system (ASES) process and solution-enhanced dispersion by SCF (SEDS) process [12,13].

A solvent for polymerization in dispersed media of acrylic and vinyl monomers [14].

This chapter will first briefly describe the basic aspects of SCFs and will then present the various SCF-based techniques for the production of drug-loaded polymer particles. For each technique, the typical physicochemical features of the particles obtained will be given in terms of particle size, drug loading, and morphology, and the factors influencing the characteristics of the SCF products will be discussed. Examples of drug delivery systems prepared using these processes will also be presented. The advantages and limitations of the processes will be pointed out and discussed, especially as regards the production yield and the scale-up issues. Finally, the potential of the polymer particles produced using SCF-based processes for therapeutic applications will be envisaged.

II. SUPERCRITICAL FLUID–BASED PROCESSES FOR THE PRODUCTION OF POLYMER MICROPARTICLES

A. Basic Aspects of Supercritical Fluids

As shown in the schematic projection of the phase diagram of a pure substance (Fig. 1), the phase transition from liquid to gas ends at the critical point. The term *supercritical fluid* strictly characterizes a substance that is above both its critical temperature and its critical pressure, but is often used improperly to describe a fluid in the relative vicinity of its critical point, mainly a fluid at a pressure over its critical pressure. In the SC state, the substance is neither a gas nor a liquid but possesses properties of both, whereby the density is liquidlike, the viscosity is gas-like, and the diffusion coefficient lies between that of a gas and a liquid [8]. One distinguishing feature of SCFs is that, though almost liquidlike in density, they possess a very high compressibility. The density behavior of carbon dioxide is shown in Fig. 2. At the critical point, the fluid compressibility diverges, as seen from the vertical slope of the critical isotherm at the critical point, resulting in an infinite rate of change in density with respect to the isothermal pressure change. The solvent power of SCF is thus highly dependent on the variations of temperature or pressure. This behavior allows the use of pressure or temperature as very sensitive parameters for tuning the properties of the fluid, and in particular its solvent power. The ratio of the experimental solubility in an SCF to the solubility predicted by assuming ideal

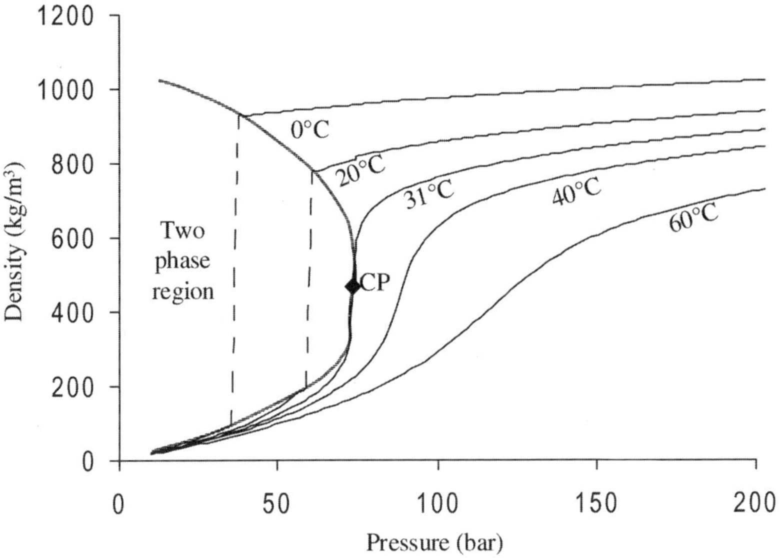

FIG. 2 Isothermal density behavior of CO_2 as a function of pressure.

gas behavior at the same pressure and temperature can be as high as 10^6 for nonvolatile solutes [15]. For polymer processing, this feature is exploited for the fractionation of polymers by programming the fluid density [16]. The properties of SCF, including the solubility and phase equilibria calculation methods, have been extensively reviewed [8,17,18].

Carbon dioxide (CO_2) is the most widely used SC solvent because it possesses easily accessible critical parameters (Table 1) and is inexpensive, inert toward most compounds, and nontoxic. Its low critical temperature allows the processing of thermolabile compounds, such as pharmaceuticals. A limitation of CO_2 in comparison with other SC solvents is its poor solvent power for highly polar, ionic, or high molecular weight compounds. Under readily achievable conditions, most of the high molecular weight polymers are not soluble in SC CO_2. Only some fluoropolymers or silicones [8] show a good solubility in pure CO_2. The addition of small amounts of a volatile cosolvent, such as acetone or ethanol, can dramatically increase the solvent power of an SCF [19]. Many therapeutic compounds or carriers are polar compounds and/or have high molecular weights, neither of these properties being conducive to high solubility in SC CO_2.

One major advantage of using SCFs is that they can be efficiently separated, by decompression, from both organic solvents or solid products, facilitating sin-

TABLE 1 Critical Temperature (T_c) and Critical Pressure (P_c) of Commonly Used Supercritical Solvents

Fluid	T_c (°C)	P_c (bar)
Ethylene	9.3	50.4
Carbon dioxide	31.1	73.8
Nitrous oxide	31.3	73.8
Ethane	32.3	48.8
Chlorodifluoromethane	96.0	51.9
Propylene	91.8	46.2
Propane	96.7	42.5
Ethanol	240.8	61.4
Water	374.1	217.7

gle-step, clean, and recycling process engineering. In the case of manufacturing techniques using pure SC solvents, a solvent-free product is recovered. This unique feature is a key point for the processing of pharmaceuticals. When SCFs are used as antisolvent, the process conditions are set up so that the SCF extracts most of the organic solvent during the particle formation. Taking advantage of their high diffusivity and solvent power, SCFs are then used for further extraction of residual solvents from the final product. The efficiency of this solvent extraction, currently used in the manufacturing of particles by SCF antisolvent techniques, is illustrated in a recent coating technology [20]. After coating tablets with an organic solution of coating agent, an SCF is contacted with the coated tablets so as to strip the solvent.

Due to their unique features and to their ability to allow the reduction of organic solvents, SCFs have been intensively used for the development of polymer particle engineering over the past 20 years.

B. Review of the Production Processes

The use of SCFs or dense gases has been recently widened to the processing of pharmaceuticals. In this chapter, we will focus on the techniques that can be used for the formation of drug-loaded polymer particles. For more general information about material processing with SCFs, particle formation, SCF extraction, or other applications of SCF, some landmark reviews or books [8,16–18,21–25] are recommended.

1. Rapid Expansion of SC Solutions

Production of particles by rapid expansion of SC solutions (RESS) relies on the fact that the solvent strength of an SCF can be dramatically reduced by altering its pressure. Depending on the nature of solutes and operating conditions, the RESS process can produce very small and nearly monodispersed particles. The first publication dealing with the formation of particles from SC solutions was published in 1879 [26], where the formation of "snow" by a sudden pressure reduction of a binary solution was reported. Surprisingly, the potentiality of processing difficult to comminute solids by decompression of an SC solution was not highlighted until the early 1980s [27].

As presented in Fig. 3, the RESS process consists of dissolving the material to be powdered in an SCF, then expanding the SC solution through a nozzle into a low-pressure chamber. Due to the simplicity of its concept, RESS allows the production of solvent-free particles within a single-step operation and can be implemented in a simple way. The SC solution is obtained by passing a preheated SCF through an extraction vessel previously loaded with the solute(s) to be processed. If the precipitation of a mixture of solutes (e.g., drug and polymer carrier) is desired, the extraction unit is loaded with multiple solutes [28], or multiple extraction vessels are used with a subsequent mixing of SC solutions [9,29]. The mixing of SC solutions from multiple extraction units makes it possible to control the content of each component in the expanded SC solution. On the other hand, the use of a single extraction vessel packed with multiple components means that the particle composition is governed by equilibrium solubilities in the extraction vessel. The expansion capillaries, or fine-diameter orifices of the nozzles, typically have a diameter ranging from 5 to 200 μm. To avoid phase separation, solute precipitation, and plugging related to cooling effects, the nozzle has to be heated. The low-pressure chamber can be set either at the atmospheric pressure, or at an intermediate pressure between

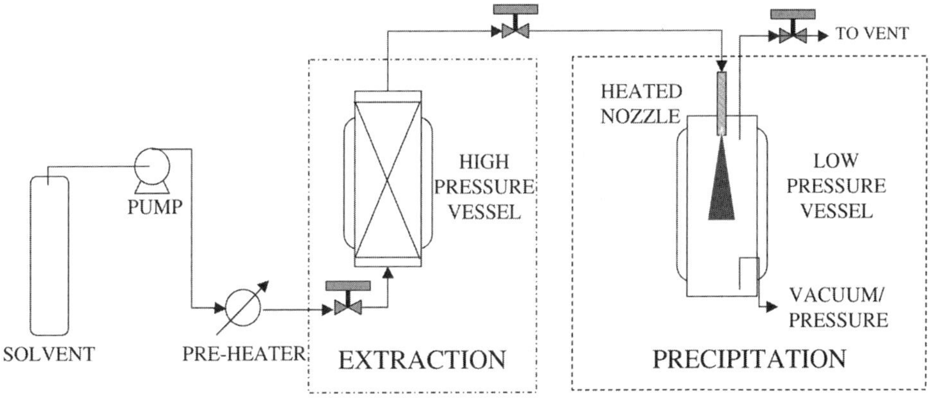

FIG. 3 Schematic representation of the RESS equipment.

atmospheric and preexpansion pressure, or even at a vacuum pressure. The main process parameters that influence the particulate product morphology were identified; these are the temperature and pressure in the extraction unit and in the precipitation vessel, the nozzle geometry and diameter, the nature of solute–solvent interaction, and the nature of the solute to be precipitated. Depending on these parameters, various solid shapes and morphologies are obtained, from fine powders to needles, or even thin films if the SC solution is expanded onto a surface [30]. Agglomeration of the discrete particles and difficulties encountered for their collection were often reported for the RESS process. The final particulate product is dry and solvent-free, which eliminates any further purification step usually needed with conventional liquid solvent–based processes. Micrometer-sized particles are typically obtained with the RESS process. However, some authors reported the production of nanometric powder of pure compounds as cholesterol or benzoic acid [31].

The flow pattern and nucleation process of RESS were investigated [9,15, 28,32–36]. If the phase change from the SC to the gas state takes place in the nozzle and in the supersonic free jet after the nozzle, the nucleation process is extremely rapid, with a particle nucleation and growth time estimated to be much less than 10^{-5} s [9]. The expansion of the SC solution leads to high supersaturation, nucleation and consequently to the formation of particles. Cooling rates of 10^9 K/s and supersaturations of about 10^5–10^8 were calculated using a modeling of the pressure and temperature changes along the expansion pathway [33].

Particle size reduction was the primary application of RESS, displaying major advantages over conventional processes, either mechanical (grinding, milling) or wet (crystallization from solution). As frequently reported [21,22,37–39], RESS has been applied to a wide variety of pure materials including pharmaceuticals, polymers, dyes, and inorganic compounds. The RESS process requires compounds to precipitate to be soluble enough in the SCF. However, most pharmaceutical compounds exhibit solubilities below 0.1 wt% under mild processing conditions [38,39].

Polymer RESS processing was described in the first comprehensive study of RESS [27]. Using propylene as the SCF, polypropylene was changed from rough-surfaced spheres of 30 μm to fibrous particles with multiple branches and high aspect ratios (about 50 μm length and 2–3 μm diameter). As shown in Table 2, various polymers were then processed using RESS. The Batelle Institute's detailed pioneer work on RESS processing of polymers [9,40,41] has demonstrated that this process can be used to produce solvent-free polymer powders with various morphologies ranging from micrometer-sized spherical particles to fibers of 100–1000 μm length. These experiments involved drastic operating conditions, such as high preexpansion temperature of 350°C and the use of flammable low molecular weight alkanes as SC solvents to overcome the low solubility of most polymers in SC CO_2.

All of these studies made it possible to foresee the opportunity offered by RESS to produce intimate mixtures of materials with a controlled morphology in a single processing step. RESS has been considered for the engineering of biodegradable polymer particles with the ultimate goal of producing drug-loaded polymer particles. However, most of the polymers used in therapeutic applications (e.g., for sustained release) are not reasonably soluble in SC CO_2.

The wide variety of bioerodible polymer particle morphologies that can be produced by RESS has been illustrated by the processing of polycaprolactones (PCLs). PCLs were found to be insoluble in SC CO_2, but were soluble in SC chlorodifluoromethane [42]. Depending on the preexpansion temperature and pressure, as well as *L/D* ratio of the orifices, various morphologies of polycaprolactone particles were obtained from fine-diameter particles to high aspect ratio fibers. Using cloud point measurements for chlorodifluoromethane/polymer systems, the product morphology was found to be dependent on the degree of saturation upstream of the expansion nozzle [32]. It was shown that a process whose operating conditions lead to a phase separation in the expansion device results in a precipitation time scale over microseconds only. The nuclei formed during the early stages of phase separation are then frozen, and submicrometer particles are produced. Conversely, the occurrence of fibers is related to a phase separation over tens of seconds in the entry region of the orifice. Thus, it appears that RESS dynamics can be used to control the shape morphology of polymer precipitates. This feature is illustrated with the precipitation of poly(L-lactic acid) (L-PLA) from a $CHClF_2$ SC solution, where tuning the operating conditions for a microsecond phase separation resulted in the formation of practically spherical particles with a diameter ranging from 0.2 to 0.6 μm [32].

The RESS processing of the poly(hydroxy acids) L-PLA, poly(D,L-lactic acid) (DL-PLA), and poly(glycolic acid) (PGA) with CO_2 as the SC solvent was also reported [28]. Commercial L-PLA ($M_w = 5500$ g/mol, $M_w/M_n = 2.0$) were found to be soluble in SC CO_2, with a solubility of 0.05 wt %. The solubility was increased up to 0.37 wt % with the addition of 1 wt % of acetone as entrainer. The polydispersity of L-PLA has a significant impact on the formation of polymeric microspheres by the RESS process. The SC solvent preferentially extracts low molecular weight L-PLA [28], thereby affecting the molecular weight and polydispersity of the processed polymer. Starting from commercial polydisperse L-PLA, the molecular weight of extracted and precipitated polymer varies during the course of the RESS experiment. The low molecular weight L-PLA were first extracted and formed a viscous low glass transition temperature precipitate. After sufficient preconditioning, the molecular weight of the extracted L-PLA was found to be 2000–3000 g/mol (starting from $M_w = 5500$ g/

TABLE 2 Polymers and Composite Drug/Polymer Particles Processed with RESS or *n*-RESS

Substrates	Supercritical fluid	Results and comments	Ref.
Polymers			
Polypropylene	Propylene	Fibrous particles with multiple branches (2- to 3-5 μm diameter, 50 μm length)	Krukonis, 1984 [27]
	Pentane	1-μm spheres or small micrometer-size fibers	Matson et al., 1987 [40]
Polystyrene	Pentane/cyclohexanol 98:2 v/v	0.3-μm particles	Smith et al., 1983 [30]
	Pentane	100- to 1000-μm fibers to 20-μm speres with increasing preexpansion temperature	Matson et al., 1987 [40]
Poly(carbosilane)	Pentane	Particles (<0.1 μm diameter) or fibers (1 μm diameter, 80–160 μm length	Matson et al., 1987 [9, 40]
Poly(phenylsulfone)	Propane	Agglomerated 0.5-μm spheres	Matson et al., 1987 [9,40]
Poly(vinyl chloride) and potassium iodide	Ethanol	7-μm spheres homogeneous in composition	Matson et al., 1987 [9]
Poly(methyl methacrylate) (PMMA)	Propane	Particles (1 μm diameter) or fibers (1 μm diameter, 100–1000 μm length)	Matson et al., 1987 [9,40]
	$CHClF_2$	Powders or fibers	Lele et al., 1992 [42]
Polycaprolactone (PCL)	$CHClF_2$	Fine-diameter powder (1–5 μm) to high aspect ratio fibers	Lele et al., 1992 [42] Lele et al., 1994 [32]
Poly(heptadecafluorodecyl acrylate)	CO_2	Particle (0.1–5 μm) and fibers	Blasig et al., 2000 [120]
Poly-L-lactic acid and poly(D,L-lactic acid) (L-PLA and DL-PLA)	CO_2	Spherical and cornflake particles (4–10 μm length)	Tom et al., 1991 [28]
L-PLA	CO_2/acetone 99:1 w/w	Particles (10–25 μm) to dendrites of up to 100 μm	Tom et al., 1991 [28]

L-PLA	$CO_2/CHClF_2$	Various operating conditions and expansion devices. Nonspherical microparticles and microspheres (<1 μm to 100 μm), dendrites (1–100 μm), and microparticles agglomerates	Tom et al., 1994 [29]
L-PLA	$CHClF_2$	0.3- to 5-μm fibers up to 1 cm length, 0.2- to 0.6-μm spherical particles	Lele et al., 1994 [32]
Poly(glycolic acid) (PGA)	CO_2	Regular morphology (oval/rectangular) 10–20 μm	Tom et al., 1993 [43]
Cellulose acetate	Pentane	0.8-μm-diameter fibers	Matson et al., 1987 [9,40]
Composite polymer particles			
DL-PLA/lovastatin	CO_2	Microspheres containing lovastatin needles	Tom et al., 1993 [43]
L-PLA/Pyrene	$CO_2/CHClF_2$	Uniform incorporation of pyrene microparticles (<5 μm) within polymer microspheres (<100 μm)	Tom et al., 1994 [29]
L-PLA/Naproxen	CO_2	Microspheres (10–90 μm) loaded with naproxen and some free naproxen particles (1–5 μm)	Kim et al., 1996 [44]
Poly(ethylene glycol)/(lipase or lyzozyme) (PEG)	CO_2/EtOH	Polymer microcapsules with protein core Particle diameter: 8–62 μm	Mishima et al., 2000 [51]
Poly(methyl methacrylate)/(lipase or lysozyme) (PMMA)	CO_2/EtOH	Microcapsules: 12–30 μm	Mishima et al., 2000 [51]
L-PLA/(lipase or lysozyme)	CO_2/EtOH	Microcapsules: 11–45 μm	Mishima et al., 2000 [51]
Poly(ethylene glycol)-poly(propylene glycol) triblock copolymer/(lipase or lysozyme) (PEG-PPG-PEG)	CO_2/EtOH	Microcapsules: 18–37 μm	Mishima et al., 2000 [51]
Aminoalkyl methacrylic copolymer/3-hydroxyflavone (Eudragit E-100)	CO_2/EtOH	Microcapsules	Mishima et al., 2000 [52]

mol, $M_w/M_n = 2$), and microparticles with diameters ranging from 10 to 25 μm were formed [28]. Usually, the poly(hydroxy acids) that can be processed using RESS are in the lower range of the commercially available molecular weights. Thus, the opportunity to control the release properties of polymer microparticles through the tuning of the polymer molecular weight is severely restricted to a narrow range of low molecular weight. This feature is one of the main limitations to the application of RESS for the production of drug-loaded polymeric particles.

The formation of drug-loaded polymer particles with tailored morphology is a major challenge for therapeutic applications. The particle size control and shape monitoring that can be achieved with the RESS process are of great interest for these applications. The first reported example of RESS-processed polymer microparticles for pharmaceutical applications dealt with the formation of composite microparticles using a commercial DL-PLA ($M_w = 5300$ g/mol, $M_w/M_n = 1.9$) and lovastatin, an anticholesterol drug ($M_w = 404$ g/mol) [43]. Experiments were carried out both with a single extraction vessel loaded with both solutes, and with two columns operating in parallel (one with both compounds and the other with lovastatin only). Using the one-column experimental setup and after a sufficient preconditioning time to extract the low molecular weight polymers, various morphologies were obtained using CO_2 as the SC solvent: these are microspheres containing single lovastatin needles, egg-shaped polymer particles enveloping lovastatin needles, polymer microspheres without protruding needles and separated lovastatin needles. If two columns were used, large microparticles containing multiple needles of lovastatin were obtained. In this study [43], it was assumed that drug crystal nucleation and growth occurred first, and that the drug crystals subsequently acted as nucleating sites for the polymer.

Using a mixture of CO_2 and $CHCLF_2$ as the SC solvent, the coprecipitation of L-PLA and the model compound pyrene was also studied [29]. The optimization of the RESS process allowed the formation of composites microspheres (<100 μm) with pyrene microparticles (<5 μm) uniformly embedded in the polymer matrix. The polymer was dissolved in $CO_2/CHClF_2$ in a first extraction column, and pyrene was dissolved in pure CO_2 in a second vessel. This experimental setup makes it possible to tune the L-PLA and pyrene concentration in the SC solution, and shows that a limited amount of pyrene can be incorporated into the microsphere (pyrene agglomerated apart from the microspheres). The most recent example [44] of RESS coprecipitation using SC CO_2 is the formation of 10- to 90-μm particles of L-PLA ($M_w = 2000$ g/mol) loaded with naproxen, a nonsteroidal anti-inflammatory drug. Composite particles consisted of a naproxen core surrounded by a thin polymer coating. These results clearly showed that the formation of composite microparticles with a low molecular weight pharmaceutical compound incorporated within a biodegradable polymer matrix can be achieved using the RESS process.

Some modifications of the RESS process were also designed to perform the coating of a drug with a polymer. One of these modifications involves the precipitation of a substrate from an SC solution onto the surface of host particles contained in the precipitation vessel. The combination of RESS with fluidized bed technology has been often reported [45–48] and, because of its good solubility in SC CO_2, paraffin was used as a coating agent in most of these studies. In addition, the coating of spherical glass beads (315 and 500 μm) with poly(vinyl chloride-*co*-vinyl acetate) (PVCVA) or hydroxypropylcellulose (HPC) was also recently reported with this modified RESS process [49]. However, HPC was not soluble enough in SC CO_2, leading to the use of acetone as a cosolvent. This modified RESS coating process could find applications in the pharmaceutical area as an alternative to conventional wet chemistry or dry particle coating methods, where the use of large amounts of organic solvents, or high shear and compressive forces, may lead to the loss of activity of bioactive compounds. However, the low solubility of most polymers of pharmaceutical interest restricts the use of this process.

The formation of polymer coated microparticles was also investigated with a second modification of RESS, i.e., the rapid expansion from SC solution with a nonsolvent (*n*-RESS) [50,51]. *n*-RESS was designed to overcome the solubility issue and the agglomeration of particles often observed for the basic RESS process applied to polymers. The key feature of *n*-RESS is the use of a cosolvent raising the polymer solubility into the SC fluid but acting as a nonsolvent of the polymer at atmospheric pressure, avoiding particle agglomeration. Using *n*-RESS, the formation of microcapsules of 3-hydroxyflavone coated by a polymer was obtained by spraying a suspension of 3-hydroxyflavone in an SC solution (CO_2/ethanol) of the aminoalkyl methacrylic copolymer Eudragit E-100 [52]. The *n*-RESS process, where the core material is suspended in the SC solution of polymer, was also applied to the formation of protein microcapsules. Lipases and lysozyme particles were coated with various polymers, such as poly(ethylene glycol) or poly(L-lactic acid), without particle agglomeration and with a control of the particle size by tuning the polymer feed composition [51]. As a conclusion, the recently reported *n*-RESS process permits coating of preformed drug particles that are nonsoluble in SC CO_2 with slightly polar or high molecular weight polymers that are not soluble enough in pure CO_2 to be processed using the basic RESS process.

From this review, it can be concluded that, when the materials to be processed meet the solubility requirements, RESS offers the unique possibility of producing solvent-free composite microparticles through a single-step operation and with monitoring of the particle size and shape. The scaling up of the RESS process to a large production stage requires a multinozzle system or the use of a porous sintered disc, which leads to greater difficulties in controlling particle size distribution. Moreover, the collection of very small particles at such a large scale is complex, as known for any process leading to micrometer-sized particles.

The low solubility of most compounds used in therapeutic applications in pure SC CO_2 remains the major limitation of RESS, where both the drug and the polymer carrier have to be dissolved prior to expansion. When RESS is used for polymer processing, the solubility issue restricts its application to low molecular weight polymers, narrowing the field of potential therapeutic applications. Nevertheless, in a process screening for the production of polymer particles, RESS and its modifications should always be considered first, as they offer many advantages.

2. Particles from Gas-Saturated Solutions

While RESS is based on the decompression of an SC solution to produce solid particles, particles from gas-saturated solutions (PGSS) [10,53] and related processes involve the decompression of a solute-rich liquid phase in which the SCF is dissolved (Fig. 4). Although the solubility of polymers in SC CO_2 is extremely low, the solubility of CO_2 in many polymers is appreciable. This sorption of CO_2 leads to a dramatic decrease in the glass transition temperature (T_g) of polymers and a subsequent swelling of the polymer. For instance, because of the dissolution of CO_2, the T_g of polystyrene can be lowered by up to 50°C [54]. In addition, the viscosity of the obtained polymer melt has been reported to be reduced by several orders of magnitude [55]. The first step of the PGSS process (Fig. 4) involves the dissolution of an SCF, generally CO_2, in a melted or liquid-

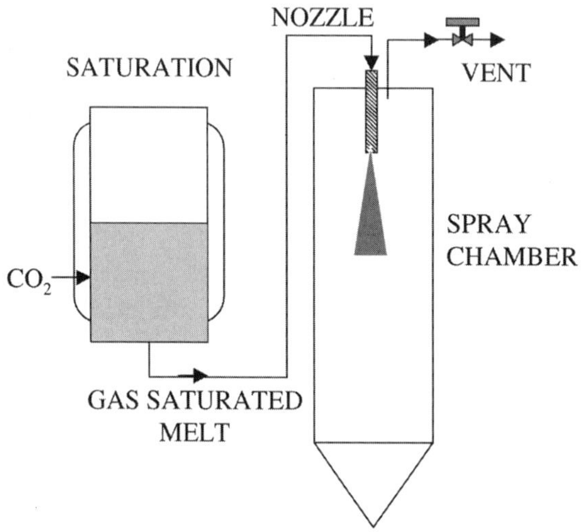

FIG. 4 Schematic representation of the PGSS equipment.

suspended material to form a so-called gas-saturated solution or suspension. This gas-saturated solution is further expanded through a nozzle to form fine particles due to the sharp temperature decrease caused by fluid expansion. The PGSS process thus allows production of particles from compounds that are not soluble in SC CO_2, especially high molecular weight polymers. PGSS was mostly applied to powder coating manufacture and application, with the Unicarb process from Union Carbide [56,57], a similar process from Nordson [58], and the VAMP process from Ferro [59]. These applications take advantage of the properties of SC CO_2 as a viscosity-reducing fluid and as a softening, swelling, and plasticizing agent of polymers. These processes are proven to operate at large scale in very large production units already used for organic-based dry powder coatings [60].

The basic PGSS process was patented for the preparation of particles or powders [53], with further investigations leading to the formation of PEG particles, whose size typically ranges from 150 to 400 μm [61]. PGSS was also applied to the processing of nifedipine, a poorly water-soluble calcium antagonist drug [10,62]. In order to increase the drug dissolution rate, pure nifedipine particles were reduced to a size of 15 μm by dissolution of SC CO_2 in the melted drug (>170°C) and subsequent expansion, which causes supersaturation and fine-particle precipitates. To avoid the degradation of the drug caused by high-temperature operating conditions used to melt the drug, a lower melting point (about 55°C) mixture of nifedipine/PEG 80:20 (w/w) was processed by PGSS [62]. Porous coprecipitates of nonregular shape showing higher dissolution rates of nifedipine were formed. Similar results were also obtained for felopidine/PEG 4000 composite microparticles [63]. The PGSS processing of the hypolidemic drug fenofibrate allowed neither micronization of the starting material, nor obtaining of higher dissolution rates of drug/PEG composites [63].

The original PGSS process, which has so far been very scarcely applied to the processing of therapeutic polymer systems, was modified to form microspheres from a suspension of a core active substance in a gas-saturated polymer carrier [64]. This process, called polymer liquefaction using SC solution (PLUSS), involves the swelling and liquefaction of the polymer carrier much below its glass transition or melting temperature by dissolution of dense CO_2. The suspension of the drug in the liquefied polymer is then atomized into a low-pressure vessel, leading to microcapsules with the drug core particles embedded in the polymer matrix, as claimed for the processing of a vaccine with polycaprolactone [64].

PGSS is already widely used at large scale and allows the processing of a very wide range of products (liquid solutions or suspensions) with many advantages over the other SCF-based processes, such as a medium operating pressure, a small volume of high-pressure equipment, a low fluid consumption, or the ability to process a wide range of compounds. However, using PGSS, the particle size and morphology is difficult to monitor and control, and the feasibility of submicrometer particles has not yet been shown. For the preparation of polymer composites containing a thermally sensitive component, the PGSS process might be limited to relatively low-melting polymers that are strongly plasticized by SC CO_2. These features may restrict the applications of PGSS to therapeutic applications involving drug-loaded microparticles.

3. Antisolvent Processes (GAS, SAS, PCA, SEDS, ASES)

Liquid antisolvent processes are widely used to precipitate, purify, or micronize various substrates. Generally speaking, starting from an organic solution, the addition of a miscible antisolvent induces supersaturation and precipitation of the solute into microparticles. Attributes of a good antisolvent include high diffusivity, low viscosity, and high solubility within the solvent, all of which are consistent with the properties of SC CO_2. One additional advantage of SC antisolvents is that they are easily removed by a simple pressure decrease, whereas the removal of liquid antisolvent requires a complex and costly downstream elimination procedure. Moreover, the reduction of the consumption of organic solvent is a major driving force for the development of new pharmaceutical processing technologies. SCF antisolvent processes have thus been proposed as alternatives to liquid antisolvent processes. These processes are particularly appropriate to the preparation of drug-loaded polymer microparticles, with polymers or drugs that are difficult to dissolve into SCF.

Various SC antisolvent processes have been described [65], from the primary gas antisolvent (GAS) concept to the numerous modifications of the spray processes, namely, supercritical anti-solvent (SAS), precipitation by compressed antisolvent (PCA), aerosol solvent extraction system (ASES), and solution-enhanced dispersion by supercritical fluids (SEDS). As recently pointed out [21], this variety of operating modes of SC antisolvent processes results in a complex intellectual property situation. From a process point of view, these techniques can be divided into two classes depending on their operating mode.

The pioneer SC antisolvent techniques were designed as batch processes operating with liquid solutions. These processes involve the expansion of a batch of liquid solution in a high-pressure vessel by mixing with a dense gas acting as the antisolvent (Fig. 5). The high-pressure vessel is partly filled with the liquid solution, then SC CO_2 is pumped into the vessel to induce the precipitation in the liquid-rich phase. The addition of the SC antisolvent induces supersaturation and precipitation of the solute that was dissolved in the liquid organic solvent. It was emphasized that the antisolvent feed rate, the precipitation temperature, and the pressure are the key parameters to control the particle size [11]. At the end of the precipitation step, the particles have to be washed with the antisolvent to eliminate the liquid organic solvent before depressurization

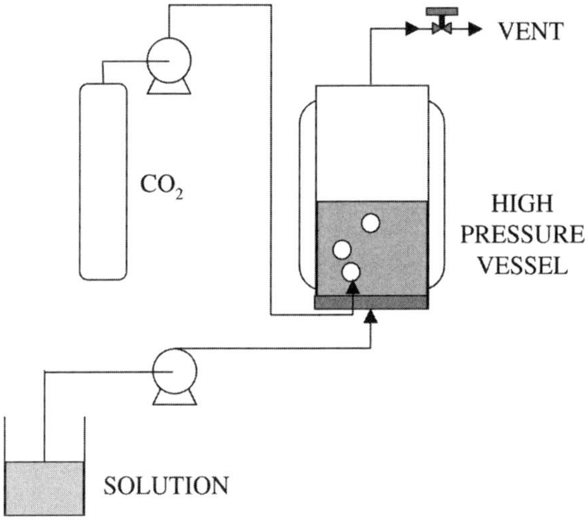

FIG. 5 Schematic representation of the GAS equipment.

and collection of particles. The washing of a powder is advantageous with an SCF due to the low surface tension, high diffusivity, and low viscosity of the antisolvent. This process has been called gas antisolvent (GAS) [11].

The other SC antisolvent processes operate in a continuous or semicontinuous way. The liquid solution is sprayed through a nozzle into compressed CO_2 at the top of a high-pressure vessel (Fig. 6). SC CO_2 is also fed in the high-pressure vessel via a high-pressure pump. The liquid solution is pumped and sprayed at a pressure higher than the vessel pressure to form small liquid droplets. The SCF dissolves in the liquid droplets, causing a large volume expansion, reduction of the liquid solvent power, and the consequent formation of small particles. These processes allow the formation of submicrometer-sized particles with a very narrow size distribution. After production of particles, pure SC CO_2 is allowed to flow through the vessel to remove residual solvents from the microparticles. The design of the liquid injection device plays a key role in this kind of process [65]. Various nozzles including standard capillary nozzles, ultrasonic dispersion devices [66–68], or coaxial nozzles [69] have been used to spray the liquid solution into the SCF. These spray processes have been called precipitation with compressed antisolvent (PCA) [70], supercritical antisolvent (SAS), or aerosol solvent extraction system (ASES) [71].

A modification of the continuous SC antisolvent processes is claimed to use the SCF both for its antisolvent properties and as a spray enhancer [13]. Called supercritical enhanced dispersion by supercritical fluid (SEDS), this process

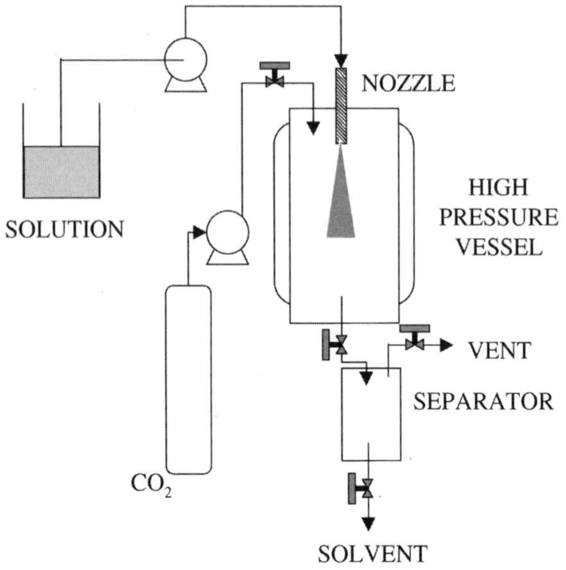

FIG. 6 Schematic representation of the SAS/PCA/ASES equipment.

(Fig. 7) involves the use of a mixing twin-fluid nozzle to blend the SCF and the liquid solution before they enter the precipitation vessel. Due to the high velocity of the SCF, very small droplets and, therefore, very fine particles are formed. Nozzles with three coaxial passages [72] are also used to precipitate water-soluble material such as proteins, which can not be processed with the other SC antisolvent process due to the nonsolubility of water in SC CO_2. Using this nozzle, a solution of the active substance in a first solvent (e.g., water) is cointroduced with a second solvent, which is miscible with both the first one and the SCF. The dispersion of the active substance solution and the second solvent occurs simultaneously with the extraction of these two solvents by the SCF. This setup also allows one to keep fragile molecules, such as therapeutic proteins, in a favorable aqueous phase until rapid precipitation occurs, thus reducing the possibility of denaturation.

Various parameters have an influence on the morphology and size of particles produced by SC antisolvent processes and, more particularly, the flow rates of SCF and solutions, the pressure and the temperature, the solution concentration, and the spraying velocity. Conflicting observations about the effect of process conditions on the particle size and morphology were reported [65]. A comprehensive description of antisolvent precipitation should include contributions from thermodynamics, hydrodynamics, mass transfer, and precipitation kinetics,

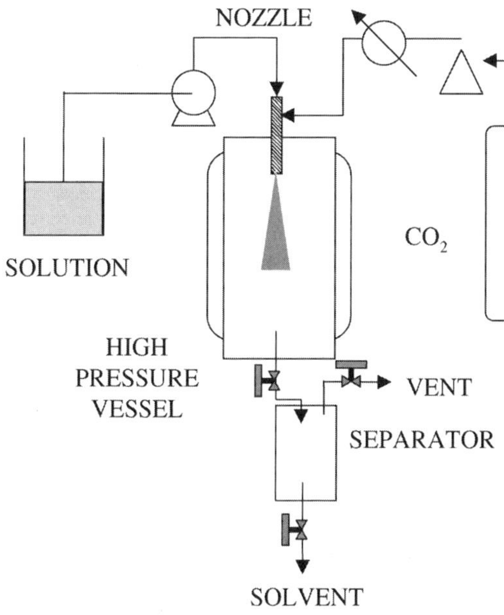

FIG. 7 Schematic representation of the SEDS equipment.

with many unresolved problems as regards the dispersion or mixing in such an environment, the achievement of a local supersaturation environment, followed by the nonequilibrium process of nucleation and surface integration. Thus, the modeling of SC antisolvent processes, although of major interest for the control of particle properties and scale-up purposes, has great complexity. Several recent studies have compiled or addressed the modeling and experimental investigation of the process parameters of antisolvent processes [65,73–80]. A discussion of the relationship between the various particle morphologies obtained and the volume expansion levels of the solvent, at which the particles are generated, was provided [65]. A recent study of the SEDS precipitation of a model drug from an ethanol solution has shown that, above the critical point of the mixture solvent/CO_2 (where a complete miscibility between ethanol and CO_2 occurs), the precipitation behavior is governed by the mixing and homogeneous nucleation [77]. This behavior is similar to a liquid solution crystallization rather than to a nucleation and postnucleation growth within the shrinking ethanol-rich droplets. A variety of precipitate morphologies have been found to result from the spray-based SC antisolvent processes; these are microspheres, fibers, hollow microspheres, and interconnected networks [81,82].

As shown in Table 3, various polymers have been precipitated by SC antisol-

TABLE 3 Polymers and Drug-Loaded Polymer Particles Processed by Supercritical Antisolvent Techniques

Substrate	Process	Solvent supercritical fluid	Results and observations	Ref.
Polymers				
Polystyrene	ASES	Toluene CO_2	Microspheres: 0.1–20 μm	Dixon et al., 1993 [70]
Polystyrene	SEDS	Toluene CO_2	Particle size: 0.5 μm	Hanna et al., 1998 [121]
Ester of alginic acid (ALAFF)	GAS	DMSO CO_2	Particle size: 0.8 μm	Pallado et al., 1996 [83]
Ester of pectinic acid			Particle size: 0.7 μm	[83]
Hyaluronic acid esters:				
HYAFF 11 (benzyl ester)			Particle size: 0.6 μm	
HYAFF 7 (ethyl ester)			Particle size: 1 μm	
HYAFF-11-p75 (partial benzyl ester)			Particle size: 0.8 μm	[83]
Cross-linked polymer of hyaluronic acid			Particle size: 0.6 μm	[83]
Dextran	ASES	DMSO CO_2	Aggregated 0.1- to 0.2-μm particles	Reverchon et al., 1999 [122] Reverchon et al., 2000 [87]
Inulin	ASES	DMSO CO_2	Microspheres, diameter <5 μm	Reverchon et al., 1999 [122]
L-PLA	ASES	CH_2Cl_2 CO_2	Microspheres of mean diameter 1–10 μm	Bleich, 1993 [85]
L-PLA	GAS ASES	CH_2CL_2 CO_2	Spheres 0.6–1.4 μm	Randolph et al., 1993 [66]

L-PLA	ASES	CH_2Cl_2 CO_2	Spherical and nonagglomerated particles 1–5 μm	Bodmeier et al., 1995 [86]
L-PLA	ASES	CH_2Cl_2/MeOH CO_2	Mean particle diameter: 12.3 μm	Bitz et al., 1996 [115]
L-PLA	ASES	CH_2Cl_2 CO_2	Agglomerated and spherical particles, mean particle size 2 μm	Ruchatz et al., 1997 [92]
L-PLA	ASES	CH_2Cl_2 CO_2	Depending on CO_2 density: nonporous 6-μm microparticles to 50-μm particles with cracks and holes	Thies and Muller, 1998 [123]
L-PLA	SEDS	CH_2Cl_2/acetone/ isopropanol CO_2	Irregular microparticles 0.5–5.3 μm	Ghaderi et al., 1999 [94]
L-PLA	SEDS	CH_2Cl_2/acetone/ isopropanol $CO_2 + N_2$	Almost spherical and irregular particles 7–10 μm	Ghaderi et al., 2000 [95]
L-PLA	ASES	CH_2Cl_2 CO_2	Well-defined independent spherical particles: mean diameter 1.4 μm	Reverchon et al., 2000 [87]
L-PLA	ASES	CH_2Cl_2 CO_2	<2-μm discrete microspheres	Tu et al., 2002 [79]
L-PLA	SEDS	CH_2Cl_2 CO_2	Agglomerated particles Mean particle diameter: 3.3–25.7 μm	Rantakyla et al., 2002 [74]
DL-PLA	SEDS	Acetone/hexane/ethyl acetate CO_2	Discrete and agglomerated porous particles 105–216 μm	Ghaderi et al., 1999 [94]
Poly(DL-lactide-co-glycolide) DL-PLGA	SEDS	CH_2Cl_2/acetone/ isopropanol CO_2	DL-PLGA inherent viscosity: 0.63 dL/g Porous and irregular particles 158–173 μm DL-PLGA inherent viscosity: 0.78 dL/g Discrete spherical particles 126–140 μm	Ghaderi et al., 1999 [94]

TABLE 3 *(Continued)*

Substrate	Process	Solvent supercricial fluid	Results and observations	Ref.
Comb poly(vinyl alcohol)-L-PLA Comb poly(vinyl alcohol)-D,L-PLA Comb poly(vinyl alcohol)-D,L-PLGA	ASES	CH_2Cl_2 CO_2	Various morphology from glassy and broken eggshell structures to free-flowing powders of microspheres, depending on the degree of crystallinity of the comb polyesters	Breitenbach et al., 2000 [91]
Polycaprolactone	SEDS	CH_2Cl_2/acetone/isopropanol CO_2	Large and very irregular porous particles 27–212 μm	Ghaderi et al., 1999 [94]
Polycaprolactone	SEDS	Acetone/ethyl acetate/isopropanol $CO_2 + N_2$	Discrete and agglomerated almost spherical microparticles: 25–85 μm	Ghaderi et al., 2000 [95]
Poly(β-hydroxybutyric acid)	ASES	CH_2Cl_2 CO_2	Spherical particles with a mean diameter <10 μm	Bleich et al., 1993 [85]
Poly(hydroxypropylmethacrylamide) (HPMA)	ASES	DMSO CO_2	Spherical microparticles Mean diameter: 0.15 μm	Reverchon, 2000 [122]
Drug-loaded polymer carriers				
HYAFF/calcitonin	GAS	DMSO CO_2	Particle size: 0.5–1 μm	Pallado et al., 1996 [83]
HYAFF/granulocyte-macrophage colony-stimulating factor (GM-CSF)	GAS	DMSO CO_2	Particle size: 0.5–1 μm	Pallado et al., 1996 [83]
Hydroxypropylcellulose/salmeterol xinafoate (HPC)	SEDS	Acetone CO_2	Fine white powder	Hannah et al., 1994 [13]

DL-PLGA (50:50)/hydrocortisone	SEDS	Acetone/ethyl acetate/isopropanol $CO_2 + N_2$	Discrete and almost spherical microparticles Mean diameter 5.6–13 μm Entrapment efficiency: 22%	Ghaderi et al., 2000 [95]
DL-PLGA (50:50)/)estriol or BSA)	ASES	TFE/CH_2Cl_2 CO_2	Spherical and agglomerated particles, mean diameter 20 μm Most of the drug is located at the surface of microparticules Drug loading 56.71% (estriol), 48.19% (BSA)	Engwicht, 1999 [89]
Three block copolymer β-poly-L-lactide-co-D,L-lactide-co-glycolide/estriol	ASES	TFE/CH_2Cl_2 CO_2	Spherical and agglomerated particles, mean diameter 10 μm Drug loading 55.68% (estriol), 52.98% (BSA)	Engwicht, 1999 [89]
L-PLA/clonidine	ASES	CH_2Cl_2 CO_2	100-μm hollow spheres	Muller and Fischer, 1991 [71]
L-PLA/hyoscine butylbromide L-PLA/indomethacin L-PLA/piroxicam L-PLA/thymopentin	ASES	CH_2Cl_2 (+MeOH) CO_2	Partially agglomerated pherical particles Mean diameter: 3–15 μm Drug loading: 0.5–19%	Bleich et al., 1994, 1996 [88], [12]
L-PLA/chlorpheniramine maleate	ASES	CH_2Cl_2 CO_2	1- to 5-μm spherical particles Theoretical drug loading: 10% Experimental drug loading: 3.73%	Bodmeier et al., 1995 [86]
L-PLA/indomethacin	ASES	CH_2Cl_2 CO_2	1- to 5-μm spherical particles Theoretical drug loading: 10% Experimental drug loading: 0.73%	Bodmeier et al., 1995 [86]

TABLE 3 *(Continued)*

Substrate	Process	Solvent supercricial fluid	Results and observations	Ref.
L-PLA/tetracosactide	ASES	CH_2Cl_2/MeOH CO_2	Mean particle diameter: 10.1 μm Theoretical drug loading: 1% w/w Entrapment efficiency of the peptide: 66.1%	Bitz et al., 1996 [115]
L-PLA/naproxen	ASES	Acetone CO_2	Particle size: 5 μm	Chou, 1997 [124]
L-PLA/Gentamicin L-PLA/rifampin L-PLA/naltrexone	SEDS	CH_2Cl_2 CO_2	0.2- to 1-μm-diameter spherical particles Gentamicin and naltrexone were dissolved in CH_2Cl_2 by hydrophobic ion pairing with AOT Drug loading from 0.5% to 37.4% Drug loading efficiency: 10% for non-ion-paired drug, up to approximately 100% for ion-paired drug	Falk, 1997 [67]
L-PLA/chymotrypsin (dioctyl sulfosuccinate hydrophobic salt)	ASES	CH_2Cl_2 CO_2	Particle size: 1–2 μm Drug loading efficiency > 80%	Elvassore, 2000 [98]
L-PLA/insulin	ASES	CH_2Cl_2/DMSO CO_2	Particle size: 1–5 μm	Elvassore, 2000 [98]
L-PLA/insulin–lauric acid conjugate	ASES	CH_2Cl_2 CO_2	Particle size: 1–5 μm	Elvassore, 2000 [98]
L-PLA/lysozyme	ASES	CH_2Cl_2/DMSO CO_2	Particle size: 1–2 μm	Elvassore, 2000 [98]

L-PLA/diuron	ASES	CH_2Cl_2 CO_2	Spherical particles of L-PLA entrapping small crystals of the herbicide diuron 1–5 μm	Taki et al., 2001 [125]
L-PLA/Parahydroxybenzoic acid (*p*-HBA)	ASES	CH_2Cl_2/MeOH CO_2	Irregular *p*-HBA particles encrusted with agglomerated L-PLA micro-spheres Drug loading: 3.8–7.7% Drug loading efficiency: 5.7–9.2%	Tu et al., 2002 [79]
L-PLA/lysozyme	ASES	CH_2Cl_2/DMSO CO_2	Clusters of polymer microspheres and protein nanospheres Drug loading: 3.7% Drug loading efficiency: 12.4–14.6%	Tu et al., 2002 [79]
Polyvinylpyrrolidone/ hydrocortisone (PVP)	ASES	EtOH CO_2	Irregular composite microparticles	Corrigan et al., 2000 [116]

vent processes, including high molecular weight polymers which are not processable by RESS. Only few examples of processing of pharmaceutically acceptable polymers by the GAS batch process were reported. The formation of loaded polysaccharide particles was claimed for the GAS process [83,84]. Due to their ability to control the particle morphology and composition, and to produce micrometer or submicrometer particles, the antisolvent spray-based processes (PCA, SAS, ASES, and SEDS) have received higher attention in recent years. However, the reported applications of these processes are mainly dedicated to the production of fine powders from pure materials, with the aim to control their properties by changing the particle size distribution. To form composite particles, the active substance and the carrier have to be dissolved in a liquid phase and sprayed in the antisolvent. In the rest of this section, the major features of these spray processes, as well as their impact on the polymer or composite particle morphology and on the drug loading, will be described.

As summarized in Table 3, several polymers were processed using the ASES process. To be successfully processed by SC antisolvent processes, the polymers must fulfill some requirements as regards their size, crystallinity, and glass transition temperature. One prerequisite is that the polymer solution could be atomized into droplets and that these droplets could harden into microparticles. Various pharmaceutically acceptable polymers, such as amorphous low molecular weight poly(D,L-lactic acid) ($M_w = 28{,}000$ g/mol) are extracted or plasticized by SCF and hence are not usually suitable for particle production by theses processes [85]. In addition, high molecular weight amorphous poly(D,L-lactide) ($M_w = 236{,}000$ g/mol, $M_w = 110{,}000$ g/mol) or poly(D,L-lactide-*co*-glycolide) ($M_w = 98{,}000$ g/mol) agglomerated in the presence of SC CO_2 [85,86]. Polycaprolactones were extracted during SAS experiments [86,87]. Polymers with low glass transition temperature, such as the commercial PLGA copolymers used for controlled release applications, can soften in the precipitation vessel and lead to agglomeration, adhesion to the precipitation vessel wall, or formation of films. The tendency of particles to form agglomerates is greater when they are produced at a temperature higher than the T_g of the polymer [88]. This behavior is enhanced by the possible high concentration of organic solvent in the gas phase, which may further depress the low T_g of these polymers. In addition, it is also worth noting that the glass transition or melting temperature of a polymer can be dramatically decreased due to the sorption of CO_2 into the polymer. To overcome the softening of PLGA 50:50 microparticles due to their low glass transition temperature, a three-block copolymer composed of L-lactide/D,L-lactide and glycolide was synthesized and further assessed with the ASES process [89]. The higher crystallinity and better thermal behavior of the three-block copolymer allowed formation of smaller microparticles with a lower agglomeration tendency. The presence of semicrystalline microdomains in block copolymers makes these compounds more suitable than statistical copolymers for

ASES processing. Following this track, the synthesis of other copolymers of lactic and glycolic acid [90] or comblike semicrystalline polyesters [91] designed for SC antisolvent processing was recently investigated.

Due to the low T_g restriction and polymer softening described above, the semicrystalline high molecular weight L-PLA (up to 100 000 g/mol) remains the biodegradable polymer that was most often processed successfully with SC antisolvent processes, as summarized in Table 3. Using ASES, the size of loaded L-PLA particles was found to be independent of the gas density or pressure, but dependent on the nature of the drug–polymer system [12,85,88]. Using an experimental design to explore the influence of various production conditions of the ASES process, it was also found that the operating conditions did not have a major influence on the size and morphology of microparticles [92]. Furthermore, in a recent study, the morphology of L-PLA microparticles processed by SAS also showed a limited dependence on such process parameters as pressure, temperature, and liquid solution concentration [87]. This confirms a characteristic behavior of the high molecular weight polymers, hardly influenced by the process conditions, while the morphology and size of particles formed from low molecular weight compounds is often highly dependent on the process conditions [65,93].

The SEDS process is claimed to take advantage of both the antisolvent properties and the "spray enhancer" feature of SCF. Using a nozzle having three coaxial passages, two separated solutions of salmeterol xinafaote and of hydroxypropylcellulose were cointroduced and mixed in SC CO_2 [13], resulting in the formation of a "fine white powder" of drug included in the polymer matrix. Using SEDS, the preparation of biodegradable polymer microparticles was further investigated [94]. DL-PLA and PCL particles were found to agglomerate due the plasticizing effect of SC CO_2. The formation of large but discrete PLGA microparticles was achieved, and the morphology was not found to be influenced by the operating pressure and temperature. However, the process yield was about 50% due to the adsorption of material to vessel walls. Semicrystalline L-PLA was also processed using SEDS, leading to the formation of smaller particles than those obtained from amorphous polymer particle [94]. The control of particle size and morphology has been improved using a combination of N_2 ($T_c = -147.1°C$, $P_c = 33.9$ bar) and CO_2, due to the lower solubility of polymers in CO_2-N_2 mixtures [95]. PLGA discrete microparticles with a diameter less than 10 μm were successfully processed using this modification and optimized conditions. Then, hydrocortisone was entrapped in DL-PLGA with an entrapment efficiency of 22%, after washing the particles with ethanol [95]. Since no discrete crystals of hydrocortisone were observed, it was concluded that the unentrapped drug was loosely adsorbed onto the surface of the microparticles. This modified SEDS process also allowed the formation of discrete, almost spherical particles of polycaprolactones, in spite of their low T_g.

The engineering of drug-loaded particles is one of the most promising applications of the SC antisolvent techniques. However, for drugs which are soluble in the dense fluid, the high solubility in the external organic solvent/CO_2 phase during the particle formation leads to poor drug loading efficiency [86,88]. With a water-insoluble drug, such as indomethacin, the experimental drug loading in L-PLA microparticles was found to be 0.73% for a theoretical drug loading of 10% [86]. Conversely, under the same operating conditions, the use of the water-soluble drug chlorpheniramine maleate led to an increase of the experimental drug loading up to 3.73%.

Due to use of a chlorinated solvent to dissolve polymers such as L-PLA, the basic SC antisolvent processes (PCA, SAS, ASES, SEDS) are hardly suitable to produce particles loaded with water-soluble drugs or large biomolecules, as peptides or proteins. As described above in this section, the SEDS process using a nozzle with three coaxial passages can be considered as well adapted to produce composite microparticles from polymers and drugs having different solubility behaviors [72]. One of the proposed approaches to encapsulate water-soluble ionic compounds within L-PLA is to apply the hydrophobic ion pairing method, which allows fine dispersion of charged biomolecules in organic solvents [96]. It consists of pairing the charged moieties of the drug with oppositely charged amphiphilic compounds, effectively increasing the drug solubility in organic solvents. The efficacy of this approach was illustrated by the encapsulation of both water-soluble and non-water-soluble drugs in L-PLA [67]. By spraying through a sonicated spray nozzle a CH_2Cl_2 solution of L-PLA and rifampin, which is an antibiotic highly soluble in CH_2Cl_2, composite microspheres were formed with a drug loading efficiency of only approximately 10%. This is because during the particle formation rifampin remains in the SC CO_2 and CH_2Cl_2 mixture. Under the same experimental conditions, the drug loading efficiency was reported to be approximately 100% with the ion-paired water-soluble and ionic drugs gentamicin and naltrexone. In spite of their high drug loadings, these particles showed a lower amount of drug located at their surface than non-ion-paired preparations. The release profile for composite particles formed from ion-paired drugs displayed little burst effect and indicated a matrix-controlled diffusion of the drug from the microspheres [67]. An optimization of the process conditions was also performed and resulted in the production of gentamicin/L-PLA microparticles with a residual CH_2Cl_2 content in the order of 1–3 ppm [97]. Thus, this promising concept allows production of controlled-release particles containing drugs that are nonsoluble in the polymer solvent. In the same way, hydrophobic ion pairing, conjugation of the protein to acylic moieties, and optimization of the solubility in solvent mixtures were also considered to produce L-PLA microparticles containing insulin, lysozyme, or chymotrypsin [98].

The coprecipitation of a model drug, *para*-hydroxybenzoic acid (*p*-HBA), and of lysozyme with L-PLA was also investigated using the ASES process and a multiple nozzle assembly [79]. More precisely, a nozzle with three coaxial passages was designed to cointroduce, mix, and spray the drug and the polymer solution. The drug solution is fed via the inner nozzle, the polymer solution via the first outer nozzle, while SC CO_2 is fed in a coaxial direction via the second outer nozzle. Using this device and operating conditions chosen to favor the formation of small and discrete particles from *p*-HBA or L-PLA alone, coprecipitation of *p*-HBA and L-PLA resulted in the formation of a fibrous network of drug and polymer mixture with low values of the encapsulation efficiency. This demonstrates that a limitation of SC antisolvent processes could be the earlier precipitation of the polymer, making the encapsulation of the drug uneasy and uneffective. Moreover, the precipitated *p*-HBA particles were large, which prevents any entrapment within smaller L-PLA microparticles. On the contrary, the lysozyme protein precipitated as nanospheres. The coprecipitation of lysozyme and L-PLA thus resulted in the formation of clusters of polymer microspheres and protein nanospheres with a higher encapsulation efficiency.

Due to their capacity to process almost any kind of compound and to control the morphology, size, and composition of particles, antisolvent processes seem particularly well-adapted to the production of drug delivery systems. These processes have proven to be very efficient for the production of micrometer and submicrometer particles of pure materials. All the studies reported are limited to the proof-of-concept step, dealing with production of milligrams at most. As for the RESS process, a viable scale-up of SC antisolvent processes will only be feasible when the difficult issue of continuous harvesting of submicrometer particles at high yield is resolved [99]. Acting as spray processes with medium throughput and medium to large high-pressure equipment, the scale-up of the SC antisolvent techniques can be commercially successful only for high-value materials. While offering the ability to produce microparticles with a low residual solvent content, they suffer from their large consumption of organic solvents during the process. Moreover, the industrial property is very complex, with many patents pending that could restrict the development of these processes.

In conclusion, polymer processing with the ultimate goal of the production of drug-loaded microparticles is one of the most challenging areas for the SC antisolvent processes. The glass transition temperature issue and its depression in the presence of SCF restricts the range of biodegradable polymers that are processable with SC antisolvents. Highly variable drug loading efficiencies are often obtained for these processes, depending on the operating conditions and on the nature of the drug–polymer system. Although some recent and promising solutions have been proposed, the processing of small water-soluble drugs or large biomolecules, such as proteins, remains a challenge for SC antisolvent processes. These processes also need a thorough investigation and modeling of the influence of operating conditions, including the design of the nozzle, prior to be successfully applied to a wide range of polymers and drugs. Although the formation of composite particles will require in-depth experimental and theoretical investigations, SC antisolvent processes will likely have a bright future in the field of drug delivery systems.

4. Polymerization in SCF

All of the processes described above are based on the use of preformed polymers, which are chosen to be either soluble or nonsoluble in an SCF, depending on the process constraints. However, SC fluids, especially SC CO_2, have also been investigated for their solvent or dispersing properties in the preparation of polymers, through a polymerization process. In this case, the polymer is prepared in situ in the SCF, and the polymerization reaction either can take place in a homogeneous medium or give rise to microparticles dispersed in the SCF. The former process requires the monomer and the polymer chains formed to be soluble in the SCF; for instance, homogeneous free-radical polymerization reactions of fluorinated monomers were earlier reported by the group of DeSimone [100,101]. Conversely, the latter process, which is a dispersion polymerization process, leads to polymer chains that are not soluble in the SCF and instead precipitate to form microparticles [102]. More precisely, the polymerization reaction is initiated homogeneously, and the resulting growing chains phase separate into primary particles to form stable nuclei. After this nucleation step, polymerization occurs within the polymer particles, following the kinetics of bulk polymerization [103]. The primary particles become stabilized by amphiphatic molecules dissolved in the continuous SC phase, which physically adsorb on the particles and prevent particle agglomeration. The amphiphatic molecule is usually a polymeric surfactant that sterically stabilizes the nucleated polymer particles. The solvent selected as the reaction medium, which is the SCF in the present case, must be a good solvent for the monomer, the initiator, and the amphiphatic steric stabilizer, but a poor solvent or a nonsolvent for the polymer being formed [104]. Monodispersed or nearly monodispersed latex particles, with diameters ranging from the submicrometer to tens of micrometers, can be produced through a dispersion polymerization process in an SCF. The first successful results were reported for the methyl methacrylate monomer by the DeSimone group [14]. Since this pioneering work, many papers have been published about the dispersion homo- or copolymerization of various monomers: styrene [105–108], vinyl acetate and ethylene [109], acrylonitrile [110], in SC CO_2. The poly(vinyl acetate) particle formation mechanism in CO_2 has been investigated in situ using high-pressure turbidimetry to monitor the volume fraction of the dispersed phase, the particle size, and the number density during the polymerization [109].

The dispersion polymerization process in SC CO_2 classically consists of successive steps as follows:

1. The reactor is charged with the initiator and the appropriate amount of steric stabilizer.
2. The reactor is purged with a flow of argon prior to the addition of monomers.
3. The reactor is pressurized to about 80 bar by injection of CO_2 using a syringe pump.
4. Under magnetic stirring, the temperature is raised to the decomposition temperature of the initiator, typically 65°C when 2,2′-azobisisobutyronitrile (AIBN) is used.
5. Additional CO_2 is slowly injected into the system until the desired temperature and pressure are reached.
6. The reaction is allowed to proceed under stirring for periods of time that typically vary between 4 and 24 h.
7. At the end of the reaction, the vessel is quickly cooled in an ice/water bath. CO_2 is gently vented through a pressure let-down valve into an organic solvent (methanol, *n*-heptane, etc.), so as to collect any residual reactants and sprayed-out polymer particles.

At the present time, this process has only been run at the laboratory scale in small-volume reactors, typically about 10 to 35 mL. No study on the scale-up of the process has been published so far.

The initiators that are used for dispersion polymerization are soluble in SC CO_2. These are classically: AIBN, diisobutyryl peroxide (Trigonox 187-C30), and a highly fluorinated azobis initiator obtained by treatment of 4,4′-azobis-4-cyanopentanoyl chloride with 1,1,2,2-tetrahydroperfluorodecanol, whose decomposition is analogous to that of AIBN [14].

The key to dispersion polymerization in SC CO_2 has been the development of effective polymeric surfactants, with a highly CO_2-soluble segment that can sterically stabilize the latexes formed. A wide variety of polymeric stabilizers has been designed and synthesized on the basis of their known solvency in CO_2 and the criteria for the steric stabilizer effectiveness [111]. They consist of a "CO_2-philic" group, which is highly soluble in SC CO_2 and extends out from the particle surface to act as a steric barrier, and a "CO_2-phobic" part, which functions as an anchoring unit that can either physically adsorb or become chemically grafted onto the growing particles. Amorphous or low-melting fluoropolymers and polysiloxanes are soluble in CO_2, while other polymers are insoluble in CO_2. For this reason, the CO_2-philic segments of most of the steric stabilizers used for dispersion polymerization in SC CO_2 consist of either fluorocarbon or polysiloxane chains. Both homopolymers or diblock copolymers have been reported (Table 4). The former ones act as effective stabilizers in protect-

TABLE 4 Homo- and Block Copolymers Used as Steric Stabilizers for Dispersion Polymerization in SC CO_2

Steric stabilizers	Nature of the polymer formed during dispersion polymerization	Ref.
Homopolymers		
Poly(1,1-dihydroperfluorooctylacrylate)	Poly(methyl methacrylate) (PMMA)	DeSimone et al., 1994 [14]
(PFOA)	PVAc	Hsiao et al., 1995 [103]
		Canelas et al., 1998 [109]
Poly(dimethylsiloxane)		Shaffer et al., 1996 [126]
(PDMS)	PVAc, Poly(vinyl acetate-*co*-ethylene) (EVA)	Canelas et al., 1998 [109]
Vinyl-terminated PDMS macromonomer	PVAc	Canelas et al., 1998 [109]
Poly(perfluorooctylethyleneoxymethyl styrene)	PS	Lacroix-Desmazes et al., 2000 [127]
(PFOS)		
Diblock copolymers		
Poly(1,1-dihydroperfluorooctylacrylate-*b*-vinyl acetate) (PFOA-*b*-PVAc)	PVAc, EVA	Canelas et al., 1998 [109]
Poly(dimethylsiloxane-*b*-vinyl acetate) (PDMS-*b*-PVAc)	PVAc	Canelas et al., 1998 [109]
Poly(styrene-*b*-dimethylsiloxane) (PS-*b*-PDMS)	Polystyrene	Canelas et al., 1997 [106]
Poly(dimethyl siloxane-*b*-methacrylic acid) (PDMS-*b*-PMAA)	PMMA	Yates et al., 1999 [128]
Poly(styrene-*b*-perfluoroethyleneoxymethyl styrene) (PS-*b*-PFOS)	PS	Lacroix-Desmazes et al., 2000 [127]
Poly(styrene-*b*-1,1,2,2-tetrahydroperfluorodecyl acrylate) (PS-*b*-PFDA)	PS	Lacroix-Desmazes et al., 2000 [127]
Acrylate-terminated PDMS macromonomer (PDMS-mMA)	Poly(methyl methacrylate-*co*-*N*,*N*-dimethylacrylamide) (P(MMA-*co*-DMA)	Filardo et al., 2002 [112]
	Poly(methyl methacrylate-*co*-1-vinyl-2-pyrrolidone) [P(MMA-*co*-VP)]	

ing latex particles from coagulation due to their high solubility and surface-active properties. However, the grafting of these stabilizers to the surface of the particles to enhance their anchoring and stabilizing efficacy has also been investigated using vinyl-terminated macromonomers, such as vinyl-terminated poly(dimethylsiloxane) (PDMS) chains [106]. In the latter situation, the CO_2-phobic segment of the diblock steric stabilizers is currently a polyvinylic or polyacrylic chain that is chosen to physically adsorb on the particle surface, and can be either hydrophilic or hydrophobic, depending on the nature of the polymer particles to be stabilized [14].

The applicability of the dispersion polymerization process to the production of polymer microparticles for specific applications in the therapeutic area has been recently investigated [112]. More particularly, the opportunity offered by this process to incorporate an active drug in polymer microparticles has just been described. However, the use of this process for production of drug-loaded polymer microparticles remains very scarce. It might be related to the nature of the monomers used for preparation of the microparticles, which could be considered as a limitation in itself. However, it is worth noting that some acrylic and vinylic homo- and copolymers are approved from a pharmaceutical regulatory point of view and currently used for coating drugs that are then administered by the oral route. The aim of the coating is classically to get either a taste-masking effect or a sustained-release profile in the gastrointestinal tract. These effects could be envisaged to be obtained by polymer particles produced using a dispersion polymerization process in SC CO_2 [112]. The major advantage would be the high purity of the polymers, which would display very low residual monomer and initiator contents, and hence very low toxicity. However, the main limitation of the dispersion polymerization process in view of applications in the therapeutic area remains the scalability of the process at the pilot and industrial scales, which is far from established at the present time.

III. APPLICATIONS: PHARMACEUTICAL ISSUES

At the present time, the main pharmaceutical applications of SCF that have been developed at the industrial scale are currently related to extraction, refinement, and fractionation of pharmacologically active compounds [23]. However, beside these well-known applications, significant advances are in progress in the field of drug delivery. For these applications, the main issue is the incorporation of drugs within polymer particles to achieve well-defined functions. It encompasses the improvement of the efficacy of the drug by enhancing drug absorption, the limitation of the side effects such as toxicity, and the sustained release of the drug to provide comfort to the patient and restrict the number of doses per day. Polymer micro- and nanoparticles are especially designed to achieve these specific goals as follows:

Taste masking of drugs with bad taste, for oral administration in tablets or syrups

Modified release of the drugs for both the oral and parenteral routes, achieved by sustained-release microparticles, for example

Absorption enhancement, by controlling the particle size and the crystallinity of the drug, by enhancing the dissolution rate of poorly soluble drugs, or by combining drugs with absorption enhancers

Drug targeting for intravenous administration, using nanoparticles that can carry drugs to solid tumors or to specific organs or tissues, avoiding capture by the macrophages

The opportunities offered by the SCF-based production processes reviewed above to achieve these specific goals are discussed below in terms of the actual potential of these techniques, their advantages, and their limitations.

A. Taste Masking

Taste masking of active drugs appears to be a key issue of the development of new oral or liquid dosage forms, such as orodispersible tablets or pediatric suspensions. A large number of drugs exhibit a bad taste, which can result in the low observance of the treatment by the patient. This is especially true for antibiotics, which are very bitter compounds, but anti-infectious drugs or drugs for the treatment of the central nervous system disorders may also require a taste-masking formulation for oral administration. The classical process for coating drugs, so as to get a taste masking effect, consists of the spray coating of an organic polymer solution onto the drug particles suspended in a fluidized bed. For this process, the batch size at the industrial scale is typically 200–500 kg.

Taste masking requires a continuous and regular coating of the drug, which should result in a microcapsule structure, i.e., a drug core surrounded by a polymer shell able to completely isolate the drug from the saliva and the papilla. Conversely, almost all of the SCF-based processes described above lead to microspheres, i.e., matrix systems in which the drug is uniformly distributed in the particles, a part of it being unavoidably located at the surface of the particles and not protected from the saliva and the papilla. For this reason, only a few SCF-based processes using polymers have been proposed for a taste masking application [64,113]. Moreover, these coating processes have only been reported at the present time at the laboratory scale, either with lipid or polymer coating materials [114]. Although these processes may look promising, their scalability from the lab to the industrial plant for production of the large amounts of coated drugs required for daily oral administration has yet to be demonstrated and worked out.

B. Modified Release

Among the potential applications of the SCF-based processes in the field of drug delivery systems, the production of polymer particles for controlled, modified, or sustained release of drugs was the first application reported in the literature [25]. For all processes the polymers used are mainly the biodegradable polyesters, such as poly(lactic acid) (PLA) and poly(lactic acid-*co*-glycolic acid) (PLGA), which can be administered by the parenteral route in subcutaneous depot formulations for sustained release of the drug over weeks or months. A few studies have also been carried out with methacrylic polymers or even PEG block copolymers [51]. Both the RESS [25,22] and antisolvent [12,22, 25,51,67] processes have been considered for this purpose, depending on the solubility of both the drug and the polymer in the SCF. Dispersion polymerization in SC CO_2 has also been recently reported for sustained-release formulations of ibuprofen to be administered by the oral route [112]. The in vitro release profile shows a sustained release of the drug over 24 h in an aqueous phase at pH 6.8 and 37°C.

RESS has been applied to a rather limited number of drugs (Table 2), since it requires a significant solubility of the drug in the SCF. This is clearly the main limitation of the process. Moreover, it appears that the recrystallization of the drug, which occurs upon reduction of solvent density and desolvation from the SC solution, cannot be straightforwardly controlled, as shown for lovastatin—DL-PLA particles, for instance [43]. For this drug, under specific conditions, the polymer DL-PLA precipitates as round-shaped particles, whereas the drug crystallizes as protruding needles whose length is much longer than the diameter of the polymer particles. This morphology does not make it possible to finely control the release kinetics of the drug, which is not completely embedded within the particles. Experimental conditions can be adjusted to restrict the size of the crystalline needles and get a higher coating factor.

Despite its limitations, RESS remains an interesting process because it does not require any organic solvent for production of drug-loaded polymer microspheres. Furthermore, it offers the ability to control the size and morphology of the resulting particles. The key issue is the control of coprecipitation of the drug and polymer and the restriction of the crystal growth of the drug by the polymer coating.

Although the process can be implemented in a rather simple equipment, the scalability of the process from the lab to a pharmaceutically approved industrial plant for production of pilot or industrial batches of injectable, or even oral, sustained-release microspheres, has still to be demonstrated. More particularly, the control of the particle size distribution upon scaling up and the particle collection remain key issues. In addition, the low solubility of most of the interesting drugs and pharmaceutically acceptable polymers in SC CO_2 could make the RESS process hardly profitable at the industrial scale, since the CO_2/drug ratio will remain very high.

Antisolvent processes, including GAS [83], ASES [12,115], and SEDS [13,67,95], have been extensively described for the production of drug-loaded polymer microparticles with a controlled-release behavior. The pharmaceutical applications of these particles are mainly injectable and inhalable products. The polymers used are the biodegradable polyesters PLA and PLGA, which can be approved for these applications. However, it has been shown that PLGA is not well adapted to the process because it is plasticized by SC CO_2, which induces a strong agglomeration of the particles during the production step and before collection. The main interest of the antisolvent processes is related to the fact that they lead to particles with a mean size typically ranging between 0.5 and 20 μm and a rather narrow particle size distribution. Moreover, the flexibility of these processes makes it possible to finely adjust the particle size distribution so that it will be well adapted to a specific way of administration; this is particularly true for inhalation, which requires microparticles in the 2- to 5-μm range, for instance. Many drugs, including both small chemical entities such as naproxen, gentamicin, piroxicam, naltrexone, hydrocortisone, and large biomolecules such as calcitonin, insulin, and granulocyte-macrophage colony-stimulating factor have been incorporated into these polymer particles using antisolvent processes. However, very few data have been reported as regards the in vitro release kinetics of the drugs [67,88,98,116]. In addition, the results show that a large amount of the drug is located near or even at the surface of the particles, which results in a fast release typically ranging from about 50% to 80% of the encapsulated drug over the first 10–20 h. The remaining drug seems to stay within the polymer matrix for a long time and not to be released [88]. More frequently, physicochemical data of the particles obtained are only reported. For this reason, the concept of controlled-release drug-loaded microparticles prepared using the SCF-based antisolvent processes is not fully proven at the present time. The most convincing results in this field could probably be the data published by Falk et al., which exhibit in vitro release periods ranging between 16 and 21 days for PLA microspheres containing ion-paired drugs, i.e., drugs ionically paired with oppositely charged surfactants capable of effectively increasing the apparent molecule solubility in the organic solvent [67]. In this case, it appears clear that the ion pairing method makes it possible to get a higher dispersion degree of the drug in the polymer matrix and to increase the interactions between the drug and the polymer, thus decreasing the release rate.

The main limitation of the SCF-based antisolvent processes remains the use of large amounts of organic solvents to dissolve the polymer, since the concentration of the initial organic solvent is rather low, typically a few weight percent. Moreover, the production rate for polymer microparticles should remain rather low even at the industrial scale, estimated at several kilograms per day, since the feed rate of the organic solution to the nozzle is limited. The scaling up will require a set of nozzles working in parallel; this will make it difficult to control the simultaneous precipitation of the particles in the column from the different nozzles and the separation of the particles from the solvent phase, as well as the homogeneity of the particle size of the powder collected. Finally, like all solvent-based processes, the SCF-based antisolvent processes induce residual organic solvents remaining in the final microparticles, which may be higher than the conforming limits for pharmaceutical applications and require additional drying and solvent extraction under vacuum for long periods of time [115].

C. Absorption Enhancement

The absorption of a drug administered by the oral route is directly related to its apparent solubility in the intestinal fluid and its transport ability through the intestinal membrane. Many different and complex transport mechanisms are involved in the passage through the membrane, which include both passive and active transport, as well as efflux effects related to the P-glycoprotein. The passage of the drug may be transcellular or paracellular, depending on the features of the drug, such as the size of the molecule and its hydrophilic or lipophilic character [117]. Whatever the mechanism, the passage requires the molecules to be solubilized prior to cross the membrane. For poorly soluble drugs, one of the key issues to increase absorption is then to increase the solubilization rate of the molecule in the intestinal fluid. This parameter can be increased by restricting the crystallinity of the drug and reducing the size of the particles to increase the specific surface area of the drug particles. Then, for this purpose, polymers can also be used to prepare drug–polymer coprecipitates, in which the crystallinity and particle size of the drug will be restricted. Moreover, the polymer can be smartly chosen to act as a permeation enhancer, i.e., to favor the crossing of the membrane by the drug molecule. RESS experiments have been reported about micronization of pharmaceutical substances so as to improve the absorption of poorly soluble drugs [118]. However, most of these studies have dealt with the drug alone, without including it in a polymer particle. Conversely, the PGSS process has been applied to the production of drug-loaded polymer particles which exhibit a dramatically enhanced in vitro dissolution rate in comparison with unprocessed drug particles. The drugs processed were nifedipine and felodipine, which were comicronized with the hydrophilic low-mass PEG polymer (M_w = 4000 g/mol) [62,63]. The amount of drug dissolved in an aqueous phase after 1 h was increased by a factor 5–10 for the PGSS-processed coprecipitates, compared to the initial drug particles and even to the drug–polymer mixture of the same composition.

The PGSS process is already used for particle design at the industrial scale for the production of powder coating formulations and paints, which makes a big difference with other processes only at the development stage at the present time. The simplicity of this solvent-free process, associated with the low processing costs (estimated to be less than about 1 Euro/kg) and the wide range of products that can be processed, should provide real opportunities for the development of the PGSS applications in the pharmaceutical field. Feasibility has been shown at the laboratory scale for the increase of dissolution rate of drugs; there is no doubt that this process will also find interesting applications in the production of polymer microparticles for controlled release of large biomolecules [53].

IV. CONCLUSIONS AND PERSPECTIVES

SCF-based processes have great potential for the production of polymer particles that have promising applications in the therapeutic area, specifically in drug delivery systems. In this field, the SCF-based processes make it possible to address the key issues of drug delivery, such as controlled release and enhancement of drug absorption. Depending on the solubility of the drug and the polymer, the SCF can be used either as solvent, or as antisolvent of both or only one of these compounds. A wide variety of SCF-based processes has been reported in the literature for polymer particle engineering. They make it possible to produce micrometer and submicrometer polymer particles with a narrow particle size distribution. This is a very versatile tool, since the features of the particles obtained, such as morphology, size distribution, internal structure, and residual solvent concentration, can be tailored for a specific application. For instance, the in vitro dissolution rate of a drug embedded in polymer microparticles can be considerably increased when both the drug and the polymer are processed together in an SCF, using the PGSS process. Of course, the choice of the polymer is crucial, but the process itself is actually the key to this stimulating result, which should result in a strong absorption enhancement of the drug administered by the oral route. Sustained-release systems based on drug-loaded polymer microparticles are also being developed at the present time, using RESS and antisolvent processes or even dispersion polymerization; these systems need to gain further insight in the internal structure of the particles and distribution of the drug within the polymer matrix, to get a better understanding of the release profile obtained and to adjust the production conditions.

In addition to the applications already foreseen, new developments in the pharmaceutical area will appear soon, with the production of drug-loaded polymer nanoparticles. These nanoparticles can be intravenously injected and used as carriers in the bloodstream for drug targeting. Their main interest is related to their ability to be extravasated through the capillaries of the porous endothelia of solid tumors, leading to an accumulation of the drug in the tumors [119]. This passive targeting should result in a higher efficiency of the drug and a reduction of toxicity, since the biodistribution can be modified. To get a long enough circulation time in the bloodstream and prevent their fast elimination by the mononuclear phagocytes, these nanoparticles have to be coated with appropriate hydrophilic polymers to prevent the adsorption of plasma proteins. The production of the nanoparticles together with their coating with hydrophilic polymers, such as a PEG of appropriate molecular weight, could likely be performed using an SCF-based process, such as PGSS or SC antisolvent spray processes using three-way coaxial nozzles (SEDS, ASES). However, the design of specific polymers with appropriate solubility in SC CO_2, adsorption behavior on polymer particles, and hydrophilic character seems to be a prerequisite with strong challenges. In a more prospective way, active targeting of drug would require the surface functionalization of the nanoparticles to allow molecular recognition of a receptor at the delivery site. This functionalization is usually carried out by chemical grafting of biomolecules (such as peptides) at the surface of the particles; this reaction could be envisaged in an SCF-based process, provided that the nanoparticles are generated using a dispersion polymerization process.

Numerous patent applications have been published regarding SCF-based processes for the production of polymer particles. That makes the intellectual property situation rather complex and intricate, and could limit the development of applications, especially for SC antisolvent processes. Although no system has been developed to the industrial scale in the pharmaceutical field, scalability has already been proven for at least one production process in the powder coating area. This makes these processes very promising and leads to the conclusion that they have an enormous potential in the field of pharmaceutical applications.

REFERENCES

1. Tinsley-Bown, A.M.; Fretwell, R.; Dowsett, A.B.; Davis, S.L.; Farrar, G.H. Formulation of poly(D,L-lactic-co-glycolic acid) microparticles for rapid plasmid DNA delivery. J. Controlled Rel. **2000**, *66*, 229–241.
2. Ruiz, J.M.; Tissier, B.; Benoit, J.P. Microencapsulation of peptide: a study of phase separation of poly(D,L-lactic acid-co-glycolic acid) copolymers 50:50 by silicone oil. Int. J. Pharmacol. **1989**, *49*, 69–77.
3. Peracchia, M.T.; Vauthier, C.; Passirani, C.; Couvreur, P.; Labarre, D. Complement consumption by poly(ethylene glycol) in different conformations chemically coupled to poly(isobutyl 2-cyanoacrylate) nanoparticles. Life Sci. **1997**, *61*, 749–761.
4. Thies, C. Formation of degradable drug loaded microparticles by in-liquid drying method. In *Microcapsules and Nanoparticles in Medicine and Pharmacy*; Donbrow, M., Ed.; CRC Press: Boca Raton, FL, 1992; 47–71.
5. Benoit, J.P.; Marchais, H.; Rolland, H.; Vande Velde, V. Biodegradable microspheres: advances in production technology. In *Microencapsulation: Methods and Industrial Applications*; Benita, S., Ed.; Marcel Dekker: New York, 1996; 35–72.

6. Sah, H. Protein instability toward organic solvent/water emulsification: implications for protein microencapsulation into microspheres. PDA J. Pharm. Sci. Technol. **1999**, *53*, 3–10.
7. Rothen-Weinhold, A.; Besseghir, K.; Vuaridel, E.; Sublet, E.; Oudry, N.; Gurny, R. Stability of somatostatin analogue in biodegradable implants. Int. J. Pharmacol. **1999**, *178*, 213–221.
8. McHugh, M.A.; Krukonis, V.J. *Supercritical Fluid Extraction: Principles and Practice*, 2nd Ed.; Butterworth-Heinemann: Boston, 1994.
9. Matson, D.W.; Fulton, J.L.; Petersen, R.C.; Smith, R.D. Rapid expansion of supercritical fluid solutions: solute formation of powders, thin films and fibers. Ind. Eng. Chem. Res. **1987**, *26*, 2298–2306.
10. Weidner, E.; Knez, Z.; Novak, Z. PGSS—A new process for powder generation. In Proceedings of the 3rd International Symposium on Supercritical Fluids, Vol. 3; Strasbourg, 1994; 229–234.
11. Gallagher, P.M.; Coffey, M.P.; Krukonis, V.J. Gas anti-solvent recrystallization of RDX: formation of ultra-fine particles of a difficult-to-comminute explosive. J. Supercrit. Fluids **1992**, *5*, 130–142.
12. Bleich, J.; Müller, B.W. Production of drug loaded microparticles by the use of supercritical gases with the aerosol solvent extraction system (ASES) process. J. Microencapsulation **1996**, *13*, 131–139.
13. Hanna, M.; York, P. Method and Apparatus for the Formation of Particles. Patent WO 95/01221, 1994.
14. DeSimone, J.M.; Maury, E.E.; Menceloglu, Y.Z.; McClain, J.B.; Romack, T.J.; Combes, J.R. Dispersion polymerizations in supercritical carbon dioxide. Science **1994**, *265*, 356–359.
15. Debenedetti, P.G. Homogeneous nucleation in supercritical fluids. AIChE J. **1990**, *36*, 1289–1297.
16. Cooper, A.I. Polymer synthesis and processing using supercritical carbon dioxide. J. Mater. Chem. **2000**, *10*, 207–234.
17. Brunner, G. *Gas Extraction*; Springer-Verlag: New York, 1994.
18. Marr, R.; Gamse, T. Use of supercritical fluids for different processes including new developments—a review. Chem. Eng. Process **2000**, *39*, 19–28.
19. Dobbs, J.M.; Wong, J.M.; Lahiere, R.J.; Johnston, K.P. Modification of supercritical fluid phase behavior using polar cosolvents. Ind. Eng. Chem. Res. **1987**, *26*, 56–65.
20. Heit, L.T.; Clevenger, J.M. Supercritical Fluid and Near Critical Gas Extraction of Organic Solvents from Formed Articles. Patent EP 0,412,053, 1991.
21. Jung, J.; Perrut, M. Particle design using supercritical fluids: literature and patent survey. J. Supercrit. Fluids **2001**, *20*, 179–219.
22. Tom, J.W.; Debenedetti, P.G. Particle formation with supercritical fluids—a review. J. Aerosol Sci. **1991**, *5*, 555–584.
23. Kaiser, C.S.; Rompp, H.; Schmidt, P.C. Pharmaceutical applications of supercritical carbon dioxide. Pharmazie **2001**, *56*, 907–926.
24. Kompella, U.B.; Koushik, K. Preparation of drug delivery sytems using supercritical fluid technology. Crit. Rev. Ther. Drug Carrier Syst. **2001**, *18*, 173–199.

25. Debenedetti, P.G.; Tom, J.W.; Yeo, S.D.; Lim, G.B. Application of supercritical fluids for the production of sustained delivery devices. J. Controlled Rel. **1993**, *24*, 27–44.
26. Hannay, J.B.; Hogarth, J. On the solubility of solids in gases. Proc. Royal Soc. London **1879**, *29*, 324.
27. Krukonis, V. Supercritical fluid nucleation of difficult to comminute solids. In Proceedings of the AIChE Annual Meeting, San Francisco, Nov 1984, paper 140F.
28. Tom, J.W.; Debenedetti, P.G. Formation of bioerodible microspheres and microparticles by rapid expansion of supercritical solutions. Biotechnol. Prog. **1991**, *7*, 403–411.
29. Tom, J.W.; Debenedetti, P.G.; Jerome, R. Precipitation of poly(L-lactic acid) and composite poly(L-lactic acid)-pyrene particles by rapid expansion of supercritical solutions. J. Supercrit. Fluids **1994**, *7*, 9–29.
30. Smith, R.D.; Wash, R. Supercritical Fluid Molecular Spray Film Deposition and Powder Formation. US Patent 4,582,731, 1986.
31. Domingo, C.; Berends, E.; van Rosmalen, G.M. Precipitation of ultrafine organic crystals from the rapid expansion of supercritical solutions over a capillary and a frit nozzle. J. Supercrit. Fluids **1997**, *10*, 39–55.
32. Lele, A.K.; Shine, A.D. Effect of RESS dynamics on polymer morphology. Ind. Eng. Chem. Res. **1994**, *33*, 1476–1485.
33. Türk, M. Formation of small organic particles by RESS: experimental and theoretical investigations. J. Supercrit. Fluids **1999**, *15*, 79–89.
34. Türk, M. Influence of thermodynamic behaviour and solute properties on homogeneous nucleation in supercritical solutions. J. Supercrit. Fluids **2000**, *18*, 169–184.
35. Sun, X.Y.; Wang, T.J.; Wang, Z.W.; Jin, Y. The characteristics of coherent structures in the rapid expansion flow of the supercritical carbon dioxide. J. Supercrit. Fluids, **2002**, *24*, 231–237.
36. Franklin, R.K.; Edwards, J.R.; Chernyak, Y.; Gould, R.; Henon, F.; Carbonell, R.G. Formation of perfluoropolyether coatings by the rapid expansion of supercritical solutions (RESS). Ind. Eng. Chem. Res. **2002**, *41*, 6127.
37. Knutson, B.L.; Debenedetti, P.G.; Tom, J.W. In *Microparticulate Systems for the Delivery of Proteins and Vaccines*; Cohen, S., Bernstein, H., Eds.; Drugs and the Pharmaceutical Sciences Series Vol. 77. Marcel Dekker: New York, 1996; 89–125.
38. Subramaniam, B.; Rajewski, R.A.; Snavely, K. Pharmaceutical processing with supercritical carbon dioxide. J. Pharm. Sci. **1997**, *86*, 885–890.
39. Phillips, E.M.; Stella, V.J. Rapid expansion from supercritical solutions: application to pharmaceutical processes. Int. J. Pharmacol. **1993**, *94*, 1–10.
40. Matson, D.W.; Petersen, R.C.; Smith, R.D. Production of powders and films from supercritical solutions. J. Mater. Sci. **1987**, *22*, 1919–1928.
41. Petersen, R.C.; Matson, D.W.; Smith, R.D. The formation of polymer fibers from the rapid expansion of supercritical fluid solutions. Polym. Eng. Sci. **1987**, *27*, 1693–1697.
42. Lele, A.K.; Shine, A.D. Morphology of polymers precipitated from a supercritical antisolvent. AIChE J. **1992**, *38*, 742–752.

43. Tom, J.W.; Lim, G.; Debenedetti, P.G.; Prud'homme, R.K. Applications of supercritical fluids in the controlled release of drugs. In *Supercritical Fluid Engineering Science*; Brennecke, J.F., Kiran, E., Eds.; ACS Symp. Series 514; ACS: Washington, DC, 1993; 238–257.
44. Kim, J.H.; Paxton, T.E.; Tomasko, D.L. Microencapsulation of naproxen using rapid expansion of supercritical solutions. J. Supercrit. Fluids **1996**, *7*, 9–29.
45. Krause, H.; Niehaus, M.; Teipel, U. Verfahren Zum Mikroverkapseln Von Partikeln. Patent DE 19,711,393, 1998.
46. Tsustumi, A.; Nakamoto, S.; Mineo, T.; Yoshida, K. A novel fluidized bed coating of fine particles by rapid expansion of supercritical fluid solutions. Powder Technol. **1995**, *85*, 275–278.
47. Wang, T.J.; Tsutsumi, A.; Hasegawa, H.; Mineo, T. Mechanism of particle coating granulation with RESS process in a fluidized bed. Powder Technol. **2001**, *118*, 229–235.
48. Schreiber, R.; Vogt, C.; Werther, J.; Brunner, G. Fluidized bed coating at supercritical fluid conditions. J. Supercrit. Fluids **2002**, *24*, 137–151.
49. Wang, Y.; Wei, D.; Dave, R.; Pfeffer, R.; Sauceau, M.; Letourneau, J.J.; Fages, J. Extraction and precipitation particle coating using supercritical CO_2. Powder Technol. **2002**, *127*, 32–44.
50. Mishima, K.; Matsuyama, K.; Uchiyama, H.; Ide, M. Microcoating of flavone and 3-hydroxyflavone with polymer using supercritical carbon dioxide. The 4th International Symposium on Supercritical Fluids, Sendai, Japan, 1997; 267–220.
51. Mishima, K.; Matsuyama, K.; Tanabe, D.; Yamauchi, S.; Young, T.J.; Johnston, K.P. Microencapsulation of proteins by rapid expansion of supercritical solution with a nonsolvent. AIChE J. **2000**, *46*, 857–865.
52. Mishima, K.; Matsuyama, K.; Yamauchi, S.; Izumi, H.; Furodono, D. Novel control of crystallinity and coating thickness of polymeric microcapsules by cosolvency of supercritical solution. Proceedings of the 5th International Symposium on Supercritical Fluids, Atlanta, Georgia, 2000.
53. Weidner, E.; Knez, Z.; Novak, Z. Process for Preparing Particles or Powders. Patent EP 0,744,992, 1995.
54. Wissinger, R.G.; Paulatis, M.E. Glass transitions in polymer/CO_2 mixtures at elevated pressures. J. Polym. Sci. B **1991**, *29*, 631–633.
55. Gulari, E.; Manke, C.W. Rheological properties of thermoplastics modified with supercritical gases. Proceedings of the 5th International Symposium on Supercritical Fluids, Atlanta, Georgia, 2000.
56. Bok, H.F.; Hoy, K.L.; Nielsen, K.A. Methods and Apparatus for Obtaining a Feathered Spray When Spraying Liquids by Airless Techniques. US Patent 5,057,342, 1989.
57. Nielsen, K.A.; Argyropoulos, J.N.; Wagner, B.E. Method for Producing Coating Powders Catalysts and Drier Water-Borne Coatings by Spraying Compositions with Compressed Fluids. US Patent 5,716,558, 1998.
58. Hastings, D.R.; Hendricks, J.A. Method and Apparatus for Spraying a Liquid Coating Containing Supercritical Fluid or Liquified Gas. Patent EP 0,421,796, 1990.

59. Mandel, F.S.; Green, C.D.; Scheibelhoffer, A.S. Method of Preparing Coating Materials. US Patent 5,548,004, 1996.
60. Mandel, F.S.; Wang, J.D. Manufacturing of specialty materials in supercritical fluid carbon dioxide. Inorg. Chim. Acta **1999**, *294*, 214–223.
61. Weidner, E.; Steiner, R.; Knez, Z. Powder generation from polyethylene glycols with compressible fluids. In *High Pressure Chemical Engineering*; Rudolf von Rohr, P., Trepp, C., Eds.; Elsevier Science: Amsterdam, 1996.
62. Sencar-Bozic, P.; Scric, S.; Knez, Z.; Kerc, J. Improvement of nifedipine dissolution characteristics using supercritical CO_2. Int. J. Pharmacol. **1997**, *148*, 123–130.
63. Kerc, J.; Srcic, S.; Knez, Z.; Sencar-Bozic, P. Micronization of drugs using supercritical carbon dioxide. Int. J. Pharmacol. **1999**, *182*, 33–39.
64. Shine, A.; Gelb, J. Microencapsulation Process Using Supercritical Fluids. Patent WO 98/15348, 1997.
65. Reverchon, E. Supercritical antisolvent precipitation of micro- and nano-particles. J. Supercrit. Fluids **1999**, *15*, 1–21.
66. Randolph, T.W.; Randolph, A.D.; Mebes, M.; Yeung, S. Sub-micrometer-sized biodegradable particles of poly(L-lactic acid) via the gas antisolvent spray precipitation process. Biotechnol. Prog. **1993**, *9*, 429–435.
67. Falk, R.; Randolph, T.W.; Meyer, J.D.; Kelly, R.M.; Manning, M.C. Controlled release of ionic compounds from poly (L-lactide) microspheres produced by precipitation with a compressed antisolvent. J. Controlled Rel. **1997**, *44*, 77–85.
68. Subramaniam, B. Micronization and coating applications with supercritical carbon dioxide. Proceedings of the AAPS Annual Meeting, Pharm. Sci. 1, S-615, 1998.
69. Mawson, S.; Kanakia, S.; Johnston, K. Coaxial nozzle for control of particle morphology in precipitation with a compressed fluid antisolvent. J. Appl. Polym. Sci. **1997**, *64*, 2105–2118.
70. Dixon, D.J.; Johnston, K.P.; Bodmeier, R.A. Polymeric materials formed by precipitation with a compressed fluid anti-solvent. AIChE J. **1993**, *39*, 127–139.
71. Muller, B.W.; Fischer, W. Verfahren zur Herstellung einer mindestens einen Wirkstoff und einen Träger umfassenden Zubereitung. Patent DE 3,744,329, 1989.
72. Hanna, M.; York, P. Method and Apparatus for the Formation of Particles. Patent WO 96/00610, 1995.
73. Werling, J.O.; Debenedetti, P.G. Numerical modelling of mass transfer in the supercritical antisolvent process. J. Supercrit. Fluids **1999**, *16*, 167–181.
74. Rantakylä, M.; Jäntti, M.; Aaltonen, O.; Hurme, M. The effect of initial drop size in the supercritical antisolvent precipitation (SAS) technique. J. Supercrit. Fluids, in press, **2002**, *24*, 251–263.
75. Shekunov, B.Y.; Hanna, M.; York, P. Crystallization process in turbulent supercritical flows. J. Cryst. Growth **1999**, *198*, 1345.
76. Shekunov, B.Y.; Baldyga, J.; York, P. Particle formation by mixing with supercritical antisolvent at high Reynolds numbers. Chem. Eng. Sci. **2001**, *56*, 2421–2433.
77. Bristow, S.; Shekunov, T.; Shekunov, B.Y.; York, P. Analysis of the supersaturation and precipitation process with supercritical CO_2. J. Supercrit. Fluids **2001**, *21*, 257–271.

78. Lengsfeld, C.S.; Delplanque, J.P.; Barocas, V.H.; Randolph, T.W. Mechanism governing microparticle morphology during precipitation by a compressed antisolvent: atomization vs nucleation and growth. J. Phys. Chem. B **2000**, *104*, 2775.
79. Sze Tu, L.; Dehghani, F.; Foster, N.R. Micronisation and microencapsulation of pharmaceuticals using a carbon dioxide antisolvent. Powder Technol. **2002**, *126*, 134–149.
80. de la Fuente Badilla, J.C.; Peters, C.J.; de Swaan Arons, J. Volume expansion in relation to the gas-antisolvent process. J. Supercrit. Fluids **2000**, *17*, 13–23.
81. Dixon, D.J.; Johnston, K.P. Formation of microporous polymer fibers and oriented fibrils by precipitation with a compressed fluid antisolvent. J. Appl. Polym. Sci. **1993**, *50*, 1929.
82. Dixon, D.J.; Luna-Barcenas, G.; Johnston, K.P. Microcellular microspheres and microballoons by precipitation with a vapour-liquid compressed fluid antisolvent. Polymer **1994**, *35*, 3998.
83. Pallado, P.; Benedetti, L.; Callegaro, L. Patent WO 96/29998, 1996.
84. Bertucco, A.; Pallado, P.; Benedetti, L. Formation of biocompatible polymer microspheres for controlled drug delivery by a supercritical antisolvent technique. In *High Pressure Chemical Engineering*; Rudolf von Rohr, P., Trepp, C., Eds.; Elsevier Science: Amsterdam, 1996.
85. Bleich, J.; Muller, B.W.; Waßmus, W. Aerosol solvent extraction system—a new microparticle production technique. Int. J. Pharmacol. **1993**, *97*, 111–117.
86. Bodmeier, R.B.; Wang, H.; Dixon, D.J.; Mawson, S.; Johnston, K.P. Polymeric microspheres prepared by spraying into compressed carbon dioxide. J. Pharm. Res. **1995**, *12*, 1211–1217.
87. Reverchon, E.; Della Porta, G.; De Rosa, I.; Subra, P.; Letourneur, D. Supercritical antisolvent micronization of some biopolymers. J. Supercrit. Fluids **2000**, *18*, 239–245.
88. Bleich, J.; Kleinebudde, P.; Muller, B.W. Influence of gas density and pressure on microparticles produced with the ASES process. Int. J. Pharmacol. **1994**, *106*, 77–84.
89. Engwicht, A.; Girreser, U.; Muller, B.W. Critical properties of lactide-co-glycolide for the use in microparticle preparation by the aerosol solvent extraction system. Int. J. Pharmacol. **1999**, *185*, 61–72.
90. Engwicht, A.; Girreser, U.; Muller, B.W. Characterization of co-polymers of lactic and glycolic acid for supercritical fluid processing. Biomaterials **2000**, *21*, 1587–1593.
91. Breitenbach, A.; Mohr, D.; Kissel, T. Biodegradable semi-crystalline comb polyesters influence the microsphere production by means of a supercritical fluid extraction technique (ASES). J. Controlled Rel. **2000**, *63*, 53–68.
92. Ruchatz, F.; Kleinebudde, P.; Muller, B.W. Residual solvents in biodegradable microparticles. Influence of process parameters on the residual solvent in microparticles produced by the aerosol solvent extraction system (ASES) process. J. Pharm. Sci. **1997**, *86*, 101–105.
93. Shekunov, B.Y.; York, P. Crystallization processes in pharmaceutical technology and drug delivery design. J. Crystal Growth **2000**, *211*, 122–136.

94. Ghaderi, R.; Artursson, P.; Carlfors, J. Preparation of microparticles using solution-enhanced dispersion by supercritical fluid (SEDS). Pharm. Res. **1999**, *16*, 676–681.
95. Ghaderi, R.; Artursson, P.; Carlfors, J. A new method for preparing biodegradable microparticles and entrapment of hydrocortisone in DL-PLG microparticles using supercritical fluids. Eur. J. Pharm. Sci. **2000**, *10*, 1–9.
96. Powers, M.E.; Matsuura, J.; Manning, M.C.; Shefter, E. Enhanced solubility of proteins and peptides in nonpolar solvents through hydrophobic ion pairing. Biopolymers **1993**, *33*,927–932.
97. Falk, R.F.; Randolph, T.W. Process variable implications for residual solvent removal and polymer morphology in the formation of gentamycin-loaded poly(L-lactide) microparticles. Pharm. Res. **1998**, *15*, 1233–1237.
98. Elvassore, N.; Bertucco, A.; Caliceti, P. Production of protein–polymer microcapsules by supercritical antisolvent techniques. Proceedings of the 5th International Symposium on Supercritical Fluids, Atlanta, Georgia, 2000.
99. Thiering, R.; Dehghani, F.; Foster, N.R. Current issues relating to antisolvent micronisation techniques and their extension to industrial scales. J. Supercrit. Fluids **2001**, *21*, 159–177.
100. DeSimone, J.M.; Guan, Z.; Elsbernd, C.S. Synthesis of fluoropolymers in supercritical carbon dioxide. Science **1992**, *257*, 945–947.
101. Combes, J.R.; Guan, Z.; DeSimone, J.M. Homogeneous free radical polymerizations in supercritical carbon dioxide. 2. Telomerization of vinylidene fluoride. Macromolecules **1994**, *27*, 865–867.
102. Barrett, K.E.J. *Dispersion Polymerization in Organic Media*. John Wiley & Sons: New York, 1975.
103. Hsiao, Y.L.; Maury, E.E.; DeSimone, J.M.; Mawson, S.; Johnston, K.P. Dispersion polymerization of methyl methacrylate stabilized with poly(1,1-dihydroperfluorooctyl acrylate) in supercritical carbon dioxide. Macromolecules **1995**, *28*, 8159–8166.
104. Kawaguchi, S.; Winnik, M.A.; Ito, K. Dispersion copolymerization of n-butyl methacrylate with poly(ethylene oxide) macromonomers in methanol-water. Comparison of experiment with theory. Macromolecules **1995**, *28*, 1159–1166.
105. Canelas, D.A.; Betts, D.E.; DeSimone, J.M. Dispersion polymerization of styrene in supercritical carbon dioxide: the importance of effective surfactants. Macromolecules **1996**, *29*, 2818–2821.
106. Canelas, D.A.; DeSimone, J.M. Dispersion polymerization of styrene in carbon dioxide stabilized with poly(styrene-b-dimethylsiloxane). Macromolecules **1997**, *30*, 5673–5682.
107. Shiho, H.; DeSimone, J.M. Dispersion polymerization of styrene in supercritical carbon dioxide utilizing random copolymers including fluorinated acrylate for preparing micron-size polystyrene particles. J. Polym. Sci. A Polym. Chem. **2000**, *38*, 1146–1153.
108. Shiho, H.; DeSimone, J.M. Preparation of micron-size polystyrene particles in supercritical carbon dioxide. J. Polym. Sci. A Polym. Chem. **1999**, *37*, 2429–2437.

109. Canelas, D.A.; Betts, D.E.; DeSimone, J.M.; Yates, M.Z.; Johnston, K.P. Poly(vinyl acetate) and poly(vinyl acetate-coethylene) latexes via dispersion polymerization in carbon dioxide. Macromolecules **1998**, *31*, 6794–6805.
110. Shiho, H.; DeSimone, J.M. Dispersion polymerization of acrylonitrile in supercritical carbon dioxide. Macromolecules **2000**, *33*, 1565–1569.
111. Napper, D.H. *Polymeric stabilization of colloidal dispersions*; Academic Press: New York, 1983.
112. Filardo, G.; Giaconia, A.; Iaia, V.; Galia, A. Dispersion copolymerisation of methyl methacrylate and hydrophilic vinyl monomers in supercritical carbon dioxide. Proceedings of the 4th International Symposium on High Pressure Process Technology and Chemical Engineering. Venice, Italy, 2002.
113. Benoit, J.P.; Richard, J.; Thies, C. Method for preparing microcapsules comprising active materials coated with a polymer and novel microcapsules in particular obtained according to the method. Patent WO 96/13136, 1998.
114. Richard, J.; Pech, B.; Thies, C.; Ribeiro Dos Santos, M.I.; Benoit, J.P. Preparation and characterization of sustained-release microcapsules obtained using supercritical fluid technology. Proceedings of the 7th ISASF Conference, Antibes, 2000; 143–146.
115. Bitz, C.; Doelker, E. Influence of the preparation method on residual solvents in biodegradable microspheres. Int. J. Pharmacol. **1996**, *131*, 171–181.
116. Corrigan, O.I.; Crean, A.M. Comparative physicochemical properties of hydrocortisone-PVP composites prepared using supercritical carbon dioxide by the GAS anti-solvent recrystallization process, by coprecipitation and by spray drying. Int. J. Pharmacol. **2002**, *245*, 75–82.
117. Daugherty, A.L.; Mrsny, R.J. Transcellular uptake mechanisms of the intestinal epithelial barrier. Part 1. Pharm. Sci. Technol. Today **1999**, *2*, 144–151.
118. Türk, M.; Hils, P.; Helfgen, B.; Schaber, K.; Martin, H.J.; Wahl, M.A. Micronization of pharmaceutical substances by the rapid expansion of supercritical solutions (RESS): a promising method to improve bioavailability of poorly soluble pharmaceutical agents. J. Supercrit. Fluids **2002**, *22*, 75–84.
119. Gabizon, A.; Catane, R.; Uziely, B.; Kaufman, B.; Safra, T.; Cohen, R.; Martin, F.; Huang, A.; Barenholz, Y. Prolonged circulation time and enhanced accumulation in malignant exudates of doxorubicin encapsulated in polyethylene-glycol coated liposomes. Cancer Res. **1994**, *54*, 987–992.
120. Blasig, A.; Norfolk, C.W.; Weber, M.; Thies, M.C. Processing polymers by RESS: the effect of concentration on product morphology. Proceedings of the 5th International Symposium on Supercritical Fluids, Atlanta, Georgia, 2000.
121. Hanna, M.; York, P. Method and apparatus for the formation of particles. Patent WO 98/36825, 1998.
122. Reverchon, E.; de Rosa, I.; Della Porta, G. Effect of process parameters on the supercritical anti-solvent precipitation of microspheres of natural polymers. GVC-Fachausschub High Pressure Chemical Engineering, March 3–5, Karlsruhe, Germany, 1999; 251–258.
123. Thies, J.; Muller, B.W. Size controlled production of biodegradable microparticles with supercritical gases. Eur. J. Pharm. Biopharm. **1998**, *45*, 67–74.

124. Chou, Y.H.; Tomasko, D.L. GAS crystallization of polymer-pharmaceutical composite particles. Proceedings of the 4th International Symposium on Supercritical Fluids, Sendai, Japan, 1997; 55–57.
125. Taki, S.; Badens, E.; Charbit, G. Controlled release system formed by supercritical anti-solvent coprecipitation of a herbicide and a biodegradable polymer. J. Supercrit. Fluids **2001**, *21*, 61–70.
126. Shaffer, K.A.; Jones, T.A.; Canelas, D.A.; DeSimone, J.M. Dispersion polymerizations in carbon dioxide using siloxane-based stabilizers. Macromolecules **1996**, *29*, 2704–2706.
127. Lacroix-Desmazes, P.; Young, J.L.; Desimone, J.M.; Boutevin, B. Synthesis and application of surfactants as stabilizers for polymerizations in supercritical CO_2. Proceedings of the 7th ISASF Conference, Antibes, 2000; 451–457.
128. Yates, M.Z.; Li, G.; Shim, J.J.; Maniar, S.; Johnston, K.P.; Lim, K.T.; Webber, S. Ambidextrous surfactants for water-dispersible polymer powders from dispersion polymerization in supercritical CO_2. Macromolecules **1999**, *32*, 1018–1026.

Index